W9-CRB-318

Springer Series in Electrophysics
Volume 20

Edited by Günter Ecker

Springer Series in Electrophysics

Editors: Günter Ecker   Walter Engl   Leopold B. Felsen

Volume 1 **Structural Pattern Recognition** By T. Pavlidis

Volume 2 **Noise in Physical Systems** Editor: D. Wolf

Volume 3 **The Boundary-Layer Method in Diffraction Problems**
By V. M. Babič, N. Y. Kirpičnikova

Volume 4 **Cavitation and Inhomogeneities** in Underwater Acoustics
Editor: W. Lauterborn

Volume 5 **Very Large Scale Integration (VLSI)**
Fundamentals and Applications
Editor: D. F. Barbe 2nd Edition

Volume 6 **Parametric Electronics** An Introduction
By K.-H. Löcherer, C. D. Brandt

Volume 7 **Insulating Films on Semiconductors**
Editors: M. Schulz, G. Pensl

Volume 8 **Fundamentals of Ocean Acoustics**
By L. Brekhovskikh, Y. P. Lysanov

Volume 9 **Principles of Plasma Electrodynamics**
By A. F. Alexandrov, L. S. Bogdankevich, A. A. Rukhadze

Volume 10 **Ion Implantation Techniques**
Editors: H. Ryssel, H. Glawischnig

Volume 11 **Ion Implantation: Equipment and Techniques**
Editors: H. Ryssel, H. Glawischnig

Volume 12 **VLSI Technology**
Fundamentals and Applications Editor: Y. Tarui

Volume 13 **Physics of Highly Charged Ions**
By R. K. Janev, L. P. Presnyakov, V. P. Shevelko

Volume 14 **Plasma Physics for Physicists**
By I. A. Artsimovich, R. Z. Sagdeev

Volume 15 **Relativity and Engineering** By J. Van Bladel

Volume 16 **Electromagnetic Induction Phenomena**
By D. Schieber

Volume 17 **Concert Hall Acoustics** By Y. Ando

Volume 18 **Planar Circuits** for Microwaves and Lightwaves
By T. Okoshi

Volume 19 **Physics of Shock Waves in Gases and Plasmas**
By M. A. Liberman, A. L. Velikovich

Volume 20 **Kinetic Theory of Particles and Photons**
Theoretical Foundations of Non-LTE Plasma Spectroscopy
By J. Oxenius

Volume 21 **Picosecond Electronics and Optoelectronics**
Editors: G. A. Mourou, D. M. Bloom, C.-H. Lee

Joachim Oxenius

Kinetic Theory of Particles and Photons

Theoretical Foundations of Non-LTE Plasma Spectroscopy

With 40 Figures

Springer-Verlag
Berlin Heidelberg New York Tokyo

Dr. Joachim Oxenius

Université Libre de Bruxelles, Service de Chimie Physique II, Code Postal no. 231, Campus Plaine U. L. B., Boulevard du Triomphe, B-1050 Bruxelles, Belgium

ISBN 3-540-15809-X Springer-Verlag Berlin Heidelberg New York Tokyo
ISBN 0-387-15809-X Springer-Verlag New York Heidelberg Berlin Tokyo

Library of Congress Cataloging in Publication Data. Oxenius, Joachim, 1930- Kinetic theory of particles and photons. (Springer series in electrophysics ; v. 20) Bibliography: p. Includes index. 1. Plasma spectroscopy. 2. Thermodynamics. 3. Gases, Kinetic theory of. 4. Photons. I. Title. II. Title: non-LTE plasma spectroscopy. III. Series. QC718.5.S6094 1986 539.7 85-17259

Typsetting: K+V Fotosatz GmbH, 6124 Beerfelden
Offset printing: Beltz Offsetdruck, 6944 Hemsbach/Bergstr.
Bookbinding: J. Schäffer OHG, 6718 Grünstadt.
2153/3130-543210

To my astrophysical friends
in Paris and elsewhere

Preface

Many laboratory and astrophysical plasmas show deviations from local thermodynamic equilibrium (LTE). This monograph develops non-LTE plasma spectroscopy as a kinetic theory of particles and photons, considering the radiation field as a photon gas whose distribution function (the radiation intensity) obeys a kinetic equation (the radiative transfer equation), just as the distribution functions of particles obey kinetic equations. Such a unified approach provides clear insight into the physics of non-LTE plasmas.

Chapter 1 treats the principle of detailed balance, of central importance for understanding the non-LTE effects in plasmas. Chapters 2, 3 deal with kinetic equations of particles and photons, respectively, followed by a chapter on the fluid description of gases with radiative interactions. Chapter 5 is devoted to the H theorem, and closes the more general first part of the book.

The last two chapters deal with more specific topics. After briefly discussing optically thin plasmas, Chap. 6 treats non-LTE line transfer by two-level atoms, the line profile coefficients of three-level atoms, and non-Maxwellian electron distribution functions. Chapter 7 discusses topics where momentum exchange between matter and radiation is crucial: the approach to thermal equilibrium through interaction with blackbody radiation, radiative forces, and Compton scattering.

A number of appendices have been added to make the book self-contained and to treat more special questions. In particular, Appendix B contains an introductory discussion of atomic line profile coefficients.

The notation conforms with general usage in this field. Specifically, the radiation intensity I_ν is used instead of the photon distribution function ϕ, $h\nu$ and kT are written rather than $\hbar\omega$ and $k_B T$, etc.

My sincere thanks are due to my friends Ivan Hubený and Eduardo Simonneau for their essential contributions to various aspects of the theory of non-LTE line transfer. I should also like to thank Alain Omont and Dietrich Voslamber for their comments on parts of the manuscript, and Hans-Werner Drawin for numerical data on optically thin plasmas.

I want to express my gratitude to Professors I. Prigogine and R. Balescu of the Université Libre de Bruxelles for their kind hospitality, and to Professor D. Palumbo of Euratom for his friendly encouragement.

I also thank P. Kinet and E. Kerckx for their technical assistance.

J. Oxenius

Contents

1. Thermal Equilibrium and Detailed Balance

This chapter treats the thermodynamic equilibrium of gases and radiation. According to the principle of detailed balance (Sect. 1.2), any reaction taking place within a thermal gas is exactly counterbalanced by its inverse reaction. Section 1.3 provides a proof of this principle for the special case of binary collisions. In Sect. 1.4, detailed balance is used to derive the thermal distribution functions encountered in plasma spectroscopy. As the principle of detailed balance is based on the reciprocity relation of the transition probabilities between quantum states, the validity of this latter relation is discussed in Sect. 1.5. Finally, starting from detailed balance and the thermal distribution functions, Sect. 1.6 derives explicit forms of the reciprocity relation, relating an atomic quantity (cross section, transition probability) which pertains to a specific reaction, to the corresponding quantity of the inverse reaction.

1.1 Introductory Remarks

This monograph deals with the statistical foundations of plasma spectroscopy. The plasmas we have here in mind contain atoms or atomic ions besides free electrons and bare nuclei; molecules and molecular ions may also be present, but are not discussed in this book. Typical examples of such plasmas are found in gas discharges and stellar atmospheres. Many of the radiation phenomena we shall study, such as spectral lines or recombination continua, involve atoms or atomic ions. We concentrate entirely upon the statistical aspects of plasma spectroscopy, assuming the physics of the various radiation and collision processes to be known. Moreover, our approach favors fundamental concepts rather than specific applications.

The systems to be studied are dilute gases composed of atoms, ions, and electrons interacting with one another and with a radiation field described in terms of a photon gas. Such a simplifying picture, which treats the photons as just another particle species, is quite sufficient for describing the chaotic light of laboratory and astrophysical plasmas. From this point of view, theoretical plasma spectroscopy is the kinetic theory of interacting particles and photons, where the behavior of each species within the plasma is governed by a kinetic equation, the kinetic equation for the photons being nothing but the well-known equation of radiative transfer.

Note that although we are concerned with plasma spectroscopy, we shall not discuss plasma radiation. According to this confusing but generally accepted terminology, plasma radiation refers to collective radiation processes in plasmas where dispersion effects due to particle correlations are important, the prominent example being the emission and absorption of bremsstrahlung with frequencies near the plasma frequency. In contrast, the radiation processes studied in plasma spectroscopy are due to single, uncorrelated particles such that the total emission and absorption of the plasma is obtained by simple summation of the contributions of all individual particles. For the purposes of plasma spectroscopy, the particles of a low-density plasma are thus considered uncorrelated and in well-defined one-particle states.

The most important physical aspect of low-density plasmas is that the collision frequencies among (material) particles may not be sufficiently high to establish local thermodynamic equilibrium (LTE). One speaks of LTE at a given point of a plasma if all distribution functions of the (material) particles are thermal distributions corresponding to a unique temperature. In contrast, in a non-LTE situation, either different particle species have different temperatures, or, usually the more important case, at least one of the particle distribution functions is nonthermal. More explicitly, the occupation numbers of bound atomic levels do not follow a Boltzmann distribution, or an ionization degree deviates from the Saha value, or a velocity distribution is non-Maxwellian.

It should be noted that the notions LTE and non-LTE refer only to the material particles and not to the photons, because a strictly thermal radiation field, namely an isotropic Planck distribution of the photons, is hardly ever found in open systems from which photons escape. On the other hand, this loss of photons through boundaries is one of the main causes of the existence of non-LTE states in low-density plasmas. Other causes can be rapid temporal variations of physical quantities like temperature or density, or transport processes due to steep spatial gradients of such quantities. Furthermore, collective plasma effects due to the long-range Coulomb interaction, like micro-instabilities or plasma turbulence, are connected with non-Maxwellian velocity distributions of the charged particles, especially of the electrons. These effects are important for plasma spectroscopy (since the various collision rates depend on the velocity distributions), but, as they belong to plasma physics in the proper sense of the term, they are not discussed in this book.

This chapter treats the state of complete thermal equilibrium of a gas including photons in some detail. Many of the results obtained are used throughout the book. In particular, a thorough understanding of the principle of detailed balance leads to a fuller comprehension of non-LTE states in plasmas.

1.2 The Principle of Detailed Balance

Let us consider a partially ionized hydrogen plasma, including photons, in complete thermal equilibrium. In this system, many different processes take place, e.g., collisional excitation of H atoms by electrons,

$$\mathrm{H}(1s)+\mathrm{e}\rightarrow\mathrm{H}(3p)+\mathrm{e}\ ,$$

radiative recombination of protons with electrons,

$$\mathrm{H}^{+}+\mathrm{e}\rightarrow\mathrm{H}(2p)+h\nu\ ,$$

absorption of photons by excited H atoms,

$$\mathrm{H}(2s)+h\nu\rightarrow\mathrm{H}(4p)\ ,$$

elastic electron-electron collisions,

$$\mathrm{e}+\mathrm{e}\rightarrow\mathrm{e}+\mathrm{e}\ ,$$

and many others.

Statistical thermodynamics enables us to calculate the thermal distribution functions determining the mean occupation numbers of single quantum states of the particles present, but it reveals nothing about the reactions that actually take place in the system. However, the state of thermal equilibrium of a dilute gas can be described in still another, more dynamic way by the principle of detailed balance, which states that in thermal equilibrium any particular reaction $i\rightarrow f$ is exactly counterbalanced by its inverse reaction $f\rightarrow i$. Here a particular reaction is characterized by an initial state i and a final state f, each comprising a group of many neighboring single quantum states $|i\rangle$ and $|f\rangle$, respectively. Take, for example, the reaction

$$\mathrm{H}(2p;\ \boldsymbol{V},d^3V)+\mathrm{e}(\boldsymbol{v},d^3v)\rightarrow\mathrm{H}(3d;\ \boldsymbol{V}',d^3V')+\mathrm{e}(\boldsymbol{v}',d^3v')\ .$$

Here the initial state i consists of all quantum states with an H atom in the $2p$ state and with a velocity in the range $(\boldsymbol{V},d^3V)$, and a free electron with a velocity in the range $(\boldsymbol{v},d^3v)$. The final state f is defined completely analogously. [A velocity $\boldsymbol{w}$ is in the range $(\boldsymbol{v},d^3v)$ if

$$v_i-\tfrac{1}{2}dv_i<w_i<v_i+\tfrac{1}{2}dv_i\quad(i=x,y,z)\ .$$

Likewise, an energy ε is said to be in the range (E,dE) if $E-dE/2<\varepsilon<E+dE/2$.]

Let us denote by $[i\rightarrow f]$ the reaction rate, that is, the number of reactions $i\rightarrow f$ that take place per unit time in the considered volume. Then we can state the principle of detailed balance:

Principle of Detailed Balance [1.1]. In a dilute gas in thermal equilibrium, each reaction $i\rightarrow f$ is exactly counterbalanced by its inverse reaction $f\rightarrow i$ such that there is equality of the reaction rates,

$$[i\rightarrow f]=[f\rightarrow i]\ , \tag{1.2.1}$$

for *all* pairs of states i,f. Equation (1.2.1) is a necessary and sufficient condition for thermal equilibrium of a dilute gas.

Proofs of the principle of detailed balance are given in Sects. 1.3, 4, and from a different point of view in Chap. 5. Here the following comments give some necessary background.

(1) The balance in question is indeed detailed, as the above example shows. In thermal equilibrium, there is not just global balancing $H(2p)+e \rightleftarrows H(3d)+e$, but balancing in every small velocity range of the particles involved.

(2) As shown below, the validity of the principle of detailed balance depends on the validity of the quantum mechanical reciprocity relation

$$w(i \rightarrow f) = w(f \rightarrow i) \quad , \tag{1.2.2}$$

where $w(i \rightarrow f)$ is the transition probability per unit time for the transition $|i\rangle \rightarrow |f\rangle$ between single quantum states of the states i and f. However, the reciprocity relation (1.2.2) does not hold in full generality, so that the principle of detailed balance (1.2.1) cannot be valid in full generality either. In other words, the principle of detailed balance is not an exact law of nature but only a (usually very good) approximation. The validity of (1.2.2) is discussed in Sect. 1.5.

(3) The density of the gases considered is supposed to be so low that one-particle quantum states are well defined and that the state of the gas as a whole is specified by the occupation numbers of these one-particle states.

1.3 Proof of the Principle of Detailed Balance for a Special Case

To become more familiar with the principle of detailed balance, let us consider its application to a reaction of the type

$$A+B \rightarrow C+D \ . \tag{1.3.1}$$

Here the letters $A, B, \ldots$ denote not only the species of the particles, but also their internal and translational states. For example, for the reaction $H(2p)+e \rightarrow H(3d)+e$ considered in the previous section, $A = H(2p; V, d^3V)$, etc.

First of all we need an explicit expression for the rate of reaction (1.3.1). Let $N(P)$ be the number of particles P (in the sense just discussed) actually present, and $G(P)$ the degree of degeneracy of the state P, that is, the number of single quantum states $|P\rangle$ it contains. Explicit expressions for $N(P)$ and $G(P)$ are given in the next section, but it may be helpful at this point to note that for free particles both $N(P)$ and $G(P)$ are proportional to the velocity range considered. Thus, for $A = H(2p; V, d^3V)$, $N(A)$ and $G(A)$ are both proportional to d^3V.

Clearly, the rate of reaction (1.3.1) is proportional to the product $N(A)N(B)$. Moreover, it is proportional to the number of reaction channels available, and hence to the number of final quantum states $G(C)G(D)$. The reaction rate can therefore be written as

$$[AB \rightarrow CD] = w(AB \rightarrow CD)N(A)N(B)G(C)G(D) \ , \tag{1.3.2}$$

where $w(AB \rightarrow CD)$ is the average transition probability per unit time between single quantum states.

As an example, consider again the process $\mathrm{H}(2p)+\mathrm{e} \rightarrow \mathrm{H}(3d)+\mathrm{e}$, supposing the velocities of the four particles to have fixed specified values compatible with momentum and energy conservation. The reaction then still comprises a great number of different channels. Since the $2p$ configuration gives rise to the levels $2\,^2P_{3/2}$ and $2\,^2P_{1/2}$ having 4 and 2 states, respectively, and taking the two orientations of the proton spin into account, $\mathrm{H}(2p)$ is seen to contain $2\times(4+2)=12$ quantum states, which together with the two spin states of the incoming electron lead to 24 quantum states of the initial state i. As the final state f has $2\times2\times(6+4)=40$ quantum states (because $3\,^2D_{5/2}$ and $3\,^2D_{3/2}$ have 6 and 4 states, respectively), there is a total of $24\times40=960$ channels for this reaction. If, using an obvious notation, one writes the reaction rate in the form

$$[\mathrm{H, e} \rightarrow \mathrm{H', e'}] = w(\mathrm{H, e} \rightarrow \mathrm{H', e'})N(\mathrm{H})N(\mathrm{e})G(\mathrm{H'})G(\mathrm{e'}) \ ,$$

the symbol $w(\mathrm{H,e} \rightarrow \mathrm{H',e'})$ stands for the mean value of the 960 transition probabilities just mentioned. It must be noted that using a mean transition probability to calculate a reaction rate supposes that all single quantum states of the initial state i have equal occupation probabilities, and that the single quantum states of the final state f are not observed individually.

Reaction rate (1.3.2) is valid only in the so-called classical limit where the mean occupation numbers of single quantum states are small, $N(P)/G(P) \ll 1$. For larger values of $N(P)/G(P)$, however, quantum corrections must be taken into account. The right-hand side of (1.3.2) must then be multiplied for each final boson B by a factor of

$$1+\frac{N(\mathrm{B})}{G(\mathrm{B})} \ , \tag{1.3.3a}$$

and for each final fermion F by a factor of

$$1-\frac{N(\mathrm{F})}{G(\mathrm{F})} \ . \tag{1.3.3b}$$

The boson factor (1.3.3a) can be understood in the following way. According to quantum field theory, the transition rate into a final boson state $|f\rangle$ is proportional to $1+N(f)$, where $N(f)$ is the number of bosons already present in state $|f\rangle$. The rate is thus composed of a spontaneous 1 and a stimulated $N(f)$ part. Indeed, acting on an N-boson state, the creation operator $a^\dagger$ produces a state with $N+1$ bosons according to $a^\dagger|N\rangle = (N+1)^{1/2}|N+1\rangle$. Hence, $|\langle N+1|a^\dagger|N\rangle|^2 = N+1$. For another derivation of the factor $1+N(f)$, see [1.2]. For a group of G neighboring quantum states, the transition rate is therefore proportional to

$$\sum_{f=1}^{G}[1+N(f)] = G+\sum_{f=1}^{G}N(f) = G+N = G\left(1+\frac{N}{G}\right),$$

where N is the occupation number of the whole group.

The fermion factor (1.3.3b) is an immediate consequence of Pauli's exclusion principle which states that for fermions a single quantum state can be occupied by only one particle. As a result, if N states are already occupied in a group of G neighboring quantum states $|f\rangle$, the transition rate is proportional to the number of unoccupied states

$$G-N = G\left(1-\frac{N}{G}\right).$$

The "attraction" of identical bosons and the "repulsion" of identical fermions are genuine quantum effects which cannot be understood in classical terms. They are a consequence of the symmetry (antisymmetry) of the total wave function of a system of identical bosons (fermions).

For example, taking quantum corrections into account, the rate of the reaction $\mathrm{H}(2p)+\mathrm{e}\rightarrow\mathrm{H}(3d)+\mathrm{e}$ is given by

$$[\mathrm{H},\mathrm{e}\rightarrow\mathrm{H}',\mathrm{e}'] = w(\mathrm{H},\mathrm{e}\rightarrow\mathrm{H}',\mathrm{e}')N(\mathrm{H})N(\mathrm{e})$$

$$\times G(\mathrm{H}')\left[1+\frac{N(\mathrm{H}')}{G(\mathrm{H}')}\right]G(\mathrm{e}')\left[1-\frac{N(\mathrm{e}')}{G(\mathrm{e}')}\right],$$

because an electron is a fermion, and an H atom, being composed of two fermions, is a boson. (Recall that a composite particle such as a nucleus, an atom, or a molecule behaves as a boson (fermion) if it is composed of an even (odd) number of fermions (protons, neutrons, electrons). An instructive discussion of the limits of this approximation can be found in [1.3].)

The rest of this section takes the quantum factors (1.3.3) into consideration. However, it should be realized that the only quantum correction of interest to us is the stimulated emission of photons, while quantum correlations in dilute material gases are always negligible.

Considering a special case, we are now ready to show that detailed balance is a necessary and sufficient condition for thermal equilibrium of a gas. Following *Oster* [1.4], consider the elastic scattering reaction

$$A(E_A^{\mathrm{i}})+B(E_B^{\mathrm{i}})\rightarrow A(E_A^{\mathrm{f}})+B(E_B^{\mathrm{f}}) \tag{1.3.4}$$

taking place in an isotropic gas which contains two different particle types A and B. Here E_A^{i}, E_B^{i} are the initial, and E_A^{f}, E_B^{f} the final kinetic energies of the particles. Conservation of energy requires

$$E_A^{\mathrm{i}} + E_B^{\mathrm{i}} = E_A^{\mathrm{f}} + E_B^{\mathrm{f}} \tag{1.3.5}$$

or, introducing explicitly the energy W transferred from particle B to particle A during the collision,

$$E_A^{\mathrm{f}} = E_A^{\mathrm{i}} + W, \quad E_B^{\mathrm{f}} = E_B^{\mathrm{i}} - W. \tag{1.3.6}$$

We first want to show that detailed balance is a sufficient condition for thermal equilibrium, i.e., that the thermal distribution functions are obtained from (1.2.1). Writing out explicitly the reaction rates by means of (1.3.2) and the quantum corrections (1.3.3), detailed balance between reaction (1.3.4) and its inverse reaction requires that

$$\begin{aligned} w(i\to f)\,N_A(E_A^{\mathrm{i}})\,N_B(E_B^{\mathrm{i}})\,G_A(E_A^{\mathrm{f}})\,[1+\alpha_A z_A(E_A^{\mathrm{f}})] \\ \cdot\, G_B(E_B^{\mathrm{f}})\,[1+\alpha_B z_B(E_B^{\mathrm{f}})] \\ = w(f\to i)\,N_A(E_A^{\mathrm{f}})\,N_B(E_B^{\mathrm{f}})\,G_A(E_A^{\mathrm{i}})\,[1+\alpha_A z_A(E_A^{\mathrm{i}})] \\ \cdot\, G_B(E_B^{\mathrm{i}})\,[1+\alpha_B z_B(E_B^{\mathrm{i}})]\ . \end{aligned} \tag{1.3.7}$$

Here

$$z(E) = \frac{N(E)}{G(E)} \tag{1.3.8}$$

is the mean occupation number of a quantum state of energy E, and

$$\alpha = \begin{cases} +1 & \text{for bosons ,} \\ -1 & \text{for fermions .} \end{cases} \tag{1.3.9}$$

Note that owing to the assumed isotropy of the gas, the occupation numbers depend only on the energy, $N = N(E)$. On account of the reciprocity relation (1.2.2), the transition probabilities $w(i\to f)$ and $w(f\to i)$ cancel out in (1.3.7). Defining now

$$\chi(E) = \frac{z(E)}{1+\alpha z(E)} \tag{1.3.10}$$

and making use of (1.3.6), the balance equation (1.3.7) becomes

$$\frac{\chi_A(E_A^{\mathrm{i}})}{\chi_A(E_A^{\mathrm{i}}+W)} = \frac{\chi_B(E_B^{\mathrm{i}}-W)}{\chi_B(E_B^{\mathrm{i}})} = r(W)\ . \tag{1.3.11}$$

Here $r(W)$ depends only on the transferred energy W, for the first ratio in (1.3.11) is independent of particle type B (and of E_B^{i}, in particular), whereas the second ratio is independent of particle type A (and of E_A^{i}, in particular). Equation (1.3.11) shows that χ_A and χ_B obey the same functional equation

$$r(W)\,\chi(E+W) = \chi(E) \ . \tag{1.3.12}$$

To solve this equation, $r(W)$ and $\chi(E+W)$ are expanded in power series with respect to W,

$$r(W) = \sum_{n=0}^{\infty} \frac{r^{(n)}(0)}{n!} W^n \ , \quad \chi(E+W) = \sum_{n=0}^{\infty} \frac{\chi^{(n)}(E)}{n!} W^n \ , \tag{1.3.13}$$

where $r^{(n)} = d^n r/dW^n$, $\chi^{(n)} = d^n\chi/dW^n$. Inserting these power series in (1.3.12) and equating the coefficients of W^0, W^1, $W^n\,(n \geqslant 2)$ on both sides of this equation yields

$$r(0) = 1 \ , \tag{1.3.14a}$$

$$r(0)\,\chi'(E) + r'(0)\,\chi(E) = 0 \ , \tag{1.3.14b}$$

$$r(0)\frac{\chi^{(n)}(E)}{n!} + \frac{r'(0)}{1!}\,\frac{\chi^{(n-1)}(E)}{(n-1)!} + \dots + \frac{r^{(n)}(0)}{n!}\,\chi(E) = 0 \ . \tag{1.3.14c}$$

Taking (1.3.14a) into account, the solution of the differential equation (1.3.14b) is given by

$$\chi(E) = \zeta \mathrm{e}^{-\beta E} \ , \tag{1.3.15}$$

where $\beta = r'(0)$ is a constant which is positive so that $\chi(E) \to 0$ and hence $z(E) \to 0$ if $E \to \infty$, and ζ is another constant. From (1.3.12) one now gets

$$r(W) = \mathrm{e}^{\beta W} \ , \tag{1.3.16}$$

and (1.3.14c) is readily seen to be fulfilled for all $n \geqslant 2$. From (1.3.10,15) follows finally that

$$z(E) = \frac{1}{\zeta^{-1}\mathrm{e}^{\beta E} - \alpha} \ . \tag{1.3.17}$$

Thus, for Bose-Einstein particles ($\alpha = +1$) detailed balance (1.2.1) leads to the distribution function

$$z_{\mathrm{BE}}(E) = \frac{1}{\zeta^{-1}\mathrm{e}^{\beta E} - 1} \ , \tag{1.3.18a}$$

and for Fermi-Dirac particles ($\alpha = -1$) to the distribution function

$$z_{\mathrm{FD}}(E) = \frac{1}{\zeta^{-1}\mathrm{e}^{\beta E} + 1} \ . \tag{1.3.18b}$$

One immediately recognizes the familiar thermal occupation numbers of single quantum states, which in the limit $\zeta \to 0$ (that is, in the limit of vanishing occupation numbers, $z \to 0$) reduce to the common classical limit of Maxwell-Boltzmann particles,

$$z_{\mathrm{MB}}(E) = \zeta \mathrm{e}^{-\beta E} . \tag{1.3.18c}$$

As is well known from statistical thermodynamics, the constants β and ζ are related to temperature T and chemical potential μ by

$$\beta = \frac{1}{kT} , \quad \zeta = \mathrm{e}^{\mu/kT} . \tag{1.3.19}$$

Note that because of (1.3.11,16), β is independent of the particle type, but in general $\zeta_A \neq \zeta_B$, as required on physical grounds, see (1.3.19).

We have thus shown that detailed balance of the reactions considered is a sufficient condition for thermal equilibrium. On the other hand, it is easy to show that detailed balance is valid in thermal equilibrium where the particle types A and B have distribution functions (1.3.17) corresponding to a common temperature, $\beta_A = \beta_B = \beta$. One merely has to check that (1.3.17) fulfills the balance equation (1.3.7). This is indeed the case, provided that the reciprocity relation (1.2.2) holds. It follows, then, that detailed balance is also a necessary condition for thermal equilibrium.

The same demonstration is also applicable to identical particles. Here detailed balance of the reactions

$$A(E) + A(E') \rightleftarrows A(E+W) + A(E'-W)$$

leads to, see (1.3.11),

$$\frac{\chi(E)}{\chi(E+W)} = \frac{\chi(E'-W)}{\chi(E')} = r(W) , \tag{1.3.20}$$

where again $r(W)$ depends only on W but not on E and E', the remaining steps of the proof being unchanged.

Let it just be mentioned that the principle of detailed balance can be proved similarly without assuming isotropy as we did above [1.4]. The equivalence of thermal equilibrium and detailed balance is again discussed in Chap. 5 in the framework of the H theorem for matter and radiation.

1.4 Derivation of Thermal Distribution Functions from Detailed Balance

Using the method of the preceding section, that is, starting from the requirement of detailed balance, we now derive some well-known thermal distribution functions of particles and photons, namely, the Maxwell, Boltzmann, Saha, and Planck distributions. In other words, we show that detailed balance is a sufficient condition for these distributions. On the other hand, that detailed balance is likewise a necessary condition for thermal equilibrium can easily be shown by inserting the distribution functions mentioned in the detailed balance equations, which are found to be fulfilled identically provided that the reciprocity relation (1.2.2) for the transition probabilities holds and that all distributions correspond to the same temperature.

1.4.1 Maxwell Distribution

Though the thermal distribution of kinetic energy has already been treated in the preceding section, here we derive the Maxwell distribution anew by considering detailed balance of elastic collisions between particles of species A and P that have kinetic energies in well-defined ranges,

$$A(E, dE) + P(\bar{E}, d\bar{E}) \rightleftarrows A(E', dE') + P(\bar{E}', d\bar{E}') \ , \tag{1.4.1a}$$

or, making use of energy conservation,

$$A(E, dE) + P(\bar{E}, d\bar{E}) \rightleftarrows A(E+W, dE') + P(\bar{E}-W, d\bar{E}') \ . \tag{1.4.1b}$$

Here and in the rest of this section we consider for simplicity only isotropic gases and radiation fields. Since the translational degrees of freedom form a quasicontinuum, we introduce densities per unit energy $N(E)$ and $G(E)$ such that $N(E)\,dE$ is the number of A particles with kinetic energies in the range (E, dE), and $G(E)\,dE$, the number of quantum states corresponding to this range, is given by the number of translational quantum states multiplied by the internal degree of degeneracy (the statistical weight) of an A particle. The corresponding quantities of the P particles are $\bar{N}(\bar{E})$ and $\bar{G}(\bar{E})$. [Notice that $N(E)$ and $G(E)$ now denote densities per unit energy, whereas in Sect. 1.3 they denoted numbers.] According to (1.3.2), detailed balance of the reactions (1.4.1b) requires

$$\begin{aligned} &N(E)\bar{N}(\bar{E})\,G(E+W)\,\bar{G}(\bar{E}-W) \\ &\quad = N(E+W)\,\bar{N}(\bar{E}-W)\,G(E)\,\bar{G}(\bar{E}) \ , \end{aligned} \tag{1.4.2}$$

since the product $dE\,d\bar{E}\,dE'\,d\bar{E}'$ drops out and the transition probabilities $w(i \to f)$ and $w(f \to i)$ cancel each other because of the reciprocity relation (1.2.2). Here we have assumed that the classical limit applies to the A and P particles, so that no quantum factors (1.3.3) appear in (1.4.2). Introducing again the mean occupation number of a quantum state of energy E, see (1.3.8),

$$z(E) = \frac{N(E)\,dE}{G(E)\,dE} = \frac{N(E)}{G(E)} \ , \tag{1.4.3}$$

we obtain from (1.4.2) in analogy to Sect. 1.3, see (1.3.11),

$$\frac{z(E)}{z(E+W)} = \frac{\bar{z}(\bar{E}-W)}{\bar{z}(\bar{E})} = r(W) \ . \tag{1.4.4}$$

Now, in Sect. 1.3 we proved that the functional equation

$$\frac{z(E)}{z(E+W)} = \frac{z(E-W)}{z(E)} = r(W) \tag{1.4.5a}$$

has the solution

$$z(E) = \zeta \mathrm{e}^{-\beta E} \ , \quad r(W) = \mathrm{e}^{\beta W} \ . \tag{1.4.5b}$$

Applying this theorem to (1.4.4) and using (1.4.3) gives

$$N(E)\,dE = \zeta G(E)\,\mathrm{e}^{-\beta E} dE \ . \tag{1.4.6}$$

To determine the quantities $G(E)$ and ζ, note that the number of translational quantum states of a μ space element $d^3r d^3p$ is given by

$$G_{\mathrm{trans}} = \frac{d^3r\, d^3p}{h^3} \ , \tag{1.4.7}$$

where d^3r and d^3p are elements of configuration and momentum space, respectively, and h is Planck's constant. Equation (1.4.7) follows from counting the number of vibration modes in a cubic cavity of length L with periodic boundary conditions. The allowed wave vectors are $(k_x, k_y, k_z) = (2\pi/L)(n_x, n_y, n_z)$, with $n_{x,y,z} = 0, \pm 1, \pm 2, \ldots$, so that a volume element d^3k in wave-number space contains $(L/2\pi)^3 d^3k$ modes. On account of the de Broglie relation $\boldsymbol{p} = (h/2\pi)\boldsymbol{k}$ between momentum $\boldsymbol{p}$ and wave vector $\boldsymbol{k}$, the corresponding number of momentum states in the volume L^3 is $L^3 d^3p/h^3$, which is (1.4.7). This derivation shows that (1.4.7) applies to particles and photons alike, since the de Broglie relation holds in both cases.

If V is the volume considered,

$$G(E)\,dE = g V \frac{4\pi p^2 dp}{h^3} = g V \frac{2^{5/2}\pi m^{3/2}}{h^3} E^{1/2} dE \ , \tag{1.4.8}$$

where p is the modulus of the momentum, and

$$E = \frac{p^2}{2m} \ , \quad dp = \left(\frac{m}{2E}\right)^{1/2} dE \ . \tag{1.4.9}$$

Here m is the particle mass, and g is the statistical weight of the particle state in question. Using (1.4.6, 8) the total number of particles N is

$$N = \int_0^\infty N(E)\, dE = \zeta g V \frac{2^{5/2} \pi m^{3/2}}{h^3} \int_0^\infty E^{1/2} \mathrm{e}^{-\beta E} dE$$

$$= \zeta g V \frac{2^{5/2} \pi m^{3/2}}{h^3} \frac{\pi^{1/2}}{2 \beta^{3/2}} , \tag{1.4.10}$$

and hence

$$\zeta = \frac{1}{g} n \lambda^3 , \quad \text{where} \tag{1.4.11}$$

$$n = \frac{N}{V} \tag{1.4.12}$$

is the number density, and

$$\lambda = \left(\frac{\beta h^2}{2 \pi m} \right)^{1/2} = \frac{h}{(2 \pi m k T)^{1/2}} \tag{1.4.13}$$

is the thermal de Broglie wavelength of the particles. Here we have used $\beta = 1/kT$, see (1.3.19). One sees from (1.4.11) that the classical limit $\zeta \ll 1$, see Sect. 1.3, is characterized by the fact that the mean number of particles in a de Broglie volume λ^3 is small compared to unity.

Inserting (1.4.8, 11, 13) in (1.4.6) finally gives the *Maxwell distribution* for the translational degrees of freedom,

$$f^{\mathrm{M}}(E)\, dE = \frac{2 E^{1/2}}{\pi^{1/2} (kT)^{3/2}} \mathrm{e}^{-E/kT} dE , \tag{1.4.14}$$

written in the form of a normalized distribution function defined by

$$f(E) = \frac{N(E)}{N} , \quad \int_0^\infty f(E)\, dE = 1 . \tag{1.4.15}$$

1.4.2 Boltzmann Distribution

To derive the Boltzmann distribution for bound atomic levels, we consider detailed balance of excitation and de-excitation collisions of free particles P with atoms A of infinite mass that are at rest,

$$A(E_1) + P(\bar{E}, d\bar{E}) \rightleftarrows A(E_2) + P(\bar{E} - E_{21}, d\bar{E}') . \tag{1.4.16}$$

Here E_1 and E_2 are the energies of the lower and upper atomic levels, so that $E_{21} = E_2 - E_1$ is the excitation energy of the considered atomic transition (Fig. 1.1). From (1.4.16) it follows as usual that

$$N_1 \bar{N}(\bar{E})\, g_2 \bar{G}(\bar{E} - E_{21}) = N_2 \bar{N}(\bar{E} - E_{21})\, g_1 \bar{G}(\bar{E}) , \tag{1.4.17}$$

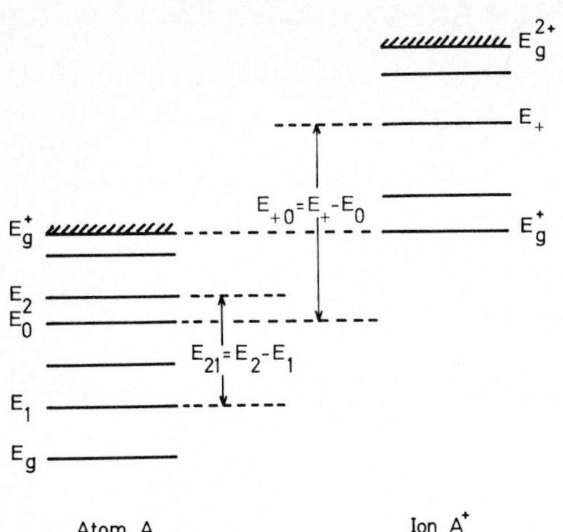

Fig. 1.1. Energy level diagram of atom A and ion A^+. E_1, E_2 denote energies of lower and upper levels of atom A, and E_0, E_+, energies of levels of atom A and ion A^+, respectively, used throughout Chaps. 1–3. The ground state energies of A, A^+, A^{2+} are E_g, E_g^+, E_g^{2+}

where N_1, N_2 are the occupation numbers, and g_1, g_2 the statistical weights of the atomic levels, or

$$\frac{N_1/g_1}{N_2/g_2} = \frac{\bar{z}(\bar{E}-E_{21})}{\bar{z}(\bar{E})} = r(E_{21}) = e^{\beta E_{21}} \ , \tag{1.4.18}$$

using (1.4.5).

In terms of number densities $n = N/V$, the *Boltzmann distribution* for the excitation degrees of freedom is thus

$$\left(\frac{n_2}{n_1}\right)^{\mathrm{B}} = \frac{g_2}{g_1} e^{-E_{21}/kT} \ . \tag{1.4.19}$$

1.4.3 Saha Distribution

Turning now to the Saha distribution for thermal ionization equilibrium, we consider detailed balance of ionization collisions and three-body recombinations,

$$A(E_0) + P(\bar{E}, d\bar{E}) \rightleftarrows A^+(E_+) + \mathrm{e}(E, dE) + P(\bar{E} - E_{+0} - E, d\bar{E}') \ , \tag{1.4.20}$$

where P is a free particle and e an electron. The atom A and the ion A^+ are considered to be of infinite mass and at rest; moreover, they are assumed to be in well-defined internal states of energies E_0 and E_+, respectively. Generally, A^+ denotes the next higher stage of ionization of the species A. For instance, if A = He, Li^+, H^-, then $A^+ = He^+$, Li^{2+}, H. The energy difference $E_{+0} = E_+ - E_0$ is the usual ionization energy of the atom A only when both E_0 and E_+ are the corresponding ground state energies E_g and E_g^+ (Fig. 1.1).

Generalizing in a straightforward manner the reaction rate (1.3.2) to processes of the type $A + B \rightleftarrows C + D + E$, (1.4.20) gives

$$N_0\bar{N}(\bar{E})\,g_+\,G_e(E)\,\bar{G}(\bar{E}-E_{+0}-E)$$
$$=N_+N_e(E)\,\bar{N}(\bar{E}-E_{+0}-E)\,g_0\,G_e(E)\ , \qquad (1.4.21)$$

where N_0, N_+ are the occupation numbers, and g_0, g_+ the statistical weights of the bound levels under consideration. This equation can be written in the form

$$\frac{N_0/g_0}{N_+/g_+}\,\frac{1}{z_e(E)}=\frac{\bar{z}(\bar{E}-E_{+0}-E)}{\bar{z}(\bar{E})}=r(E_{+0}+E)=e^{\beta(E_{+0}+E)}\ , \qquad (1.4.22)$$

again using (1.4.5). Now, since the electron distribution cannot depend on the energies E_0 and E_+, we may conclude from (1.4.22) that

$$z_e(E)=\zeta_e e^{-\beta E}=\tfrac{1}{2}n_e\lambda_e^3 e^{-\beta E}\ . \qquad (1.4.23)$$

Here the constant ζ_e has been determined by (1.4.11) with $g_e=2$, owing to the electron spin 1/2, and

$$\lambda_e=\frac{h}{(2\pi m_e kT)^{1/2}} \qquad (1.4.24)$$

is the thermal de Broglie wavelength of free electrons, m_e being the electron mass, see (1.4.13).

Inserting (1.4.23) in (1.4.22) now gives the *Saha distribution* for the ionization degrees of freedom,

$$\left(\frac{n_e n_+}{n_0}\right)^S=\frac{2g_+}{g_0}\lambda_e^{-3}e^{-E_{+0}/kT}\ , \qquad (1.4.25)$$

written in terms of number densities $n=N/V$. Owing to the approximation of infinite atom and ion masses, the de Broglie wavelength λ_e of free electrons appears in (1.4.25). The correct Saha equation is obtained from (1.4.25) by replacing there the electron mass m_e by the reduced mass $\mu=m_+m_e/(m_++m_e)$, m_+ being the ion mass, see Appendix E, (E.2.13).

Clearly, the Saha equation governs not only the thermal distributions of atoms and positive ions but also those of atoms and negative ions. For example, the thermal number densities of H^- ions and H atoms in the ground state obey the Saha equation

$$\frac{n_e n_H}{n_{H^-}}=4\lambda_e^{-3}e^{-\varepsilon/kT}\ ,$$

since $2g_H/g_{H^-}=4$, and where $\varepsilon=0.75$ eV is the detachment energy of an H^- ion.

1.4.4 Planck Distribution

To derive the thermal distribution function of the photons from detailed balance, two points must be observed. First, stimulated photon processes must be taken into account, since they determine the low-frequency (or Rayleigh-Jeans) part of the Planck function. And second, one must consider processes by which the photon number is changed (that is, emissions and absorptions as opposed to scattering processes), because otherwise the chemical potential of the photon gas cannot be determined.

Therefore let us consider detailed balance of emission and absorption of photons by free particles P in the field of particles Q which are of infinite mass and hence at rest. For example, for ordinary bremsstrahlung, P is an electron and Q is a positive ion. Detailed balance requires that

$$Q+P(\bar{E},d\bar{E}) \rightleftarrows Q+P(\bar{E}-h\nu,d\bar{E}')+\gamma(\nu,d\nu) \ , \tag{1.4.26}$$

where we used the fact that the stationary particle Q can take up only momentum but not energy. In (1.4.26), $\gamma(\nu,d\nu)$ denotes a photon with frequency in the range $(\nu,d\nu)$. Let $N(\nu)\,d\nu$ be the number of photons, and $G(\nu)\,d\nu$ the number of photon states in this range, and let, furthermore,

$$z(\nu)=\frac{N(\nu)\,d\nu}{G(\nu)\,d\nu}=\frac{N(\nu)}{G(\nu)} \tag{1.4.27}$$

denote the mean occupation number of a photon state of frequency ν. Detailed balance (1.4.26) then leads to

$$\bar{N}(\bar{E})\,\bar{G}(\bar{E}-h\nu)\,G(\nu)\,[1+z(\nu)]=\bar{N}(\bar{E}-h\nu)\,N(\nu)\,\bar{G}(\bar{E}) \ , \tag{1.4.28}$$

where the factor $[1+z(\nu)]$ accounts for spontaneous and induced emission of photons, see (1.3.3a). In a now familiar way, one obtains from (1.4.28)

$$\frac{1+z(\nu)}{z(\nu)}=\frac{\bar{z}(\bar{E}-h\nu)}{\bar{z}(\bar{E})}=r(h\nu)=\mathrm{e}^{\beta h\nu} \ , \tag{1.4.29}$$

and hence

$$z(\nu)=\frac{1}{\mathrm{e}^{\beta h\nu}-1} \ , \quad \text{or} \tag{1.4.30}$$

$$N(\nu)\,d\nu=G(\nu)\frac{1}{\mathrm{e}^{\beta h\nu}-1}\,d\nu \ . \tag{1.4.31}$$

Comparison with (1.3.18a) shows that for photons $\zeta=1$. Since, according to (1.3.19), $\zeta=\exp(\mu/kT)$, we obtain the well-known result that the chemical potential μ of the photon gas is zero, due to the nonconservation of the photon number.

The number of photon states in the frequency range $(\nu, d\nu)$ follows from (1.4.7), which is also valid for photons:

$$G(\nu)\,d\nu = 2\,V\frac{4\,\pi\,p^2 dp}{h^3} = V\frac{8\,\pi\,\nu^2}{c^3}\,d\nu \tag{1.4.32}$$

because the photon momentum is $p = h\,\nu/c$. In (1.4.32), the factor 2 is due to the two polarization states of a photon, and V is again the volume considered.

From (1.4.31,32) we get the *Planck distribution*

$$n(\nu)\,d\nu = \frac{8\,\pi\,\nu^2}{c^3}\,\frac{1}{\mathrm{e}^{h\nu/kT}-1}\,d\nu \tag{1.4.33}$$

in terms of the photon density per unit frequency and unit volume

$$n(\nu) = \frac{N(\nu)}{V}\ . \tag{1.4.34}$$

Let us now rewrite the blackbody distribution (1.4.33) in terms of the specific intensity of radiation which is the quantity commonly used for describing a radiation field. The *specific intensity* $I_\nu(\boldsymbol{n}) \equiv I_\nu(\boldsymbol{n},\boldsymbol{r},t)$ of an unpolarized radiation field is defined as follows (Fig. 1.2). At point $\boldsymbol{r}$ and time t, consider a beam of unpolarized light in the frequency range $(\nu, d\nu)$ passing through the fixed area dF into the solid angle $d\Omega$ whose direction is characterized by the unit vector $\boldsymbol{n}$. If ϑ is the angle between the direction of the beam and the normal of the area dF, then the radiant energy dE transported by the beam through dF during the time dt is

$$dE = I_\nu(\boldsymbol{n})\,d\nu\,d\Omega\,dF\cos\vartheta\,dt\ . \tag{1.4.35}$$

This relation defines the specific intensity $I_\nu(\boldsymbol{n})$ which, essentially, is the photon distribution function (Sect. 3.1). The radiant energy (1.4.35) fills the volume

$$dV = dF\cos\vartheta\;c\,dt \tag{1.4.36}$$

since all photons of the beam move in the same direction with the same speed c, and corresponds to

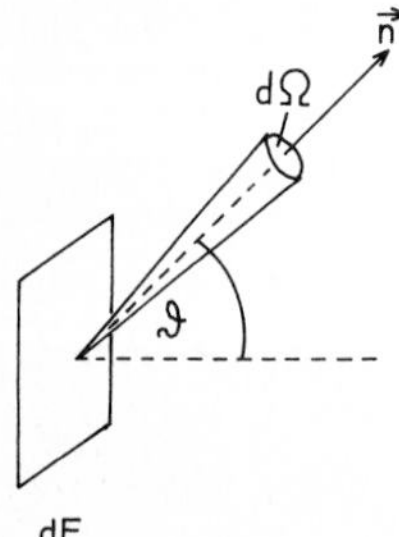

Fig. 1.2. Geometrical quantities used for defining the specific intensity of a radiation field

$$dN = \frac{1}{h\nu} I_\nu(\boldsymbol{n})\, d\nu\, d\Omega\, dF \cos\vartheta\, dt \tag{1.4.37}$$

photons since each photon carries the same energy $h\nu$. Hence, a light beam $I_\nu(\boldsymbol{n})\, d\nu\, d\Omega$ has the energy density

$$\frac{dE}{dV} = \frac{1}{c} I_\nu(\boldsymbol{n})\, d\nu\, d\Omega\ , \tag{1.4.38}$$

the photon density

$$\frac{dN}{dV} = \frac{1}{h\nu c} I_\nu(\boldsymbol{n})\, d\nu\, d\Omega\ , \tag{1.4.39}$$

and the photon flux density

$$c\frac{dN}{dV} = \frac{1}{h\nu} I_\nu(\boldsymbol{n})\, d\nu\, d\Omega\ . \tag{1.4.40}$$

On the other hand, analogous to (1.4.7, 32), the number of photon states $(\nu, d\nu; \boldsymbol{n}, d\Omega)$ is given by

$$dG = 2\frac{dV\, d^3p}{h^3} = \frac{2\nu^2}{c^3} d\nu\, d\Omega\, dV\ , \tag{1.4.41}$$

using the element of momentum space

$$d^3p = p^2 dp\, d\Omega = \frac{h^3\nu^2}{c^3} d\nu\, d\Omega \tag{1.4.42}$$

since $p = h\nu/c$. Thus, a light beam of specific intensity $I_\nu(\boldsymbol{n})$ corresponds to a mean occupation number of a photon state

$$z_\nu(\boldsymbol{n}) \equiv \frac{dN}{dG} = \frac{c^2}{2h\nu^3} I_\nu(\boldsymbol{n})\ . \tag{1.4.43}$$

We now return to the special case of blackbody radiation. In an isotropic radiation field, the relation between the quantity $n(\nu)$, (1.4.34), and the specific intensity I_ν is, according to (1.4.39),

$$\frac{dN}{dV} = n(\nu)\, d\nu = \frac{4\pi}{h\nu c} I_\nu d\nu\ ,$$

recalling that $n(\nu)$ here refers to the solid angle 4π, or

$$n(\nu) = \frac{4\pi}{h\nu c} I_\nu\ . \tag{1.4.44}$$

Inserting this expression in (1.4.33) and writing as usual $B_\nu(T)$ for the specific intensity of a blackbody radiation field of temperature T, we obtain the *Planck function* in its familiar form

$$B_\nu(T) = \frac{2h\nu^3}{c^2} \frac{1}{e^{h\nu/kT}-1} . \tag{1.4.45}$$

In the foregoing, we have tacitly assumed that the refractive index is equal to unity for all frequencies considered. This assumption is made throughout this book.

1.5 Validity of the Reciprocity Relation $w(i \to f) = w(f \to i)$

In the preceding sections it was shown that the reciprocity relation (1.2.2) is an essential condition for the validity of the principle of detailed balance. Furthermore, the reciprocity relation also plays a crucial part in the proof of the H theorem (Chap. 5). It is therefore worthwhile to investigate the validity of this relation in more detail.

A very general statement about the reciprocity relation is provided by first-order perturbation theory in quantum mechanics. Here the transition rate $w(i \to f)$ is related to the matrix element $H_{fi} = \langle f | H_{\text{int}} | i \rangle$ of the interaction Hamiltonian H_{int} through

$$w(i \to f) = \frac{2\pi}{\hbar} |H_{fi}|^2 \delta(E_f - E_i) . \tag{1.5.1}$$

Here $\hbar = h/2\pi$, $|i\rangle$ and $|f\rangle$ are eigenstates of the unperturbed Hamiltonian H_0 such that $H_0|i\rangle = E_i|i\rangle$ and $H_0|f\rangle = E_f|f\rangle$, and the delta function ensures conservation of energy. Since H_{int} is a Hermitian operator, $H_{if} = H_{fi}^*$, thus $|H_{if}|^2 = |H_{fi}|^2$, and hence, see (1.2.2),

$$w(i \to f) = w(f \to i) . \tag{1.5.2}$$

The same argument applies more generally to the first nonvanishing contribution of the perturbation series. For example, if the first-order term vanishes, $H_{fi} = 0$, the transition probability is still given by (1.5.1) provided that there H_{fi} is replaced by the second-order term

$$H_{fi}^{(2)} = \sum_n \frac{H_{fn} H_{ni}}{E_i - E_n} . \tag{1.5.3}$$

Here the summation is over all intermediate states $|n\rangle$ with $H_0|n\rangle = E_n|n\rangle$, for which $E_n \neq E_i$. Since all E_k are real and, in addition, $E_f = E_i$ because of the delta function in (1.5.1), then $H_{if}^{(2)} = H_{fi}^{(2)*}$, again leading to (1.5.2).

It follows, then, that the reciprocity relation (1.2.2) is valid for all radiative processes, for these are accurately described by the first nonvanishing order of perturbation theory. Likewise, it holds for all collision processes for which the Born approximation is valid.

Perturbation theory, however, is not sufficient to ensure the validity of the reciprocity relation (1.2.2) for more general collision processes. The general framework for describing collision processes is provided by the scattering or S matrix. This approach applies to any kind of collision, and hence, in particular, to the three collision types considered in this book: elastic, inelastic, and rearrangement collisions (the latter referring to general reactions in which electrons and nuclei are conserved, e.g., ionization collisions).

In loose terms, the S matrix may be viewed as the time evolution operator of the system of colliding particles. If at time $t = -\infty$ (before the collision) the system was in the initial state $|i\rangle$, at time $t = +\infty$ (after the collision) it will be in the state

$$|F\rangle = S|i\rangle \ . \tag{1.5.4}$$

According to the general rules of quantum mechanics, the probability for the transition $|i\rangle \to |f\rangle$ into a specific final state $|f\rangle$ is proportional to the square of the modulus of the projection of $|f\rangle$ onto $|F\rangle$:

$$\langle f|F\rangle = \langle f|S|i\rangle = S_{fi} \ . \tag{1.5.5}$$

Equation (1.5.5) is the S matrix element between the considered states $|i\rangle$ and $|f\rangle$. These are taken to be free particle states since at $t = \pm\infty$ the particles are far apart and hence noninteracting. One now defines the transition or T matrix through

$$S_{fi} = \delta_{fi} - 2\pi i T_{fi} \delta(E_f - E_i) \ . \tag{1.5.6}$$

Here δ_{fi} contributes only to those cases where no transition takes place ($|f\rangle = |i\rangle$), whereas the second term, containing the T matrix element, accounts for true collisions that change the state of the system ($|f\rangle \neq |i\rangle$). For $|f\rangle \neq |i\rangle$, the transition rate $w(i\to f)$ is in terms of the T matrix element given by [1.5, 6]

$$w(i\to f) = \frac{2\pi}{\hbar} |T_{fi}|^2 \delta(E_f - E_i) \ . \tag{1.5.7}$$

Note that in contrast to the perturbation expression (1.5.1) which has the same form, (1.5.7) is an exact result.

We now return to the question of the validity of the reciprocity relation (1.2.2) for general collision processes. To this end, symmetry properties must be considered. Since in this book we are concerned only with electromagnetic interactions, the Hamiltonian of a system of colliding particles is invariant under space reflection. Moreover, we assume that it is also invariant under time reversal. This latter assumption requires in general the absence of an external magnetic field and/or a negligible dependence of the Hamiltonian on magnetic effects, for time reversal symmetry relates a specific state of the system in magnetic field $\boldsymbol{B}$ to the time-reversed state in the reversed field $-\boldsymbol{B}$ rather than in $\boldsymbol{B}$.

It is pertinent now to ask whether the invariance of the Hamiltonian with respect to space reflection and time reversal ensures the validity of the reciprocity relation $w(i\to f)=w(f\to i)$. In classical physics, application of space reflection (parity) P to a given configuration corresponds to performing the substitution $\boldsymbol{r}\to-\boldsymbol{r}$ for the position vectors, leaving the time t unchanged; linear momenta $\boldsymbol{p}$ and angular momenta $\boldsymbol{j}$ then change according to $\boldsymbol{p}\to-\boldsymbol{p}$ and $\boldsymbol{j}\to\boldsymbol{j}$. On the other hand, application of time reversal T to a given configuration corresponds to substituting $t\to-t$, leaving $\boldsymbol{r}$ unchanged; linear and angular momenta then change according to $\boldsymbol{p}\to-\boldsymbol{p}$ and $\boldsymbol{j}\to-\boldsymbol{j}$. Thus, formally,

$$
\begin{aligned}
P:\ t &\to t \qquad & T:\ t &\to -t\\
\boldsymbol{r} &\to -\boldsymbol{r} & \boldsymbol{r} &\to \boldsymbol{r}\\
\boldsymbol{p} &\to -\boldsymbol{p} & \boldsymbol{p} &\to -\boldsymbol{p}\\
\boldsymbol{j} &\to \boldsymbol{j} & \boldsymbol{j} &\to -\boldsymbol{j}\,.
\end{aligned}
\tag{1.5.8}
$$

Turning to quantum mechanics, two theorems about S matrix elements are stated below without proof which, however, are easily interpreted in view of the classical substitution rules (1.5.8). To this end, the state of a free particle is specified by the linear momentum $\boldsymbol{p}$ and by the usual quantum numbers j and m corresponding to the internal angular momentum (spin) of the particle. Thus, a free particle state is $|a,\boldsymbol{p},j,m\rangle$, where a stands for all remaining quantum numbers of the particle. Note that the classical substitution $\boldsymbol{j}\to-\boldsymbol{j}$ corresponds to the quantum-mechanical substitution $j,m\to j,-m$. It should also be observed that the spin of a particle contains in general contributions from internal orbital motions. For instance, the $3\,^2P_{3/2}$ state of an H atom has "spin" $j=3/2$ that is partly due to the orbital angular momentum $l=1$ (P state) and partly due to the spin $s=1/2$ of the electron.

In the following $\boldsymbol{p}_i(\boldsymbol{p}_f)$ denotes the whole set of linear momenta of all incoming (outgoing) particles, and the quantum numbers a,j,m are to be interpreted analogously. Then the following two theorems hold [1.5, 6].

(1) If the Hamiltonian is invariant under space reflection P, it is

$$
\begin{aligned}
&\langle a_f,\boldsymbol{p}_f,j_f,m_f|\,S\,|a_i,\boldsymbol{p}_i,j_i,m_i\rangle\\
&\quad=\langle a_f,-\boldsymbol{p}_f,j_f,m_f|\,S\,|a_i,-\boldsymbol{p}_i,j_i,m_i\rangle\,.
\end{aligned}
\tag{1.5.9}
$$

That is, the S matrix element between two states is equal to the S matrix element between the space-reflected states. In (1.5.9), invariance of the quantum numbers a_i and a_f under P has been assumed.

(2) If the Hamiltonian is invariant under time reversal T, it is

$$\langle a_f, p_f, j_f, m_f | S | a_i, p_i, j_i, m_i \rangle$$
$$= \pm \langle a_i, -p_i, j_i, -m_i | S | a_f, -p_f, j_f, -m_f \rangle . \qquad (1.5.10)$$

That is, apart from a phase factor ± 1 which is irrelevant in the following considerations, the S matrix element between two states is equal to the S matrix element between the time-reversed states if, in agreement with the naive picture of time reversal, the initial and final states are interchanged. In (1.5.10), invariance of the quantum numbers a_i and a_f under T has been assumed.

Let us now consider the following transitions:

(1) $|a_i, p_i, j_i, m_i\rangle \to |a_f, p_f, j_f, m_f\rangle$ $\quad (S)$,

(2) $|a_f, -p_f, j_f, -m_f\rangle \to |a_i, -p_i, j_i, -m_i\rangle$ $\quad (S_T)$,

(3) $|a_f, p_f, j_f, -m_f\rangle \to |a_i, p_i, j_i, -m_i\rangle$ $\quad (S_{TP})$,

(4) $|a_f, p_f, j_f, m_f\rangle \to |a_i, p_i, j_i, m_i\rangle$ $\quad (S_{\text{inv}})$.

Reaction (1) is the reaction $|i\rangle \to |f\rangle$ under consideration. Reaction (2) follows from (1) by applying the symmetry operation T, (3) follows from (2) by applying P or, equivalently, from (1) by applying TP. Finally, reaction (4) is the inverse reaction of the original one, $|f\rangle \to |i\rangle$. The corresponding four S matrix elements are denoted by S, S_T, S_{TP}, and S_{inv}, respectively. The important point to notice is that the inverse reaction (4) cannot be obtained from the direct reaction (1) by applying the symmetry operations P and T.

If the Hamiltonian of the system is invariant under P and T, then through (1.5.9, 10)

$$|S| = |S_T| = |S_{TP}|, \quad \text{whereas in general} \qquad (1.5.11)$$

$$|S| \neq |S_{\text{inv}}| . \qquad (1.5.12)$$

Consequently, since the transition probability is proportional to $|S|^2$,

$$w(i \to f) \neq w(f \to i) \qquad (1.5.13)$$

even if the Hamiltonian is P- and T-invariant.

Inspecting reactions (3,4) above shows that they are identical if $m_i = m_f = 0$. In particular, if all incoming and outgoing particles are spinless ($j = 0$), reactions (3,4) coincide. In this case, and if the Hamiltonian is P- and T-invariant, $|S_{TP}| = |S_{\text{inv}}|$, and the reciprocity relation (1.2.2) holds. This result is the quantum analogue to Boltzmann's statement that detailed balance in a gas obtains only when the colliding particles are spherically symmetric.

The foregoing result suggests that the reciprocity relation (1.2.2) may be generally valid for unpolarized particles, provided the Hamiltonian is *P*- and *T*-invariant. This is indeed the case, as can be seen as follows. For unpolarized particles, the magnetic sublevels $m = j, j-1, \ldots, -j$ corresponding to given a, p, j are occupied with equal probability. Thus, considering as initial and final states of unpolarized particles the whole groups of magnetic sublevels,

$$w(i \to f) \propto \frac{1}{g_i g_f} \sum_{m_i} \sum_{m_f} |S_{fi}|^2 , \tag{1.5.14}$$

where m_i and m_f stand for the whole sets of magnetic quantum numbers of all initial and final particles, and $g_i = \prod (2j_i+1)$, $g_f = \prod (2j_f+1)$ are the corresponding statistical weights. One observes that in (1.5.14), $(1/g_i)$ times the sum over m_i is the statistical average over the equally populated initial sublevels, that the sum over the final sublevels m_f must be carried out because the polarization of the outgoing particles is not observed, and that the factor $1/g_f$ appears since $w(i \to f)$ is defined as the transition probability between single quantum states. Now,

$$\frac{1}{g_f g_i} \sum_{m_f} \sum_{m_i} |S_{TP}|^2 = \frac{1}{g_f g_i} \sum_{m_f} \sum_{m_i} |S_{\text{inv}}|^2 \tag{1.5.15}$$

because the two sums include exactly the same terms, merely in a different order. However, $|S_{TP}| = |S|$ on account of (1.5.11), and hence

$$\frac{1}{g_i g_f} \sum_{m_i} \sum_{m_f} |S|^2 = \frac{1}{g_f g_i} \sum_{m_f} \sum_{m_i} |S_{\text{inv}}|^2 , \tag{1.5.16}$$

which, in view of (1.5.14), leads to the reciprocity relation for unpolarized particles, $w(i \to f) = w(f \to i)$.

To sum up, the reciprocity relation $w(i \to f) = w(f \to i)$ is not generally valid, not even when the Hamiltonian of the system is invariant under space reflection and time reversal. However, it holds in the framework of perturbation theory, and thus applies to radiative processes and to collisions in the Born approximation. The reciprocity relation holds moreover for all collision processes involving unpolarized particles, provided that the Hamiltonian is *P*- and *T*- invariant.

In this book it is always assumed that particles and photons are unpolarized, and that the reciprocity relation $w(i \to f) = w(f \to i)$ is valid for all processes considered.

1.6 Explicit Forms of the Reciprocity Relation $w(i \to f) = w(f \to i)$

The reciprocity relation (1.2.2) relates an atomic quantity that determines the rate of a specific process (for example, a cross section or an Einstein coefficient) to the corresponding quantity of the inverse process. In this section the most important of these relations are derived for inelastic collision processes and for radiative processes, while the corresponding relations for elastic processes, such as elastic collisions of particles or Thomson scattering of photons, are discussed at the appropriate places below. These relations are derived through the principle of detailed balance, by writing down the rate equations in question and using the explicit forms of the thermal distribution functions given in Sect. 1.4. This procedure is very simple and straightforward, while the relations so found are, of course, independent of this method and could be derived just as well directly from quantum mechanics.

1.6.1 Collisional Excitation and De-Excitation

Let us consider detailed balance between collisional excitation of an atom in a lower level 1 by an electron in a well-defined velocity range, and the corresponding collisional de-excitation of an atom in a higher level 2,

$$A(E_1) + \mathrm{e}(\boldsymbol{v}, d^3v) \rightleftarrows A(E_2) + \mathrm{e}(\boldsymbol{v}', d^3v') , \tag{1.6.1}$$

where E_1, E_2 are the energies of the atomic levels (Fig. 1.1). Here and in the following, the approximation of stationary atoms is made for all kinds of electron-atom collisions, that is, the mass of the atom is considered to be infinite, so that the relative velocity between electron and atom can be replaced by the electron velocity, and the recoil of the atom can be ignored. In this approximation, conservation of energy requires

$$\tfrac{1}{2} m_e v^2 = \tfrac{1}{2} m_e v'^2 + E_{21} , \tag{1.6.2}$$

$$d(\tfrac{1}{2} m_e v^2) = d(\tfrac{1}{2} m_e v'^2) , \quad v\, dv = v'\, dv' . \tag{1.6.3}$$

Here, $E_{21} = E_2 - E_1$ is the excitation energy, and dv is related to d^3v through

$$d^3v = v^2 dv\, d\Omega , \tag{1.6.4}$$

where $d\Omega$ is the element of solid angle.

Detailed balance (1.6.1) implies the following relation which expresses the equality of the two reaction rates per unit volume,

$$\begin{aligned} &n_1 v n_e f_e(\boldsymbol{v}) v^2 dv\, d\Omega\, q_{12}(\boldsymbol{v}; \boldsymbol{v}')\, d\Omega' \\ &\quad = n_2 v' f_e(\boldsymbol{v}') v'^2 dv'\, d\Omega'\, q_{21}(\boldsymbol{v}'; \boldsymbol{v})\, d\Omega . \end{aligned} \tag{1.6.5}$$

Here n_1 and n_2 are the densities of atoms in the lower and upper level, n_e is the electron density, and $f_e(\boldsymbol{v})$ is the normalized velocity distribution of the electrons. Furthermore, $q_{12}(\boldsymbol{v}; \boldsymbol{v}')$ and $q_{21}(\boldsymbol{v}'; \boldsymbol{v})$ are the differential cross sections of the considered excitation and de-excitation collisions, which for unpolarized particles may also be written as $q_{12}(v, \vartheta)$ and $q_{21}(v', \vartheta)$, respectively, because they depend only on the modulus of the relative velocity before the collision and on the scattering angle ϑ defined by $\boldsymbol{v} \cdot \boldsymbol{v}' = v v' \cos\vartheta$. Recall that the number of reactions per unit volume and unit time is given by the product of the density of scattering centers, the incoming flux density, and the cross section [thus, for example, $n_1 \times v n_e f_e(\boldsymbol{v}) d^3 v \times q_{12} d\Omega'$].

According to the principle of detailed balance, (1.6.5) is valid in thermodynamic equilibrium for arbitrary ranges of the electron velocities. Inserting therefore in (1.6.5) for $f_e(\boldsymbol{v})$ the Maxwell distribution (1.4.14), written in terms of velocity $\boldsymbol{v}$ rather than kinetic energy E [$E = mv^2/2$, $dE = mv\,dv$, $f(\boldsymbol{v}) 4\pi v^2 dv = f(E)\,dE$],

$$f^{\mathrm{M}}(\boldsymbol{v}) = f^{\mathrm{M}}(v) = \left(\frac{m}{2\pi kT}\right)^{3/2} \mathrm{e}^{-mv^2/2kT} , \tag{1.6.6a}$$

$$\int f^{\mathrm{M}}(\boldsymbol{v})\, d^3 v = 1 , \tag{1.6.6b}$$

inserting for n_2/n_1 the Boltzmann distribution (1.4.19), and using (1.6.2, 3), one obtains

$$g_1 v^2 q_{12}(v, \vartheta) = g_2 v'^2 q_{21}(v', \vartheta) , \tag{1.6.7}$$

where g_1, g_2 are the statistical weights of the atomic levels, and v and v' are connected by (1.6.2).

Equation (1.6.7) is an example of the explicit form of the reciprocity relation $w(i \to f) = w(f \to i)$, relating the differential cross section for electronic excitation to that of de-excitation. Note that the temperature and the particle densities have dropped out, as they should.

Given a differential cross section $q(v, \vartheta)$, the corresponding total cross section $Q(v)$ is defined by

$$Q(v) = \int q(v, \vartheta)\, d\Omega = 2\pi \int_0^\pi q(v, \vartheta) \sin\vartheta\, d\vartheta . \tag{1.6.8}$$

From (1.6.7) the *Klein-Rosseland relation* for the total cross sections Q_{12} and Q_{21} then follows immediately. Written in terms of kinetic energies $E = m_e v^2/2$ and $E' = m_e v'^2/2$,

$$g_1 E Q_{12}(E) = g_2 E' Q_{21}(E') , \tag{1.6.9}$$

where

$$E = E' + E_{21} \tag{1.6.10}$$

on account of (1.6.2).

Above we have discussed only inelastic collisions with electrons, using the approximation of stationary atoms. Inelastic collisions with heavy particles are considered in Appendix E.

1.6.2 Collisional Ionization and Three-Body Recombination

Here we consider detailed balance between collisional ionization of an atom A by electron impact, and the corresponding three-body recombination of the ion A^+,

$$A(E_0)+\mathrm{e}(\boldsymbol{v},d^3v)\rightleftarrows A^+(E_+)+\mathrm{e}(\boldsymbol{v}',d^3v')+\mathrm{e}(\boldsymbol{v}'',d^3v'') \ , \tag{1.6.11}$$

where E_0 and E_+ are the energies of the bound levels (Fig. 1.1). Read from left to right, the reaction may be collisional ionization of an atom or positive ion (e.g., $\mathrm{He}^+ + \mathrm{e}\to \mathrm{He}^{2+}+\mathrm{e}+\mathrm{e}$), or collisional detachment of a negative ion (e.g., $\mathrm{H}^- + \mathrm{e}\to\mathrm{H}+\mathrm{e}+\mathrm{e}$). Assuming again the heavy particles A and A^+ to be stationary, conservation of energy requires

$$\tfrac{1}{2}m_\mathrm{e}v^2=\tfrac{1}{2}m_\mathrm{e}v'^2+\tfrac{1}{2}m_\mathrm{e}v''^2+E_{+0} \ , \tag{1.6.12}$$

where $E_{+0}=E_+-E_0$ is the ionization energy.

The rate equation for the balanced reactions (1.6.11) can be written as

$$\begin{aligned} &n_0 v n_\mathrm{e} f_\mathrm{e}(\boldsymbol{v})\,d^3v\,\omega_{0+}(\boldsymbol{v};\boldsymbol{v}',\boldsymbol{v}'')d^3v'\,d^3v'' \\ &\quad = n_+ v' n_\mathrm{e} f_\mathrm{e}(\boldsymbol{v}')d^3v'\,v''\,n_\mathrm{e} f_\mathrm{e}(\boldsymbol{v}'')d^3v''\,\omega_{+0}(\boldsymbol{v}',\boldsymbol{v}'';\boldsymbol{v})\,d^3v \ . \end{aligned} \tag{1.6.13}$$

Here, n_0, n_+, n_e are the particle densities, and $f_\mathrm{e}(\boldsymbol{v})$ is the normalized electron velocity distribution, so that $v n_\mathrm{e} f_\mathrm{e}(\boldsymbol{v})\,d^3v$ is the flux density of electrons with velocities in the range $(\boldsymbol{v},d^3v)$ impinging on a particle at rest. The precise meaning of the quantities ω_{0+} and ω_{+0} follows from (1.6.13). The left-hand side of this equation is the number of ionization collisions per unit time and unit volume due to electrons $\mathrm{e}(\boldsymbol{v},d^3v)$ such that, after the collision, the two outgoing electrons have velocities in the ranges $(\boldsymbol{v}',d^3v')$ and $(\boldsymbol{v}'',d^3v'')$, respectively, and the right-hand side is the rate of the inverse three-body recombinations per unit volume. In view of (1.6.12), $\omega_{0+}(\boldsymbol{v};\boldsymbol{v}',\boldsymbol{v}'')$ and $\omega_{+0}(\boldsymbol{v}',\boldsymbol{v}'';\boldsymbol{v})$ both contain the delta function

$$\delta(\tfrac{1}{2}m_\mathrm{e}v^2-\tfrac{1}{2}m_\mathrm{e}v'^2-\tfrac{1}{2}m_\mathrm{e}v''^2-E_{+0})$$

as a factor [but not the delta function $\delta(m_\mathrm{e}\boldsymbol{v}-m_\mathrm{e}\boldsymbol{v}'-m_\mathrm{e}\boldsymbol{v}'')$, since the heavy particle can take up any amount of momentum].

Inserting now into (1.6.13) the Maxwell distribution (1.6.6) for $f_\mathrm{e}(\boldsymbol{v})$, the Saha distribution (1.4.25) for n_+n_e/n_0, and using (1.6.12), one obtains

$$g_0 v\,\omega_{0+}(\boldsymbol{v};\boldsymbol{v}',\boldsymbol{v}'')=2g_+\frac{m_\mathrm{e}^3}{h^3}v'v''\,\omega_{+0}(\boldsymbol{v}',\boldsymbol{v}'';\boldsymbol{v}) \ , \tag{1.6.14}$$

where g_0, g_+ are the statistical weights of the bound levels, and h is Planck's constant.

In analogy to the passage from differential to total cross sections, total quantities Ω_{0+} and Ω_{+0} may be introduced in the following manner: Ω_{0+} is obtained from ω_{0+} by summing over all directions of the two outgoing electrons,

$$\Omega_{0+}(v;v',v'') = \iint \omega_{0+}(\boldsymbol{v};\boldsymbol{v}',\boldsymbol{v}'')\,d\Omega'\,d\Omega'' \ , \tag{1.6.15}$$

while Ω_{+0} is obtained from ω_{+0} by summing over the direction of the outgoing electron and averaging over the relative orientations of the two velocities of the incoming electrons. This is clearly equivalent to

$$\Omega_{+0}(v',v'';v) = \iint \omega_{+0}(\boldsymbol{v}',\boldsymbol{v}'';\boldsymbol{v})\,\frac{d\Omega'\,d\Omega''}{4\pi} \ . \tag{1.6.16}$$

The total quantities Ω in terms of kinetic energies $E = m_e v^2/2$ are defined by

$$\Omega_{0+}(v;v',v'')\,v'^2 dv'\,v''^2 dv'' = \Omega_{0+}(E;E',E'')\,dE'\,dE'' \qquad \text{or}$$

$$\Omega_{0+}(v;v',v'') = \frac{m_e^3}{2(E'E'')^{1/2}}\,\Omega_{0+}(E;E',E'') \ , \tag{1.6.17}$$

and by

$$\Omega_{+0}(v',v'';v)\,v^2 dv = \Omega_{+0}(E',E'';E)\,dE \qquad \text{or}$$

$$\Omega_{+0}(v',v'';v) = \frac{m_e^{3/2}}{(2E)^{1/2}}\,\Omega_{+0}(E',E'';E) \ . \tag{1.6.18}$$

Because of the energy relation, see (1.6.12),

$$E = E' + E'' + E_{+0} \ , \tag{1.6.19}$$

$\Omega_{0+}(E;E',E'')$ and $\Omega_{+0}(E',E'';E)$ both contain the delta function $\delta(E-E'-E''-E_{+0})$ as a factor.

Using (1.6.15 – 19), one readily obtains from (1.6.14) the *Fowler relation*

$$g_0 E\,\Omega_{0+}(E;E',E'') = \frac{16\pi m_e}{h^3}\,g_+ E' E''\,\Omega_{+0}(E',E'';E) \ , \tag{1.6.20}$$

which can also be derived directly from the rate equation for isotropic electron distributions

$$\begin{aligned} n_0 \left(\frac{2E}{m_e}\right)^{1/2} & n_e f_e(E)\,dE\,\Omega_{0+}(E;E',E'')\,dE'\,dE'' \\ = n_+ & \left(\frac{2E'}{m_e}\right)^{1/2} n_e f_e(E')\,dE' \\ & \cdot \left(\frac{2E''}{m_e}\right)^{1/2} n_e f_e(E'')\,dE''\,\Omega_{+0}(E',E'';E)\,dE \end{aligned} \tag{1.6.21}$$

by here substituting the Maxwell and Saha distributions, and using the energy relation (1.6.19).

The total cross section for collisional ionization Q_{0+} is related to Ω_{0+} by

$$Q_{0+}(E) = \iint \Omega_{0+}(E;E',E'')\,dE'\,dE'' \ , \tag{1.6.22}$$

where the integration is over all distinct outgoing electron pairs (E',E'').

It should be noted that Ω_{+0} (or ω_{+0}) for three-body recombination cannot be derived via a reciprocity relation from the ionization cross section Q_{0+}, which is the quantity usually measured or calculated. This quantity Ω_{+0} (or ω_{+0}) is needed, for example, when recombination rates in plasmas with non-Maxwellian electron velocity distributions are to be calculated.

Collisional ionization by heavy particles (rather than by electrons) and the corresponding three-body recombination are considered in Appendix E.

1.6.3 Autoionization and Radiationless Capture

Atoms (or ions) with two or more bound electrons have doubly excited levels, that is, energy levels belonging to configurations with two excited electrons. If the energy of a doubly excited level is higher than the ground state energy of the ion A^+, radiationless *autoionization* can occur, yielding an ion A^+ and a free electron. The inverse process is *radiationless capture* of a free electron by an ion, yielding a doubly excited atom. *Dielectronic recombination* occurs when radiationless capture is followed by a radiative transition producing a singly excited atom (Sect. 2.3.5). Consider, for instance, the following reactions:

$$\mathrm{He}^+(1s) + \mathrm{e} \rightleftarrows \mathrm{He}(2p, 34d) \to \mathrm{He}(1s, 34d) + h\nu \ .$$

Here a He^+ ion in its ground state $1s$ captures a free electron into a state with a high principal quantum number ($n = 34$) under simultaneous excitation of the bound electron ($1s\to 2p$) (radiationless capture). The doubly exçited He atom so formed either decomposes again into an electron-ion pair (autoionization), or the captured electron remains in its state, while the other electron jumps back into the ground state ($2p\to 1s$) under emission of a photon, thereby producing a singly excited He atom (dielectronic recombination).

Let us now consider detailed balance between autoionization of a doubly excited atom A^{**} and the inverse radiationless capture of a free electron with kinetic energy in the range (E, dE) by the ion A^+ (Fig. 1.3),

$$A^{**}(E', dE') \rightleftarrows A^+(E_+) + \mathrm{e}(E, dE) \ . \tag{1.6.23}$$

Here the continuously varying energy E' of the doubly excited atom accounts for the broadening of the level due to its short lifetime, the latter resulting from autoionization and radiative transitions to lower lying levels. More pre-

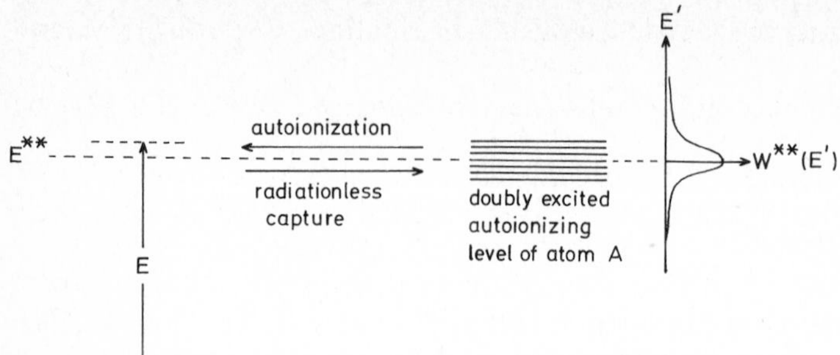

Fig. 1.3. Autoionization and radiationless capture. On the left the energy continuum E of free electrons above an energy level E_+ of ion A^+, on the right an autoionizing energy level of atom A with a level profile $w^{**}(E')$ centered about the energy E^{**}

cisely, we define the normalized level profile $w^{**}(E')$, with $\int w^{**}(E')\,dE' = 1$, such that $w^{**}(E')\,dE'$ is the probability of finding the doubly excited atom in the energy range (E', dE') (Fig. 1.3). For some purposes, the level broadening can be ignored, in which case the level profile is approximated by a delta function, $w^{**}(E') = \delta(E' - E^{**})$, where E^{**} is the energy of the level center. For the reactions (1.6.23), conservation of energy requires

$$E^{**} \simeq E' = E_+ + E\ , \quad dE' = dE\ . \tag{1.6.24}$$

The balanced reactions (1.6.23) imply the following rate equation:

$$n^{**} A_a w^{**}(E')\,dE' = n_+ \left(\frac{2E}{m_e}\right)^{1/2} n_e f_e(E)\,dE\,Q_{rc}(E)\ . \tag{1.6.25}$$

Here n^{**} is the density of atoms in level E^{**}, A_a is the transition probability per unit time for autoionization, and $Q_{rc}(E)$ is the cross section of an ion in level E_+ for radiationless capture of an electron with kinetic energy E. Inserting in (1.6.25) the Maxwell distribution (1.4.14) for $f_e(E)$, the Saha distribution (1.4.25) with $E_{+0} = E_+ - E^{**} = -E$ for $n_e n_+/n^{**}$, and using (1.6.24), one obtains

$$g^{**} A_a w^{**}(E') = \frac{16\pi m_e}{h^3} g_+ E Q_{rc}(E)\ , \tag{1.6.26}$$

where g^{**} and g_+ are the statistical weights of the bound levels. Equation (1.6.26) relates the cross section for radiationless capture to the transition probability for autoionization and to the level profile of the autoionizing level.

1.6.4 Line Emission and Absorption

We now turn to radiative processes, and consider first detailed balance of photon emission and absorption in a spectral line due to stationary atoms,

$$A(E_1) + h\nu \rightleftarrows A(E_2)\ . \tag{1.6.27}$$

Conservation of energy then requires

$$h\nu = E_{21} \;, \tag{1.6.28}$$

$E_{21} = E_2 - E_1$ being the excitation energy of the atomic transition (Fig. 1.1).

For a thermal radiation field, the rate equation corresponding to (1.6.27) can be written as

$$n_1 B_{12} B_\nu(T) = n_2 [A_{21} + B_{21} B_\nu(T)] \;, \tag{1.6.29}$$

expressing the fact that the absorption rate is equal to the sum of the rates of spontaneous and stimulated emissions. Here $B_\nu(T)$ is the Planck function (1.4.45), which is isotropic and whose intensity may be considered constant over the small frequency range of a spectral line, and A_{21}, B_{21}, B_{12} are, respectively, the Einstein coefficients for spontaneous emission, stimulated emission, and absorption. Note that B_{12} and B_{21} are defined here such that the corresponding transition rates per unit volume are given by the product of the particle density, the specific intensity, and the B coefficient (also called the Einstein-Milne coefficient). (The B coefficients defined originally by *Einstein,* and still sometimes used in the literature, differ from the Einstein-Milne coefficients as defined here by the factor $c/4\pi$.)

The ratio of stimulated to spontaneous emissions is generally given by the mean occupation number of the considered photon state [see (1.3.3a)], so that the detailed balance equation (1.6.29) may be written alternatively as

$$n_1 B_{12} B_\nu(T) = n_2 A_{21} \left[1 + \frac{c^2}{2h\nu^3} B_\nu(T) \right] \tag{1.6.30}$$

on account of (1.4.43). Comparison of (1.6.29, 30) shows that

$$\frac{A_{21}}{B_{21}} = \frac{2h\nu^3}{c^2} \;, \tag{1.6.31}$$

where ν is the frequency of the spectral line. On the other hand, inserting into (1.6.29) the Boltzmann ratio (1.4.19) for n_2/n_1 and the explicit expression of the Planck function, (1.4.45), and using (1.6.28, 31), one obtains

$$g_1 B_{12} = g_2 B_{21} \;, \tag{1.6.32}$$

where g_1 and g_2 are the statistical weights of the atomic levels. Equations (1.6.31, 32) are the so-called *Einstein relations.*

Notice that

$$\frac{B_\nu(T)}{1 + \dfrac{c^2}{2h\nu^3} B_\nu(T)} = \frac{2h\nu^3}{c^2} \mathrm{e}^{-h\nu/kT} \equiv B_\nu^{\mathrm{W}}(T) \tag{1.6.33}$$

is the *Wien function,* that is, the Planck function in the low-temperature limit $[\exp(h\nu/kT) \gg 1]$, where stimulated emissions are negligible compared to spontaneous ones. The detailed balance equation (1.6.30) may therefore be written in the equivalent form

$$n_1 B_{12} B_\nu^{\mathrm{W}}(T) = n_2 A_{21} \ . \tag{1.6.34}$$

This means that the correct relation between B_{12} and A_{21} may also be obtained by neglecting stimulated emissions altogether, and using the Wien function instead of the Planck function.

The Einstein coefficients A_{21}, B_{21}, B_{12} are defined for an atomic transition $1 \leftrightarrow 2$. In particular, A_{21} is the total transition probability per unit time for a transition $2 \to 1$ due to spontaneous emission of a single photon, independent of the direction and the exact frequency of the emitted photon. Furthermore, it should be observed that the rates per unit volume of absorptions and stimulated emissions are given by $n_1 B_{12} I$ and $n_2 B_{21} I$ only when the radiation field, in the rest frame of the atom, is isotropic and of constant intensity I within the frequency range of the spectral line. Expressions for these rates valid in general radiation fields of specific intensity $I_\nu(\boldsymbol{n})$ are given in Sects. 2.4.1 and 3.2.1.

Comparison of (1.6.29, 30) shows that the introduction of a B coefficient for stimulated emission is quite unnecessary. Indeed, such a coefficient is usually introduced (for historical reasons) only for atomic bound-bound transitions, but not for free-bound or free-free transitions, Sects. 1.6.5, 6. Moreover, the consideration of multiphoton processes, Sect. 1.6.7, shows that the whole concept of B coefficients for stimulated emission is rather artificial. In any case, it is easier to take account of stimulated emissions in the manner exemplified by (1.6.30) rather than by (1.6.29).

1.6.5 Photoionization and Radiative Recombination

We consider detailed balance between photoionization of an atom and radiative recombination of an ion with an electron:

$$A(E_0) + \gamma(\nu, d\nu; \boldsymbol{n}, d\Omega) \rightleftarrows A^+(E_+) + \mathrm{e}(\boldsymbol{v}, d^3v) \ . \tag{1.6.35}$$

Here $\gamma(\nu, d\nu; \boldsymbol{n}, d\Omega)$ denotes a photon with frequency in the range $(\nu, d\nu)$ and direction in the range $(\boldsymbol{n}, d\Omega)$, where $\boldsymbol{n}$ and $d\Omega$ are a unit vector and an element of solid angle, respectively. Read from left to right, reaction (1.6.35) may be photoionization of an atom or positive ion (e.g., $\mathrm{H} + h\nu \to \mathrm{H}^+ + \mathrm{e}$) or photodetachment of a negative ion (e.g., $\mathrm{H}^- + h\nu \to \mathrm{H} + \mathrm{e}$). For stationary heavy particles A and A^+, conservation of energy requires

$$h\nu = \tfrac{1}{2} m_\mathrm{e} v^2 + E_{+0} \ , \tag{1.6.36}$$

with $E_{+0} = E_+ - E_0$ being the ionization energy in question (Fig. 1.1), from which follows

$$h\,d\nu = d(\tfrac{1}{2} m_\mathrm{e} v^2) = m_\mathrm{e} v\,dv \ . \tag{1.6.37}$$

The balanced reactions (1.6.35) lead to the following rate equation:

$$n_0 \frac{1}{h\nu} I_\nu(\boldsymbol{n})\, d\nu\, d\Omega\, \sigma_{0+}(\nu,\vartheta)\, d\Omega_v$$

$$= n_+ v n_e f_e(\boldsymbol{v})\, v^2 dv\, d\Omega_v \sigma_{+0}(v,\vartheta)\, d\Omega \left[1 + \frac{c^2}{2h\nu^3} I_\nu(\boldsymbol{n})\right] , \quad (1.6.38)$$

where stimulated radiative recombinations have been taken into account. Here $\sigma_{0+}(\nu,\vartheta)$ and $\sigma_{+0}(v,\vartheta)$ are the differential cross sections for photoionization and spontaneous radiative recombination, respectively. (We denote by σ cross sections of radiative processes, i.e., processes involving at least one photon in the initial or final state, and by q, Q cross sections of collision processes.) Furthermore, ϑ denotes the angle between the photon direction and the electron velocity, i.e., $\boldsymbol{n}\cdot\boldsymbol{v} = v\cos\vartheta$, and the elements of solid angle $d\Omega$ and $d\Omega_v$ refer to the directions of the photon and the electron, respectively. Recall that $(I_\nu/h\nu)\,d\nu\,d\Omega$ is the flux density of photons, see (1.4.40), and $v n_e f_e d^3v$ that of the electrons.

Substituting the Planck function, the Maxwell and the Saha distributions in (1.6.38) gives rise to

$$g_0 (h\nu)^2 \sigma_{0+}(\nu,\vartheta) = g_+ m_e c^2 m_e v^2 \sigma_{+0}(v,\vartheta) , \quad (1.6.39)$$

where g_0 and g_+ are the statistical weights of the bound levels. In view of (1.6.33), (1.6.39) may also be obtained by substituting the Wien function for I_ν and neglecting stimulated recombinations. Integrating with respect to solid angle gives the *Milne relation* for the total cross sections, which in terms of kinetic energy $E = m_e v^2/2$ reads

$$g_0 (h\nu)^2 \sigma_{0+}(\nu) = 2 g_+ m_e c^2 E \sigma_{+0}(E) , \quad (1.6.40)$$

total radiative cross sections being defined by

$$\sigma(\nu) = \int \sigma(\nu,\vartheta)\, d\Omega = 2\pi \int_0^\pi \sigma(\nu,\vartheta) \sin\vartheta\, d\vartheta . \quad (1.6.41)$$

1.6.6 Free-Free Emission and Absorption

Here we consider detailed balance of free-free emission (bremsstrahlung) and free-free absorption (inverse bremsstrahlung) of an electron in the field of a heavy, stationary particle P (an ion or a neutral atom),

$$P + e(\boldsymbol{v}, d^3v) + \gamma(\nu, d\nu; \boldsymbol{n}, d\Omega) \rightleftarrows P + e(\boldsymbol{v}', d^3v') , \quad (1.6.42)$$

where conservation of energy requires

$$\tfrac{1}{2} m_e v^2 + h\nu = \tfrac{1}{2} m_e v'^2 . \quad (1.6.43)$$

The corresponding rate equation can be written as

$$n_P n_e f_e(\boldsymbol{v})\, d^3v\, I_\nu(\boldsymbol{n})\, d\nu\, d\Omega\, \beta_\nu(\boldsymbol{n}, \boldsymbol{v}; \boldsymbol{v}')\, d^3v'$$

$$= n_P n_e f_e(\boldsymbol{v}')\, d^3v'\, \alpha_\nu(\boldsymbol{v}'; \boldsymbol{n}, \boldsymbol{v})\, d\nu\, d\Omega\, d^3v \left[1 + \frac{c^2}{2h\nu^3} I_\nu(\boldsymbol{n})\right] , \qquad (1.6.44)$$

where stimulated bremsstrahlung has been taken into account. Here n_P is the density of heavy particles P that give rise to the free-free transitions of the electrons. Equation (1.6.44) defines α_ν and β_ν precisely, which correspond to spontaneous free-free emission and free-free absorption, thus being analogous to the Einstein coefficients A_{21} and B_{12} (Sect. 1.6.4). The left-hand side of (1.6.44) is the number of free-free absorptions of photons $\gamma(\nu, d\nu; \boldsymbol{n}, d\Omega)$ by electrons $e(\boldsymbol{v}, d^3v)$ per unit time and unit volume such that the outgoing electron has a velocity in the range $(\boldsymbol{v}', d^3v')$, and the right-hand side is the sum of the inverse spontaneous and stimulated free-free emissions.

Note that in (1.6.44), the reaction rates are calculated by incoherent summation of the individual one-electron emission and absorption processes. This means that collective radiation effects of the electrons are not considered, thereby restricting the frequencies in question to values well above the plasma frequency. The plasma frequency in cgs units is given by (e being the elementary charge) $\nu_p = (2\pi)^{-1}(4\pi n_e e^2/m_e)^{1/2}$.

Substituting the Planck function and the Maxwell distribution into (1.6.44) leads in a by now familiar way to

$$\beta_\nu(\boldsymbol{n}, \boldsymbol{v}; \boldsymbol{v}') = \frac{c^2}{2h\nu^3} \alpha_\nu(\boldsymbol{v}'; \boldsymbol{n}, \boldsymbol{v}) , \qquad (1.6.45)$$

which is analogous to $B_{12} = (g_2/g_1)(c^2/2h\nu^3)A_{21}$, see (1.6.31, 32).

Because of (1.6.43), α_ν and β_ν both contain the delta function

$$\delta(\tfrac{1}{2} m_e v'^2 - \tfrac{1}{2} m_e v^2 - h\nu)$$

as a factor, but not the delta function $\delta(m_e\boldsymbol{v}' - m_e\boldsymbol{v} - [h\nu/c]\boldsymbol{n})$ since the heavy particle P can absorb any amount of momentum.

1.6.7 Two-Photon Emission and Absorption

An excited atom may decay by simultaneous emission of two photons. For instance, the following two-photon decays are important in astrophysics:

$$\mathrm{H}(2s) \rightarrow \mathrm{H}(1s) + h\nu' + h\nu'' ,$$

$$\mathrm{He}(2\,^1S) \rightarrow \mathrm{He}(1\,^1S) + h\nu' + h\nu'' ,$$

$$\mathrm{He}(2\,^3S) \rightarrow \mathrm{He}(1\,^1S) + h\nu' + h\nu'' .$$

The inverse process, two-photon absorption, also exists but is usually negligible, except in high-intensity laser light which is not considered in this book. Nevertheless, we shall derive the relation between the Einstein coefficients for two-photon emission and absorption, respectively, to gain some insight into the physics involved.

Let

$$A_{21}(\nu',\boldsymbol{n}';\nu'',\boldsymbol{n}'')d\nu'\,d\Omega'\,d\nu''\,d\Omega''$$

denote the transition probability per unit time for simultaneous spontaneous emission of two photons $\gamma'\equiv\gamma(\nu',d\nu';\boldsymbol{n}',d\Omega')$ and $\gamma''\equiv\gamma(\nu'',d\nu'';\boldsymbol{n}'',d\Omega'')$, respectively, in an atomic transition from an upper level 2 to a lower level 1. Since conservation of energy requires

$$h\nu'+h\nu''=E_{21}\ , \tag{1.6.46}$$

where $E_{21}=E_2-E_1$ is the energy difference of the atomic levels, $A_{21}(\nu',\boldsymbol{n}';\nu'',\boldsymbol{n}'')$ is proportional to the delta function $\delta(E_{21}-h\nu'-h\nu'')$, so that it depends effectively on only one of the two photon frequencies. Moreover, it depends on the photon directions $\boldsymbol{n}'$ and $\boldsymbol{n}''$ only through the angle ϑ between them ($\boldsymbol{n}'\cdot\boldsymbol{n}''=\cos\vartheta$). The total transition probability for spontaneous two-photon emission is hence

$$\begin{aligned}A_{21}&=\iiiint A_{21}(\nu',\boldsymbol{n}';\nu'',\boldsymbol{n}'')d\nu'\,d\Omega'\,d\nu''\,d\Omega''\\&=4\pi\int_0^{E_{21}/h}d\nu\int_0^{\pi}2\pi\sin\vartheta\,d\vartheta\,A_{21}(\nu,\vartheta)\ .\end{aligned} \tag{1.6.47}$$

For example, for a hydrogen atom, $A_{2s,1s}=8.2\ \mathrm{s}^{-1}$.

On the other hand, we define $B_{12}(\nu',\boldsymbol{n}';\nu'',\boldsymbol{n}'')$ such that

$$B_{12}(\nu',\boldsymbol{n}';\nu'',\boldsymbol{n}'')I_{\nu'}(\boldsymbol{n}')d\nu'\,d\Omega'\,I_{\nu''}(\boldsymbol{n}'')d\nu''\,d\Omega''$$

is the transition probability for simultaneous absorption of two photons γ' and γ'', respectively, in a radiation field of specific intensity $I_\nu(\boldsymbol{n})$.

Let us now consider detailed balance of two-photon emissions and absorptions,

$$A(E_1)+\gamma'+\gamma''\rightleftarrows A(E_2)\ . \tag{1.6.48}$$

The corresponding rate equation is

$$\begin{aligned}&n_1B_{12}(\nu',\boldsymbol{n}';\nu'',\boldsymbol{n}'')I_{\nu'}(\boldsymbol{n}')d\nu'\,d\Omega'\,I_{\nu''}(\boldsymbol{n}'')d\nu''\,d\Omega''\\&=n_2A_{21}(\nu',\boldsymbol{n}';\nu'',\boldsymbol{n}'')d\nu'\,d\Omega'\,d\nu''\,d\Omega''\\&\quad\cdot\left[1+\frac{c^2}{2h\nu'^3}I_{\nu'}(\boldsymbol{n}')\right]\left[1+\frac{c^2}{2h\nu''^3}I_{\nu''}(\boldsymbol{n}'')\right]\ ,\end{aligned} \tag{1.6.49}$$

taking stimulated emission into account. On substituting the Planck function and the Boltzmann distribution into this equation, one obtains

$$B_{12}(\nu',\boldsymbol{n}';\nu'',\boldsymbol{n}'') = \frac{g_2}{g_1}\frac{c^2}{2h\nu'^3}\frac{c^2}{2h\nu''^3}A_{21}(\nu',\boldsymbol{n}';\nu'',\boldsymbol{n}'') \ , \qquad (1.6.50)$$

which is the two-photon analogue to the Einstein relation $B_{12} = (g_2/g_1)(c^2/2h\nu^3)A_{21}$ for one-photon processes, see (1.6.31, 32).

Inspecting the right-hand side of (1.6.49) shows that four different two-photon emission processes occur: one doubly spontaneous emission, two mixed stimulated-spontaneous emissions, and one doubly stimulated emission. Hence, three different B_{21} coefficients for stimulated emission have to be distinguished. (For n-photon emissions there are 2^n-1 different B_{21} coefficients [1.7].) This is quite cumbersome. As already mentioned at the end of Sect. 1.6.4, it is much more natural (and much easier) to treat the stimulated emissions as indicated in (1.6.30, 38, 44, 49), without introducing any quantities analogous to a B_{21} coefficient.

It must be emphasized, however, that deriving reciprocity relations as outlined above is open to the criticism that induced two-photon processes (absorptions and stimulated emissions) are nonlinear, and that deviations from detailed balance are expected in this domain. The main aim of this discussion was to show that stimulated emissions are best treated without defining any special coefficients for them.

2. Kinetic Equations of Particles

This chapter treats some aspects of the kinetic theory of gases required in plasma spectroscopy. Section 2.1 recalls, in simple terms, the basic notions of the kinetic theory (distribution function, kinetic equation, collision term). Section 2.2 discusses the most important elastic collision terms, namely, the Boltzmann and Fokker-Planck terms. Inelastic collision terms corresponding to collisional excitation, de-excitation, ionization, and three-body recombination are derived in Sect. 2.3, and collision terms for radiative processes corresponding to bound-bound, free-bound, and free-free transitions in Sect. 2.4.

2.1 Kinetic Equations

Equilibrium and nonequilibrium states of a gas are microscopically described by the distribution functions of the various components it is composed of. The *distribution function* $F(\boldsymbol{r}, \boldsymbol{v}, t)$ of a specific particle type (for example, H atoms in the excited $2p$ state) is defined such that $F(\boldsymbol{r}, \boldsymbol{v}, t)\, d^3r\, d^3v$ is the average or most probable number of particles of the type considered that at time t are in the volume element $(\boldsymbol{r}, d^3r)$ and whose velocities are in the range $(\boldsymbol{v}, d^3v)$. *Particle type* refers to atoms or ions in a well-defined internal state, and *particle species,* to a chemical species. Thus, the species H (atomic hydrogen) is composed of the particle types H$(1s)$, H$(2s)$, H$(2p)$, etc. The local number density n of a given particle type is therefore related to F by

$$n(\boldsymbol{r}, t) = \int F(\boldsymbol{r}, \boldsymbol{v}, t)\, d^3v \ . \tag{2.1.1}$$

It is often convenient to indicate explicitly the number density. Then

$$F(\boldsymbol{r}, \boldsymbol{v}, t) = n(\boldsymbol{r}, t) f(\boldsymbol{r}, \boldsymbol{v}, t) \ , \tag{2.1.2}$$

so introducing the velocity distribution f normalized to unity,

$$\int f(\boldsymbol{r}, \boldsymbol{v}, t)\, d^3v = 1 \ . \tag{2.1.3}$$

To be mentioned in passing is that despite the fact that the description of a gas by means of distribution functions F is very detailed, nevertheless some physical phenomena are not covered by it, for example fluctuations, or correlations between particles.

The time evolution of a distribution function is governed by its *kinetic equation*. To derive this equation, let us introduce the six-dimensional μ space in which each particle is represented by a point $\boldsymbol{x} \equiv (\boldsymbol{r}, \boldsymbol{v})$. Now consider a fixed, small volume $d^6x \equiv d^3r\, d^3v$ in this space. By definition of the distribution function F, the number of particles at time t in this volume is $F(\boldsymbol{x}, t)\, d^6x$. In the course of time this number can change for two different reasons. First, it can change because of the streaming of the particles, which move on their trajectories with velocities $\dot{\boldsymbol{x}} \equiv (\dot{\boldsymbol{r}}, \dot{\boldsymbol{v}}) = (\boldsymbol{v}, \boldsymbol{K}/m)$, where

$$\boldsymbol{K} = m\gamma + q\boldsymbol{E} + \frac{q}{c}\,\boldsymbol{v} \times \boldsymbol{B} \tag{2.1.4}$$

is the force on the particle (of mass m and electric charge q) due to gravitational, electric, and magnetic fields. (Contributions to the force $\boldsymbol{K}$ due to higher electric and magnetic moments are neglected.) And second, it can change because of the collisions between particles, whereby some particles are scattered out of the volume d^6x, while others are scattered into it from outside. Therefore

$$\frac{\partial F}{\partial t} = -\nabla^{\mu} \cdot (\dot{\boldsymbol{x}} F) + \left(\frac{\delta F}{\delta t}\right)_{\text{coll}} . \tag{2.1.5}$$

Here the first term on the right is the negative divergence of the particle flow (which, by Gauss' theorem, is equal to the net flow of particles into the volume d^6x through its surface), where $\nabla^{\mu} \equiv (\partial/\partial \boldsymbol{r}, \partial/\partial \boldsymbol{v})$. Explicitly

$$\begin{aligned}\nabla^{\mu} \cdot (\dot{\boldsymbol{x}} F) &\equiv \frac{\partial}{\partial \boldsymbol{r}} \cdot (\dot{\boldsymbol{r}} F) + \frac{\partial}{\partial \boldsymbol{v}} \cdot (\dot{\boldsymbol{v}} F) \\ &= \frac{\partial}{\partial \boldsymbol{r}} \cdot (\boldsymbol{v} F) + \frac{\partial}{\partial \boldsymbol{v}} \cdot \left(\frac{\boldsymbol{K}}{m} F\right) \\ &= \boldsymbol{v} \cdot \frac{\partial F}{\partial \boldsymbol{r}} + \frac{\boldsymbol{K}}{m} \cdot \frac{\partial F}{\partial \boldsymbol{v}}\end{aligned} \tag{2.1.6}$$

since

$$\frac{\partial}{\partial \boldsymbol{v}} \cdot \boldsymbol{K} = 0$$

on account of (2.1.4). Equation (2.1.5) may therefore be written as

$$\frac{\partial F}{\partial t} + \boldsymbol{v} \cdot \frac{\partial F}{\partial \boldsymbol{r}} + \frac{\boldsymbol{K}}{m} \cdot \frac{\partial F}{\partial \boldsymbol{v}} = \left(\frac{\delta F}{\delta t}\right)_{\text{coll}} , \tag{2.1.7}$$

which is the standard form of a kinetic equation.

Strictly speaking, (2.1.7) as it stands is nothing but the definition of the collision term $(\delta F/\delta t)_{\text{coll}}$, which has to be specified in terms of physical quantities like distribution functions and cross sections, otherwise the equation is meaningless. Generally, the right-hand side of (2.1.7) is a *collision term* only if it has the following three properties:

(1) The term $(\delta F/\delta t)_{\text{coll}}$ depends only on one-particle distribution functions (as opposed to many-particle distribution functions). In analogy to the one-particle distribution function $F(\boldsymbol{r}, \boldsymbol{v}, t)$, one defines the two-particle distribution function $F^{(2)}$ as follows: $F^{(2)}(\boldsymbol{r}_1, \boldsymbol{v}_1; \boldsymbol{r}_2, \boldsymbol{v}_2; t)\, d^3r_1\, d^3v_1\, d^3r_2 d^3v_2$ is the number of particle pairs at time t such that one particle is in $(\boldsymbol{r}_1, d^3r_1; \boldsymbol{v}_1, d^3v_1)$ and the other in $(\boldsymbol{r}_2, d^3r_2; \boldsymbol{v}_2, d^3v_2)$. If the particles are uncorrelated, then $F^{(2)}(1,2) = F(1)F(2)$. In general, however, $F^{(2)}(1,2) = F(1)F(2) + G^{(2)}(1,2)$. Property (1) means, more precisely, that $(\delta F/\delta t)_{\text{coll}}$ does not depend on the pair correlation function $G^{(2)}$ or on higher-order correlation functions $G^{(3)}, \ldots$.

(2) The term $(\delta F/\delta t)_{\text{coll}}$ is Markovian. This means that it depends at time t only on the instantaneous values of the distribution functions or their derivatives with respect to $\boldsymbol{v}$, but not on the past history of the system through memory terms like $\int_{-\infty}^{t} \ldots F(t')\, dt'$.

(3) The term $(\delta F/\delta t)_{\text{coll}}$ gives rise to an irreversible approach to thermal equilibrium.

It should be noted that particle interactions enter the kinetic equation (2.1.7) not only via the collision term $(\delta F/\delta t)_{\text{coll}}$, but also via the force $\boldsymbol{K}$. Indeed, all three fields contributing to the force (2.1.4) are in general composed of external and so-called self-consistent parts,

$$\gamma = \gamma^{\text{ext}} + \gamma^{\text{self}} , \tag{2.1.8a}$$

$$\boldsymbol{E} = \boldsymbol{E}^{\text{ext}} + \boldsymbol{E}^{\text{self}} , \tag{2.1.8b}$$

$$\boldsymbol{B} = \boldsymbol{B}^{\text{ext}} + \boldsymbol{B}^{\text{self}} . \tag{2.1.8c}$$

Self-consistent electric and magnetic fields are ubiquitous in plasma physics, self-consistent gravitational fields are encountered, for instance, in all calculations of stellar evolution. For quasi-static forces, i.e., neglecting retardation effects, the self-consistent gravitational, electric, and magnetic fields (in cgs units) are given by

$$\gamma^{\text{self}}(\boldsymbol{r}, t) = -G \int \rho(\boldsymbol{r}', t) \frac{\boldsymbol{r} - \boldsymbol{r}'}{|\boldsymbol{r} - \boldsymbol{r}'|^3}\, d^3r' , \tag{2.1.9a}$$

$$\boldsymbol{E}^{\text{self}}(\boldsymbol{r}, t) = \int \tau(\boldsymbol{r}', t) \frac{\boldsymbol{r} - \boldsymbol{r}'}{|\boldsymbol{r} - \boldsymbol{r}'|^3}\, d^3r' \tag{2.1.9b}$$

$$\boldsymbol{B}^{\text{self}}(\boldsymbol{r},t)=\frac{1}{c}\int \boldsymbol{j}(\boldsymbol{r}',t)\times\frac{\boldsymbol{r}-\boldsymbol{r}'}{|\boldsymbol{r}-\boldsymbol{r}'|^3}\,d^3r' \;, \tag{2.1.9c}$$

where G is the gravitational constant, and the mass density ρ, the charge density τ, and the current density $\boldsymbol{j}$ are determined by the distribution functions of all gas components a through

$$\rho(\boldsymbol{r}',t)=\sum_a \int m_a F_a(\boldsymbol{r}',\boldsymbol{v},t)\,d^3v \;, \tag{2.1.10a}$$

$$\tau(\boldsymbol{r}',t)=\sum_a \int q_a F_a(\boldsymbol{r}',\boldsymbol{v},t)\,d^3v \;, \tag{2.1.10b}$$

$$\boldsymbol{j}(\boldsymbol{r}',t)=\sum_a \int q_a \boldsymbol{v} F_a(\boldsymbol{r}',\boldsymbol{v},t)\,d^3v \;. \tag{2.1.10c}$$

Thus, the distribution function F at point $\boldsymbol{r}$ is, via the self-consistent force $\boldsymbol{K}^{\text{self}}$, affected by the distribution functions F_a at all other points $\boldsymbol{r}'$ of the system considered.

At first sight, the decomposition of particle interactions into self-consistent fields on the one hand, and collisions on the other, may seem rather ill-defined [especially when considering small distances $|\boldsymbol{r}-\boldsymbol{r}'|$ in (2.1.9)], even if the self-consistent fields are spatially and temporally smooth, whereas the term collision suggests a somewhat irregular behavior. However, closer examination reveals that this decomposition is quite plausible insofar as it turns out that self-consistent fields do not give rise to irreversible behavior which, according to property (3) stated above, is characteristic of collisions (see Sect. 5.1.2).

We now return to the collision term of the kinetic equation:

$$\left(\frac{\delta F}{\delta t}\right)_{\text{coll}}=\left(\frac{\delta F}{\delta t}\right)_{\text{el}}+\left(\frac{\delta F}{\delta t}\right)_{\text{inel}}+\left(\frac{\delta F}{\delta t}\right)_{\text{rad}} \;. \tag{2.1.11}$$

Here the terms on the right-hand side correspond to elastic collisions with particles, to inelastic collisions with particles, and to radiative interactions, respectively, which are discussed in turn in the following.

2.2 Elastic Collision Terms

The derivation of elastic collision terms and the study of their properties are major topics in the kinetic theory of gases. In this section we collect some of the main results, referring for their derivation and thorough discussion to the literature [2.1 – 5]. However, to make this book more self-contained, we discuss in simple terms some aspects of the Boltzmann and Fokker-Planck equations in the Appendices C and D, respectively.

2.2.1 Boltzmann Collision Terms

Elastic collisions of electrically neutral particles with neutral and charged particles, on the one hand, and of charged particles with neutral particles, on the other, are described by Boltzmann collision terms. Here the collisions are due to short-range interactions (as opposed to the long-range Coulomb interaction between charged particles), so that there is usually a large momentum transfer in a single collision (strong collisions). In dilute gases, only binary collisions (involving just two particles) have to be taken into account. Hence the following physical picture applies: each particle moves on an unperturbed trajectory until, by a sudden collision with another particle, it is scattered into a new trajectory on which it moves till the next collision, and so on.

In such a situation the *Boltzmann collision term* applies which, for particles of type a, reads

$$\left(\frac{\delta F_a}{\delta t}\right)_{\mathrm{B}} = \sum_b \iint [F_a(\boldsymbol{v}_1)F_b(\boldsymbol{v}_1') - F_a(\boldsymbol{v})F_b(\boldsymbol{v}')]\, w q_{ab}(w,\vartheta)\, d\Omega\, d^3v' \quad (2.2.1)$$

where $F_a \equiv F_a(\boldsymbol{r},\boldsymbol{v},t)$, and all distribution functions on the right-hand side refer to the same space-time point $(\boldsymbol{r},t)$: $F_a(\boldsymbol{v}_1) \equiv F_a(\boldsymbol{r},\boldsymbol{v}_1,t)$, etc. In (2.2.1), $q_{ab}(w,\vartheta)$ is the differential cross section for the elastic collision $a(\boldsymbol{v}) + b(\boldsymbol{v}')$ $\rightarrow a(\boldsymbol{v}_1) + b(\boldsymbol{v}_1')$, or briefly $(\boldsymbol{v},\boldsymbol{v}' \rightarrow \boldsymbol{v}_1,\boldsymbol{v}_1')$, w is the modulus of the relative velocity $\boldsymbol{w} = \boldsymbol{v} - \boldsymbol{v}'$, ϑ the angle between the relative velocities before and after the collision (i.e., $\boldsymbol{w}\cdot\boldsymbol{w}_1 = w w_1 \cos\vartheta$, where $\boldsymbol{w}_1 = \boldsymbol{v}_1 - \boldsymbol{v}_1'$), and $d\Omega = \sin\vartheta\, d\vartheta\, d\varphi$, the element of solid angle. If the particles a are neutral, the summation in (2.2.1) is over all components of the gas (including, of course, particle type a itself), whereas if the particles a are charged, the summation is over only the neutral components. (See Appendix C for a brief discussion of the Boltzmann equation of a one-component gas.)

Collision term (2.2.1) is seen to be the sum of a creation term and a destruction term. The destruction term is due to all collisions $(\boldsymbol{v},\boldsymbol{v}' \rightarrow \boldsymbol{v}_1,\boldsymbol{v}_1')$ that destroy a particles of the considered velocity $\boldsymbol{v}$. On the other hand, the creation term is due to all inverse collisions $(\boldsymbol{v}_1,\boldsymbol{v}_1' \rightarrow \boldsymbol{v},\boldsymbol{v}')$ that create a particles of velocity $\boldsymbol{v}$. Notice that elastic collisions leave the modulus of the relative velocity unchanged ($w = w_1$), and that the cross sections for the direct and inverse collisions, respectively, are equal, which is a special case of the reciprocity relation (1.2.2) [see the text after (E.1.15)]. Consequently, in (2.2.1),

$$w q_{ab}(w,\vartheta) \equiv w q_{ab}(\boldsymbol{v},\boldsymbol{v}';\boldsymbol{v}_1,\boldsymbol{v}_1') = w_1 q_{ab}(\boldsymbol{v}_1,\boldsymbol{v}_1';\boldsymbol{v},\boldsymbol{v}') \ .$$

An interesting special case arises if one of the collision partners has a much smaller mass than the other. Electron collisions with neutral atoms in a weakly ionized plasma are the most prominent example. More precisely, let us consider a situation with so low an electron density that for the electron gas collisions with atoms are dominant, and where, on the other hand, the atoms have a Maxwellian distribution function of temperature T owing to frequent atom-atom collisions. Here an initially arbitrary electron velocity distribution quickly becomes isotropic as a result of elastic collisions with the (almost stationary) atoms, while its approach to a Maxwell distribution is rather slow because of the small energy transfer between electron and atom in a single collision. Then, the Boltzmann collision term for the isotropic electron distribution function $F_e(v)$ takes the form [2.6,7]

$$\left(\frac{\delta F_e}{\delta t}\right)_B^{e0} = \frac{m_e}{m_0}\frac{1}{v^2}\frac{\partial}{\partial v}\left(v^3 \gamma_{e0}(v)\left[F_e(v)+\frac{kT}{m_e v}\frac{\partial F_e}{\partial v}\right]\right). \qquad (2.2.2)$$

Here m_e and m_0 are the electron and atom masses, respectively. Furthermore

$$\gamma_{e0}(v) = n_0 v \int_0^{\pi} q_{e0}(v,\vartheta)(1-\cos\vartheta)\, 2\pi \sin\vartheta \, d\vartheta \qquad (2.2.3)$$

is the effective collision frequency for momentum transfer for an electron of velocity v. [If, in a collision with a stationary atom, an electron of initial velocity $\boldsymbol{v}$ is deflected by the angle ϑ, it transfers a momentum to the atom whose component in the $\boldsymbol{v}$ direction is $m_e v(1-\cos\vartheta)$.] The density of the atoms is n_0, and $q_{e0}(v,\vartheta)$ is the differential cross section. Collision term (2.2.2) is valid for electron velocities well above the thermal velocity of the atoms, $v \gg (kT/m_0)^{1/2}$. For a Maxwellian electron distribution function of temperature T, $F_e(v) = F_e^M(v)$, collision term (2.2.2) vanishes identically, and hence can be accepted as a reasonable approximation even for electron velocities $v \lesssim (kT/m_0)^{1/2}$.

2.2.2 Fokker-Planck Collision Terms

Elastic collisions of electrically charged particles with other charged particles differ significantly from the collisions discussed in the preceding section. Indeed, in contrast to binary collisions involving at least one neutral particle, a charged particle in a plasma interacts simultaneously with a great number of other charged particles owing to the long range of the Coulomb interaction. As mentioned in Sect. 2.1, the Coulomb interaction of a charged particle with the surrounding plasma can be split into a reversible interaction with the self-consistent field, and an irreversible interaction due to collisions. Here we are concerned only with collisions. The following provides an appropriate physical picture: a charged particle interacts (collides) simultaneously with numerous other charged particles such that any single collision, regarded separately, leads to only a small momentum transfer between the collision partners (weak collisions). In such a situation one can no longer speak of unperturbed trajectories suddenly interrupted by collisions, rather, the particle trajectories slowly change in a quasi-continuous manner. This type of motion is called

Brownian motion. In a plasma there are, of course, also close encounters between charged particles, i.e., strong collisions, but usually their contribution to the collision term is negligible compared to that of the weak distant encounters. (See Appendix D for an elementary discussion of Brownian motion and the Fokker-Planck equation [2.8].)

The collision term describing weak elastic collisions between charged particles is the *Fokker-Planck collision term* which, for the charged particle type a, reads

$$\left(\frac{\delta F_a}{\delta t}\right)_{\text{FP}} = -\frac{\partial}{\partial \boldsymbol{v}} \cdot \left(\left\langle\frac{\Delta \boldsymbol{v}}{\Delta t}\right\rangle_a F_a\right) + \frac{1}{2}\frac{\partial^2}{\partial \boldsymbol{v}\,\partial \boldsymbol{v}} : \left(\left\langle\frac{\Delta \boldsymbol{v}\,\Delta \boldsymbol{v}}{\Delta t}\right\rangle_a F_a\right) \tag{2.2.4}$$

where $F_a \equiv F_a(\boldsymbol{r}, \boldsymbol{v}, t)$. The dot denotes the scalar product of two vectors, and the double dot the contracted product of two tensors,

$$\boldsymbol{a} \cdot \boldsymbol{b} = \sum_{i=1}^{3} a_i b_i \ , \quad \overleftrightarrow{\boldsymbol{A}} : \overleftrightarrow{\boldsymbol{B}} = \sum_{i,j=1}^{3} A_{ij} B_{ji} \ .$$

The vector $\langle \Delta \boldsymbol{v}/\Delta t\rangle_a$ and the tensor $\langle \Delta \boldsymbol{v}\,\Delta \boldsymbol{v}/\Delta t\rangle_a$ describe, respectively, dynamical friction and diffusion in velocity space of an a particle of velocity $\boldsymbol{v}$ in the plasma considered. They are given explicitly by

$$\left\langle\frac{\Delta \boldsymbol{v}}{\Delta t}\right\rangle_a = \frac{4\pi q_a^4}{m_a^2} \ln \Lambda \frac{\partial}{\partial \boldsymbol{v}} \sum_b \left(\frac{q_b}{q_a}\right)^2 \frac{m_a + m_b}{m_b} \int \frac{F_b(\boldsymbol{v}')}{|\boldsymbol{v} - \boldsymbol{v}'|} d^3 v' \ , \tag{2.2.5a}$$

$$\left\langle\frac{\Delta \boldsymbol{v}\,\Delta \boldsymbol{v}}{\Delta t}\right\rangle_a = \frac{4\pi q_a^4}{m_a^2} \ln \Lambda \frac{\partial^2}{\partial \boldsymbol{v}\,\partial \boldsymbol{v}} \sum_b \left(\frac{q_b}{q_a}\right)^2 \int F_b(\boldsymbol{v}')\,|\boldsymbol{v} - \boldsymbol{v}'|\,d^3 v' \ , \tag{2.2.5b}$$

where $F_b(\boldsymbol{v}') \equiv F_b(\boldsymbol{r}, \boldsymbol{v}', t)$. Here m_a, m_b are the masses, and q_a, q_b the charges (in cgs units) of particles of type a and b, respectively, and the summation is over all charged components of the gas (including particle type a itself). The so-called plasma parameter Λ is essentially the number of charged particles in a Debye sphere. For example, for a proton-electron plasma of electron density (= proton density) n_e and temperature T,

$$\Lambda = 24\pi n_e L_D^3 \tag{2.2.6}$$

with the Debye length

$$L_D = \left(\frac{kT}{8\pi n_e e^2}\right)^{1/2} \ . \tag{2.2.7}$$

An important special case of a Fokker-Planck collision term concerns the collisions of high-velocity electrons with the bulk of the electron gas that has an (almost) Maxwellian velocity distribution. For not too low electron densities, the thermalization time of free electrons due to electron-electron collisions is rather short, so that they quickly approach a Maxwell distribution. However, it may happen that the tail of the distribution function is perturbed by inelastic collision processes with atoms, and therefore deviates from a Maxwellian. The corresponding collision term for an isotropic distribution of suprathermal electrons, $F_e(v)$ with $v \gg (kT_e/m_e)^{1/2}$, interacting with the bulk of thermal electrons of temperature T_e and density n_e, is given by [2.4]

$$\left(\frac{\delta F_e}{\delta t}\right)_{\mathrm{FP}}^{\mathrm{ee}} = \frac{4\pi n_e e^4}{m_e} \ln \Lambda \frac{1}{v^2} \frac{\partial}{\partial v}\left(F_e(v) + \frac{kT_e}{m_e v}\frac{\partial F_e}{\partial v}\right) . \quad (2.2.8)$$

Equation (2.2.8) has the same structure as (2.2.2), for the Coulomb cross section q_{ee} is proportional to v^{-4} (Rutherford formula), hence the collision frequency $\gamma_{ee} \propto v q_{ee} \propto v^{-3}$, so that $v^3 \gamma_{ee}(v) = \text{const}$. Likewise, for a Maxwell distribution of temperature T_e, $F_e(v) = F_e^{\mathrm{M}}(v)$, collision term (2.2.8) vanishes identically.

2.2.3 General Properties of Elastic Collision Terms

Elastic collision terms, which in general are sums of Boltzmann and Fokker-Planck terms,

$$\left(\frac{\delta F_a}{\delta t}\right)_{\mathrm{el}} = \left(\frac{\delta F_a}{\delta t}\right)_{\mathrm{B}} + \left(\frac{\delta F_a}{\delta t}\right)_{\mathrm{FP}} , \quad (2.2.9)$$

have the following general properties.

(1) If the distribution function is positive at time t_0, $F_a(\boldsymbol{r}, \boldsymbol{v}, t_0) \geqslant 0$, it remains so for all later times $t > t_0$, $F_a(\boldsymbol{r}, \boldsymbol{v}, t) \geqslant 0$. Clearly, this property is necessary for F_a to be interpreted in terms of a probability density.

(2) Since elastic collisions conserve mass, momentum, and kinetic energy, then

$$\int m_a \left(\frac{\delta F_a}{\delta t}\right)_{\mathrm{el}} d^3v = 0 , \quad (2.2.10a)$$

$$\sum_a \int m_a \boldsymbol{v} \left(\frac{\delta F_a}{\delta t}\right)_{\mathrm{el}} d^3v = 0 , \quad (2.2.10b)$$

$$\sum_a \int \frac{1}{2} m_a v^2 \left(\frac{\delta F_a}{\delta t}\right)_{\mathrm{el}} d^3v = 0 . \quad (2.2.10c)$$

Whereas the mass of each particle type is conserved separately, conservation of momentum and kinetic energy holds, of course, only for the sum of all gas components. Note that (2.2.10) are local conservation laws valid for every space-time point $(\boldsymbol{r}, t)$, that is, physically, in every small volume d^3r during a small time dt.

(3) The H theorem is valid,

$$\left(\frac{\delta H}{\delta t}\right)_{\text{el}} \leqslant 0 \ , \quad \text{where} \tag{2.2.11}$$

$$H = \sum_a \int F_a \ln F_a d^3v \ , \tag{2.2.12}$$

$(\delta/\delta t)_{\text{el}}$ denoting the rate of change due to elastic collisions. Moreover, $(\delta H/\delta t)_{\text{el}} = 0$ if and only if all distribution functions F_a are Maxwell distributions corresponding to the same temperature T. As shown in Sect. 5.1.2, H is essentially the negative entropy density, so that $(\delta H/\delta t)_{\text{el}}$ is proportional to the negative entropy production. The H theorem (2.2.11) thus expresses the fact that as a result of elastic collisions, the distribution functions change irreversibly until thermal equilibrium is reached and the entropy production becomes zero.

In Appendix C, properties (1 – 3) are proved for the Boltzmann collision term of a one-component gas.

2.3 Inelastic Collision Terms

Inelastic collisions comprise excitation and de-excitation collisions as well as rearrangement collisions corresponding to reactions like ionization, dissociation, or collisional three-body recombination. This section considers only inelastic collisions of electrons with atoms or ions, for they are usually the most important ones. Moreover, we shall make the simplifying assumptions that velocity changes of atoms due to inelastic collisions with electrons can be ignored, and that the relative velocity between electron and atom may be replaced by the electron velocity. The inelastic collision terms are then very simple creation and destruction terms which can be formulated immediately. Only energy conservation must be taken into account, whereas momentum conservation need not be considered.

Sometimes in very weakly ionized plasmas, inelastic collisions with heavy particles become important. Collision terms due to heavy particles are more complicated than the electronic collision terms discussed here, because the momentum transfer between colliding particles must be treated correctly. However, the explicit form of inelastic collision terms due to heavy-particle collisions is hardly of practical interest because in situations where such collisions play a role, elastic collisions among heavy particles are usually so frequent that all velocity distributions are Maxwellian. Hence, only the rate coefficients corresponding to Maxwellian velocity distributions are required to determine the number densities of the various excitation and ionization states in such a weakly ionized plasma.

For atoms in a given bound level i, the inelastic collision term is

$$\left(\frac{\delta F_i}{\delta t}\right)_{\text{inel}} = \sum_{k>i}\left(\frac{\delta F_i}{\delta t}\right)^{ik}_{\text{inel}} + \sum_{j<i}\left(\frac{\delta F_i}{\delta t}\right)^{ji}_{\text{inel}} + \sum_{+}\left(\frac{\delta F_i}{\delta t}\right)^{i+}_{\text{inel}} + \sum_{-}\left(\frac{\delta F_i}{\delta t}\right)^{-i}_{\text{inel}} . \tag{2.3.1}$$

Here the sums are over all higher levels k ($>i$) and all lower levels j ($<i$) of the considered particle species, over all bound levels "+" of the species of the next higher stage of ionization, and over all bound levels "−" of the species of the next lower stage of ionization. Likewise, the inelastic collision term for free electrons is the sum of the contributions of all bound-bound ("1,2") and all free-bound ("0,+") transitions of all particle species present,

$$\left(\frac{\delta F_e}{\delta t}\right)_{\text{inel}} = \sum_{(1,2)}\left(\frac{\delta F_e}{\delta t}\right)^{12}_{\text{inel}} + \sum_{(0,+)}\left(\frac{\delta F_e}{\delta t}\right)^{0+}_{\text{inel}} . \tag{2.3.2}$$

Of particular importance in plasma spectroscopy is the total production rate of particles of type a (for example, of atoms in level i, or free electrons) per unit time and unit volume due to inelastic collisions, which is given by

$$\left(\frac{\delta n_a}{\delta t}\right)_{\text{inel}} = \int\left(\frac{\delta F_a}{\delta t}\right)_{\text{inel}} d^3v . \tag{2.3.3}$$

2.3.1 Collisional Excitation and De-Excitation (Atoms)

Let us consider excitation collisions of free electrons with atoms,

$$A_1(\boldsymbol{v}) + \mathrm{e}(\boldsymbol{v}') \rightarrow A_2(\boldsymbol{v}) + \mathrm{e}(\boldsymbol{v}'') , \tag{2.3.4}$$

and the inverse de-excitation collisions. Here 1 and 2 denote the lower and upper atomic levels. Energy conservation requires

$$\tfrac{1}{2} m_e v'^2 = \tfrac{1}{2} m_e v''^2 + E_{21} , \tag{2.3.5}$$

where $E_{21} = E_2 - E_1$ is the excitation energy of the transition considered. Note that in our approximation the atom in level 2 has the same velocity $\boldsymbol{v}$ as the atom in level 1.

The corresponding inelastic collision terms for the atomic distribution functions $F_1 \equiv F_1(\boldsymbol{r}, \boldsymbol{v}, t)$ and $F_2 \equiv F_2(\boldsymbol{r}, \boldsymbol{v}, t)$ are given by

$$\left(\frac{\delta F_1}{\delta t}\right)^{12}_{\text{inel}} = -\left(\frac{\delta F_2}{\delta t}\right)^{12}_{\text{inel}} = -F_1(\boldsymbol{v})\, n_e C_{12} + F_2(\boldsymbol{v})\, n_e C_{21} . \tag{2.3.6}$$

Here n_e is the electron density, and the rate coefficients for excitation and de-excitation collisions are

$$C_{12} = \iint v' f_e(\boldsymbol{v}') q_{12}(v', \vartheta) d^3v' d\Omega \ , \tag{2.3.7a}$$

$$C_{21} = \iint v'' f_e(\boldsymbol{v}'') q_{21}(v'', \vartheta) d^3v'' d\Omega \ , \tag{2.3.7b}$$

where $f_e(\boldsymbol{v}) \equiv f_e(\boldsymbol{r}, \boldsymbol{v}, t)$ is the normalized velocity distribution of free electrons, q_{12} and q_{21} the differential cross sections, and $d\Omega = \sin\vartheta \, d\vartheta \, d\varphi$ the element of solid angle, ϑ denoting the angle between $\boldsymbol{v}'$ and $\boldsymbol{v}''$, i.e., $\boldsymbol{v}' \cdot \boldsymbol{v}'' = v' v'' \cos\vartheta$. Notice that according to our approximation the relative electron-atom velocity has been replaced in (2.3.7) by the electron velocity. Furthermore, recall that the cross sections q_{12} and q_{21} are related to each other by (1.6.7).

For isotropic electron distributions, the integrations over $d\Omega$ in (2.3.7) can be performed, yielding total cross sections Q_{12} and Q_{21}, (1.6.8). In terms of kinetic energy $E = m_e v^2/2$,

$$C_{12} = \int_{E_{21}}^{\infty} \left(\frac{2E'}{m_e}\right)^{1/2} f_e(E') Q_{12}(E') dE' \ , \tag{2.3.8a}$$

$$\begin{aligned} C_{21} &= \int_{0}^{\infty} \left(\frac{2E''}{m_e}\right)^{1/2} f_e(E'') Q_{21}(E'') dE'' \\ &= \frac{g_1}{g_2} \int_{E_{21}}^{\infty} \left(\frac{2E'}{m_e}\right)^{1/2} \left(\frac{E'}{E'-E_{21}}\right)^{1/2} f_e(E'-E_{21}) Q_{12}(E') dE' \ , \end{aligned} \tag{2.3.8b}$$

where $f_e(E) dE = f_e(v) 4\pi v^2 dv$, and where in (2.3.8b) the Klein-Rosseland relation (1.6.9) has been used, with $E'' = E' - E_{21}$.

The rate of change of the atomic number densities follows from (2.3.6) by integrating over all velocities,

$$\left(\frac{\delta n_1}{\delta t}\right)_{\text{inel}}^{12} = -\left(\frac{\delta n_2}{\delta t}\right)_{\text{inel}}^{12} = -n_1 n_e C_{12} + n_2 n_e C_{21} \ . \tag{2.3.9}$$

If the electron distribution function is Maxwellian of temperature T_e, (1.4.14), it follows from (2.3.8) that the rate coefficients for excitation and de-excitation collisions are related by

$$C_{21} = \frac{g_1}{g_2} e^{E_{21}/kT_e} C_{12} \ . \tag{2.3.10}$$

This relation can also be derived by the following simple reasoning. In thermal equilibrium of temperature T_e there is detailed balance, and hence, a fortiori, global balance $n_1 n_e C_{12} = n_2 n_e C_{21}$, see (2.3.9), and the atomic populations follow a Boltzmann distribution $n_2/n_1 = (g_2/g_1) \exp(-E_{21}/kT_e)$. Combining these two relations gives (2.3.10) at once.

2.3.2 Collisional Ionization and Three-Body Recombination (Atoms)

Consider the following ionization collision between an atom of velocity $\boldsymbol{v}$ and an electron of kinetic energy E,

$$A(\boldsymbol{v})+\mathrm{e}(E)\rightarrow A^{+}(\boldsymbol{v})+\mathrm{e}(E')+\mathrm{e}(E'') \; , \tag{2.3.11}$$

and the inverse three-body recombination. The atom A and the ion A^+ are supposed to be in well-defined internal states of energies E_0 and E_+, respectively. Clearly,

$$E=E'+E''+E_{+0} \; , \tag{2.3.12}$$

where $E_{+0}=E_+-E_0$ is the ionization energy of the transition considered. In (2.3.11) it was again assumed that the atom and ion have the same velocity $\boldsymbol{v}$. For simplicity, we consider only isotropic electron distributions in the following.

The corresponding collision terms for the distribution functions $F_0\equiv F_0(\boldsymbol{r},\boldsymbol{v},t)$ of the atoms and $F_+\equiv F_+(\boldsymbol{r},\boldsymbol{v},t)$ of the ions are given by

$$\left(\frac{\delta F_0}{\delta t}\right)^{0+}_{\text{inel}}=-\left(\frac{\delta F_+}{\delta t}\right)^{0+}_{\text{inel}}=-F_0(\boldsymbol{v})\,n_\mathrm{e}M_{0+}+F_+(\boldsymbol{v})\,n_\mathrm{e}^2N_{+0} \tag{2.3.13}$$

with the rate coefficients (Sect. 1.6.2)

$$\begin{aligned} M_{0+} &= \iiint \left(\frac{2E}{m_\mathrm{e}}\right)^{1/2} f_\mathrm{e}(E)\,\Omega_{0+}(E;E',E'')\,dE\,dE'dE'' \\ &= \int_{E_{+0}}^{\infty}\left(\frac{2E}{m_\mathrm{e}}\right)^{1/2} f_\mathrm{e}(E)\,Q_{0+}(E)\,dE \; , \end{aligned} \tag{2.3.14a}$$

$$\begin{aligned} N_{+0} &= \iiint \left(\frac{2E'}{m_\mathrm{e}}\right)^{1/2}\left(\frac{2E''}{m_\mathrm{e}}\right)^{1/2} f_\mathrm{e}(E')f_\mathrm{e}(E'')\,\Omega_{+0}(E',E'';E)\,dE'dE''dE \\ &= \frac{g_0}{2g_+}\,\frac{h^3}{2^{5/2}\pi m_\mathrm{e}^{3/2}}\iiint\left(\frac{2E}{m_\mathrm{e}}\right)^{1/2}\left(\frac{E}{E'E''}\right)^{1/2} f_\mathrm{e}(E')f_\mathrm{e}(E'') \\ &\quad\cdot\,\Omega_{0+}(E;E',E'')\,dE\,dE'dE'' \; . \end{aligned} \tag{2.3.14b}$$

In (2.3.14a) the total cross section for collisional ionization Q_{0+} has been introduced, see (1.6.22), and in (2.3.14b) the Fowler relation (1.6.20) has been used. Again, relative electron-atom velocities have been approximated by electron velocities.

It should be realized that in integrals like

$$\iint \Omega_{0+}(E;E',E'')\,dE'dE'' \quad \text{or} \quad \iint \Omega_{+0}(E',E'';E)\,dE'dE'' ,$$

the summation is over all distinct electron pairs (E',E''). It is therefore advantageous in such integrals to write always *ordered* pairs with $E'<E''$, say.

The rate of change of the number densities of the atoms and ions follows by integrating (2.3.13) over all velocities $\boldsymbol{v}$,

$$\left(\frac{\delta n_0}{\delta t}\right)^{0+}_{\text{inel}} = -\left(\frac{\delta n_+}{\delta t}\right)^{0+}_{\text{inel}} = -n_0 n_e M_{0+} + n_+ n_e^2 N_{+0} . \tag{2.3.15}$$

If the electrons have a Maxwell distribution of temperature T_e, (2.3.14) shows that the rate coefficients for collisional ionization and three-body recombination are related by

$$N_{+0} = \frac{g_0}{2g_+}\,\lambda_e^3 e^{E_{+0}/kT_e} M_{0+} , \tag{2.3.16}$$

where λ_e is the thermal de Broglie wavelength of the electrons, see (1.4.24). This relation can also be derived by considering thermal equilibrium at temperature T_e. Here the global balance is $n_0 n_e M_{0+} = n_+ n_e^2 N_{+0}$, (2.3.15), and the Saha distribution $n_+ n_e/n_0 = (2g_+/g_0)\lambda_e^{-3}\exp(-E_{+0}/kT_e)$ applies. Equation (2.3.16) follows from these two relations.

2.3.3 Collisional Excitation and De-Excitation (Electrons)

The preceding two sections discussed the inelastic collision terms for the distribution functions of the atoms (or ions). We now turn to the corresponding collision terms for the electron distribution function, and consider first excitation and de-excitation collisions, as in Sect. 2.3.1.

Electrons of velocity $\boldsymbol{v}$ are destroyed by excitation collisions

$$A_1 + e(\boldsymbol{v}) \rightarrow A_2 + e(\boldsymbol{v}') , \tag{2.3.17}$$

and by de-excitation collisions

$$A_2 + e(\boldsymbol{v}) \rightarrow A_1 + e(\boldsymbol{v}'') , \tag{2.3.18}$$

and they are created by the corresponding inverse reactions. Reaction (2.3.18) takes place for all velocities $\boldsymbol{v}$, whereas reaction (2.3.17) requires $v > (2E_{21}/m_e)^{1/2}$, where E_{21} is the excitation energy of the atomic transition. Clearly,

$$v' = \left(v^2 - \frac{2E_{21}}{m_e}\right)^{1/2} , \quad v'' = \left(v^2 + \frac{2E_{21}}{m_e}\right)^{1/2} . \tag{2.3.19}$$

In our approximation, the atoms are treated as if they were at rest.

Accordingly, the inelastic collision term for the electron distribution function $F_e \equiv F_e(\boldsymbol{r}, \boldsymbol{v}, t)$ is given by

$$\left(\frac{\delta F_e}{\delta t}\right)_{\text{inel}}^{12} = -n_1 v F_e(\boldsymbol{v}) Q_{12}(v) + n_2 \frac{v'^2}{v} \int F_e(\boldsymbol{v}') q_{21}(v', \vartheta) \, d\Omega$$

$$- n_2 v F_e(\boldsymbol{v}) Q_{21}(v) + n_1 \frac{v''^2}{v} \int F_e(\boldsymbol{v}'') q_{12}(v'', \vartheta) \, d\Omega \ . \quad (2.3.20)$$

Here the first and second terms describe, respectively, the excitation collisions (2.3.17) and the inverse de-excitation collisions, and the third and fourth terms the de-excitation collisions (2.3.18) and the inverse excitation collisions, respectively. Note that $Q_{12}(v) = q_{21}(v', \vartheta) = 0$ if $v < (2E_{21}/m_e)^{1/2}$, $\boldsymbol{v}'$ being defined by (2.3.17).

The second term of the right-hand side of (2.3.20) is derived as follows. In the volume d^3r at $\boldsymbol{r}$, the production rate of electrons with velocity in the range $(\boldsymbol{v}, d^3v)$ due to de-excitation collisions with atoms A_2 is given by

$$\frac{\delta F_e}{\delta t} d^3r \, d^3v = n_2 d^3r \iint' v' F_e(\boldsymbol{v}') q_{21}(v', \vartheta) \, d^3v' d\Omega \ ,$$

that is, by the product of the number of atoms $n_2 d^3r$, the flux density of electrons $v' F_e d^3v'$, and the cross section $q_{21} d\Omega$. The dash at the integral sign signifies that the integration d^3v' must be performed only over those $\boldsymbol{v}'$ that lead to a final velocity in the considered range $(\boldsymbol{v}, d^3v)$. For a fixed angle ϑ between the velocities $\boldsymbol{v}$ and $\boldsymbol{v}'$, i.e., $\boldsymbol{v} \cdot \boldsymbol{v}' = v v' \cos\vartheta$, and using spherical coordinates in velocity space,

$$d^3v' = \frac{d^3v'}{d^3v} d^3v = \frac{v'^2 dv' d\omega'}{v^2 dv \, d\omega} d^3v = \frac{v'}{v} d^3v \ . \quad (2.3.21)$$

Here we have used the fact that $v' dv' = v \, dv$ through (2.3.19). Furthermore, since a fixed ϑ means that the relative orientation of $\boldsymbol{v}$ and $\boldsymbol{v}'$ is held fixed, the elements of solid angle are equal, $d\omega' = d\omega$. [See also (E.1.13).] Thus, after cancellation of $d^3r \, d^3v$, we get the second term of (2.3.20), where $\boldsymbol{v}'$ denotes the vector of length v', given by (2.3.19), that has polar and azimuthal angles ϑ and φ with respect to the final velocity $\boldsymbol{v}$ as polar axis, and $d\Omega = \sin\vartheta \, d\vartheta \, d\varphi$. The fourth term in (2.3.20) is derived analogously.

If the electrons have an isotropic velocity distribution, the differential cross sections in (2.3.20) can be integrated to yield total cross sections. In terms of kinetic energy $E = m_e v^2/2$,

$$\left(\frac{\delta F_e}{\delta t}\right)^{12}_{\text{inel}} = -n_1\left(\frac{2E}{m_e}\right)^{1/2} F_e(E)Q_{12}(E) + n_2\left(\frac{2E'}{m_e}\right)^{1/2} F_e(E')Q_{21}(E')$$

$$-n_2\left(\frac{2E}{m_e}\right)^{1/2} F_e(E)Q_{21}(E) + n_1\left(\frac{2E''}{m_e}\right)^{1/2} F_e(E'')Q_{12}(E'') \; , \tag{2.3.22}$$

where $F_e(E)dE = F_e(v)4\pi v^2 dv$, and $E' = E - E_{21}$, $E'' = E + E_{21}$. Of course, (2.3.22) can be formulated directly.

Integrating (2.3.20) over all velocities $\boldsymbol{v}$ and using (2.3.21) or integrating (2.3.22) over all energies E leads to

$$\left(\frac{\delta n_e}{\delta t}\right)^{12}_{\text{inel}} = 0 \tag{2.3.23}$$

for arbitrary densities n_1 and n_2, as it should.

2.3.4 Collisional Ionization and Three-Body Recombination (Electrons)

Finally, let us derive the inelastic collision term for the electron distribution function for the free-bound transitions considered in Sect. 2.3.2. For simplicity, we suppose that the velocity distribution of the electrons is isotropic.

Electrons of kinetic energy E are destroyed by ionization collisions

$$A + e(E) \rightarrow A^+ + e(E') + e(E'') \; , \tag{2.3.24}$$

and by three-body recombinations

$$A^+ + e(E) + e(E_1') \rightarrow A + e(E_1'') \; , \tag{2.3.25}$$

and they are created by the corresponding inverse reactions. Reaction (2.3.25) takes place for all kinetic energies E, but reaction (2.3.24) requires $E > E_{+0}$, where E_{+0} is the ionization energy in question. Clearly,

$$E = E' + E'' + E_{+0} \; , \quad E_1'' = E + E_1' + E_{+0} \; . \tag{2.3.26}$$

Atoms A and ions A^+ are considered to be stationary.

The inelastic collision term for the isotropic electron distribution function $F_e \equiv F_e(\boldsymbol{r}, E, t)$ is thus given by

$$\left(\frac{\delta F_e}{\delta t}\right)^{0+}_{\text{inel}}$$

$$= -n_0\left(\frac{2E}{m_e}\right)^{1/2} F_e(E)\,Q_{0+}(E)$$

$$+n_+\iint\left(\frac{2E'}{m_e}\right)^{1/2}\left(\frac{2E''}{m_e}\right)^{1/2} F_e(E')F_e(E'')\,\Omega_{+0}(E',E'';E)\,dE'dE''$$

$$-n_+\left(\frac{2E}{m_e}\right)^{1/2} F_e(E)\iint\left(\frac{2E_1'}{m_e}\right)^{1/2} F_e(E_1')\,\Omega_{+0}(E,E_1';E_1'')\,dE_1'dE_1''$$

$$+n_0\iint\left(\frac{2E_1''}{m_e}\right)^{1/2} F_e(E_1'')\,\Omega_{0+}(E_1'';E,E_1')\,dE_1''dE_1' \ . \tag{2.3.27}$$

Here the first and second terms on the right-hand side describe, respectively, the collisional ionizations (2.3.24) and the inverse three-body recombinations, and the third and fourth terms the three-body recombinations (2.3.25) and the inverse collisional ionizations, respectively. Note that $Q_{0+}(E) = \Omega_{+0}(E',E'';E) = 0$ if $E<E'+E''+E_{+0}$.

The rate of change of the electron density due to the free-bound transition considered is obtained by integrating (2.3.27) over all kinetic energies E. Writing $F_e(E) = n_e f_e(E)$, the first two terms yield

$$-n_0 n_e M_{0+} + n_+ n_e^2 N_{+0}$$

according to (2.3.14). However, some care is required when integrating the remaining two terms. Using the notation of ordered pairs (E',E'') as proposed in Sect. 2.3.2 after (2.3.14), integration over all E of the third term of (2.3.27) yields the following *two* terms, with $d^3E \equiv dE\,dE_1'\,dE_1''$,

$$-\int\ldots\Omega_{+0}(E,E_1';E_1'')\,d^3E - \int\ldots\Omega_{+0}(E_1',E;E_1'')\,d^3E \ .$$

Likewise, the fourth term yields

$$+\int\ldots\Omega_{0+}(E_1'';E,E_1')\,d^3E + \int\ldots\Omega_{0+}(E_1'';E_1',E)\,d^3E \ .$$

Thus, the contribution of the last two terms in (2.3.27) is

$$-2n_+ n_e^2 N_{+0} + 2n_0 n_e M_{0+}$$

on account of (2.3.14). Hence

$$\left(\frac{\delta n_e}{\delta t}\right)_{\text{inel}}^{0+} = n_0 n_e M_{0+} - n_+ n_e^2 N_{+0} \ , \tag{2.3.28}$$

in agreement with (2.3.15).

2.3.5 Dielectronic Recombination

As discussed in Sect. 1.6.3, dielectronic recombination proceeds via radiationless capture of a free electron by an ion, giving rise to a doubly excited atom, followed by emission of a photon in a radiative transition leading to a singly excited atom. To be definite, suppose that the ion is in the ground state. Then the following chain of reactions occurs (Fig. 2.1, Sect. 1.6.3),

$$A^+(E_g^+) + e(E) \rightleftarrows A(E^{**}) \rightarrow A(E^*) + h\nu \ , \tag{2.3.29}$$

where

$$E_g^+ + E = E^{**} \ , \quad h\nu = E^{**} - E^* \ . \tag{2.3.30}$$

As indicated in (2.3.29), the doubly excited atom $A(E^{**})$ formed through radiationless capture of an electron is either split again into an electron-ion pair by autoionization, or it emits a photon, thus leading to dielectronic recombination.

In the following we shall calculate the rate of change of the electron (and ion) density due to dielectronic recombination, (2.3.29), for the limiting case of very low density, so that inelastic collisions as well as all radiative processes can be ignored with the exception of spontaneous emissions (optically thin plasma).

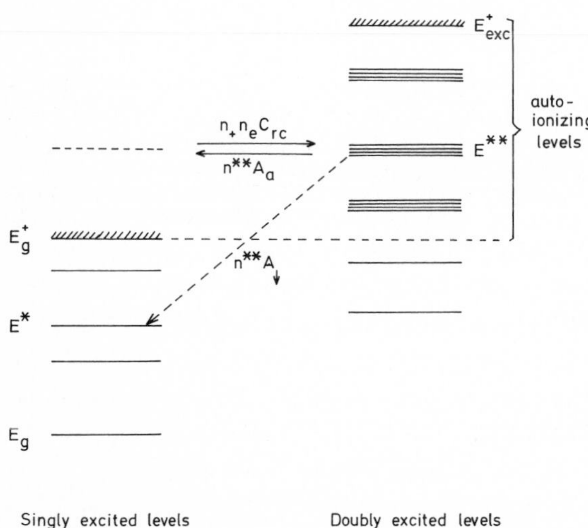

Fig. 2.1. Energy level diagram for dielectronic recombination in a low-density plasma. Reaction rates per unit time and unit volume are indicated. Doubly excited autoionizing levels are strongly broadened owing to their short lifetimes. E_g = ground-state energy of atom A, E_g^+ = ground-state energy of ion A^+, E_{exc}^+ = energy of an excited level of ion A^+

Owing to the very short lifetime of an autoionizing level, its population is determined even in nonstationary situations by the instantaneous steady-state value

$$n^{**}=\frac{n_+ n_e C_{rc}}{A_a + A_\downarrow}\,, \tag{2.3.31}$$

which is valid in the limit of low density where only the three reactions indicated in (2.3.29) (or in Fig. 2.1) contribute. Here n_+ denotes the density of ions in the ground state $A^+(E_g^+)$, A_a is the transition probability for autoionization of the doubly excited atom, $A_\downarrow$ is the Einstein coefficient (transition probability) for the spontaneous radiative transition $A(E^{**})\rightarrow A(E^*)+h\nu$, and

$$C_{rc}=\int v f_e(\boldsymbol{v})\, Q_{rc}(v)\, d^3v \tag{2.3.32}$$

is the rate coefficient for radiationless capture of a free electron, Q_{rc} being the corresponding cross section. The rate of dielectronic recombination via the considered chain of reactions is hence

$$\left(\frac{\delta n_e}{\delta t}\right)^{\text{diel}}=\left(\frac{\delta n_+}{\delta t}\right)^{\text{diel}}=-n^{**}A_\downarrow=-n_+ n_e C_{rc}\frac{A_\downarrow}{A_a+A_\downarrow}\,. \tag{2.3.33}$$

If the electrons have a Maxwellian velocity distribution of temperature T_e, the rate coefficient C_{rc} can be expressed in terms of the transition probability for autoionization A_a:

$$\begin{aligned} C_{rc} &= \int\left(\frac{2E'}{m_e}\right)^{1/2} f_e^M(E')\, Q_{rc}(E')\, dE' \\ &\simeq \left(\frac{2E}{m_e}\right)^{1/2} f_e^M(E)\int Q_{rc}(E')\, dE' = A_a\frac{g^{**}}{2g_+}\lambda_e^3 e^{-E/kT_e}\,, \end{aligned} \tag{2.3.34}$$

where $f_e^M(E)$ is the normalized Maxwell distribution in terms of kinetic energy $E=m_e v^2/2$, (1.4.14), and λ_e the thermal de Broglie wavelength of electrons (1.4.24). In the first step we have used the fact that the cross section $Q_{rc}(E')$ is appreciably different from zero only for energies in the immediate neighborhood of $E=E^{**}-E_g^+$, and in the second step use has been made of

$$\int Q_{rc}(E')\, dE'=\frac{g^{**}}{2g_+}\,\frac{h^3}{8\pi m_e E}A_a\,, \tag{2.3.35}$$

which follows from the reciprocity relation (1.6.26), g^{**} and g_+ being the statistical weights of the levels $A(E^{**})$ and $A^+(E_g^+)$, respectively. Relation (2.3.34) can also be derived by considering thermal equilibrium at temperature

T_e. Here there is detailed balance, and hence also global balance $n_+ n_e C_{rc} = n^{**} A_a$. Inserting here the Saha distribution (1.4.25) with $E_{+0} = E_g^+ - E^{**} = -E$ gives again (2.3.34).

According to (2.3.33, 34), the rate of dielectronic recombination via the considered reactions in a Maxwellian electron gas of temperature T_e is given by

$$\left(\frac{\delta n_e}{\delta t}\right)^{\text{diel}} = \left(\frac{\delta n_+}{\delta t}\right)^{\text{diel}} = -n_+ n_e \frac{A_a A_\downarrow}{A_a + A_\downarrow} \frac{g^{**}}{2g_+} \lambda_e^3 e^{-E/kT_e} , \qquad (2.3.36)$$

where $E = E^{**} - E_g^+$. The total recombination rate of an atom is, of course, an appropriate sum of expressions like (2.3.36) [2.9, 10].

2.4 Collision Terms Due to Radiative Processes

This section is devoted to a discussion of collision terms due to radiative bound-bound, free-bound, and free-free atomic transitions, neglecting recoil effects due to emission and absorption of photons. Momentum exchange between particles and photons is considered in Chap. 7; in particular, examples of collision terms with recoil effects taken into account are to be found in Sect. 7.3 regarding radiative forces, and in Sect. 7.4 regarding Compton scattering.

When deriving collision terms corresponding to radiative processes, it must be remembered that the measured or calculated values of atomic quantities like cross sections or Einstein coefficients refer to atoms at rest. Therefore, in principle all quantities of the radiation field must be transformed from the laboratory frame to the rest frame of the particle. Denoting quantities in the atomic rest frame by the superscript $^{\circ}$, this transformation changes the frequency of a photon from ν to ν° (Doppler effect), its direction from $\boldsymbol{n}$ to $\boldsymbol{n}^{\circ}$ (aberration), and the specific intensity from I to I°. The explicit formulas are derived in Appendix A. According to (A.1.27, 28, 36), for non-relativistic particle velocities $\boldsymbol{v}$

$$\nu^{\circ} = \nu\left(1 - \frac{1}{c}\boldsymbol{n}\cdot\boldsymbol{v}\right), \qquad (2.4.1a)$$

$$\boldsymbol{n}^{\circ} = \boldsymbol{n}\left[1 + \mathrm{O}\left(\frac{v}{c}\right)\right], \qquad (2.4.1b)$$

$$I^{\circ} = I\left[1 + \mathrm{O}\left(\frac{v}{c}\right)\right], \qquad (2.4.1c)$$

where in the last two equations correction terms of the order v/c have not been written down explicitly. Furthermore, the photon flux density, see (1.4.40), is transformed according to

$$\frac{1}{h\nu^{\text{o}}} I^{\text{o}}(\nu^{\text{o}},\boldsymbol{n}^{\text{o}})\,d\nu^{\text{o}}\,d\Omega^{\text{o}} = \frac{\nu^{\text{o}}}{\nu}\,\frac{1}{h\nu} I(\nu,\boldsymbol{n})\,d\nu\,d\Omega$$

$$= \left(1-\frac{1}{c}\boldsymbol{n}\cdot\boldsymbol{v}\right)\frac{1}{h\nu} I(\nu,\boldsymbol{n})\,d\nu\,d\Omega \tag{2.4.2}$$

on account of $d\nu^{\text{o}}/d\nu = \nu^{\text{o}}/\nu$, $d\Omega^{\text{o}}/d\Omega = (\nu/\nu^{\text{o}})^2$, and $I^{\text{o}}/I = (\nu^{\text{o}}/\nu)^3$, see (A.1.13, 19, 26).

The reaction rates for absorption, stimulated emission, and scattering are proportional to the flux density of the incoming photons. According to (2.4.2), replacing the flux density in the rest frame of the particle by the flux density in the laboratory frame gives rise to a relative error of the order of v/c which for all practical purposes can be neglected. The same applies to the photon direction and to the specific intensity. Concerning the photon frequency, two cases have to be distinguished. For continuum radiation due to free-bound and free-free transitions, the frequency of a photon in the atom's rest frame may simply be replaced by the laboratory frequency, because here all cross sections are slowly varying functions of the frequency. In contrast, for line radiation due to bound-bound transitions, the Doppler effect must be taken into account in all quantities that vary rapidly as functions of frequency, such as cross sections, line profile coefficients, or redistribution functions.

To sum up, for nonrelativistic particle velocities the following approximations will be used:

$$\nu^{\text{o}} \simeq \begin{cases} \nu - \dfrac{\nu_0}{c}\boldsymbol{n}\cdot\boldsymbol{v} & \text{(bb)}\ , \\[2ex] \nu & \text{(fb and ff)}\ , \end{cases} \tag{2.4.3a}$$

$$\boldsymbol{n}^{\text{o}} \simeq \boldsymbol{n}\ , \tag{2.4.3b}$$

$$I^{\text{o}} \simeq I\ , \tag{2.4.3c}$$

$$I^{\text{o}}(\nu^{\text{o}},\boldsymbol{n}^{\text{o}})\,d\nu^{\text{o}}\,d\Omega^{\text{o}} \simeq I_\nu(\boldsymbol{n})\,d\nu\,d\Omega\ , \tag{2.4.3d}$$

$$\frac{1}{h\nu^{\text{o}}} I^{\text{o}}(\nu^{\text{o}},\boldsymbol{n}^{\text{o}})\,d\nu^{\text{o}}\,d\Omega^{\text{o}} \simeq \frac{1}{h\nu} I_\nu(\boldsymbol{n})\,d\nu\,d\Omega\ . \tag{2.4.3e}$$

In (2.4.3a), bb, fb, ff stand for bound-bound, free-bound, and free-free transitions. Note that in the bound-bound case, the frequency appearing in the correction term has been replaced by the constant frequency $\nu_0 \equiv \nu_0^{\text{o}}$ corresponding to the line center in the atomic rest frame.

2.4.1 Line Emission and Absorption

Let us first consider radiative bound-bound transitions between a lower atomic level 1 and a higher level 2. Atoms in level 1 are destroyed by absorption of photons,

$$A_1(\boldsymbol{v})+h\nu\rightarrow A_2(\boldsymbol{v}) \ , \tag{2.4.4}$$

and they are created by inverse emissions. Since recoil effects are neglected, the velocity of the atom is not changed by these processes. Clearly,

$$h\nu\simeq h\nu_0=E_{21} \ , \tag{2.4.5}$$

where $E_{21}=E_2-E_1$ is the excitation energy of the bound-bound transition, and $\nu_0\equiv\nu_0^{\rm o}$ is the frequency of the line center as measured in the atomic rest frame.

The corresponding collision terms for the atomic distribution functions $F_1\equiv F_1(\boldsymbol{r},\boldsymbol{v},t)$ and $F_2\equiv F_2(\boldsymbol{r},\boldsymbol{v},t)$ are in this approximation given by

$$\begin{aligned}\left(\frac{\delta F_1}{\delta t}\right)^{12}_{\rm rad} = -\left(\frac{\delta F_2}{\delta t}\right)^{12}_{\rm rad} &= -F_1(\boldsymbol{v})B_{12}\iint I_\nu(\boldsymbol{n})\,\alpha_{12}(\xi)\,d\nu\frac{d\Omega}{4\pi}\\ &+F_2(\boldsymbol{v})A_{21}\iint\left[1+\frac{c^2}{2h\nu^3}I_\nu(\boldsymbol{n})\right]\eta_{21}(\xi,\boldsymbol{n})\,d\nu\frac{d\Omega}{4\pi} \ ,\end{aligned} \tag{2.4.6}$$

where the first term describes absorptions, and the second, spontaneous and stimulated emissions. Here the approximations (2.4.3) have been used according to which only the photon frequency appearing in the atomic absorption and emission profiles α_{12} and η_{21} is transformed from the laboratory frame to the rest frame of the atom, $\nu\rightarrow\nu^{\rm o}\simeq\xi$, with

$$\xi\equiv\xi(\nu,\boldsymbol{n};\boldsymbol{v}) = \nu-\frac{\nu_0}{c}\boldsymbol{n}\cdot\boldsymbol{v} \ . \tag{2.4.7}$$

In (2.4.6), B_{12} and A_{21} are the Einstein coefficients for absorption and spontaneous emission, respectively, and stimulated emissions are taken into account through

$$\frac{c^2}{2h\nu^3}I_\nu(\boldsymbol{n})\simeq\frac{c^2}{2h\nu_0^3}I_\nu(\boldsymbol{n}) = \frac{B_{21}}{A_{21}}I_\nu(\boldsymbol{n}) \ , \tag{2.4.8}$$

where $I_\nu(\boldsymbol{n})\equiv I_\nu(\boldsymbol{n},\boldsymbol{r},t)$ is the specific intensity of the radiation field, and B_{21} the Einstein coefficient for stimulated emission, (1.6.31).

The atomic absorption profile is defined such that $\alpha_{12}(\xi)\,d\xi$ is the probability that an atom in level 1 absorbs through a radiative transition $1\rightarrow2$ a photon $(\xi,d\xi)$ if irradiated with light whose intensity is constant over the frequency range of the considered spectral line. The atomic emission profile is defined such that $\eta_{21}(\xi,\boldsymbol{n})\,d\xi\,d\Omega/4\pi$ is the probability that an atom in level 2 emits a photon $(\xi, d\xi; \boldsymbol{n}, d\Omega)$ provided that a spontaneous radiative transition $2\to1$ takes place. Accordingly, these profiles are normalized, see (B.1.1),

$$\int \alpha_{12}(\xi)\,d\xi = \iint \eta_{21}(\xi,\boldsymbol{n})\,d\xi\frac{d\Omega}{4\pi} = 1 \ . \tag{2.4.9}$$

A more detailed discussion of the atomic profile coefficients α_{12} and η_{21} can be found in Appendix B.

Using (2.4.8, 9), the collision term (2.4.6) can be written as

$$\left(\frac{\delta F_1}{\delta t}\right)^{12}_{\text{rad}} = -\left(\frac{\delta F_2}{\delta t}\right)^{12}_{\text{rad}} = -F_1(\boldsymbol{v})B_{12}I_{12}(\boldsymbol{v}) + F_2(\boldsymbol{v})\left[A_{21}+B_{21}I_{21}(\boldsymbol{v})\right] \ , \tag{2.4.10}$$

where, see (B.2.20,3.3),

$$I_{12}(\boldsymbol{v}) = \iint I_\nu(\boldsymbol{n})\,\alpha_{12}(\xi)\,d\nu\frac{d\Omega}{4\pi}, \tag{2.4.11}$$

$$I_{21}(\boldsymbol{v}) = \iint I_\nu(\boldsymbol{n})\,\eta_{21}(\xi,\boldsymbol{n})\,d\nu\frac{d\Omega}{4\pi} \ . \tag{2.4.12}$$

The rate of change of the corresponding number densities follows immediately by integrating (2.4.10) over all velocities $\boldsymbol{v}$, yielding

$$\left(\frac{\delta n_1}{\delta t}\right)^{12}_{\text{rad}} = -\left(\frac{\delta n_2}{\delta t}\right)^{12}_{\text{rad}} = -n_1B_{12}J_{12}+n_2(A_{21}+B_{21}J_{21}) \ , \tag{2.4.13}$$

where

$$J_{12} = \int f_1(\boldsymbol{v})I_{12}(\boldsymbol{v})\,d^3v = \iint I_\nu(\boldsymbol{n})\,\varphi_\nu^{12}(\boldsymbol{n})\,d\nu\frac{d\Omega}{4\pi} \ , \tag{2.4.14}$$

$$J_{21} = \int f_2(\boldsymbol{v})I_{21}(\boldsymbol{v})\,d^3v = \iint I_\nu(\boldsymbol{n})\,\psi_\nu^{21}(\boldsymbol{n})\,d\nu\frac{d\Omega}{4\pi} \ . \tag{2.4.15}$$

Here we have introduced the so-called *absorption and emission profiles* of the spectral line, which are the appropriate average values of the atomic profiles α_{12} and η_{21},

$$\varphi_\nu^{12}(\boldsymbol{n}) = \int f_1(\boldsymbol{v})\,\alpha_{12}(\xi)\,d^3v \ , \tag{2.4.16}$$

$$\psi_\nu^{21}(\boldsymbol{n}) = \int f_2(\boldsymbol{v})\,\eta_{21}(\xi,\boldsymbol{n})\,d^3v \ , \tag{2.4.17}$$

and which are normalized according to

$$\iint \varphi_\nu^{12}(\boldsymbol{n})\, d\nu \frac{d\Omega}{4\pi} = \iint \psi_\nu^{21}(\boldsymbol{n})\, d\nu \frac{d\Omega}{4\pi} = 1 \; . \tag{2.4.18}$$

In contrast to the profiles α_{12} and η_{21} that refer to an atom, the profile coefficients φ_ν and ψ_ν refer to a small volume of the gas at the space-time point $(\boldsymbol{r}, t)$. Note that even for an isotropic profile α_{12} as assumed above, the absorption profile φ_ν will be anisotropic if the velocity distribution of absorbing atoms $f_1(\boldsymbol{v})$ is anisotropic.

2.4.2 Photoionization and Radiative Recombination (Atoms)

In radiative free-bound transitions, atoms A in an internal state of energy E_0 are destroyed by photoionizations that yield an ion A^+ in an internal state of energy E_+ and a free electron,

$$A(\boldsymbol{v}) + h\nu \rightarrow A^+(\boldsymbol{v}) + \mathrm{e}(\boldsymbol{v}_\mathrm{e}) \;, \tag{2.4.19}$$

and they are created by the inverse radiative recombinations. Since momentum exchange of the heavy particles with photons and electrons is neglected, the atom and the ion in (2.4.19) have the same velocity $\boldsymbol{v}$. On the other hand, energy conservation requires in the approximation of stationary atoms and ions that

$$h\nu = \tfrac{1}{2} m_\mathrm{e} v_\mathrm{e}^2 + E_{+0} \tag{2.4.20}$$

with ionization energy $E_{+0} = E_+ - E_0$.

The corresponding collision terms for the distribution functions $F_0 \equiv F_0(\boldsymbol{r}, \boldsymbol{v}, t)$ and $F_+ \equiv F_+(\boldsymbol{r}, \boldsymbol{v}, t)$ are given by

$$\begin{aligned} \left(\frac{\delta F_0}{\delta t}\right)_{\mathrm{rad}}^{0+} &= -\left(\frac{\delta F_+}{\delta t}\right)_{\mathrm{rad}}^{0+} = -F_0(\boldsymbol{v}) \iint \frac{1}{h\nu} I_\nu(\boldsymbol{n})\, \sigma_{0+}(\nu)\, d\nu\, d\Omega \\ &\quad + F_+(\boldsymbol{v}) \iint d\nu\, d\Omega \left[1 + \frac{c^2}{2h\nu^3} I_\nu(\boldsymbol{n})\right] \int d^3 v_\mathrm{e} v_\mathrm{e} F_\mathrm{e}(\boldsymbol{v}_\mathrm{e}) \\ &\quad \cdot \sigma_{+0}(v_\mathrm{e}, \vartheta) \frac{1}{v_\mathrm{e}^2} \delta\left(v_\mathrm{e} - \left[\frac{2}{m_\mathrm{e}}(h\nu - E_{+0})\right]^{1/2}\right) . \end{aligned} \tag{2.4.21}$$

The first term on the right-hand side describes photoionizations, $\sigma_{0+}(\nu)$ being the total cross section for photoionization, recalling that $I_\nu\, d\nu\, d\Omega / h\nu$ is the flux density of photons, (1.4.40). Notice that according to our approximation (2.4.3), no transformation from the laboratory frame to the atomic rest frame has been carried out here for continuum radiation. On the other hand, the second term describes spontaneous and induced recombinations. In the approximation of stationary ions, $v_e F_e(\boldsymbol{v}_e)\,d^3v_e$ is the flux density of incoming electrons in the velocity range $(\boldsymbol{v}_e, d^3v_e)$, and $\sigma_{+0}(v_e, \vartheta)$ is the differential cross section for spontaneous radiative recombination, where ϑ denotes the angle between the electron velocity and the direction of the emitted photon $(\boldsymbol{n}\cdot\boldsymbol{v}_e = v_e\cos\vartheta)$. The delta function (with the normalization factor v_e^{-2} because of $d^3v_e \equiv v_e^2\,dv_e\,d\Omega_e \equiv v_e^2 \sin\vartheta\,dv_e d\vartheta\,d\varphi$) ensures that for a given photon frequency ν only those electrons contribute to recombination whose kinetic energy has the value required by energy conservation, (2.4.20).

The rate of change of the corresponding number densities follows from (2.4.21) by integration over all velocities $\boldsymbol{v}$,

$$\left(\frac{\delta n_0}{\delta t}\right)^{0+}_{\rm rad} = -\left(\frac{\delta n_+}{\delta t}\right)^{0+}_{\rm rad} = -n_0 \iint \frac{1}{h\nu} I_\nu(\boldsymbol{n})\,\sigma_{0+}(\nu)\,d\nu\,d\Omega$$

$$+ n_+ n_e \iint d\nu\,d\Omega \left[1 + \frac{c^2}{2h\nu^3} I_\nu(\boldsymbol{n})\right] v_e \int f_e(\boldsymbol{v}_e)\,\sigma_{+0}(v_e, \vartheta)\,d\Omega_e\,, \tag{2.4.22a}$$

where

$$v_e \equiv \left[\frac{2}{m_e}(h\nu - E_{+0})\right]^{1/2}, \qquad \cos\vartheta \equiv \frac{\boldsymbol{n}\cdot\boldsymbol{v}_e}{v_e}\,. \tag{2.4.22b}$$

Here, $\boldsymbol{v}_e$ is the vector of length v_e that has polar and azimuthal angles ϑ and φ in a coordinate system whose polar axis is given by the photon direction $\boldsymbol{n}$, so that $d\Omega_e = \sin\vartheta\,d\vartheta\,d\varphi$. On the other hand, the solid angle $d\Omega$ corresponds to the photon direction $\boldsymbol{n}$.

Finally, a useful form of the rate coefficient for spontaneous radiative recombinations R_{+0} is obtained when the electrons have a Maxwellian velocity distribution of temperature T_e. Clearly, in terms of kinetic energy $E = m_e v_e^2/2$,

$$R_{+0} = \int_0^\infty \left(\frac{2E}{m_e}\right)^{1/2} f_e^{\rm M}(E)\,\sigma_{+0}(E)\,dE\,, \tag{2.4.23}$$

where $\sigma_{+0}(E)$ is the total cross section for spontaneous recombination. Inserting here the Maxwell distribution (1.4.14) and using the Milne relation (1.6.40) gives

$$R_{+0} = \frac{g_0}{g_+}\left(\frac{2}{\pi}\right)^{1/2} \frac{h^3 e^{E_{+0}/kT_e}}{c^2(m_e kT_e)^{3/2}} \int_{\nu_{+0}}^\infty \nu^2 e^{-h\nu/kT_e}\sigma_{0+}(\nu)\,d\nu \tag{2.4.24}$$

in terms of the total cross section for photoionization $\sigma_{0+}(\nu)$, and where $\nu_{+0} \equiv E_{+0}/h$.

This result can also be derived by considering thermal equilibrium at temperature T_e, where radiative recombinations are balanced by photoionizations, $n_+ n_e R_{+0} = n_0 P_{0+}$. However, since only spontaneous recombinations are considered, the rate coefficient for photoionization P_{0+} has to be calculated using the Wien function, which is the Planck function in the limit of negligible induced photon processes [see the discussion after (1.6.33)], that is,

$$P_{0+} = \int_{\nu_{+0}}^{\infty} \frac{4\pi}{h\nu} B_\nu^W(T_e)\,\sigma_{0+}(\nu)\,d\nu \;,$$

since the total flux density of incoming photons $(\nu, d\nu)$ is $(4\pi/h\nu)B_\nu^W(T_e)\,d\nu$. Using the explicit form of the Wien function, (1.6.33), and the Saha distribution (1.4.25) for the ratio $n_0/n_+ n_e$, (2.4.24) again results.

2.4.3 Photoionization and Radiative Recombination (Electrons)

Let us now discuss the collision term for the electron distribution function, corresponding to the radiative free-bound transitions considered in the preceding section. Destruction of electrons of velocity $\boldsymbol{v}$ is due to recombinations

$$A^+ + e(\boldsymbol{v}) \rightarrow A + h\nu \;, \tag{2.4.25}$$

and their creation is due to inverse photoionizations. Here the ion $A^+ \equiv A^+(E_+)$ and the atom $A \equiv A(E_0)$ are considered as stationary particles, and

$$\tfrac{1}{2} m_e v^2 + E_{+0} = h\nu \;, \tag{2.4.26}$$

with $E_{+0} = E_+ - E_0$.

The corresponding collision term for the electron distribution function $F_e \equiv F_e(\boldsymbol{r}, \boldsymbol{v}, t)$ is now

$$\begin{aligned}\left(\frac{\delta F_e}{\delta t}\right)^{0+}_{\text{rad}} &= -n_+ v F_e(\boldsymbol{v}) \iint d\nu\, d\Omega\, \sigma_{+0}(v, \vartheta)\left[1 + \frac{c^2}{2h\nu^3} I_\nu(\boldsymbol{n})\right] \\ &\quad \cdot \delta\left(\nu - \frac{1}{h}\left[\frac{1}{2} m_e v^2 + E_{+0}\right]\right) \\ &\quad + n_0 \frac{m_e}{hv} \iint d\nu\, d\Omega \frac{1}{h\nu} I_\nu(\boldsymbol{n})\, \sigma_{0+}(\nu, \vartheta) \\ &\quad \cdot \delta\left(\nu - \frac{1}{h}\left[\frac{1}{2} m_e v^2 + E_{+0}\right]\right) \;.\end{aligned} \tag{2.4.27}$$

The first term, which describes the destruction of electrons $e(\boldsymbol{v})$ due to spontaneous and stimulated recombinations, is easily interpreted. Here $\sigma_{+0}(v, \vartheta)$ is the differential cross section for spontaneous recombination, ϑ being the angle between the electron velocity and the direction of the emitted photon,

$$\cos\vartheta = \frac{\boldsymbol{n}\cdot\boldsymbol{v}}{v} \ . \tag{2.4.28}$$

The delta function ensures that only photons with the correct frequency, as required by energy conservation (2.4.26), contribute to stimulated recombination.

The second term on the right-hand side of (2.4.27) is derived as follows. The number of electrons $e(\boldsymbol{v}, d^3v)$ created at time t in the volume $(\boldsymbol{r}, d^3r)$ by photoionizations is given by

$$\frac{\delta F_e}{\delta t} d^3r\, d^3v = n_0 d^3r \iint' d\nu\, d\Omega \frac{1}{h\nu} I_\nu(\boldsymbol{n})\, \sigma_{0+}(v, \vartheta)\, d\Omega_v \ , \tag{2.4.29}$$

that is, by the product of the number of atoms $n_0 d^3r$, the flux density of photons $I_\nu\, d\nu\, d\Omega/h\nu$, and the cross section $\sigma_{0+}\, d\Omega_v$ where the solid angle $d\Omega_v$ corresponds to the outgoing electrons. The dash at the integral signs signifies that the integrations $d\nu\, d\Omega$ must be performed only over those ν and $\boldsymbol{n}$ that lead to a final electron velocity in the range $(\boldsymbol{v}, d^3v)$ considered. However,

$$d\nu\, d\Omega_v = \frac{d\nu\, d\Omega_v}{d^3v} d^3v = \frac{d\nu\, d\Omega_v}{v^2 dv\, d\Omega_v} d^3v = \frac{m_e}{hv} d^3v \tag{2.4.30}$$

since $d\nu = m_e v\, dv/h$ through (2.4.26). Equation (2.4.29) can hence be written as

$$\frac{\delta F_e}{\delta t} = n_0 \frac{m_e}{hv} \int \frac{1}{h\nu} I_\nu(\boldsymbol{n})\, \sigma_{0+}(v, \vartheta)\, d\Omega \ , \tag{2.4.31a}$$

where ν stands for

$$\nu \equiv \frac{1}{h}\left(\frac{1}{2} m_e v^2 + E_{+0}\right) \ . \tag{2.4.31b}$$

Clearly, (2.4.31) is equivalent to the last term of (2.4.27).

The rate of change of the electron density is, of course,

$$\left(\frac{\delta n_e}{\delta t}\right)^{0+}_{\text{rad}} = \left(\frac{\delta n_+}{\delta t}\right)^{0+}_{\text{rad}} \ , \tag{2.4.32}$$

where $(\delta n_+/\delta t)^{0+}_{\text{rad}}$ is given by (2.4.22).

2.4.4 Bremsstrahlung and Inverse Bremsstrahlung (Electrons)

As the last case we treat radiative free-free transitions of electrons in the field of heavy particles (ions or atoms), that is, we consider the collision term for the electron distribution function due to free-free emission (bremsstrahlung)

$$P+\mathrm{e}(\boldsymbol{v}',d^3v')\rightarrow P+\mathrm{e}(\boldsymbol{v},d^3v)+\gamma(\nu,d\nu;\boldsymbol{n},d\Omega)\ , \tag{2.4.33}$$

and the inverse free-free absorption (also called inverse bremsstrahlung), where P is a heavy particle assumed as stationary. The photon frequency is supposed to be sufficiently above the plasma frequency so that collective plasma effects are negligible (Sect. 1.6.6). Of course,

$$\tfrac{1}{2}m_\mathrm{e}v'^2=\tfrac{1}{2}m_\mathrm{e}v^2+h\nu\ . \tag{2.4.34}$$

In Sect. 1.6.6 two quantities α_ν and β_ν were introduced which determine the rates of free-free emissions and absorptions, respectively. As already mentioned there, they both contain a delta function on account of the energy relation (2.4.34). However, for our present purposes it is advantageous to define new quantities $\tilde{\alpha}_\nu$ and $\tilde{\beta}_\nu$ that no longer contain delta functions. To this end, the rate of spontaneous free-free emissions is written in the following two equivalent ways, see (1.6.44),

$$\begin{aligned}&n_PF_\mathrm{e}(\boldsymbol{v}')d^3v'\,\alpha_\nu(\boldsymbol{v}';\boldsymbol{n},\boldsymbol{v})\,d\nu\,d\Omega\,d^3v\\&\quad=n_PF_\mathrm{e}(\boldsymbol{v}')d^3v'\tilde{\alpha}_\nu(\boldsymbol{v}';\boldsymbol{n},\boldsymbol{e})\,d\nu\,d\Omega\,d\Omega_v\ ,\end{aligned} \tag{2.4.35a}$$

with

$$\begin{aligned}&v\equiv(v'^2-\varepsilon)^{1/2}\ ,\quad \varepsilon=\frac{2h\nu}{m_\mathrm{e}}\ ,\\&\boldsymbol{e}=\frac{\boldsymbol{v}}{v}\ ,\qquad d^3v=v^2dv\,d\Omega_v\ .\end{aligned} \tag{2.4.35b}$$

The rate of free-free absorptions is, see (1.6.44),

$$\begin{aligned}&n_PF_\mathrm{e}(\boldsymbol{v})d^3v\,I_\nu(\boldsymbol{n})\,d\nu\,d\Omega\,\beta_\nu(\boldsymbol{n},\boldsymbol{v};\boldsymbol{v}')\,d^3v'\\&\quad=n_PF_\mathrm{e}(\boldsymbol{v})d^3v\,I_\nu(\boldsymbol{n})\,d\nu\,d\Omega\,\tilde{\beta}_\nu(\boldsymbol{n},\boldsymbol{v};\boldsymbol{e}')\,d\Omega_v'\ ,\end{aligned} \tag{2.4.36a}$$

with

$$\begin{aligned}&v'\equiv(v^2+\varepsilon)^{1/2}\ ,\quad \varepsilon=\frac{2h\nu}{m_\mathrm{e}}\ ,\\&\boldsymbol{e}'=\frac{\boldsymbol{v}'}{v'}\ ,\qquad d^3v'=v'^2dv'\,d\Omega_v'\ .\end{aligned} \tag{2.4.36b}$$

Inserting the factors $\delta(v-[v'^2-\varepsilon]^{1/2})\,dv$ and $\delta(v'-[v^2+\varepsilon]^{1/2})\,dv'$ in the right-hand sides of (2.4.35a,36a), respectively, gives the following relations between the old α_ν, β_ν and the new $\tilde{\alpha}_\nu$, $\tilde{\beta}_\nu$,

$$v^2\alpha_\nu(\boldsymbol{v}';\boldsymbol{n},\boldsymbol{v}) = \tilde{\alpha}_\nu(\boldsymbol{v}';\boldsymbol{n},\boldsymbol{e})\,\delta(v-[v'^2-\varepsilon]^{1/2}) \ , \tag{2.4.37a}$$

$$v'^2\beta_\nu(\boldsymbol{n},\boldsymbol{v};\boldsymbol{v}') = \tilde{\beta}_\nu(\boldsymbol{n},\boldsymbol{v};\boldsymbol{e}')\,\delta(v'-[v^2+\varepsilon]^{1/2}) \ . \tag{2.4.37b}$$

On the other hand, from

$$\delta(v'-[v^2+\varepsilon]^{1/2})\,dv' = \delta(v-[v'^2-\varepsilon]^{1/2})\,dv \tag{2.4.38}$$

together with

$$v'dv' = v\,dv \ , \tag{2.4.39}$$

which follows from (2.4.35b or 36b), one gets

$$\delta(v'-[v^2+\varepsilon]^{1/2}) = \frac{v'}{v}\,\delta(v-[v'^2-\varepsilon]^{1/2}) \ . \tag{2.4.40}$$

Substituting the preceding equation into (2.4.37b), and using the reciprocity relation (1.6.45) gives in view of (2.4.37a) the new reciprocity relation

$$\tilde{\beta}_\nu(\boldsymbol{n},\boldsymbol{v};\boldsymbol{e}') = \frac{v'}{v}\,\frac{c^2}{2h\nu^3}\,\tilde{\alpha}_\nu(\boldsymbol{v}';\boldsymbol{n},\boldsymbol{e}) \ . \tag{2.4.41}$$

Of course, this equation can also be derived directly from the principle of detailed balance by substituting in the right-hand sides of (2.4.35a,36a) for F_e the Maxwell distribution (1.6.6a), and for I_ν the Wien function (1.6.33), and equating the expressions so obtained.

After these preliminary remarks we now turn to the derivation of the collision term. For this, observe that electrons of velocity $\boldsymbol{v}$ are created by the processes

$$P+\mathrm{e}(\boldsymbol{v}')\rightarrow P+\mathrm{e}(\boldsymbol{v})+h\nu \ , \tag{2.4.42a}$$

$$P+\mathrm{e}(\boldsymbol{v}'')+h\nu\rightarrow P+\mathrm{e}(\boldsymbol{v}) \ , \tag{2.4.42b}$$

and destroyed by the processes

$$P+\mathrm{e}(\boldsymbol{v})\rightarrow P+\mathrm{e}(\boldsymbol{v}'')+h\nu \ , \tag{2.4.42c}$$

$$P+\mathrm{e}(\boldsymbol{v})+h\nu\rightarrow P+\mathrm{e}(\boldsymbol{v}') \ . \tag{2.4.42d}$$

Conservation of energy requires

$$\tfrac{1}{2}\,m_e v'^2-\tfrac{1}{2}\,m_e v^2 = h\nu = \tfrac{1}{2}\,m_e v^2-\tfrac{1}{2}\,m_e v''^2 \ . \tag{2.4.43}$$

The net production rate of electrons $e(\boldsymbol{v}, d^3v)$, taking stimulated bremsstrahlung into account, is accordingly given by

$$\frac{\delta F_e}{\delta t} d^3v = \int n_P F_e(\boldsymbol{v}') d^3v' \tilde{\alpha}_\nu(\boldsymbol{v}'; \boldsymbol{n}, \boldsymbol{e}) \left[1 + \frac{c^2}{2h\nu^3} I_\nu(\boldsymbol{n})\right] d\nu \, d\Omega \, d\Omega_v$$

$$+ \int n_P F_e(\boldsymbol{v}'') d^3v'' I_\nu(\boldsymbol{n}) d\nu \, d\Omega \, \tilde{\beta}_\nu(\boldsymbol{n}, \boldsymbol{v}''; \boldsymbol{e}) \, d\Omega_v$$

$$- \int n_P F_e(\boldsymbol{v}) d^3v \, \tilde{\alpha}_\nu(\boldsymbol{v}; \boldsymbol{n}, \boldsymbol{e}'') \left[1 + \frac{c^2}{2h\nu^3} I_\nu(\boldsymbol{n})\right] d\nu \, d\Omega \, d\Omega_v''$$

$$- \int n_P F_e(\boldsymbol{v}) d^3v I_\nu(\boldsymbol{n}) d\nu \, d\Omega \, \tilde{\beta}_\nu(\boldsymbol{n}, \boldsymbol{v}; \boldsymbol{e}') \, d\Omega_v' \,, \tag{2.4.44}$$

where

$$\boldsymbol{v}' \equiv v' \boldsymbol{e}' \quad \text{with} \quad v' \equiv \left(v^2 + \frac{2h\nu}{m_e}\right)^{1/2} , \tag{2.4.45a}$$

$$\boldsymbol{v}'' \equiv v'' \boldsymbol{e}'' \quad \text{with} \quad v'' \equiv \left(v^2 - \frac{2h\nu}{m_e}\right)^{1/2} . \tag{2.4.45b}$$

To remove any restrictions for the regions of integration, we substitute in the first line of (2.4.44)

$$d^3v' d\Omega_v = v'^2 dv' \, d\Omega_v' \, d\Omega_v = \frac{v'}{v} d\Omega_v' \, d^3v \,, \tag{2.4.46a}$$

and in the second line

$$d^3v'' d\Omega_v = v''^2 dv'' d\Omega_v'' d\Omega_v = \frac{v''}{v} d\Omega_v'' d^3v \,, \tag{2.4.46b}$$

where $v \, dv = v' dv' = v'' dv''$ and $d^3v = v^2 dv \, d\Omega_v$ have been used. This finally gives the collision term for the electron distribution function $F_e \equiv F_e(\boldsymbol{r}, \boldsymbol{v}, t)$ due to radiative free-free transitions

$$\left(\frac{\delta F_e}{\delta t}\right)_{\text{rad}}^{\text{ff}} = n_P \iiint d\nu \, d\Omega \, d\Omega_v' \frac{v'}{v} F_e(\boldsymbol{v}') \tilde{\alpha}_\nu(\boldsymbol{v}'; \boldsymbol{n}, \boldsymbol{e}) \left[1 + \frac{c^2}{2h\nu^3} I_\nu(\boldsymbol{n})\right]$$

$$+ n_P \iiint d\nu \, d\Omega \, d\Omega_v'' \frac{v''}{v} F_e(\boldsymbol{v}'') I_\nu(\boldsymbol{n}) \tilde{\beta}_\nu(\boldsymbol{n}, \boldsymbol{v}''; \boldsymbol{e})$$

$$- n_P \iiint d\nu \, d\Omega \, d\Omega_v'' F_e(\boldsymbol{v}) \tilde{\alpha}_\nu(\boldsymbol{v}; \boldsymbol{n}, \boldsymbol{e}'') \left[1 + \frac{c^2}{2h\nu^3} I_\nu(\boldsymbol{n})\right]$$

$$- n_P \iiint d\nu \, d\Omega \, d\Omega_v' F_e(\boldsymbol{v}) I_\nu(\boldsymbol{n}) \tilde{\beta}_\nu(\boldsymbol{n}, \boldsymbol{v}; \boldsymbol{e}') \,. \tag{2.4.47}$$

We now want to rewrite this collision term for the limiting case of soft photons, assuming in addition an isotropic situation [2.11]. For isotropy, $F_e(\boldsymbol{v}) = F_e(v)$ and $I_\nu(\boldsymbol{n}) = I_\nu$, and (2.4.47) takes the form

$$\left(\frac{\delta F_e}{\delta t}\right)^{\rm ff}_{\rm rad} = \frac{16\pi^2 n_P}{v}\int d\nu\left([v'\tilde{\alpha}_\nu(v')F_e(v') - v\,\tilde{\alpha}_\nu(v)F_e(v)]\right.$$
$$+[v'\tilde{\alpha}_\nu(v')F_e(v') - v\,\tilde{\alpha}_\nu(v)F_e(v)]\frac{c^2}{2h\nu^3}I_\nu$$
$$\left.+[v\,\tilde{\alpha}_\nu(v)F_e(v'') - v'\tilde{\alpha}_\nu(v')F_e(v)]\frac{c^2}{2h\nu^3}I_\nu\right), \tag{2.4.48}$$

where the reciprocity relation (2.4.41) has been employed to express $\tilde{\beta}_\nu$ in terms of $\tilde{\alpha}_\nu$, and where $\tilde{\alpha}_\nu(v)$ denotes $\tilde{\alpha}_\nu(\boldsymbol{v};\boldsymbol{n},\boldsymbol{e}'')$ averaged over directions,

$$\tilde{\alpha}_\nu(v) = \iint \alpha_\nu(\boldsymbol{v};\boldsymbol{n},\boldsymbol{e}'')\frac{d\Omega\, d\Omega_v''}{(4\pi)^2}\,. \tag{2.4.49}$$

On the other hand, the limiting case of soft photons means that the photon energy is small compared to the mean kinetic energy of the electrons. For non-relativistic plasmas, usually only this limiting case is important, as for higher frequencies the cross section for free-free emission (and hence the quantity $\tilde{\alpha}_\nu$) becomes very small [2.12]. In this limiting case,

$$\frac{2h\nu}{m_e} = v'^2 - v^2 \simeq 2v(v'-v)\,, \qquad \frac{2h\nu}{m_e} = v^2 - v''^2 \simeq 2v(v-v'')\,,$$

or

$$v'-v \simeq v - v'' \simeq \frac{h\nu}{m_e v}\,. \tag{2.4.50}$$

Thus,

$$v'\tilde{\alpha}_\nu(v')F_e(v') - v\,\tilde{\alpha}_\nu(v)F_e(v) \simeq \frac{h\nu}{m_e v}\frac{\partial}{\partial v}[v\,\tilde{\alpha}_\nu(v)F_e(v)]\,, \tag{2.4.51}$$

$$v'\tilde{\alpha}_\nu(v')[F_e(v') - F_e(v)] - v\,\tilde{\alpha}_\nu(v)[F_e(v) - F_e(v'')]$$
$$\simeq \frac{h\nu}{m_e}\left[\tilde{\alpha}_\nu(v')\left(\frac{\partial F_e}{\partial v}\right)_{v'} - \tilde{\alpha}_\nu(v)\left(\frac{\partial F_e}{\partial v}\right)_v\right]$$
$$\simeq \frac{h^2\nu^2}{m_e^2 v}\frac{\partial}{\partial v}\left[\tilde{\alpha}_\nu(v)\frac{\partial F_e}{\partial v}\right]. \tag{2.4.52}$$

Inserting these two equations into (2.4.48) yields

$$\left(\frac{\delta F_e}{\delta t}\right)^{ff}_{rad} = \frac{16\pi^2 h}{m_e v^2} n_P \int d\nu \left(\nu \frac{\partial}{\partial v} [v \tilde{\alpha}_\nu(v) F_e(v)] \right.$$

$$\left. + \frac{c^2}{2m_e \nu} \frac{\partial}{\partial v} \left[\tilde{\alpha}_\nu(v) \frac{\partial F_e}{\partial v} \right] I_\nu \right). \tag{2.4.53}$$

This derivation shows clearly how derivatives of the distribution function with respect to velocity emerge in collision terms due to weak collisions (as, for instance, in Fokker-Planck collision terms).

Let us check the case of thermal equilibrium. For a Maxwell distribution (1.6.6a),

$$\frac{\partial F_e^M}{\partial v} = -\frac{m_e v}{kT} F_e^M . \tag{2.4.54}$$

On the other hand, in the limiting case $h\nu/kT \ll 1$, the Planck function (1.4.45) reduces to the *Rayleigh-Jeans function*

$$B_\nu^{RJ}(T) = \frac{2kT\nu^2}{c^2} , \tag{2.4.55}$$

which corresponds to the classical limit as Planck's constant h has dropped out. Inserting (2.4.54) and $I_\nu = B_\nu^{RJ}(T)$ in (2.4.53) gives

$$\left(\frac{\delta F_e}{\delta t}\right)^{ff}_{rad} = 0 \tag{2.4.56}$$

as required.

3. The Kinetic Equation of Photons

This chapter considers the radiative transfer equation as the kinetic equation of the photon gas. Section 3.1 introduces the transfer equation and defines the phenomenological emission and absorption coefficients and the source function. Section 3.2 discusses emission-absorption processes corresponding to bound-bound, free-bound, and free-free transitions, and Sect. 3.3 scattering by stationary and moving particles, respectively.

3.1 The Equation of Radiative Transfer

As discussed in Chap. 2, the distribution function $F(\boldsymbol{r}, \boldsymbol{v}, t)$ of a specific type of material particles obeys the kinetic equation (2.1.7). This equation can be written as

$$\frac{\partial F}{\partial t} + \boldsymbol{v} \cdot \frac{\partial F}{\partial \boldsymbol{r}} + \frac{\boldsymbol{K}}{m} \cdot \frac{\partial F}{\partial \boldsymbol{v}} = \left(\frac{\delta F}{\delta t}\right)_+ - \left(\frac{\delta F}{\delta t}\right)_- , \tag{3.1.1}$$

where the collision term has been decomposed into a creation term $(\delta F/\delta t)_+$ and a destruction term $(\delta F/\delta t)_-$ whose meanings are obvious.

Turning now to the photons, their distribution function $\phi(\boldsymbol{r}, \boldsymbol{p}, t)$ is defined analogously to that of material particles in the following way: $\phi(\boldsymbol{r}, \boldsymbol{p}, t)\, d^3r\, d^3p$ is the average or most probable number of photons at time t in the volume element $(\boldsymbol{r}, d^3r)$ that have momenta in the range $(\boldsymbol{p}, d^3p)$. Recall that in this book only unpolarized radiation fields are considered, so that there is no need to introduce an internal variable for photon polarization.

In analogy to (3.1.1) the kinetic equation of the photon gas is

$$\frac{\partial \phi}{\partial t} + c\boldsymbol{n} \cdot \frac{\partial \phi}{\partial \boldsymbol{r}} = \left(\frac{\delta \phi}{\delta t}\right)_+ - \left(\frac{\delta \phi}{\delta t}\right)_- , \tag{3.1.2}$$

where the terms on the right-hand side describe the creation and destruction of photons with momentum $\boldsymbol{p}$. In contrast to (3.1.1), there is no force term in (3.1.2), owing to the fact that the photon velocity $c\boldsymbol{n}$ is constant, $\boldsymbol{n}$ being again the unit vector in the direction of propagation ($\boldsymbol{n} \equiv \boldsymbol{p}/p$).

Usually the kinetic equation for photons is written in terms of the specific intensity $I_\nu \equiv I_\nu(\boldsymbol{n},\boldsymbol{r},t)$ rather than in terms of the distribution function ϕ. The relation between these two quantities is easily established by observing that the number density of photons $(\nu, d\nu; \boldsymbol{n}, d\Omega) \equiv (\boldsymbol{p}, d^3p)$ can be written in the two following ways, see (1.4.39),

$$\phi(\boldsymbol{r},\boldsymbol{p},t)\,d^3p = \frac{1}{h\nu c} I_\nu(\boldsymbol{n},\boldsymbol{r},t)\,d\nu\,d\Omega \ . \tag{3.1.3}$$

Using

$$\boldsymbol{p} = \frac{h\nu}{c}\boldsymbol{n} \ , \quad d^3p = p^2 dp\,d\Omega = \frac{h^3\nu^2}{c^3}\,d\nu\,d\Omega \ , \tag{3.1.4}$$

from (3.1.3) the desired relation is immediately obtained:

$$\phi(\boldsymbol{r},\boldsymbol{p},t) = \frac{c^2}{h^4\nu^3} I_\nu(\boldsymbol{n},\boldsymbol{r},t) \ . \tag{3.1.5}$$

Thus, apart from a numerical factor, I_ν/ν^3 is the photon distribution function [see also (A.1.20–26)].

Multiplying (3.1.2) by $h^4\nu^3/c^3$ now yields

$$\frac{1}{c}\frac{\partial I_\nu}{\partial t} + \boldsymbol{n}\cdot\nabla I_\nu = \frac{1}{c}\left(\frac{\delta I_\nu}{\delta t}\right)_+ - \frac{1}{c}\left(\frac{\delta I_\nu}{\delta t}\right)_- \ , \tag{3.1.6}$$

where $\nabla \equiv \partial/\partial\boldsymbol{r}$. Here the creation term may be called the creation coefficient $e_\nu \equiv e_\nu(\boldsymbol{n},\boldsymbol{r},t)$ for photons $(\nu,\boldsymbol{n})$,

$$\frac{1}{c}\left(\frac{\delta I_\nu}{\delta t}\right)_+ = e_\nu \ . \tag{3.1.7}$$

On the other hand, for all cases considered in this book, the destruction term is proportional to the specific intensity I_ν, that is, to the number of photons $(\nu,\boldsymbol{n})$ present, so that

$$\frac{1}{c}\left(\frac{\delta I_\nu}{\delta t}\right)_- = a_\nu I_\nu \ , \tag{3.1.8}$$

where $a_\nu \equiv a_\nu(\boldsymbol{n},\boldsymbol{r},t)$ may be called the destruction coefficient for photons $(\nu,\boldsymbol{n})$. Since, according to (3.1.3),

$$\frac{1}{c}\frac{\delta I_\nu}{\delta t}\,d\nu\,d\Omega = h\nu\frac{\delta\phi}{\delta t}\,d^3p \ , \tag{3.1.9}$$

the following relations hold:

$$e_\nu \, d\nu \, d\Omega = h\nu \text{ times creation rate (per unit time and unit volume) of photons } (\nu, d\nu; \boldsymbol{n}, d\Omega) \,, \tag{3.1.10a}$$

$$a_\nu I_\nu \, d\nu \, d\Omega = h\nu \text{ times destruction rate (per unit time and unit volume) of photons } (\nu, d\nu; \boldsymbol{n}, d\Omega) \,. \tag{3.1.10b}$$

The creation and destruction terms (3.1.7, 8) consider creation and destruction of photons $(\nu, \boldsymbol{n})$ from the point of view of quantum electrodynamics. That is, e_ν and $a_\nu I_\nu$ are proportional to $|\langle n+1|a^\dagger|n\rangle|^2$ and $|\langle n-1|a|n\rangle|^2$, respectively, where $a^\dagger$ and a are the creation and destruction operators of the photon mode considered, and n is the occupation number of this mode.

However, according to (1.3.3a), photon creation (through emission or scattering) is always composed of spontaneous and stimulated parts, such that

$$e_\nu = \varepsilon_\nu \left(1 + \frac{c^2}{2h\nu^3} I_\nu\right) \tag{3.1.11}$$

on account of (1.4.43), where ε_ν describes spontaneous creation alone. Therefore the right-hand side of (3.1.6) has the following two alternative forms,

$$e_\nu - a_\nu I_\nu = \varepsilon_\nu - \kappa_\nu I_\nu \,, \tag{3.1.12}$$

with

$$\kappa_\nu = a_\nu - \frac{c^2}{2h\nu^3} \varepsilon_\nu \,. \tag{3.1.13}$$

The quantity $\varepsilon_\nu \equiv \varepsilon_\nu(\boldsymbol{n}, \boldsymbol{r}, t)$ is called the (phenomenological) *emission coefficient,* and $\kappa_\nu \equiv \kappa_\nu(\boldsymbol{n}, \boldsymbol{r}, t)$ the (phenomenological) *absorption coefficient* (or *opacity coefficient*). They are related to the coefficients e_ν and a_ν through

$$\varepsilon_\nu = \frac{e_\nu}{1 + \dfrac{c^2}{2h\nu^3} I_\nu} \,, \quad \kappa_\nu = a_\nu - \frac{e_\nu}{\dfrac{2h\nu^3}{c^2} + I_\nu} \,, \tag{3.1.14a}$$

$$e_\nu = \varepsilon_\nu \left(1 + \frac{c^2}{2h\nu^3} I_\nu\right) , \quad a_\nu = \kappa_\nu + \frac{c^2}{2h\nu^3} \varepsilon_\nu \,. \tag{3.1.14b}$$

To summarize, the coefficients e_ν and a_ν describe in the sense of quantum electrodynamics the true, physical creation and destruction (through emission, absorption, and scattering) of photons in a group of neighboring modes, whereas the coefficients ε_ν and κ_ν account phenomenologically for the net rate of change of photons $(\nu,\boldsymbol{n})$ such that, by definition, the coefficient $\varepsilon_\nu(\boldsymbol{n})$ does not depend explicitly on the specific intensity $I_\nu(\boldsymbol{n})$ of the considered photons $(\nu,\boldsymbol{n})$. [However, $\varepsilon_\nu(\boldsymbol{n})$ may depend explicitly on the intensity $I_{\nu'}(\boldsymbol{n}')$ in other photon modes $(\nu',\boldsymbol{n}')\neq(\nu,\boldsymbol{n})$ as, for example, in the case of Rayleigh or Thomson scattering.] In this latter description, absorptions and stimulated emissions are included in the same phenomenological term $\kappa_\nu I_\nu$.

In the literature on radiative transfer, the phenomenological coefficients ε_ν and κ_ν are generally used. In addition, the *source function* $S_\nu \equiv S_\nu(\boldsymbol{n},\boldsymbol{r},t)$ for photons $(\nu,\boldsymbol{n})$ is defined as the ratio of the emission to the absorption coefficient,

$$S_\nu = \frac{\varepsilon_\nu}{\kappa_\nu} \; . \tag{3.1.15}$$

So, finally, we obtain from (3.1.6) the *equation of radiative transfer* in its usual forms

$$\frac{1}{c}\frac{\partial I_\nu}{\partial t} + \boldsymbol{n}\cdot\nabla I_\nu = \varepsilon_\nu - \kappa_\nu I_\nu \tag{3.1.16a}$$

or

$$\frac{1}{c}\frac{\partial I_\nu}{\partial t} + \boldsymbol{n}\cdot\nabla I_\nu = \kappa_\nu(S_\nu - I_\nu) \; . \tag{3.1.16b}$$

The radiative transfer equation is the kinetic equation for the (strangely normalized) photon distribution function I_ν.

As for a kinetic equation for particles, the transfer equation (3.1.16) is useless unless ε_ν, κ_ν, S_ν are specified in terms of well-defined physical quantities. Recall that ε_ν, κ_ν, S_ν as defined above comprise all kinds of emission, absorption, and scattering. The emission and absorption coefficients are additive quantities,

$$\varepsilon_\nu = \sum_j \varepsilon_\nu^{(j)} \; , \quad \kappa_\nu = \sum_j \kappa_\nu^{(j)} \; , \tag{3.1.17a}$$

where each superscript j refers to a specific radiative process of a specific particle species, whereas the total source function is related to the partial source functions through

$$S_\nu = \frac{\sum_j \kappa_\nu^{(j)} S_\nu^{(j)}}{\sum_j \kappa_\nu^{(j)}} = \sum_j \frac{\kappa_\nu^{(j)}}{\kappa_\nu} S_\nu^{(j)} \; , \tag{3.1.17b}$$

where $S_\nu^{(j)} \equiv \varepsilon_\nu^{(j)}/\kappa_\nu^{(j)}$.

From a physical point of view, two great classes of radiative processes can be distinguished: emission and absorption processes, on the one hand, and scattering, on the other. According to (3.1.17),

$$\varepsilon_\nu = \varepsilon_\nu^{\mathrm{ea}} + \varepsilon_\nu^{\mathrm{sc}} \ , \quad \kappa_\nu = \kappa_\nu^{\mathrm{ea}} + \kappa_\nu^{\mathrm{sc}} \ , \tag{3.1.18a}$$

$$S_\nu = \frac{\kappa_\nu^{\mathrm{ea}} S_\nu^{\mathrm{ea}} + \kappa_\nu^{\mathrm{sc}} S_\nu^{\mathrm{sc}}}{\kappa_\nu^{\mathrm{ea}} + \kappa_\nu^{\mathrm{sc}}} \ . \tag{3.1.18b}$$

The difference between these two classes of radiation processes is clearly exhibited by comparing the respective source functions in thermal equilibrium. Section 3.2 shows that the emission-absorption source function S_ν^{ea} reduces to the Planck function,

$$S_\nu^{\mathrm{ea}} = B_\nu(T) \ , \tag{3.1.19}$$

whenever the *matter* is in local thermodynamic equilibrium LTE, that is, when the particle distribution functions are the Maxwell, Boltzmann, and Saha distributions of the same temperature T,

$$f = f^{\mathrm{M}} \ , \quad \frac{n_2}{n_1} = \left(\frac{n_2}{n_1}\right)^{\mathrm{B}} \ , \quad \frac{n_+ n_{\mathrm{e}}}{n_0} = \left(\frac{n_+ n_{\mathrm{e}}}{n_0}\right)^{\mathrm{S}} \ . \tag{3.1.20}$$

By contrast, as shown in Sect. 3.3, the scattering source function S_ν^{sc} reduces to the Planck function,

$$S_\nu^{\mathrm{sc}} = B_\nu(T) \ , \tag{3.1.21}$$

whenever the *radiation* is thermal, that is, when the specific intensity is a Planck distribution,

$$I_\nu = B_\nu(T) \ . \tag{3.1.22}$$

For Raman scattering, the validity of (3.1.21) requires in addition a Boltzmann distribution for the atomic levels.

It follows from (3.1.18b, 19, 21) that in complete thermodynamic equilibrium (for matter *and* radiation) the total source function is given by the Planck function,

$$S_\nu = B_\nu(T) \ , \tag{3.1.23}$$

as it should.

The following comments on the transfer equation (3.1.16) may be helpful.

(1) As already mentioned, the attenuation of a light beam is supposed to be linear, thus excluding from the consideration nonlinear effects due to light of very high intensity.

(2) Validity of geometrical optics is assumed. This requires the photon wavelength λ to be much smaller than all linear dimensions of the system over which gradients may occur, and the medium to be weakly absorbing, i.e., $\kappa_\nu \lambda \ll 1$.

(3) By assuming straight photon trajectories and setting the photon velocity equal to that in vacuo, c, a refractive index equal to unity has tacitly been assumed. Strictly speaking, this contradicts the existence of a non-vanishing absorption coefficient, because the refractive index and the absorption coefficient are related to each other via the Kramers-Kronig relations. However, setting the refractive index of dilute gases equal to one is usually a good approximation, even in strong spectral lines.

(4) The description of the radiation field in terms of the specific intensity is incomplete. Indeed, laser radiation and radiation emitted from a thermal source having the same intensity cannot be distinguished in this picture. In other words, coherence properties of the radiation field are not described by the specific intensity, in the same way as correlations between particles are not included in the one-particle distribution function. The laboratory and astrophysical radiation fields considered in this book are always thermal or chaotic with respect to coherence, and we neglect any coherence effects.

Despite the fact that our derivation of the transfer equation is phenomenological, simply based on the analogy between a photon gas and a gas of material particles, there is no doubt that this equation describes the radiation transport in laboratory and astrophysical plasmas adequately. Attempts to derive this equation from first principles are found in [3.1 – 5]. The transfer equation for dispersive media is discussed in [3.6 – 10], and for polarized radiation in [3.11, 12].

3.2 Emission and Absorption

For emission and absorption processes it is easiest to calculate the quantities a_ν and ε_ν that correspond to absorption and spontaneous emission, respectively. In analogy to (3.1.10a),

$$\varepsilon_\nu \, d\nu \, d\Omega = h\nu \text{ times spontaneous creation rate (per unit time and unit volume) of photons } (\nu, d\nu; \boldsymbol{n}, d\Omega) \ , \tag{3.2.1}$$

while a_ν is defined via (3.1.10b). According to (3.1.13), the phenomenological absorption coefficient is given by

$$\kappa_\nu = a_\nu - \frac{c^2}{2h\nu^3} \varepsilon_\nu \ , \tag{3.2.2}$$

and the source function (3.1.15) takes in terms of ε_ν and a_ν the form

$$S_\nu = \frac{\varepsilon_\nu}{\kappa_\nu} = \frac{2h\nu^3}{c^2} \frac{1}{\dfrac{2h\nu^3}{c^2}\dfrac{a_\nu}{\varepsilon_\nu} - 1} . \tag{3.2.3}$$

In particular, the source function reduces to the Planck function (1.4.45),

$$S_\nu = B_\nu(T) \equiv \frac{2h\nu^3}{c^2} \frac{1}{e^{h\nu/kT} - 1} , \tag{3.2.4}$$

if

$$\frac{\varepsilon_\nu}{a_\nu} = B_\nu^{\mathrm{W}}(T) \equiv \frac{2h\nu^3}{c^2} e^{-h\nu/kT} , \tag{3.2.5}$$

where $B_\nu^{\mathrm{W}}(T)$ is the Wien function (1.6.33). In this case, the absorption coefficient (3.2.2) can be written as

$$\kappa_\nu = a_\nu (1 - e^{-h\nu/kT}) . \tag{3.2.6}$$

All of the above relations apply to emission-absorption processes and to scattering alike.

3.2.1 Bound-Bound Transitions

The rate of spontaneous emission of photons $(\nu, d\nu; \boldsymbol{n}, d\Omega)$ in an atomic bound-bound transition $2 \to 1$ due to atoms in the upper level 2 with velocities in the range $(\boldsymbol{v}, d^3v)$ is given by

$$n_2 f_2(\boldsymbol{v})\, d^3v\, A_{21}\, \eta_{21}(\xi, \boldsymbol{n})\, d\nu \frac{d\Omega}{4\pi} . \tag{3.2.7}$$

Here $f_2(\boldsymbol{v})$ is the normalized velocity distribution of atoms in level 2, A_{21} the Einstein coefficient for spontaneous emission, and $\eta_{21}(\xi, \boldsymbol{n})$ the atomic emission profile (Sect. 2.4.1 and Appendix B). The quantity

$$\xi = \nu - \frac{\nu_0}{c} \boldsymbol{n} \cdot \boldsymbol{v} \tag{3.2.8}$$

is the frequency of the photon $(\nu, \boldsymbol{n})$ measured in the rest frame of an atom of velocity $\boldsymbol{v}$, $\nu_0 \equiv \nu_0^0$ being the frequency of the line center in the rest frame of the atom, see (2.4.7). Recall that approximation (2.4.3) is used, according to which only the photon frequency is transformed when passing from the laboratory frame to the atomic rest frame. From (3.2.1, 7) the emission coefficient for the transition $2 \to 1$ follows

$$\varepsilon_\nu^{21}(\boldsymbol{n}) = \frac{h\nu_0}{4\pi} n_2 A_{21} \psi_\nu^{21}(\boldsymbol{n}) \ , \tag{3.2.9}$$

where

$$\psi_\nu^{21}(\boldsymbol{n}) = \int f_2(\boldsymbol{v}) \eta_{21}(\xi, \boldsymbol{n}) d^3 v \tag{3.2.10}$$

is the emission profile, see (2.4.17). In (3.2.9), a constant energy $h\nu_0$ has been ascribed to all photons of the spectral line considered.

On the other hand, the absorption rate of photons $(\nu, d\nu; \boldsymbol{n}, d\Omega)$ by atoms in lower level 1 with velocities in the range $(\boldsymbol{v}, d^3 v)$ is given by

$$n_1 f_1(\boldsymbol{v}) d^3 v B_{12} I_\nu(\boldsymbol{n}) \alpha_{12}(\xi) d\nu \frac{d\Omega}{4\pi} \ , \tag{3.2.11}$$

where $f_1(\boldsymbol{v})$ is the normalized velocity distribution of atoms in level 1, B_{12} the Einstein coefficient for absorption, and $\alpha_{12}(\xi)$ the atomic absorption profile (Sect. 2.4.1 and Appendix B). Hence, according to (3.1.10b), the coefficient a_ν for the considered transition $1 \rightarrow 2$ is

$$a_\nu^{12}(\boldsymbol{n}) = \frac{h\nu_0}{4\pi} n_1 B_{12} \varphi_\nu^{12}(\boldsymbol{n}) \ , \tag{3.2.12}$$

where

$$\varphi_\nu^{12}(\boldsymbol{n}) = \int f_1(\boldsymbol{v}) \alpha_{12}(\xi) d^3 v \tag{3.2.13}$$

is the absorption profile, see (2.4.16).

Let us again introduce the Einstein coefficient for stimulated emission, B_{21}, which obeys the Einstein relations, see (1.6.31, 32),

$$\frac{A_{21}}{B_{21}} = \frac{2h\nu_0^3}{c^2} \ , \quad \frac{B_{12}}{B_{21}} = \frac{g_2}{g_1} \ , \tag{3.2.14}$$

g_1 and g_2 being the statistical weights of the atomic levels. From (3.2.2) now follows the *line absorption coefficient*

$$\kappa_\nu^{12}(\boldsymbol{n}) = \frac{h\nu_0}{4\pi} [n_1 B_{12} \varphi_\nu^{12}(\boldsymbol{n}) - n_2 B_{21} \psi_\nu^{21}(\boldsymbol{n})] \ , \tag{3.2.15}$$

and from (3.2.3) the *line source function*

$$S_\nu^{21}(\boldsymbol{n}) = \frac{n_2 A_{21} \psi_\nu^{21}(\boldsymbol{n})}{n_1 B_{12} \varphi_\nu^{12}(\boldsymbol{n}) - n_2 B_{21} \psi_\nu^{21}(\boldsymbol{n})} \ . \tag{3.2.16}$$

Note that the profile coefficient for stimulated emission is ψ_ν (obvious from the point of view of quantum electrodynamics), rather than φ_ν as sometimes stated in the literature.

The line source function reduces to the Planck function $S^{21} = B_{\nu_0}(T)$ if

$$\frac{n_2 A_{21} \psi_\nu^{21}}{n_1 B_{12} \varphi_\nu^{12}} = \frac{2h\nu_0^3}{c^2} e^{-h\nu_0/kT} \tag{3.2.17}$$

according to (3.2.5, 9, 12). This means, first, that one must have a Boltzmann distribution (1.4.19) for the occupation numbers,

$$\frac{n_2}{n_1} = \left(\frac{n_2}{n_1}\right)^{\mathrm{B}} \equiv \frac{g_2}{g_1} e^{-E_{21}/kT} , \tag{3.2.18}$$

as can be seen by using (3.2.14), and recalling that $h\nu_0 = E_{21} \equiv E_2 - E_1$ is the energy difference of the atomic levels. And second, one must have *complete redistribution,* that is, equality of the emission and absorption profiles,

$$\psi_\nu^{21}(\boldsymbol{n}) = \varphi_\nu^{12}(\boldsymbol{n}) . \tag{3.2.19}$$

Generally in the case of complete redistribution, the line source function (3.2.16) is independent of frequency and isotropic,

$$S_\nu^{21}(\boldsymbol{n}) = S^{21} . \tag{3.2.20}$$

3.2.2 Free-Bound Transitions

The rate of spontaneous recombinations in a specific free-bound transition due to free electrons with velocities in the range $(\boldsymbol{v}, d^3v)$ and yielding photons $(\nu, d\nu; \boldsymbol{n}, d\Omega)$ is given by

$$n_+ v n_e f_e(\boldsymbol{v}) d^3v \, \sigma_{+0}(v, \vartheta) d\Omega , \tag{3.2.21}$$

where $f_e(\boldsymbol{v})$ is the normalized electron velocity distribution, and $\sigma_{+0}(v, \vartheta)$ is the differential cross section for spontaneous recombination, ϑ being the angle between the electron velocity and the direction of the emitted photon $(\boldsymbol{n} \cdot \boldsymbol{v} = v \cos\vartheta)$. In (3.2.21), the approximation of stationary ions has been used. Using conservation of energy,

$$h\nu = \tfrac{1}{2} m_e v^2 + E_{+0} , \quad h\, d\nu = m_e v\, dv , \tag{3.2.22}$$

where $E_{+0} = E_+ - E_0$ is the energy difference of the ionic and atomic levels, and writing

$$d^3v = v^2 dv\, d\Omega_v , \tag{3.2.23}$$

one easily derives from (3.2.1, 21) the emission coefficient for the free-bound transition considered

$$\varepsilon_\nu^{+0}(\boldsymbol{n}) = n_+ n_e \frac{h^2 \nu}{m_e} v^2 \int f_e(\boldsymbol{v}) \sigma_{+0}(v, \vartheta) d\Omega_v \tag{3.2.24a}$$

with

$$v \equiv \left[\frac{2}{m_e}(h\nu - E_{+0})\right]^{1/2} . \tag{3.2.24b}$$

Equivalently, this may also be written in terms of a delta function as

$$\varepsilon_\nu^{+0}(\boldsymbol{n}) = n_+ n_e \frac{h^2\nu}{m_e} \int d\Omega_v \int dv\, \delta\left(v - \left[\frac{2}{m_e}(h\nu - E_{+0})\right]^{1/2}\right)$$

$$\cdot\, v^2 f_e(\boldsymbol{v})\, \sigma_{+0}(v, \vartheta) , \tag{3.2.25}$$

or, using (3.2.23),

$$\varepsilon_\nu^{+0}(\boldsymbol{n}) = n_+ n_e \frac{h^2\nu}{m_e} \int f_e(\boldsymbol{v})\, \sigma_{+0}(v, \vartheta)$$

$$\cdot\, \delta\left(v - \left[\frac{2}{m_e}(h\nu - E_{+0})\right]^{1/2}\right) d^3v . \tag{3.2.26}$$

On the other hand, the corresponding rate of photoionizations due to photons $(\nu, d\nu; \boldsymbol{n}, d\Omega)$ is

$$n_0 \frac{1}{h\nu} I_\nu(\boldsymbol{n})\, d\nu\, d\Omega\, \sigma_{0+}(\nu) , \tag{3.2.27}$$

where $\sigma_{0+}(\nu)$ is the total cross section for photoionization. Here we have again adopted the approximations (2.4.3) according to which, for continuum radiation, no transformations are performed when going from the laboratory frame to the rest frame of the atom. Hence, from (3.1.10b, 2.27),

$$a_\nu^{0+} = n_0 \sigma_{0+}(\nu) , \tag{3.2.28}$$

which is an evident result.

From (3.2.24 or 26) and (3.2.28), the absorption coefficient $\kappa_\nu^{0+}(\boldsymbol{n})$ and the source function $S_\nu^{+0}(\boldsymbol{n})$ of the free-bound transition considered are obtained using (3.2.2, 3), respectively.

If the electron velocity distribution is isotropic, in (3.2.24a) integration over $d\Omega_v$ can be carried out, yielding the total cross section for spontaneous recombination $\sigma_{+0}(v)$. In terms of kinetic energy $E = m_e v^2/2$, i.e., $f_e(E)\,dE = f_e(v) 4\pi v^2 dv$, (3.2.24) yields the isotropic emission coefficient

$$\varepsilon_\nu^{+0} = n_+ n_e \frac{h^2\nu}{2^{3/2}\pi m_e^{1/2}} E^{1/2} f_e(E)\, \sigma_{+0}(E) \tag{3.2.29a}$$

where

$$E \equiv h\nu - E_{+0} \ . \tag{3.2.29b}$$

In particular, if the electrons have a Maxwell distribution (1.4.14),

$$f_e(E) = f_e^M(E) \equiv \frac{2E^{1/2}}{\pi^{1/2}(kT_e)^{3/2}} e^{-E/kT_e} , \tag{3.2.30}$$

(3.2.29a) takes the form

$$\varepsilon_\nu^{+0} = n_+ n_e \frac{h^2 \nu}{2^{1/2} \pi^{3/2} m_e^{1/2} (kT_e)^{3/2}} E e^{-E/kT_e} \sigma_{+0}(E) \ . \tag{3.2.31}$$

This may also be written as

$$\varepsilon_\nu^{+0} = n_+ n_e \frac{2 m_e \nu}{h} \lambda_e^3 E e^{-E/kT_e} \sigma_{+0}(E) \tag{3.2.32}$$

in terms of the thermal de Broglie wavelength of the electrons, (1.4.24),

$$\lambda_e = \frac{h}{(2\pi m_e k T_e)^{1/2}} \ . \tag{3.2.33}$$

According to (3.2.5), the source function of the free-bound transition considered becomes the Planck function corresponding to the electron temperature T_e, $S_\nu^{+0} = B_\nu(T_e)$, if $\varepsilon_\nu^{+0}/a_\nu^{0+} = B_\nu^W(T_e)$. Using (3.2.28, 32) together with the Milne relation (1.6.40) and the energy relation (3.2.29b), it is readily shown that this implies a Saha distribution (1.4.25) for the occupation numbers,

$$\frac{n_+ n_e}{n_0} = \left(\frac{n_+ n_e}{n_0}\right)^S \equiv \frac{2 g_+}{g_0} \lambda_e^{-3} e^{-E_{+0}/kT_e} \ . \tag{3.2.34}$$

Recall that by using (3.2.32), a Maxwellian velocity distribution of the electrons has been assumed.

That only the electron temperature enters (3.2.32, 34) is due to our approximation of stationary atoms and ions. Without this approximation, the validity of $S_\nu^{+0} = B_\nu(T)$ requires Maxwell distributions for the atoms, ions, and electrons, and a Saha distribution for the number densities, where all distributions correspond to the same temperature T.

Let us now briefly discuss the emission coefficient due to *dielectronic recombination,* restricting ourselves to the optically thin case discussed in Sect. 2.3.5. Using the same notation as there, then

$$\varepsilon_\nu^{\text{diel}} d\nu \, d\Omega = h\nu_0 n^{**} A_\downarrow \psi_\nu^{**} d\nu \frac{d\Omega}{4\pi} , \tag{3.2.35}$$

where

$$h\nu_0 = E^{**} - E^* \tag{3.2.36}$$

is the photon energy. In analogy to (3.2.10), the emission profile is given by

$$\psi_\nu^{**} = \int f^{**}(\boldsymbol{v})\, w^{**}\left(\nu - \frac{\nu_0}{c}\boldsymbol{n}\cdot\boldsymbol{v}\right) d^3v \ , \tag{3.2.37}$$

where $f^{**}(\boldsymbol{v})$ is the normalized velocity distribution of the doubly excited atoms, and $w^{**}(\nu)$ is the normalized level profile of the autoionizing level, Sect. 1.6.3. In (3.2.37) it has been assumed for simplicity that the broadening of the lower (singly excited) atomic level is negligible compared to that of the higher (doubly excited) level.

If the electrons have a Maxwellian velocity distribution, the number density of doubly excited atoms is given by

$$n^{**} = \frac{n_+ n_e}{A_a + A_\downarrow} A_a \frac{g^{**}}{2g_+} \lambda_e^3 e^{-E/kT_e} \tag{3.2.38}$$

on account of (2.3.31, 34), where λ_e is the thermal de Broglie wavelength (3.2.33), and

$$E = E^{**} - E_g^+ \tag{3.2.39}$$

is the kinetic energy of the captured electron. Thus, combining (3.2.35, 38) leads to the emission coefficient

$$\varepsilon_\nu^{\text{diel}} = \frac{h\nu_0}{4\pi} n_+ n_e \frac{A_a A_\downarrow}{A_a + A_\downarrow} \frac{g^{**}}{2g_+} \lambda_e^3 e^{-E/kT_e} \psi_\nu^{**} \ . \tag{3.2.40}$$

3.2.3 Free-Free Transitions

From the definitions of the quantities α_ν and β_ν in (1.6.44), and (3.2.1, 1.10b), the coefficients ε_ν and a_ν for photons $(\nu, \boldsymbol{n})$ due to free-free transitions of electrons in the field of heavy, stationary particles P (which may be ions or neutral atoms) can be formulated immediately:

$$\varepsilon_\nu^{\text{ff}}(\boldsymbol{n}) = h\nu n_P n_e A_\nu(\boldsymbol{n}) \ , \tag{3.2.41}$$

$$a_\nu^{\text{ff}}(\boldsymbol{n}) = h\nu n_P n_e B_\nu(\boldsymbol{n}) \ , \tag{3.2.42}$$

where

$$A_\nu(\boldsymbol{n}) = \iint f_e(\boldsymbol{v}')\, \alpha_\nu(\boldsymbol{v}'; \boldsymbol{n}, \boldsymbol{v})\, d^3v\, d^3v' \ , \tag{3.2.43}$$

$$B_\nu(\boldsymbol{n}) = \iint f_e(\boldsymbol{v})\, \beta_\nu(\boldsymbol{n}, \boldsymbol{v}; \boldsymbol{v}')\, d^3v\, d^3v' \ . \tag{3.2.44}$$

The absorption coefficient $\kappa_\nu^{\mathrm{ff}}(\boldsymbol{n})$ and the source function $S_\nu^{\mathrm{ff}}(\boldsymbol{n})$ are obtained using (3.2.2, 3), respectively.

For a Maxwellian electron velocity distribution of temperature T_e,

$$f_\mathrm{e}^\mathrm{M}(v') = \mathrm{e}^{-h\nu/kT_\mathrm{e}} f_\mathrm{e}^\mathrm{M}(v) \tag{3.2.45}$$

because the velocities v' and v are related through

$$\tfrac{1}{2} m_\mathrm{e} v'^2 = \tfrac{1}{2} m_\mathrm{e} v^2 + h\nu \ . \tag{3.2.46}$$

[Note that both α_ν and β_ν in (3.2.43, 44) contain the delta function $\delta(m_\mathrm{e}v'^2/2 - m_\mathrm{e}v^2/2 - h\nu)$.] Inserting this relation in (3.2.43), and using the reciprocity relation (1.6.45), one obtains

$$\frac{A_\nu}{B_\nu} = \frac{\varepsilon_\nu^{\mathrm{ff}}}{a_\nu^{\mathrm{ff}}} = \frac{2h\nu^3}{c^2}\,\mathrm{e}^{-h\nu/kT_\mathrm{e}} \ . \tag{3.2.47}$$

In view of (3.2.5), this means that the free-free source function is a Planck function, $S_\nu^{\mathrm{ff}} = B_\nu(T_\mathrm{e})$, if the electron gas is Maxwellian.

3.3 Scattering

According to the point of view adopted in this book, scattering of light is described in terms of photon collisions with material particles. For Rayleigh and Thomson scattering, the collision is elastic, i.e., the internal state of the scattering particle is not changed by the collision, so that neglecting recoil effects, the frequencies of the incoming and outgoing photons are equal in the rest frame of the scattering particle. (Raman scattering, which is not discussed in this book, corresponds in this picture to an inelastic photon collision [3.13].) This nonresonant photon scattering is an instantaneous process, and for Rayleigh scattering by an atom, intermediate excited states of the atom are only virtual, that is, not observable. This should be contrasted with resonant photon scattering, e.g., resonance fluorescence, where the frequency of the incoming photon is in resonance with an electronic transition of the atom. Here the re-emission of the scattered photon is time-delayed with respect to the previous absorption of the incoming photon, owing to the exponential decay of the excited atomic level, and the intermediate excited states of the atom are real and can be observed, e.g., by inelastic de-excitation collisions with free electrons.

On the other hand, nonresonant light scattering is easily explained by classical physics. As an example, consider Rayleigh scattering by atoms. In a classical picture, the oscillating electric field of the incoming light wave induces in the atoms oscillating electric dipoles of the same frequency, which in turn emit electromagnetic radiation in directions other than that of the incident wave (scattering). In this picture the angular distribution of the scattered light is obtained by the *coherent* superposition of the *amplitudes* of the field strengths of the emitted secondary waves, whereas in the previous quantum picture it is determined by the *incoherent* superposition of the scattered light *intensities* (which are proportional to the square of the amplitudes). Let us see how these two conflicting pictures are reconciled [3.14].

Consider a scattering volume V (Fig. 3.1) containing many ($N \gg 1$) identical atoms, on which a plane light wave impinges with an electric field vector $\boldsymbol{E}_{\text{inc}} \propto \exp(\mathrm{i}\boldsymbol{k}\cdot\boldsymbol{r})$, where $\boldsymbol{k} = k\boldsymbol{n}$ is the wave vector pointing in the direction of the unit vector $\boldsymbol{n}$. (Here and in the following, a time dependence $\exp(-\mathrm{i}\omega t)$ is tacitly assumed.) In the jth atom at position $\boldsymbol{r}_j$, this field induces an oscillating electric dipole $\boldsymbol{d}_j \propto \exp(\mathrm{i}\boldsymbol{k}\cdot\boldsymbol{r}_j)$, which emits an electromagnetic wave whose electric field vector in the far radiation zone ($r \gg r_j$) is proportional to

$$\boldsymbol{E}_j \propto \left(\frac{g(\vartheta)}{|\boldsymbol{r}-\boldsymbol{r}_j|}\exp(\mathrm{i}k|\boldsymbol{r}-\boldsymbol{r}_j|)\right)\exp(\mathrm{i}\boldsymbol{k}\cdot\boldsymbol{r}_j) \ . \tag{3.3.1}$$

The first term is an outgoing spherical wave with angular dependence $g(\vartheta)$, and the second term is the just-mentioned phase factor of the emitting dipole $\boldsymbol{d}_j$. In (3.3.1) we have used the fact that for $r \gg r_j$ the angle between the vectors $\boldsymbol{r}-\boldsymbol{r}_j$ and $\boldsymbol{n}$ can, for all j, be replaced by the constant scattering angle ϑ (Fig. 3.1), and that the scattering is elastic, i.e., $k' = k$. Now, for $r \gg r_j$,

$$|\boldsymbol{r}-\boldsymbol{r}_j| = (r^2+2\boldsymbol{r}\cdot\boldsymbol{r}_j+r_j^2)^{1/2} \simeq (r^2+2\boldsymbol{r}\cdot\boldsymbol{r}_j)^{1/2} \simeq r-\frac{\boldsymbol{r}\cdot\boldsymbol{r}_j}{r}$$

or

$$|\boldsymbol{r}-\boldsymbol{r}_j| \simeq r-\boldsymbol{n}'\cdot\boldsymbol{r}_j \ , \tag{3.3.2}$$

where $\boldsymbol{n}'$ is the unit vector in the direction of the scattered wave, so that $\boldsymbol{k}' = k\boldsymbol{n}'$, $\boldsymbol{r} = r\boldsymbol{n}'$. From (3.3.1, 2) then

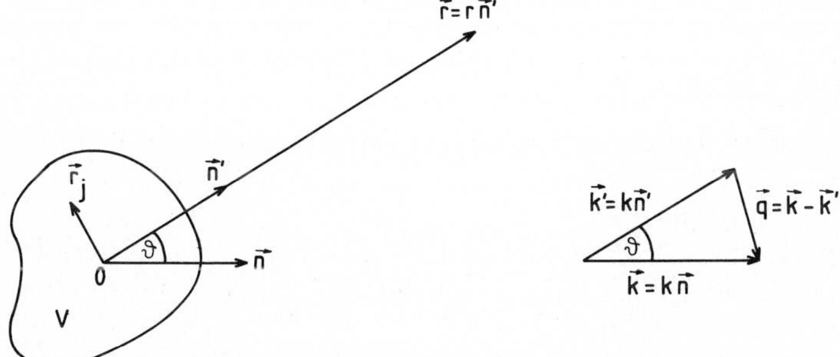

Fig. 3.1. Scattering of an electromagnetic wave from a scattering volume V. See text for details

$$E_j \propto \frac{g(\vartheta)}{r} \, \mathrm{e}^{\mathrm{i}kr} \mathrm{e}^{-\mathrm{i}\boldsymbol{k}'\cdot\boldsymbol{r}_j} \mathrm{e}^{\mathrm{i}\boldsymbol{k}\cdot\boldsymbol{r}_j}$$

or, suppressing all factors independent of j,

$$\boldsymbol{E}_j \propto \mathrm{e}^{\mathrm{i}\boldsymbol{q}\cdot\boldsymbol{r}_j} \, , \quad \boldsymbol{q} \equiv \boldsymbol{k} - \boldsymbol{k}' \; . \tag{3.3.3}$$

The total electric field strength of the scattered wave is obtained by the coherent superposition

$$\boldsymbol{E}_{\mathrm{sc}} = \sum_j \boldsymbol{E}_j \propto \mathcal{F}(\boldsymbol{q}) \equiv \sum_j \mathrm{e}^{\mathrm{i}\boldsymbol{q}\cdot\boldsymbol{r}_j} \, , \tag{3.3.4}$$

$\mathcal{F}(\boldsymbol{q})$ being called the structure factor, and the intensity of the scattered light is hence

$$I_{\mathrm{sc}} \propto |\boldsymbol{E}_{\mathrm{sc}}|^2 \propto |\mathcal{F}(\boldsymbol{q})|^2 \equiv \sum_j \sum_{j'} \exp[\mathrm{i}\boldsymbol{q}\cdot(\boldsymbol{r}_j - \boldsymbol{r}_{j'})] \; . \tag{3.3.5}$$

Up to now nothing has been assumed as regards the positions $\boldsymbol{r}_j$ of the scattering atoms. Restricting now the discussion to a gas, we observe that owing to the thermal motion of the atoms, the position vectors $\boldsymbol{r}_j = \boldsymbol{r}_j(t)$ are randomly distributed, so that in the double sum in (3.3.5) only the terms with $j = j'$ contribute, while those with $j \neq j'$ give on the average a negligible contribution. Thus $|\mathcal{F}(\boldsymbol{q})|^2 \simeq N$, which shows that the intensity of the scattered light is proportional to the number of scattering atoms,

$$I_{\mathrm{sc}} \propto N \; . \tag{3.3.6}$$

In other words, for a gas where the scattering particles are distributed at random, interference effects are negligible, and the simple picture of elastically scattered photons may be applied.

In the following, nonresonant light scattering by stationary and moving particles, respectively, is treated in terms of elastic photon collisions. As always in this book, it is assumed that the radiation field is unpolarized and that the polarization of outgoing photons is not observed.

3.3.1 Scattering by Stationary Particles

In the simplest model of a scattering gas all particles are assumed to be at rest, thus neglecting not only the recoil of the particles due to scattering of photons, but also the Doppler shift of the scattered photons. Hence, in this model, scattering is considered to be frequency-coherent in the observer's frame. That is, if the incoming and scattered photons are characterized by $(\nu, \boldsymbol{n})$ and $(\nu', \boldsymbol{n}')$, respectively, then

$$\nu' = \nu \; . \tag{3.3.7}$$

This simple model is adequate for describing Rayleigh and Thomson scattering of continuum radiation whose intensity is a slowly varying function of frequency, and even for Rayleigh scattering of spectral lines provided that the gas temperature is sufficiently low. However, owing to the high thermal velocities of free electrons, Thomson scattering of spectral lines generally requires that the thermal motion of electrons be taken into account (Sect. 3.3.2).

We write the differential scattering cross section in the form

$$\sigma(\nu, \vartheta) = g(\vartheta)\,\sigma(\nu) \ , \tag{3.3.8}$$

where $\sigma(\nu)$ denotes the total scattering cross section. For Rayleigh scattering $\sigma(\nu) \propto \nu^4$, and for Thomson scattering, in cgs units, $\sigma(\nu) = \sigma_e \equiv (8\pi/3)(e^2/m_e c^2)^2$ independent of ν. On the other hand, the phase function $g(\vartheta)$ is a function of the scattering angle ϑ of the photon (i.e., $\boldsymbol{n} \cdot \boldsymbol{n}' = \cos\vartheta$), which is symmetric and normalized,

$$g(\boldsymbol{n},\boldsymbol{n}') = g(\boldsymbol{n}',\boldsymbol{n}) \ , \tag{3.3.9}$$

$$\int g(\boldsymbol{n},\boldsymbol{n}')\,d\Omega' = \int g(\boldsymbol{n},\boldsymbol{n}')\,d\Omega = 1 \ , \tag{3.3.10}$$

where $d\Omega' = d\Omega = 2\pi \sin\vartheta\, d\vartheta$. Rayleigh and Thomson scattering correspond to dipole scattering,

$$g_1(\vartheta) = \frac{3}{16\pi}(1 + \cos^2\vartheta) \ , \tag{3.3.11a}$$

while the approximation of isotropic scattering corresponds to

$$g_0(\vartheta) = \frac{1}{4\pi} \ . \tag{3.3.11b}$$

Let us now write the creation and destruction terms of the radiative transfer equation, corresponding to scattering by stationary particles. According to (3.1.10a,b)

$$e_\nu^{\mathrm{sc}}(\boldsymbol{n}) = h\nu n_s \int \frac{1}{h\nu} I_\nu(\boldsymbol{n}')\,\sigma(\nu,\vartheta)\,d\Omega' \left[1 + \frac{c^2}{2h\nu^3} I_\nu(\boldsymbol{n})\right] , \tag{3.3.12}$$

$$a_\nu^{\mathrm{sc}} I_\nu(\boldsymbol{n}) = h\nu n_s \frac{1}{h\nu} I_\nu(\boldsymbol{n}) \int \sigma(\nu,\vartheta)\left[1 + \frac{c^2}{2h\nu^3} I_\nu(\boldsymbol{n}')\right] d\Omega' \ . \tag{3.3.13}$$

Interpreting these two expressions is straightforward if one remembers that $I_\nu(\boldsymbol{n})\,d\nu\, d\Omega/h\nu$ is the flux density of photons $(\nu, d\nu; \boldsymbol{n}, d\Omega)$, see (1.4.40). In (3.3.12, 13), n_s is the density of scattering particles, and $d\Omega' = 2\pi \sin\vartheta\, d\vartheta$ is the element of solid angle with respect to the photon direction $\boldsymbol{n}'$. Moreover, frequency coherence of scattering, (3.3.7), has been used, and stimulated scattering has been taken into account.

If (3.3.13) is subtracted from (3.3.12), thus forming the combination $e_\nu - a_\nu I_\nu$ that appears in the transfer equation (3.1.6), see (3.1.7, 8), the terms describing induced scattering cancel in this model. Of course, this does not mean that induced scattering does not exist. For example, the relaxation time of a dilute gas immersed in blackbody radiation for approaching thermal equilibrium depends on the rate of induced scattering processes (Sect. 7.2.2). On the other hand, by definition (3.1.12), $e_\nu - a_\nu I_\nu \equiv \varepsilon_\nu - \kappa_\nu I_\nu$. Hence the following quantities may be taken as emission and absorption coefficients for scattering by stationary particles, neglecting induced scattering altogether,

$$\varepsilon_\nu^{\mathrm{sc}}(\boldsymbol{n}) = n_{\mathrm{s}}\sigma(\nu) \int I_\nu(\boldsymbol{n}')\, g(\boldsymbol{n}',\boldsymbol{n})\, d\Omega' \ , \tag{3.3.14}$$

$$\kappa_\nu^{\mathrm{sc}} = n_{\mathrm{s}}\sigma(\nu) \ , \tag{3.3.15}$$

where (3.3.8,10) have been used. Finally, the corresponding source function is, from (3.1.15),

$$S_\nu^{\mathrm{sc}}(\boldsymbol{n}) = \int I_\nu(\boldsymbol{n}')\, g(\boldsymbol{n}',\boldsymbol{n})\, d\Omega' \ . \tag{3.3.16}$$

Using the normalization (3.3.10), one checks easily that

$$\int [\varepsilon_\nu^{\mathrm{sc}}(\boldsymbol{n}) - \kappa_\nu^{\mathrm{sc}} I_\nu(\boldsymbol{n})]\, d\Omega = 0 \ , \tag{3.3.17}$$

which expresses the fact that in this model the number of photons of a given frequency is conserved. On the other hand, in a blackbody radiation field, the specific intensity is $I_\nu(\boldsymbol{n}) = B_\nu(T)$, so that the scattering source function (3.3.16) reduces to the Planck function $S_\nu^{\mathrm{sc}} = B_\nu(T)$.

3.3.2 Scattering by Moving Particles

In gases of sufficiently high temperature, the Doppler shift of photons scattered by Rayleigh or Thomson scattering may be an important effect for radiation transport in spectral lines, so that here the thermal motion of the scattering particles must be considered. For example, the disappearance of photospheric Fraunhofer lines in the solar corona is caused by Thomson scattering of photons from the neighboring continuum into the spectral range of the line.

If recoil is neglected, Thomson and Rayleigh scattering are, in the particle's rest frame, coherent in frequency. That is, if in the observer's frame the incoming and scattered photons are characterized by $(\nu,\boldsymbol{n})$ and $(\nu',\boldsymbol{n}')$, respectively, both frequencies ν and ν' correspond to the same frequency ν^{o} in the rest frame of the scattering particle, so that through (A.1.27)

$$\nu^{\mathrm{o}} = \nu - \frac{\nu}{c}\boldsymbol{n}\cdot\boldsymbol{v} \ , \tag{3.3.18a}$$

$$\nu^{\text{o}} = \nu' - \frac{\nu'}{c}\boldsymbol{n}' \cdot \boldsymbol{v} \simeq \nu' - \frac{\nu}{c}\boldsymbol{n}' \cdot \boldsymbol{v} \ , \tag{3.3.18b}$$

where $\boldsymbol{v}$ is the velocity of the particle. The approximation made in (3.3.18b) is consistent with the nonrelativistic formulas used in which terms of the order v^2/c^2 are neglected. Equations (3.3.18) show that the frequencies ν and ν' are related by [valid for scattering $(\nu', \boldsymbol{n}') \rightarrow (\nu, \boldsymbol{n})$ too]

$$\nu' = \nu + \frac{\nu}{c}(\boldsymbol{n}' - \boldsymbol{n}) \cdot \boldsymbol{v} \, , \qquad \nu = \nu' + \frac{\nu'}{c}(\boldsymbol{n} - \boldsymbol{n}') \cdot \boldsymbol{v} \, . \tag{3.3.19}$$

On account of (3.1.10), the creation and destruction terms for photons $(\nu, \boldsymbol{n})$ due to scattering by particles with density n_{s} and normalized velocity distribution function $f_{\text{s}}(\boldsymbol{v})$ are given by

$$e_{\nu}^{\text{sc}}(\boldsymbol{n}) = h\nu \int d^3v \, n_{\text{s}} f_{\text{s}}(\boldsymbol{v}) \int \frac{1}{h\nu} I_{\nu'}(\boldsymbol{n}') \sigma(\nu) \, g(\boldsymbol{n}', \boldsymbol{n}) \, d\Omega' \cdot \left[1 + \frac{c^2}{2h\nu^3} I_{\nu}(\boldsymbol{n}) \right] , \tag{3.3.20}$$

$$a_{\nu}^{\text{sc}}(\boldsymbol{n}) I_{\nu}(\boldsymbol{n}) = h\nu \int d^3v \, n_{\text{s}} f_{\text{s}}(\boldsymbol{v}) \frac{1}{h\nu} I_{\nu}(\boldsymbol{n}) \cdot \int \sigma(\nu) \, g(\boldsymbol{n}, \boldsymbol{n}') \left[1 + \frac{c^2}{2h\nu^3} I_{\nu'}(\boldsymbol{n}') \right] d\Omega' \ , \tag{3.3.21}$$

where the differential cross section has been written in the form of (3.3.8), and induced scattering has been taken into account. In (3.3.20, 21) the approximations (2.4.3) have been used, which in the present context mean that the correct transformation of the frequency, (3.3.19), is applied only to the specific intensity, because it is the only quantity that can strongly vary as a function of frequency. Elsewhere

$$\nu' \simeq \nu \ , \quad d\nu' \simeq d\nu \ , \quad \sigma(\nu') \simeq \sigma(\nu) \ . \tag{3.3.22}$$

Of course, aberration effects have also been neglected. For example, the flux density of photons $(\nu', d\nu'; \boldsymbol{n}', d\Omega')$ is $I_{\nu'}(\boldsymbol{n}') \, d\nu \, d\Omega' / h\nu$.

Forming now the combination $e_{\nu} - a_{\nu} I_{\nu}$ as required for the transfer equation (3.1.6), the terms describing stimulated scattering cancel again. Therefore, we may take the following quantities as emission and absorption coefficients for scattering by moving particles,

$$\varepsilon_{\nu}^{\text{sc}}(\boldsymbol{n}) = n_{\text{s}} \sigma(\nu) \int d^3v f_{\text{s}}(\boldsymbol{v}) \int I\left(\nu + \frac{\nu}{c} [\boldsymbol{n}' - \boldsymbol{n}] \cdot \boldsymbol{v}, \boldsymbol{n}' \right) g(\boldsymbol{n}', \boldsymbol{n}) \, d\Omega' \ , \tag{3.3.23}$$

$$\kappa_{\nu}^{\text{sc}} = n_{\text{s}} \sigma(\nu) \ , \tag{3.3.24}$$

where (3.3.19) has been used in (3.3.23). According to (3.1.15), the corresponding source function is given by

$$S_\nu^{sc}(\boldsymbol{n}) = \int d^3v f_s(\boldsymbol{v}) \int I\left(\nu + \frac{\nu}{c}[\boldsymbol{n}' - \boldsymbol{n}] \cdot \boldsymbol{v}, \boldsymbol{n}'\right) g(\boldsymbol{n}', \boldsymbol{n})\, d\Omega' \ . \tag{3.3.25}$$

If all particles are at rest, $f_s(\boldsymbol{v}) = \delta(\boldsymbol{v})$, and one recovers from (3.3.23 – 25) the expressions (3.3.14 – 16).

The fact that scattering conserves the total number of photons is expressed by

$$\iint [\varepsilon_\nu^{sc}(\boldsymbol{n}) - \kappa_\nu^{sc} I_\nu(\boldsymbol{n})]\, d\nu\, d\Omega = 0 \ , \tag{3.3.26}$$

which is readily proved by noticing that in (3.3.23) the integration over ν can, for fixed $\boldsymbol{n}$, $\boldsymbol{n}'$, and $\boldsymbol{v}$, be replaced by that over ν', and using $\sigma(\nu) \simeq \sigma(\nu')$ according to (3.3.22). On the other hand, in a blackbody radiation field

$$I_{\nu'}(\boldsymbol{n}') \equiv I\left(\nu + \frac{\nu}{c}[\boldsymbol{n}' - \boldsymbol{n}] \cdot \boldsymbol{v}, \boldsymbol{n}'\right) \simeq B_\nu(T)$$

if variations of the intensity over the scattering Doppler width are neglected. According to (3.3.25), the scattering source function is then given by the Planck function $S_\nu^{sc} = B_\nu(T)$, independently of the velocity distribution of the scattering particles.

The scattering source function (3.3.25) can be written in the following alternative form

$$S_\nu^{sc}(\boldsymbol{n}) = \iint I_{\nu'}(\boldsymbol{n}')\, a(\nu', \boldsymbol{n}' \to \nu, \boldsymbol{n})\, d\nu'\, d\Omega' \tag{3.3.27}$$

with the scattering function (or scattering kernel)

$$\begin{aligned} a(\nu', \boldsymbol{n}' \to \nu, \boldsymbol{n}) &= g(\boldsymbol{n}', \boldsymbol{n}) \int f_s(\boldsymbol{v})\, \delta\left(\nu' - \nu - \frac{\nu}{c}(\boldsymbol{n}' - \boldsymbol{n}) \cdot \boldsymbol{v}\right) d^3v \\ &= g(\boldsymbol{n}', \boldsymbol{n}) \int f_s(\boldsymbol{v})\, \delta\left(\nu' - \nu - \frac{\nu'}{c}(\boldsymbol{n}' - \boldsymbol{n}) \cdot \boldsymbol{v}\right) d^3v \ , \end{aligned} \tag{3.3.28}$$

the second line being equivalent to the first in view of (3.3.18b). The scattering function is easily seen to be symmetric and normalized,

$$a(\nu', \boldsymbol{n}' \to \nu, \boldsymbol{n}) = a(\nu, \boldsymbol{n} \to \nu', \boldsymbol{n}') \ , \tag{3.3.29}$$

$$\iint a(\nu', \boldsymbol{n}' \to \nu, \boldsymbol{n})\, d\nu\, d\Omega = \iint a(\nu', \boldsymbol{n}' \to \nu, \boldsymbol{n})\, d\nu'\, d\Omega' = 1 \ . \tag{3.3.30}$$

Hence $a(\nu', \boldsymbol{n}' \to \nu, \boldsymbol{n})\, d\nu\, d\Omega$ is the probability that the scattering of an incoming photon $(\nu', \boldsymbol{n}')$ gives rise to an outgoing photon in the range

$(\nu, d\nu; \boldsymbol{n}, d\Omega)$. Equations (3.3.30) simply express conservation of the number of photons for scattering.

Let us calculate the scattering function of a gas whose velocity distribution is Maxwellian. Figure 3.2 shows that

$$|\boldsymbol{n}' - \boldsymbol{n}| = 2\sin(\vartheta/2) \; ,$$

where ϑ is the scattering angle. Choosing a coordinate system such that the z axis points in the direction of the vector $\boldsymbol{n}' - \boldsymbol{n}$ so that

$$(\boldsymbol{n}' - \boldsymbol{n}) \cdot \boldsymbol{v} = 2 v_z \sin(\vartheta/2) \; ,$$

then

$$\iiint f_{\mathrm{s}}^{\mathrm{M}}(v_x, v_y, v_z)\, \delta\left(\nu' - \nu - 2\frac{\nu}{c} v_z \sin\frac{\vartheta}{2}\right) dv_x\, dv_y\, dv_z$$

$$= \int_{-\infty}^{\infty} f_{\mathrm{s}}^{\mathrm{M}}(v_z)\, \delta\left(\nu' - \nu - 2\frac{\nu}{c} v_z \sin\frac{\vartheta}{2}\right) dv_z$$

$$= \frac{1}{2\dfrac{\nu}{c}\sin\dfrac{\vartheta}{2}} \left(\frac{m_{\mathrm{s}}}{2\pi k T_{\mathrm{s}}}\right)^{1/2} \exp\left[-\frac{m_{\mathrm{s}}}{2kT_{\mathrm{s}}}\left(\frac{\nu' - \nu}{2\dfrac{\nu}{c}\sin\dfrac{\vartheta}{2}}\right)^2\right] \; , \qquad (3.3.31)$$

m_{s} being the mass of a scattering particle, and T_{s} the gas temperature. The scattering function of a Maxwellian gas is obtained by substituting (3.3.31) into (3.3.28), which yields

$$a^{\mathrm{M}}(\nu', \boldsymbol{n}' \to \nu, \boldsymbol{n}) = \frac{g(\vartheta)}{2\pi^{1/2}\Delta\nu_{\mathrm{D}}\sin\dfrac{\vartheta}{2}} \exp\left[-\left(\frac{\nu' - \nu}{2\Delta\nu_{\mathrm{D}}\sin\dfrac{\vartheta}{2}}\right)^2\right] \qquad (3.3.32)$$

where $g(\vartheta) \equiv g(\boldsymbol{n}', \boldsymbol{n})$, and where

$$\Delta\nu_{\mathrm{D}} = \frac{\nu}{c}\left(\frac{2kT_{\mathrm{s}}}{m_{\mathrm{s}}}\right)^{1/2} \qquad (3.3.33)$$

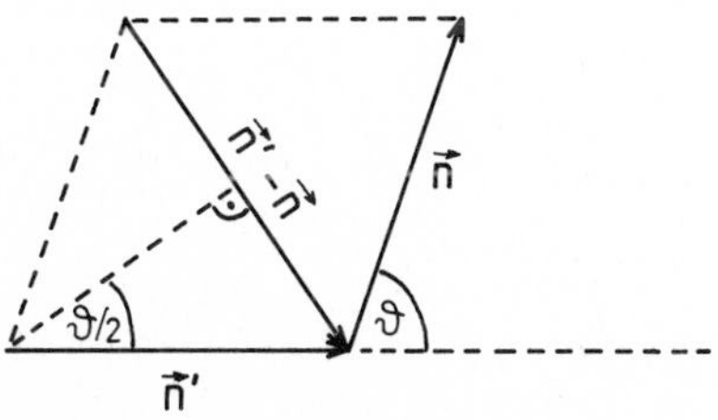

Fig. 3.2. Geometry of scattering. Unit vectors $\boldsymbol{n}'$, $\boldsymbol{n}$ define the photon directions, ϑ is the scattering angle

is the Doppler width. Because of the isotropy of the Maxwell distribution, the scattering function depends only on the scattering angle $\vartheta = \arccos(\boldsymbol{n}' \cdot \boldsymbol{n})$ rather than on the directions $\boldsymbol{n}'$ and $\boldsymbol{n}$ separately. Equation (3.3.32) shows that a monochromatic light beam is scattered without broadening in the forward direction $\vartheta = 0$, whereas maximum broadening $2\Delta\nu_D$ occurs for backward scattering $\vartheta = \pi$.

4. Moment Equations and Fluid Description

This chapter is devoted to the fluid description of gases and radiation near thermal equilibrium by a closed set of moment equations with transport coefficients. The moment equations corresponding to mass, momentum, and energy of a multi-component gas are derived in Sect. 4.1, and the moment equations corresponding to radiant momentum and energy in Sect. 4.2. Section 4.3 recalls how the fluid description of a one-component gas near thermal equilibrium in the absence of radiation is achieved by introducing transport coefficients and employing the thermal and caloric equations of state. In Sect. 4.4, the analogous problem in the presence of radiation is considered where both matter and radiation are near thermodynamic equilibrium. The radiation pressure and the coefficients of radiative heat conductivity and viscosity are derived, and the hydrodynamic equations containing these radiative quantities are obtained.

4.1 Moment Equations for Particles

The *velocity moments* of zeroth, first, second, etc. order of the distribution function $F(\boldsymbol{r},\boldsymbol{v},t)$ of material particles are defined as the tensor quantities

$$\int F(\boldsymbol{r},\boldsymbol{v},t)\,d^3v\ ,\quad \int v_iF(\boldsymbol{r},\boldsymbol{v},t)\,d^3v\ ,\quad \int v_iv_jF(\boldsymbol{r},\boldsymbol{v},t)\,d^3v\ ,\ \ldots,$$

where v_i, v_j, ... denote Cartesian components of the velocity vector $\boldsymbol{v}$. The velocity moments are in general functions of position $\boldsymbol{r}$ and time t. The equations they obey are obtained by multiplying the kinetic equation (2.1.7) for the distribution function F by, respectively, 1, v_i, v_iv_j, ... and performing the integration over all velocities. However, the equations for the velocity moment tensor of order n do not form a closed set of equations. For, on the one hand, owing to the streaming term $\boldsymbol{v}\cdot(\partial F/\partial\boldsymbol{r})$ in the kinetic equation, the moment tensor of order $n+1$ appears on the left-hand sides of the equations for the tensor of order n, while, on the other hand, their right-hand sides, which contain the collision term of the kinetic equation, cannot generally be expressed in terms of the moment tensor of order n alone. The moment equations, therefore, form an infinite set of coupled equations that, as a whole, is equivalent to the original kinetic equation. Hence, in practice, to work with moment equations, this chain has to be cut somewhere to provide a finite closed set of equations. As this closure of the moment equations corresponds to replacing a microscopic by a macroscopic description, this procedure cannot be expected to work generally. If, however, a gas is in a state near thermal equilibrium, the set of moment equations corresponding to the conserved quantities mass, momentum, and energy can be closed through the introduction of macroscopic transport coefficients, thus leading to the well-known hydrodynamic equations (Sect. 4.3) [4.1 – 7].

4.1.1 One Particle Type *a*

Let us first focus attention on one particular type of particles. Recall from Sect. 2.1 that in this book this term refers to an atom or ion in a well-defined internal state so that, for instance, hydrogen atoms $H(1s)$ and $H(2p)$ are considered as two different particle types. We characterize the particle type a by its mass m_a, its electric charge q_a (higher electric and magnetic moments are ignored), and its excitation energy E_a. By definition, free electrons (e) and free nuclei (H^+, He^{2+}) have $E_a = 0$. Then, for example, $H(1s)$, $He^+(1s)$, and $He(1\,^1S)$ have excitation energies $E_a = -13.6$ eV, -54.4 eV, and -79.0 eV, respectively.

The distribution function of particle type a, $F_a \equiv F_a(\boldsymbol{r}, \boldsymbol{v}, t)$, obeys the kinetic equation, see (2.1.7),

$$\frac{DF_a}{Dt} \equiv \frac{\partial F_a}{\partial t} + \boldsymbol{v} \cdot \frac{\partial F_a}{\partial \boldsymbol{r}} + \frac{\boldsymbol{K}_a}{m_a} \cdot \frac{\partial F_a}{\partial \boldsymbol{v}} = \left(\frac{\delta F_a}{\delta t}\right)_{\text{coll}} , \tag{4.1.1}$$

where

$$\boldsymbol{K}_a = m_a \boldsymbol{\gamma} + q_a \boldsymbol{E} + \frac{1}{c} q_a \boldsymbol{v} \times \boldsymbol{B} \equiv \boldsymbol{K}_a^0 + \frac{1}{c} q_a \boldsymbol{v} \times \boldsymbol{B} \tag{4.1.2}$$

is the force on a particle of velocity $\boldsymbol{v}$, and $\boldsymbol{K}_a^0$ denotes the velocity-independent part of this force. The gravitational, electric, and magnetic fields $\boldsymbol{\gamma}$, $\boldsymbol{E}$, and $\boldsymbol{B}$ are in general functions of $\boldsymbol{r}$ and t, and are composed of external and self-consistent parts, (2.1.8). On the other hand, the collision term in (4.1.1) is a sum of terms describing elastic and inelastic collisions, and radiative interactions, see (2.1.11).

The moment equations corresponding to mass, momentum, and energy are of special importance. Multiplying (4.1.1) respectively by 1 (instead of m_a), $m_a \boldsymbol{v}$, and $m_a v^2/2 + E_a$ and integrating over all velocities yields the moment equations

$$\int \frac{DF_a}{Dt} d^3v = \int \left(\frac{\delta F_a}{\delta t}\right)_{\text{coll}} d^3v , \tag{4.1.3}$$

$$\int m_a \boldsymbol{v} \frac{DF_a}{Dt} d^3v = \int m_a \boldsymbol{v} \left(\frac{\delta F_a}{\delta t}\right)_{\text{coll}} d^3v \ , \tag{4.1.4}$$

$$\int \left(\frac{1}{2} m_a v^2 + E_a\right) \frac{DF_a}{Dt} d^3v = \int \left(\frac{1}{2} m_a v^2 + E_a\right) \left(\frac{\delta F_a}{\delta t}\right)_{\text{coll}} d^3v \ . \tag{4.1.5}$$

When carrying out the various integrations in (4.1.3 – 5), one encounters the following macroscopic quantities of the gas of particle type a:

number density

$$n_a = \int F_a(\boldsymbol{v})\, d^3v \ , \tag{4.1.6}$$

mass density

$$\rho_a = m_a n_a \ , \tag{4.1.7}$$

mean velocity

$$\boldsymbol{u}_a = \frac{1}{n_a} \int \boldsymbol{v} F_a(\boldsymbol{v})\, d^3v \ , \tag{4.1.8}$$

momentum density

$$\boldsymbol{g}_a = m_a n_a \boldsymbol{u}_a = \rho_a \boldsymbol{u}_a \ , \tag{4.1.9}$$

mean force density

$$\boldsymbol{k}_a = \int \boldsymbol{K}_a(\boldsymbol{v}) F_a(\boldsymbol{v})\, d^3v = \boldsymbol{k}_a^0 + \frac{1}{c} q_a n_a \boldsymbol{u}_a \times \boldsymbol{B} \ , \tag{4.1.10}$$

velocity-independent force density

$$\boldsymbol{k}_a^0 = n_a \boldsymbol{K}_a^0 = m_a n_a \boldsymbol{\gamma} + q_a n_a \boldsymbol{E} \ , \tag{4.1.11}$$

pressure tensor

$$\overleftrightarrow{\boldsymbol{P}}_a = m_a \int (\boldsymbol{v} - \boldsymbol{u}_a)(\boldsymbol{v} - \boldsymbol{u}_a) F_a(\boldsymbol{v})\, d^3v \ , \tag{4.1.12}$$

heat flux density

$$\boldsymbol{Q}_a = \tfrac{1}{2} m_a \int (\boldsymbol{v} - \boldsymbol{u}_a)^2 (\boldsymbol{v} - \boldsymbol{u}_a) F_a(\boldsymbol{v})\, d^3v \ . \tag{4.1.13}$$

One further defines the following energy densities:

(total) energy density

$$e_a^{\text{tot}} = \int \left(\tfrac{1}{2} m_a v^2 + E_a\right) F_a(\boldsymbol{v})\, d^3v \ , \tag{4.1.14}$$

excitation energy density

$$e_a^{\text{exc}} = \int E_a F_a(\boldsymbol{v})\, d^3v = E_a n_a \,, \tag{4.1.15}$$

thermal energy density, where $\text{Tr}\{\overleftrightarrow{\boldsymbol{P}}\} = \sum_i P_{ii}$ denotes the trace of the tensor $\overleftrightarrow{\boldsymbol{P}}$,

$$\begin{aligned} e_a^{\text{th}} &= \int \tfrac{1}{2} m_a (\boldsymbol{v}-\boldsymbol{u}_a)^2 F_a(\boldsymbol{v})\, d^3v = \tfrac{1}{2}\,\text{Tr}\{\overleftrightarrow{\boldsymbol{P}}_a\} \\ &= e_a^{\text{tot}} - e_a^{\text{exc}} - \tfrac{1}{2} m_a u_a^2 n_a \,, \end{aligned} \tag{4.1.16}$$

kinetic energy density

$$e_a^{\text{kin}} = e_a^{\text{tot}} - e_a^{\text{exc}} - e_a^{\text{th}} = \tfrac{1}{2} m_a u_a^2 n_a \,. \tag{4.1.17}$$

Lastly, one has the following production rates (PR) per unit time and unit volume due to collisions and radiative interactions, referring to the a particle gas:

PR of particle number

$$\hat{\nu}_a = \int \left(\frac{\delta F_a}{\delta t}\right)_{\text{coll}} d^3v \,, \tag{4.1.18}$$

PR of momentum

$$\hat{\boldsymbol{\pi}}_a = \int m_a \boldsymbol{v} \left(\frac{\delta F_a}{\delta t}\right)_{\text{coll}} d^3v \,, \tag{4.1.19}$$

PR of (total) energy

$$\hat{\varepsilon}_a^{\text{tot}} = \int \left(\frac{1}{2} m_a v^2 + E_a\right)\left(\frac{\delta F_a}{\delta t}\right)_{\text{coll}} d^3v \,, \tag{4.1.20}$$

PR of excitation energy

$$\hat{\varepsilon}_a^{\text{exc}} = \int E_a \left(\frac{\delta F_a}{\delta t}\right)_{\text{coll}} d^3v = E_a \hat{\nu}_a \,, \tag{4.1.21}$$

PR of thermal energy

$$\begin{aligned} \hat{\varepsilon}_a^{\text{th}} &= \int \frac{1}{2} m_a (\boldsymbol{v}-\boldsymbol{u}_a)^2 \left(\frac{\delta F_a}{\delta t}\right)_{\text{coll}} d^3v \\ &= \hat{\varepsilon}_a^{\text{tot}} - \hat{\varepsilon}_a^{\text{exc}} - \boldsymbol{u}_a \cdot \hat{\boldsymbol{\pi}}_a + \tfrac{1}{2} m_a u_a^2 \hat{\nu}_a \,, \end{aligned} \tag{4.1.22}$$

PR of kinetic energy

$$\hat{\varepsilon}_a^{\mathrm{kin}} = \hat{\varepsilon}_a^{\mathrm{tot}} - \hat{\varepsilon}_a^{\mathrm{exc}} - \hat{\varepsilon}_a^{\mathrm{th}} = \boldsymbol{u}_a \cdot \hat{\boldsymbol{\pi}}_a - \tfrac{1}{2} m_a u_a^2 \hat{v}_a \ . \tag{4.1.23}$$

We note the following useful relations, whose derivation is straightforward,

$$m_a \int \boldsymbol{v}\boldsymbol{v} F_a(\boldsymbol{v})\, d^3v = \boldsymbol{g}_a \boldsymbol{u}_a + \overset{\leftrightarrow}{\boldsymbol{P}}_a \ , \tag{4.1.24}$$

$$\tfrac{1}{2} m_a \int v^2 F_a(\boldsymbol{v})\, d^3v = e_a^{\mathrm{th}} + e_a^{\mathrm{kin}} \ , \tag{4.1.25}$$

$$\tfrac{1}{2} m_a \int v^2 \boldsymbol{v} F_a(\boldsymbol{v})\, d^3v = (e_a^{\mathrm{th}} + e_a^{\mathrm{kin}}) \boldsymbol{u}_a + \overset{\leftrightarrow}{\boldsymbol{P}}_a \cdot \boldsymbol{u}_a + \boldsymbol{Q}_a \ . \tag{4.1.26}$$

The moment equations (4.1.3 – 5) may now be written in the form of balance equations for particle number, momentum, and energy,

$$\frac{\partial n_a}{\partial t} + \nabla \cdot (n_a \boldsymbol{u}_a) = \hat{v}_a \ , \tag{4.1.27}$$

$$\frac{\partial \boldsymbol{g}_a}{\partial t} + \nabla \cdot (\boldsymbol{g}_a \boldsymbol{u}_a + \overset{\leftrightarrow}{\boldsymbol{P}}_a) = \boldsymbol{k}_a + \hat{\boldsymbol{\pi}}_a \ , \tag{4.1.28}$$

$$\frac{\partial e_a^{\mathrm{tot}}}{\partial t} + \nabla \cdot (e_a^{\mathrm{tot}} \boldsymbol{u}_a + \overset{\leftrightarrow}{\boldsymbol{P}}_a \cdot \boldsymbol{u}_a + \boldsymbol{Q}_a) = \boldsymbol{k}_a^0 \cdot \boldsymbol{u}_a + \hat{\varepsilon}_a^{\mathrm{tot}} \ . \tag{4.1.29}$$

Indeed, since each left-hand side of (4.1.27 – 29) is the sum of the partial time derivative of a density and the divergence of the corresponding flux density (composed, in general, of a convective flow proportional to the mean velocity, and a conduction flow), it follows from Gauss' theorem that each right-hand side is the production rate per unit time and unit volume of the corresponding quantity. As expected, in addition to the various source terms due to collisions, the mean force density $\boldsymbol{k}_a$ acts as a local source of the momentum density, and the work done by the velocity-independent force density, $\boldsymbol{k}_a^0 \cdot \boldsymbol{u}_a$, acts as a source of the energy density.

In the special case where the velocity-independent force can be derived from a time-independent potential φ_a,

$$\boldsymbol{K}_a^0 = -\nabla \varphi_a \ , \quad \frac{\partial \varphi_a}{\partial t} = 0 \ , \tag{4.1.30}$$

the following relation, using (4.1.27), is readily derived:

$$\frac{\partial (n_a \varphi_a)}{\partial t} + \nabla \cdot (n_a \varphi_a \boldsymbol{u}_a) = -\boldsymbol{k}_a^0 \cdot \boldsymbol{u}_a + \varphi_a \hat{v}_a \ . \tag{4.1.31}$$

Here, the balance equation of energy (4.1.29) can be written as

$$\frac{\partial e_a'^{\text{tot}}}{\partial t} + \nabla \cdot (e_a'^{\text{tot}} \boldsymbol{u}_a + \overleftrightarrow{\boldsymbol{P}}_a \cdot \boldsymbol{u}_a + \boldsymbol{Q}_a) = \hat{\varepsilon}_a'^{\text{tot}} , \qquad (4.1.32)$$

where

$$e_a'^{\text{tot}} = e_a^{\text{tot}} + \varphi_a n_a , \quad \hat{\varepsilon}_a'^{\text{tot}} = \hat{\varepsilon}_a^{\text{tot}} + \varphi_a \hat{\nu}_a \qquad (4.1.33)$$

denote the density of total energy, including the potential energy, and its production rate due to collisions, respectively.

We now return to the energy balance (4.1.29), which we want to write in another, more useful form. To this end, the scalar product of the momentum balance (4.1.28) with the mean velocity $\boldsymbol{u}_a$ is formed, which yields the balance equation for the kinetic energy

$$\frac{\partial e_a^{\text{kin}}}{\partial t} + \nabla \cdot (e_a^{\text{kin}} \boldsymbol{u}_a + \overleftrightarrow{\boldsymbol{P}}_a \cdot \boldsymbol{u}_a) = \overleftrightarrow{\boldsymbol{P}}_a : \nabla \boldsymbol{u}_a + \boldsymbol{k}_a^0 \cdot \boldsymbol{u}_a + \hat{\varepsilon}_a^{\text{kin}} , \qquad (4.1.34)$$

observing that the velocity-dependent Lorentz force does not contribute to the right-hand side. The double dot denotes the contracted tensor product,

$$\overleftrightarrow{\boldsymbol{P}} : \nabla \boldsymbol{u} = \sum_{i,j} P_{ij} \partial u_j / \partial x_i .$$

Here the density of kinetic energy and its production rate are defined by (4.1.17, 23), and use has been made of

$$\boldsymbol{u}_a \cdot \left[\frac{\partial \boldsymbol{g}_a}{\partial t} + \nabla \cdot (\boldsymbol{g}_a \boldsymbol{u}_a) \right] = \frac{\partial e_a^{\text{kin}}}{\partial t} + \nabla \cdot (e_a^{\text{kin}} \boldsymbol{u}_a) + \frac{1}{2} m_a u_a^2 \hat{\nu}_a , \qquad (4.1.35)$$

which holds in view of (4.1.27).

Subtracting (4.1.34) from (4.1.29) gives the balance equation for the internal energy,

$$\frac{\partial e_a}{\partial t} + \nabla \cdot (e_a \boldsymbol{u}_a + \boldsymbol{Q}_a) = -\overleftrightarrow{\boldsymbol{P}}_a : \nabla \boldsymbol{u}_a + \hat{\varepsilon}_a . \qquad (4.1.36)$$

Here the internal energy density e_a and its production rate due to collisions $\hat{\varepsilon}_a$ are defined by

$$e_a = e_a^{\text{exc}} + e_a^{\text{th}} = e_a^{\text{tot}} - e_a^{\text{kin}} , \qquad (4.1.37)$$

$$\hat{\varepsilon}_a = \hat{\varepsilon}_a^{\text{exc}} + \hat{\varepsilon}_a^{\text{th}} = \hat{\varepsilon}_a^{\text{tot}} - \hat{\varepsilon}_a^{\text{kin}} . \qquad (4.1.38)$$

As can be seen from (4.1.36), the heat flux $\boldsymbol{Q}_a$ is the conduction flow corresponding to the internal energy. On the other hand, apart from the internal energy production by collisions and radiative interactions, there is a source term on the right-hand side of (4.1.36) which describes the transformation of kinetic into internal energy and vice versa by pressure forces, see also the right-hand side of (4.1.34). Energy balance is more conveniently expressed by (4.1.36) than by the original (4.1.29) because, as shown above, the balance of kinetic energy is already implicitly contained in the momentum balance equation (4.1.28).

4.1.2 One Particle Species *A*

Next we consider one particle species A composed of several particle types $a_1, a_2, \ldots$, of equal mass m_A and electric charge q_A, but differing in their excitation energies E_a. For instance, the particle species H, atomic hydrogen, is composed of the particle types H(1*s*), H(2*s*), H(2*p*), ...

Let us define

$$F_A(\boldsymbol{r}, \boldsymbol{v}, t) \equiv \sum_a F_a(\boldsymbol{r}, \boldsymbol{v}, t) \tag{4.1.39}$$

where the sum is over all particle types of the considered species. The quantity F_A is merely an abbreviation for the sum on the right-hand side of (4.1.39), but it may be interpreted as the distribution function of the species A in all expressions not containing excitation energies E_a. Then the macroscopic quantities n_A, ρ_A, $\boldsymbol{u}_A$, $\boldsymbol{g}_A$, $\boldsymbol{k}_A$, $\boldsymbol{k}_A^0$, $\overleftrightarrow{\boldsymbol{P}}_A$, $\boldsymbol{Q}_A$ are simply defined by (4.1.6 – 13), replacing a by A everywhere. Note, in particular, that in the definitions of $\overleftrightarrow{\boldsymbol{P}}_A$ and $\boldsymbol{Q}_A$ the velocity $\boldsymbol{v}-\boldsymbol{u}_A$ now appears. The following relations are easily derived:

$$n_A = \sum_a n_a \ , \quad \rho_A = \sum_a \rho_a \ , \quad n_A\boldsymbol{u}_A = \sum_a n_a\boldsymbol{u}_a \ , \quad \boldsymbol{g}_A = \sum_a \boldsymbol{g}_a \ , \tag{4.1.40}$$

$$\boldsymbol{k}_A = \sum_a \boldsymbol{k}_a \ , \quad \boldsymbol{k}_A^0 \cdot \boldsymbol{u}_A = \sum_a \boldsymbol{k}_a^0 \cdot \boldsymbol{u}_a \ . \tag{4.1.41}$$

The various energy densities are now defined by

$$e_A^{\mathrm{tot}} = \sum_a e_a^{\mathrm{tot}} \ , \tag{4.1.42}$$

$$e_A^{\mathrm{exc}} = \sum_a e_a^{\mathrm{exc}} \ , \tag{4.1.43}$$

$$\begin{aligned} e_A^{\mathrm{th}} &= \int \tfrac{1}{2} m_A (\boldsymbol{v}-\boldsymbol{u}_A)^2 F_A(\boldsymbol{v})\, d^3v = \tfrac{1}{2}\mathrm{Tr}\{\overleftrightarrow{\boldsymbol{P}}_A\} \\ &= e_A^{\mathrm{tot}} - e_A^{\mathrm{exc}} - \tfrac{1}{2} m_A u_A^2 n_A \ , \end{aligned} \tag{4.1.44}$$

$$e_A^{\mathrm{kin}} = e_A^{\mathrm{tot}} - e_A^{\mathrm{exc}} - e_A^{\mathrm{th}} = \tfrac{1}{2} m_A u_A^2 n_A \ , \tag{4.1.45}$$

and the various production rates due to collisions and radiative interactions by

$$\hat{\nu}_A = \sum_a \hat{\nu}_a \ , \tag{4.1.46}$$

$$\hat{\pi}_A = \sum_a \hat{\pi}_a \ , \tag{4.1.47}$$

$$\hat{\varepsilon}_A^{\text{tot}} = \sum_a \hat{\varepsilon}_a^{\text{tot}} \ , \tag{4.1.48}$$

$$\hat{\varepsilon}_A^{\text{exc}} = \sum_a \hat{\varepsilon}_a^{\text{exc}} \ , \tag{4.1.49}$$

$$\begin{aligned} \hat{\varepsilon}_A^{\text{th}} &= \sum_a \int \frac{1}{2} m_A (\boldsymbol{v} - \boldsymbol{u}_A)^2 \left(\frac{\delta F_a}{\delta t} \right)_{\text{coll}} d^3 v \\ &= \hat{\varepsilon}_A^{\text{tot}} - \hat{\varepsilon}_A^{\text{exc}} - \boldsymbol{u}_A \cdot \hat{\boldsymbol{\pi}}_A + \tfrac{1}{2} m_A u_A^2 \hat{\nu}_A \ , \end{aligned} \tag{4.1.50}$$

$$\begin{aligned} \hat{\varepsilon}_A^{\text{kin}} &= \hat{\varepsilon}_A^{\text{tot}} - \hat{\varepsilon}_A^{\text{exc}} - \hat{\varepsilon}_A^{\text{th}} \\ &= \boldsymbol{u}_A \cdot \hat{\boldsymbol{\pi}}_A - \tfrac{1}{2} m_A u_A^2 \hat{\nu}_A \ . \end{aligned} \tag{4.1.51}$$

By straightforward calculation, the following relations are shown to hold:

$$g_A \boldsymbol{u}_A + \overleftrightarrow{\boldsymbol{P}}_A = \sum_a (g_a \boldsymbol{u}_a + \overleftrightarrow{\boldsymbol{P}}_a) \ , \tag{4.1.52}$$

$$e_A^{\text{th}} + e_A^{\text{kin}} = \sum_a (e_a^{\text{th}} + e_a^{\text{kin}}) \ , \tag{4.1.53}$$

$$(e_A^{\text{th}} + e_a^{\text{kin}}) \boldsymbol{u}_A + \overleftrightarrow{\boldsymbol{P}}_A \cdot \boldsymbol{u}_A + \boldsymbol{Q}_A = \sum_a [(e_a^{\text{th}} + e_a^{\text{kin}}) \boldsymbol{u}_a + \overleftrightarrow{\boldsymbol{P}}_a \cdot \boldsymbol{u}_a + \boldsymbol{Q}_a] \ . \tag{4.1.54}$$

Using the various definitions and relations above, the balance equations of particle number and momentum for the particle species A are derived at once by summing (4.1.27, 28) over all particle types a,

$$\frac{\partial n_A}{\partial t} + \nabla \cdot (n_A \boldsymbol{u}_A) = \hat{\nu}_A \ , \tag{4.1.55}$$

$$\frac{\partial \boldsymbol{g}_A}{\partial t} + \nabla \cdot (\boldsymbol{g}_A \boldsymbol{u}_A + \overleftrightarrow{\boldsymbol{P}}_A) = \boldsymbol{k}_A + \hat{\boldsymbol{\pi}}_A \ . \tag{4.1.56}$$

More care is required to derive the corresponding balance equation of energy. Let us write (4.1.29) in the form

$$\begin{aligned} &\frac{\partial}{\partial t}(e_a^{\text{th}} + e_a^{\text{kin}}) + \nabla \cdot [(e_a^{\text{th}} + e_a^{\text{kin}}) \boldsymbol{u}_a + \overleftrightarrow{\boldsymbol{P}}_a \cdot \boldsymbol{u}_a + \boldsymbol{Q}_a] + \frac{\partial e_a^{\text{exc}}}{\partial t} + \nabla \cdot (e_a^{\text{exc}} \boldsymbol{u}_a) \\ &\quad = \boldsymbol{k}_a^0 \cdot \boldsymbol{u}_a + \hat{\varepsilon}_a^{\text{tot}} \ . \end{aligned} \tag{4.1.57}$$

When summing this equation over all a, one must observe that, in view of (4.1.43),

$$\sum_a \frac{\partial e_a^{\text{exc}}}{\partial t} = \frac{\partial e_A^{\text{exc}}}{\partial t} , \tag{4.1.58}$$

$$\sum_a \nabla \cdot (e_a^{\text{exc}} \boldsymbol{u}_a) = \nabla \cdot (e_A^{\text{exc}} \boldsymbol{u}_A) + \nabla \cdot \sum_a e_a^{\text{exc}} (\boldsymbol{u}_a - \boldsymbol{u}_A) . \tag{4.1.59}$$

Using furthermore (4.1.41, 48, 53, 54), one obtains from (4.1.57)

$$\frac{\partial e_A^{\text{tot}}}{\partial t} + \nabla \cdot (e_A^{\text{tot}} \boldsymbol{u}_A + \overleftrightarrow{\boldsymbol{P}}_A \cdot \boldsymbol{u}_A + \boldsymbol{Q}_A + \boldsymbol{D}_A) = \boldsymbol{k}_A^0 \cdot \boldsymbol{u}_A + \hat{\varepsilon}_A^{\text{tot}} , \tag{4.1.60}$$

where

$$\boldsymbol{D}_A = \sum_a e_a^{\text{exc}} (\boldsymbol{u}_a - \boldsymbol{u}_A) . \tag{4.1.61}$$

Thus, when dealing with a composite particle species, the diffusion current of excitation energy $\boldsymbol{D}_A$ appears in the energy balance equation.

The scalar product of (4.1.56) with $\boldsymbol{u}_A$ again gives the balance equation of kinetic energy

$$\frac{\partial e_A^{\text{kin}}}{\partial t} + \nabla \cdot (e_A^{\text{kin}} \boldsymbol{u}_A + \overleftrightarrow{\boldsymbol{P}}_A \cdot \boldsymbol{u}_A) = \overleftrightarrow{\boldsymbol{P}}_A : \nabla \boldsymbol{u}_A + \boldsymbol{k}_A^0 \cdot \boldsymbol{u}_A + \hat{\varepsilon}_A^{\text{kin}} . \tag{4.1.62}$$

Subtracting now this equation from (4.1.60) yields the balance equation for the internal energy e_A,

$$\frac{\partial e_A}{\partial t} + \nabla \cdot (e_A \boldsymbol{u}_A + \boldsymbol{J}_A) = -\overleftrightarrow{\boldsymbol{P}}_A : \nabla \boldsymbol{u}_A + \hat{\varepsilon}_A , \tag{4.1.63}$$

where

$$e_A = e_A^{\text{exc}} + e_A^{\text{th}} = e_A^{\text{tot}} - e_A^{\text{kin}} , \tag{4.1.64}$$

$$\hat{\varepsilon}_A = \hat{\varepsilon}_A^{\text{exc}} + \hat{\varepsilon}_A^{\text{th}} = \hat{\varepsilon}_A^{\text{tot}} - \hat{\varepsilon}_A^{\text{kin}} , \tag{4.1.65}$$

$$\boldsymbol{J}_A = \boldsymbol{D}_A + \boldsymbol{Q}_A . \tag{4.1.66}$$

The conduction flow of internal energy is hence the sum of the diffusion current of excitation energy and the ordinary heat current.

4.1.3 Several Particle Species

We finally turn to the moment equations for a gas that contains several particle species A, B, ..., each possibly composed of different excitation states $a_1, a_2, \ldots; b_1, b_2, \ldots$. The masses of the particles are $m_A, m_B, \ldots$, their electric charges $q_A, q_B, \ldots$, and their excitation energies $E_{a_1}, E_{a_2}, \ldots;$ $E_{b_1}, E_{b_2}, \ldots$.

The following macroscopic quantities refer to the composite gas as a whole, where the summations are over all particle species present,

$$\rho = \sum_A \rho_A \; , \quad \boldsymbol{g} = \sum_A \boldsymbol{g}_A \; , \quad \boldsymbol{u} = \frac{1}{\rho}\boldsymbol{g} \; , \tag{4.1.67}$$

$$\boldsymbol{k} = \sum_A \boldsymbol{k}_A \; , \tag{4.1.68}$$

$$\overleftrightarrow{\boldsymbol{P}} = \sum_A m_A \int (\boldsymbol{v}-\boldsymbol{u})(\boldsymbol{v}-\boldsymbol{u}) F_A(\boldsymbol{v})\, d^3v \; , \tag{4.1.69}$$

$$\boldsymbol{Q} = \sum_A \tfrac{1}{2} m_A \int (\boldsymbol{v}-\boldsymbol{u})^2 (\boldsymbol{v}-\boldsymbol{u}) F_A(\boldsymbol{v})\, d^3v \; . \tag{4.1.70}$$

The velocity $\boldsymbol{u}$, defined in (4.1.67) and used in the definitions of the pressure tensor and the heat flux density, is the mean mass velocity of the gas, in terms of which the hydrodynamic equations take their most convenient form. On the other hand, the mean (particle) velocity $\boldsymbol{u}_0$, defined by

$$\boldsymbol{u}_0 = \frac{1}{n} \sum_A n_A \boldsymbol{u}_A \; , \quad n = \sum_A n_A \; , \tag{4.1.71}$$

allows a more symmetric formulation of diffusion phenomena, but is not used in the following.

The following energy densities refer to the gas as a whole:

$$e^{\text{tot}} = \sum_A e_A^{\text{tot}} \; , \tag{4.1.72}$$

$$e^{\text{exc}} = \sum_A e_A^{\text{exc}} \; , \tag{4.1.73}$$

$$\begin{aligned} e^{\text{th}} &= \sum_A \int \tfrac{1}{2} m_A (\boldsymbol{v}-\boldsymbol{u})^2 F_A(\boldsymbol{v})\, d^3v = \tfrac{1}{2} \operatorname{Tr}\{\overleftrightarrow{\boldsymbol{P}}\} \\ &= e^{\text{tot}} - e^{\text{exc}} - \tfrac{1}{2} \rho u^2 \; , \end{aligned} \tag{4.1.74}$$

$$e^{\text{kin}} = e^{\text{tot}} - e^{\text{exc}} - e^{\text{th}} = \tfrac{1}{2} \rho u^2 \; . \tag{4.1.75}$$

Since the total mass of the gas is conserved, and since the momentum and the energy of the gas are not changed by elastic and inelastic collisions between particles, the following relations are valid:

$$\sum_A m_A \hat{\nu}_A = 0 \; , \tag{4.1.76}$$

$$\sum_A \hat{\boldsymbol{\pi}}_A = \hat{\boldsymbol{\pi}}_{\text{mat}} \; , \tag{4.1.77}$$

$$\sum_A \hat{\varepsilon}_A^{\text{tot}} = \hat{\varepsilon}_{\text{mat}} \; . \tag{4.1.78}$$

Here $\hat{\pi}_{\text{mat}}$ and $\hat{\varepsilon}_{\text{mat}}$ denote, respectively, the production rates of particle momentum and energy per unit volume, owing to momentum and energy transfer from the radiation field to the material gas via radiative interactions. In addition, one defines the following energy production rates of the total gas:

$$\hat{\varepsilon}^{\text{exc}} = \sum_A \hat{\varepsilon}_A^{\text{exc}} \ , \tag{4.1.79}$$

$$\begin{aligned}\hat{\varepsilon}^{\text{th}} &= \sum_A \sum_a \int \frac{1}{2} m_A (\boldsymbol{v}-\boldsymbol{u})^2 \left(\frac{\delta F_a}{\delta t}\right)_{\text{coll}} d^3v \\ &= \hat{\varepsilon}_{\text{mat}} - \hat{\varepsilon}^{\text{exc}} - \boldsymbol{u} \cdot \hat{\boldsymbol{\pi}}_{\text{mat}} \ ,\end{aligned} \tag{4.1.80}$$

$$\hat{\varepsilon}^{\text{kin}} = \hat{\varepsilon}_{\text{mat}} - \hat{\varepsilon}^{\text{exc}} - \hat{\varepsilon}^{\text{th}} = \boldsymbol{u} \cdot \hat{\boldsymbol{\pi}}_{\text{mat}} \ , \tag{4.1.81}$$

where (4.1.76) has been used in (4.1.80, 81).

The following relations can be verified by direct calculation,

$$\boldsymbol{g}\boldsymbol{u} + \overleftrightarrow{\boldsymbol{P}} = \sum_A (\boldsymbol{g}_A \boldsymbol{u}_A + \overleftrightarrow{\boldsymbol{P}}_A) \ , \tag{4.1.82}$$

$$e^{\text{th}} + e^{\text{kin}} = \sum_A (e_A^{\text{th}} + e_A^{\text{kin}}) \ , \tag{4.1.83}$$

$$(e^{\text{th}} + e^{\text{kin}})\boldsymbol{u} + \overleftrightarrow{\boldsymbol{P}} \cdot \boldsymbol{u} + \boldsymbol{Q} = \sum_A [(e_A^{\text{th}} + e_A^{\text{kin}})\boldsymbol{u}_A + \overleftrightarrow{\boldsymbol{P}}_A \cdot \boldsymbol{u}_A + \boldsymbol{Q}_A] \ . \tag{4.1.84}$$

Let us now derive the balance equations of mass, momentum, and energy. Multipyling (4.1.55) by the particle mass m_A and summing over all particle species A leads to the balance equation of mass

$$\frac{\partial \rho}{\partial t} + \nabla \cdot (\rho \boldsymbol{u}) = 0 \ , \tag{4.1.85}$$

where (4.1.76) has been used. The balance equation of momentum follows from (4.1.56) by summing over all species A,

$$\frac{\partial \boldsymbol{g}}{\partial t} + \nabla \cdot (\boldsymbol{g}\boldsymbol{u} + \overleftrightarrow{\boldsymbol{P}}) = \boldsymbol{k} + \hat{\boldsymbol{\pi}}_{\text{mat}} \ . \tag{4.1.86}$$

Finally, to derive the balance equation of energy, we first write (4.1.60) in the form

$$\begin{aligned}&\frac{\partial}{\partial t}(e_A^{\text{th}} + e_A^{\text{kin}}) + \nabla \cdot [(e_A^{\text{th}} + e_A^{\text{kin}})\boldsymbol{u}_A + \overleftrightarrow{\boldsymbol{P}}_A \cdot \boldsymbol{u}_A + \boldsymbol{Q}_A] \\ &+ \frac{\partial e_A^{\text{exc}}}{\partial t} + \nabla \cdot \left(\sum_a e_a^{\text{exc}} \boldsymbol{u}_a\right) = \boldsymbol{k}_A^0 \cdot \boldsymbol{u}_A + \hat{\varepsilon}_A^{\text{tot}} \ ,\end{aligned} \tag{4.1.87}$$

using (4.1.43, 61). Summation over all species A now leads to the balance equation

$$\frac{\partial e^{\text{tot}}}{\partial t}+\nabla\cdot(e^{\text{tot}}\boldsymbol{u}+\overleftrightarrow{\boldsymbol{P}}\cdot\boldsymbol{u}+\boldsymbol{Q}+\boldsymbol{D})=\sum_A \boldsymbol{k}_A^0\cdot\boldsymbol{u}_A+\hat{\varepsilon}_{\text{mat}}\ , \tag{4.1.88}$$

where

$$\boldsymbol{D}=\sum_A\sum_a e_a^{\text{exc}}(\boldsymbol{u}_a-\boldsymbol{u}) \tag{4.1.89}$$

is the total diffusion current of excitation energy relative to the mean mass velocity. On the other hand, scalar multiplication of (4.1.86) by $\boldsymbol{u}$ yields the balance equation of kinetic energy,

$$\frac{\partial e^{\text{kin}}}{\partial t}+\nabla\cdot(e^{\text{kin}}\boldsymbol{u}+\overleftrightarrow{\boldsymbol{P}}\cdot\boldsymbol{u})=\overleftrightarrow{\boldsymbol{P}}:\nabla\boldsymbol{u}+\boldsymbol{k}\cdot\boldsymbol{u}+\hat{\varepsilon}^{\text{kin}}\ , \tag{4.1.90}$$

and subtracting this equation from (4.1.88) gives the balance equation of internal energy

$$\frac{\partial e}{\partial t}+\nabla\cdot(e\boldsymbol{u}+\boldsymbol{J})=-\overleftrightarrow{\boldsymbol{P}}:\nabla\boldsymbol{u}+\sum_A\boldsymbol{k}_A^0\cdot\boldsymbol{u}_A-\boldsymbol{k}\cdot\boldsymbol{u}+\hat{\varepsilon}\ . \tag{4.1.91}$$

Here

$$e=e^{\text{exc}}+e^{\text{th}}=e^{\text{tot}}-e^{\text{kin}}=\sum_A\sum_a n_a E_a+\tfrac{1}{2}\,\text{Tr}\{\overleftrightarrow{\boldsymbol{P}}\} \tag{4.1.92}$$

is the internal energy of the gas,

$$\hat{\varepsilon}=\hat{\varepsilon}^{\text{exc}}+\hat{\varepsilon}^{\text{th}}=\hat{\varepsilon}_{\text{mat}}-\hat{\varepsilon}^{\text{kin}}=\hat{\varepsilon}_{\text{mat}}-\boldsymbol{u}\cdot\hat{\boldsymbol{\pi}}_{\text{mat}} \tag{4.1.93}$$

is its production rate due to collisions and radiative interactions, and

$$\boldsymbol{J}=\boldsymbol{D}+\boldsymbol{Q} \tag{4.1.94}$$

is the conduction flow of internal energy, composed of the diffusion current of excitation energy and the ordinary heat current.

We now want to write the two force terms on the right of (4.1.91) in a more lucid form. To this end we use the explicit expressions of the force densities, (4.1.10, 11, 41, 68) to write

$$\sum_A\boldsymbol{k}_A^0\cdot\boldsymbol{u}_A=\sum_A n_A(m_A\boldsymbol{\gamma}+q_A\boldsymbol{E})\cdot\boldsymbol{u}_A=\boldsymbol{g}\cdot\boldsymbol{\gamma}+\boldsymbol{j}\cdot\boldsymbol{E}\ , \tag{4.1.95}$$

$$\begin{aligned}\boldsymbol{k}\cdot\boldsymbol{u}&=\sum_A n_A\left(m_A\boldsymbol{\gamma}+q_A\boldsymbol{E}+\frac{1}{c}q_A\boldsymbol{u}_A\times\boldsymbol{B}\right)\cdot\boldsymbol{u}\\&=\boldsymbol{g}\cdot\boldsymbol{\gamma}+\left(\tau\boldsymbol{E}+\frac{1}{c}\boldsymbol{j}\times\boldsymbol{B}\right)\cdot\boldsymbol{u}\ .\end{aligned} \tag{4.1.96}$$

Here γ, $\boldsymbol{E}$, $\boldsymbol{B}$ are the gravitational, electric, and magnetic fields, $\boldsymbol{g}$ is the momentum density of the gas, see (4.1.67), and

$$\tau = \sum_A n_A q_A \ , \quad \boldsymbol{j} = \sum_A n_A q_A \boldsymbol{u}_A \tag{4.1.97}$$

are the electric charge and current densities, respectively. Hence

$$\sum_A \boldsymbol{k}_A^0 \cdot \boldsymbol{u}_A - \boldsymbol{k} \cdot \boldsymbol{u} = \boldsymbol{j} \cdot \boldsymbol{E} - \left(\tau \boldsymbol{E} + \frac{1}{c} \boldsymbol{j} \times \boldsymbol{B} \right) \cdot \boldsymbol{u} \ , \tag{4.1.98}$$

the gravitational terms dropping out. Here the first term on the right-hand side is the total energy transferred from the electromagnetic field to the gas (recalling that only the Coulomb force contributes because the Lorentz force leaves the energy of a particle unchanged), and the second term is the mechanical work on the gas done by the electric and magnetic forces. It follows that the difference of these two terms is the electromagnetic energy dissipated in the gas per unit time and unit volume, that is, the Joule heat $\hat{q}_{\mathrm{J}}$,

$$\boldsymbol{j} \cdot \boldsymbol{E} - \left(\tau \boldsymbol{E} + \frac{1}{c} \boldsymbol{j} \times \boldsymbol{B} \right) \cdot \boldsymbol{u} = \hat{q}_{\mathrm{J}} \ . \tag{4.1.99}$$

To illustrate this more explicitly, let us introduce, at a given space-time point of the gas, the local rest frame in which, by definition, $\boldsymbol{u}^{\mathrm{o}} = 0$, the superscript ^{o} denoting the values of physical quantities in this frame. Using the nonrelativistic transformation formulas for electromagnetic quantities (which are best obtained from the relativistic formulas, neglecting terms of order u^2/c^2 [4.8]), the electric field strength and the electric current density are given in this frame by

$$\boldsymbol{E}^{\mathrm{o}} = \boldsymbol{E} + \frac{1}{c} \boldsymbol{u} \times \boldsymbol{B} \ , \tag{4.1.100a}$$

$$\boldsymbol{j}^{\mathrm{o}} = \boldsymbol{j} - \tau \boldsymbol{u} \ . \tag{4.1.100b}$$

On the other hand, Ohm's law is valid in the local rest frame,

$$\boldsymbol{j}^{\mathrm{o}} = \overleftrightarrow{\sigma} \cdot \boldsymbol{E}^{\mathrm{o}} \ , \tag{4.1.101a}$$

where $\overleftrightarrow{\sigma}$ is the conductivity tensor. Choosing the orientation of our coordinate system such that the z axis points in the direction of the local magnetic field $\boldsymbol{B}^{\mathrm{o}}$, the conductivity tensor takes the form

$$\overleftrightarrow{\sigma} = \begin{pmatrix} \sigma_\perp & -\sigma_{\mathrm{H}} & 0 \\ \sigma_{\mathrm{H}} & \sigma_\perp & 0 \\ 0 & 0 & \sigma_\parallel \end{pmatrix} \ , \tag{4.1.101b}$$

where $\sigma_\perp$ and $\sigma_\parallel$ are the electric conductivities perpendicular and parallel to the magnetic field, and σ_{H} is the Hall conductivity [4.9]. In the absence of a magnetic field, $\boldsymbol{B}^\circ = 0$, then $\sigma_\perp = \sigma_\| = \sigma$ and $\sigma_H = 0$, and hence $\overleftrightarrow{\sigma} = \sigma\overleftrightarrow{I}$, where $\overleftrightarrow{I}$ denotes the unit tensor, so that here (4.1.101a) reduces to the familiar form of Ohm's law with a scalar conductivity σ,

$$\boldsymbol{j}^\circ = \sigma \boldsymbol{E}^\circ . \tag{4.1.101c}$$

Using (4.1.100, 101), thus

$$\begin{aligned} \boldsymbol{j}\cdot\boldsymbol{E} - \left(\tau\boldsymbol{E} + \frac{1}{c}\boldsymbol{j}\times\boldsymbol{B}\right)\cdot\boldsymbol{u} &= \boldsymbol{j}\cdot\boldsymbol{E} - \tau\boldsymbol{u}\cdot\boldsymbol{E} - \frac{1}{c}(\boldsymbol{j}\times\boldsymbol{B})\cdot\boldsymbol{u} \\ &= (\boldsymbol{j} - \tau\boldsymbol{u})\cdot\boldsymbol{E} + \boldsymbol{j}\cdot\frac{1}{c}(\boldsymbol{u}\times\boldsymbol{B}) \\ &= \boldsymbol{j}^\circ\cdot\boldsymbol{E} + \boldsymbol{j}^\circ\cdot\frac{1}{c}(\boldsymbol{u}\times\boldsymbol{B}) \\ &= \boldsymbol{j}^\circ\cdot\boldsymbol{E}^\circ = \boldsymbol{E}^\circ\cdot\overleftrightarrow{\sigma}\cdot\boldsymbol{E}^\circ \\ &= \sigma_\perp(E_\perp^\circ)^2 + \sigma_\|(E_\|^\circ)^2 , \end{aligned} \tag{4.1.102}$$

with $(E_\perp^\circ)^2 \equiv (E_x^\circ)^2 + (E_y^\circ)^2$ and $(E_\|^\circ)^2 \equiv (E_z^\circ)^2$. This is indeed the Joule heat $\hat{q}_J$. In particular, for a scalar conductivity $\sigma = \sigma_\perp = \sigma_\|$, the right-hand side of (4.1.102) reduces to $\sigma(E^\circ)^2 = (1/\sigma)(j^\circ)^2$, as it should.

We therefore obtain the balance equation for the internal energy (4.1.91) in the following convenient form,

$$\frac{\partial e}{\partial t} + \nabla\cdot(e\boldsymbol{u} + \boldsymbol{J}) = -\overleftrightarrow{\boldsymbol{P}}:\nabla\boldsymbol{u} + \hat{q}_J + \hat{\varepsilon}_{\text{mat}} - \boldsymbol{u}\cdot\hat{\boldsymbol{\pi}}_{\text{mat}} , \tag{4.1.103}$$

taking (4.1.93, 99) into account.

Equations (4.1.85, 86, 103) form the final result. In these balance equations appear the total force density $\boldsymbol{k}$ due to gravitational and electromagnetic fields, the dissipated electromagnetic energy $\hat{q}_J$ (there is no dissipation of gravitational energy), and the production rates of gas momentum and energy due to radiative interactions, $\hat{\boldsymbol{\pi}}_{\text{mat}}$ and $\hat{\varepsilon}_{\text{mat}}$.

4.2 Moment Equations for Photons

As for material gases, one can derive moment equations for the photon gas. Here one defines the moments of the radiation intensity $I_\nu(\boldsymbol{n},\boldsymbol{r},t)$ as the tensor quantities

$$\iint I_\nu(\boldsymbol{n},\boldsymbol{r},t)\,d\nu\,d\Omega\ ,\quad \iint n_i I_\nu(\boldsymbol{n},\boldsymbol{r},t)\,d\nu\,d\Omega\ ,\quad \iint n_i n_j I_\nu(\boldsymbol{n},\boldsymbol{r},t)\,d\nu\,d\Omega\ ,\ \ldots ,$$

where n_i, n_j, ... denote Cartesian components of the unit vector $\boldsymbol{n}$, and the integrations are over all frequencies ν and all directions $\boldsymbol{n}$. These moments are in general functions of position $\boldsymbol{r}$ and time t, and the equations they obey follow in an obvious way from the transfer equation (3.1.16). Like the moment equations of a material gas, the moment equations of the radiation field form an infinite set of coupled equations which, as a whole, is equivalent to the equation of radiative transfer. Of particular importance are the moment equations that express the balance of radiant energy and momentum, considered below.

First of all we derive the macroscopic quantities of the radiation field that are related to its energy and momentum. Our starting point is the energy density of a light beam $I_\nu(\boldsymbol{n})\,d\nu\,d\Omega$ which, according to (1.4.38), is given by

$$\frac{1}{c}I_\nu(\boldsymbol{n})\,d\nu\,d\Omega \ .$$

The total *energy density* of the radiation field is therefore

$$e_{\mathrm{R}} = \frac{1}{c}\iint I_\nu(\boldsymbol{n})\,d\nu\,d\Omega \ . \qquad (4.2.1)$$

On the other hand, the total net flux of radiant energy through an oriented area $\boldsymbol{s}\,dF$ (the unit vector $\boldsymbol{s}$ being the normal to the area dF) is evidently

$$\iint c\cos\vartheta\,\frac{1}{c}I_\nu(\boldsymbol{n})\,d\nu\,d\Omega = \boldsymbol{s}\cdot\iint \boldsymbol{n}\,I_\nu(\boldsymbol{n})\,d\nu\,d\Omega \ ,$$

where $\cos\vartheta \equiv \boldsymbol{s}\cdot\boldsymbol{n}$, and is hence the projection of the vector

$$\boldsymbol{f}_{\mathrm{R}} = \iint \boldsymbol{n}\,I_\nu(\boldsymbol{n})\,d\nu\,d\Omega \qquad (4.2.2)$$

on the direction $\boldsymbol{s}$. This means that the quantity $\boldsymbol{f}_{\mathrm{R}}$ is the total *energy flux density* of the radiation field.

Since the photon momentum $\boldsymbol{p} = (h\nu/c)\boldsymbol{n}$ is related to the photon energy $E = h\nu$ by $\boldsymbol{p} = (\boldsymbol{n}/c)E$, the momentum density of a light beam $I_\nu(\boldsymbol{n})\,d\nu\,d\Omega$ is given by

$$\frac{1}{c^2}\boldsymbol{n}\,I_\nu(\boldsymbol{n})\,d\nu\,d\Omega \ ,$$

and the total *momentum density* of the radiation field is hence

$$\boldsymbol{g}_{\mathrm{R}} = \frac{1}{c^2}\iint \boldsymbol{n}\,I_\nu(\boldsymbol{n})\,d\nu\,d\Omega \ . \qquad (4.2.3)$$

Comparison with (4.2.2) shows that the momentum density is related to the energy flux density by

$$\boldsymbol{g}_{\mathrm{R}} = \frac{1}{c^2}\boldsymbol{f}_{\mathrm{R}} \ , \qquad (4.2.4)$$

which is a well-known relation of classical electrodynamics. On the other hand, the total net flux of radiant momentum through an oriented area $\boldsymbol{s}\,dF$ is obviously given by

$$\iint c\cos\vartheta \frac{1}{c^2} \boldsymbol{n} I_\nu(\boldsymbol{n})\, d\nu\, d\Omega = \boldsymbol{s} \cdot \frac{1}{c} \iint \boldsymbol{n}\boldsymbol{n} I_\nu(\boldsymbol{n})\, d\nu\, d\Omega$$

because $\cos\vartheta \equiv \boldsymbol{s}\cdot\boldsymbol{n}$, and is hence the scalar product of the unit vector $\boldsymbol{s}$ with the tensor

$$\overleftrightarrow{\boldsymbol{P}}_\mathrm{R} = \frac{1}{c} \iint \boldsymbol{n}\boldsymbol{n} I_\nu(\boldsymbol{n})\, d\nu\, d\Omega \;. \tag{4.2.5}$$

In other words, the quantity $\overleftrightarrow{\boldsymbol{P}}_\mathrm{R}$ is the total momentum flux density or the *pressure tensor* of the radiation field. Since $\boldsymbol{n}$ is a unit vector, then the trace $\mathrm{Tr}\{\boldsymbol{n}\boldsymbol{n}\} = 1$, and hence

$$\mathrm{Tr}\{\overleftrightarrow{\boldsymbol{P}}_\mathrm{R}\} = e_\mathrm{R} \tag{4.2.6}$$

in view of (4.2.1).

Performing the integration $\iint d\nu\, d\Omega$ of the transfer equation (3.1.16),

$$\frac{1}{c}\,\frac{\partial I_\nu}{\partial t} + \boldsymbol{n}\cdot\nabla I_\nu = \varepsilon_\nu - \kappa_\nu I_\nu \;, \tag{4.2.7}$$

yields immediately the balance equation of radiant energy

$$\frac{\partial e_\mathrm{R}}{\partial t} + \nabla\cdot \boldsymbol{f}_\mathrm{R} = \hat{\varepsilon}_\mathrm{rad} \;, \quad \text{where} \tag{4.2.8}$$

$$\hat{\varepsilon}_\mathrm{rad} = \iint (\varepsilon_\nu - \kappa_\nu I_\nu)\, d\nu\, d\Omega \tag{4.2.9}$$

is the total production rate of radiant energy per unit time and unit volume due to the material gas characterized by the phenomenological emission and absorption coefficients $\varepsilon_\nu(\boldsymbol{n},\boldsymbol{r},t)$ and $\kappa_\nu(\boldsymbol{n},\boldsymbol{r},t)$. In the usual approximation where particle recoil and the energy change due to Doppler shifts of photons are neglected, photon scattering does not contribute to the energy production rate (4.2.9), see (3.3.26). Likewise, operating with $\iint d\nu\, d\Omega\, \boldsymbol{n}/c$ on the transfer equation (4.2.7) leads to the balance equation of radiant momentum

$$\frac{\partial \boldsymbol{g}_\mathrm{R}}{\partial t} + \nabla\cdot \overleftrightarrow{\boldsymbol{P}}_\mathrm{R} = \hat{\boldsymbol{\pi}}_\mathrm{rad} \;, \quad \text{where} \tag{4.2.10}$$

$$\hat{\boldsymbol{\pi}}_\mathrm{rad} = \frac{1}{c} \iint \boldsymbol{n}(\varepsilon_\nu - \kappa_\nu I_\nu)\, d\nu\, d\Omega \tag{4.2.11}$$

is the total production rate of radiant momentum per unit time and unit volume due to the material gas in question.

4.3 Fluid Description of a Gas

In the two preceding sections it has been shown that the balance equations for mass, momentum, and energy are special cases of moment equations that follow from the corresponding kinetic equations. For the total material gas one has the balance equations (4.1.85, 86, 103) which contain the production rates of gas momentum and energy due to radiative interactions, $\hat{\pi}_{\text{mat}}$ and $\hat{\varepsilon}_{\text{mat}}$. On the other hand, the corresponding balance equations for the radiation field (4.2.8, 10) contain the production rates of radiant momentum and energy, $\hat{\pi}_{\text{rad}}$ and $\hat{\varepsilon}_{\text{rad}}$. Since conservation of momentum and energy requires

$$\hat{\pi}_{\text{mat}} + \hat{\pi}_{\text{rad}} = 0 \ , \tag{4.3.1}$$

$$\hat{\varepsilon}_{\text{mat}} + \hat{\varepsilon}_{\text{rad}} = 0 \ , \tag{4.3.2}$$

the two sets of balance equations for matter and radiation are coupled to each other.

It should be remembered that the various quantities entering the balance equations depend on the distribution functions. So the macroscopic gas quantities ρ, $\boldsymbol{u}$, $\boldsymbol{g}$, $\overset{\leftrightarrow}{\boldsymbol{P}}$, $\boldsymbol{k}$, e, $\boldsymbol{J}$, $\hat{q}_{\mathrm{J}}$ depend on the particle distribution functions F_a, the radiative quantities e_{R}, $\boldsymbol{f}_{\mathrm{R}}$, $\boldsymbol{g}_{\mathrm{R}}$, $\overset{\leftrightarrow}{\boldsymbol{P}}_{\mathrm{R}}$ depend on the specific intensity I_ν, and the production rates $\hat{\pi}_{\text{mat}} = -\hat{\pi}_{\text{rad}}$ and $\hat{\varepsilon}_{\text{mat}} = -\hat{\varepsilon}_{\text{rad}}$ depend both on the F_a via ε_ν and κ_ν, and on I_ν, see (4.2.9, 11). Now, to know the distribution functions, and hence the macroscopic quantities, one must in principle solve either the kinetic equations or, what is equivalent, the whole infinite set of coupled moment equations.

There is one important exception, however. If the state of the system is near thermal equilibrium, one can transform the balance equations into a closed set of equations without having to consider the infinite set of the remaining moment equations. In this section we consider this question for a material gas without radiative interactions. The possibility of a fluid description of the gas requires all distribution functions to be approximately Maxwellian [4.2, 7]

$$F_a \simeq F_a^{\mathrm{M}} \quad \text{for all } a \ , \quad \text{where} \tag{4.3.3}$$

$$F_a^{\mathrm{M}}(\boldsymbol{r}, \boldsymbol{v}, t) = n_a \left(\frac{m_a}{2\pi k T}\right)^{3/2} \exp\left(-\frac{m_a |\boldsymbol{v} - \boldsymbol{u}|^2}{2kT}\right) \tag{4.3.4}$$

is a local Maxwell distribution, with $\boldsymbol{u} = \boldsymbol{u}(\boldsymbol{r}, t)$ being the local mean mass velocity and $T = T(\boldsymbol{r}, t)$ the local kinetic temperature.

To see how the fluid description of a gas by a closed set of hydrodynamic equations is achieved in situations near local thermodynamic equilibrium (LTE), let us consider the simplest case, namely, a one-component gas containing only one species of particles without radiative interactions [4.1]. Using $\boldsymbol{g} = \rho \boldsymbol{u}$, (4.1.67), and $\hat{\pi}_{\text{mat}} = \hat{\varepsilon}_{\text{mat}} = 0$, we rewrite (4.1.85, 86, 103) in the form

$$\frac{\partial \rho}{\partial t} + \nabla \cdot (\rho \boldsymbol{u}) = 0 \ , \tag{4.3.5}$$

$$\frac{\partial(\rho \boldsymbol{u})}{\partial t}+\nabla \cdot(\rho \boldsymbol{u} \boldsymbol{u}+\overleftrightarrow{\boldsymbol{P}})=\boldsymbol{k} \ , \tag{4.3.6}$$

$$\frac{\partial e}{\partial t}+\nabla \cdot(e \boldsymbol{u}+\boldsymbol{J})=-\overleftrightarrow{\boldsymbol{P}}: \nabla \boldsymbol{u}+\hat{q}_{\mathrm{J}} \ . \tag{4.3.7}$$

Although a true one-component gas has vanishing electrical conductivity, we keep the Joule heat $\hat{q}_{\mathrm{J}}$ in our formulas. Possible theoretical models are the magnetohydrodynamic approximation of a plasma, and the electron gas of a plasma with a fixed, neutralizing ion background.

Assuming the force $\boldsymbol{k}$ and the Joule heat $\hat{q}_{\mathrm{J}}$ to be known, these are only five equations for 14 unknowns (ρ, three u_i, six P_{ij}, e, three J_i). In a situation near LTE, closure of these equations is possible by the following procedure. In strict thermodynamic equilibrium, the gas is homogeneous (ρ, T, $\boldsymbol{u}=$ const), so that $\boldsymbol{J}=0$, and $\overleftrightarrow{\boldsymbol{P}}=p\overleftrightarrow{\boldsymbol{I}}$, where p is the hydrostatic pressure and $\overleftrightarrow{\boldsymbol{I}}$ the unit tensor. Allowing now for small inhomogeneities, one assumes that the correction terms to the heat current $\boldsymbol{J}$ and to the pressure tensor $\overleftrightarrow{\boldsymbol{P}}$ are linear in the local gradients. Since ∇T and $\nabla \rho$ are vectors, and $\nabla \boldsymbol{u}$ is a tensor, then

$$\boldsymbol{J}=-\zeta \nabla T-\mu \nabla \rho \quad \text{and} \tag{4.3.8}$$

$$\overleftrightarrow{\boldsymbol{P}}=p\overleftrightarrow{\boldsymbol{I}}+\overleftrightarrow{\boldsymbol{\Pi}} \ , \tag{4.3.9}$$

where the symmetric stress tensor

$$\overleftrightarrow{\boldsymbol{\Pi}}=-\eta \overset{\circ}{\overleftrightarrow{\boldsymbol{S}}}-\eta'(\nabla \cdot \boldsymbol{u})\overleftrightarrow{\boldsymbol{I}} \tag{4.3.10}$$

is linear in the velocity gradients. Here $\overset{\circ}{\overleftrightarrow{\boldsymbol{S}}}$ denotes the symmetric traceless tensor

$$\overset{\circ}{S}_{ij}=\frac{\partial u_j}{\partial x_i}+\frac{\partial u_i}{\partial x_j}-\frac{2}{3} \frac{\partial u_l}{\partial x_l} \delta_{ij} \ , \tag{4.3.11}$$

where $\partial u_l / \partial x_l \equiv \sum_l \partial u_l / \partial x_l=\nabla \cdot \boldsymbol{u}$. The phenomenological quantities ζ, μ, η, η' thus introduced are called *transport coefficients*. As shown presently, the second law, which requires the local entropy production to be positive, leads to

$$\zeta, \eta, \eta' \geqslant 0 \ , \quad \mu=0 \ . \tag{4.3.12}$$

Assuming the thermal conductivity ζ and the coefficients of shear and bulk viscosity η and η' to be known, there are now five equations for seven unknowns (ρ, three u_i, T, p, e). To eliminate two more of these, one must furthermore assume that the thermal and caloric equations of state valid in strict thermodynamic equilibrium also apply to the near-LTE state considered,

$$p=p(\rho, T) \ , \tag{4.3.13}$$

$$e=e(\rho, T) \ . \tag{4.3.14}$$

At this point a thermal Boltzmann distribution is required for the number densities $n_a = n_a(\boldsymbol{r}, t)$ that appear in the Maxwell distributions (4.3.4) if the excited particles significantly contribute to these equations of state.

To summarize: the fluid description of a one-component gas by means of the closed set of five hydrodynamic equations for the five macroscopic quantities ρ, u_x, u_y, u_z, T is possible only in situations near LTE. Furthermore, the three transport coefficients ζ, η, η' and the two equations of state (4.3.13, 14) must be known. [Two additional transport coefficients, the electrical conductivities $\sigma_\parallel$ and $\sigma_\perp$, (4.1.99, 102), enter the Joule heat $\hat{q}_J$ in (4.3.7) and so must be known, too.] These statements hold, mutatis mutandis, also for multicomponent gases.

To derive (4.3.12), we now turn to the entropy balance in a near-LTE state. Instead of densities per unit volume of internal energy and of entropy, e and s, we shall temporarily use the corresponding densities per unit mass, $e_{\rm m}$ and $s_{\rm m}$, defined by

$$e = \rho e_{\rm m} \ , \quad s = \rho s_{\rm m} \ . \tag{4.3.15}$$

It is readily shown that

$$\frac{\partial e}{\partial t} + \nabla \cdot (e\boldsymbol{u}) = \rho\left(\frac{\partial e_{\rm m}}{\partial t} + \boldsymbol{u} \cdot \nabla e_{\rm m}\right) \ , \tag{4.3.16}$$

$$\frac{\partial s}{\partial t} + \nabla \cdot (s\boldsymbol{u}) = \rho\left(\frac{\partial s_{\rm m}}{\partial t} + \boldsymbol{u} \cdot \nabla s_{\rm m}\right) \ . \tag{4.3.17}$$

In agreement with the supposed validity of the equations of state (4.3.13, 14) for near-LTE states, one likewise assumes for near-LTE states the validity of the first law in the form

$$de_{\rm m} = T\,ds_{\rm m} - p\,d\frac{1}{\rho} \ , \tag{4.3.18}$$

which, in turn, yields the Gibbs relation

$$ds_{\rm m} = \frac{1}{T}\,de_{\rm m} - \frac{p}{T\rho^2}\,d\rho \ , \tag{4.3.19}$$

and hence

$$\frac{\partial s_{\rm m}}{\partial t} = \frac{1}{T}\,\frac{\partial e_{\rm m}}{\partial T} - \frac{p}{T\rho^2}\,\frac{\partial \rho}{\partial t} \ , \tag{4.3.20a}$$

$$\boldsymbol{u} \cdot \nabla s_{\rm m} = \frac{1}{T}\boldsymbol{u} \cdot \nabla e_{\rm m} - \frac{p}{T\rho^2}\boldsymbol{u} \cdot \nabla \rho \ . \tag{4.3.20b}$$

Using (4.3.16, 17, 20), one can now write

$$\frac{\partial s}{\partial t}+\nabla\cdot(s\boldsymbol{u})=\rho\left(\frac{\partial s_{\mathrm{m}}}{\partial t}+\boldsymbol{u}\cdot\nabla s_{\mathrm{m}}\right)$$

$$=\frac{\rho}{T}\left(\frac{\partial e_{\mathrm{m}}}{\partial t}+\boldsymbol{u}\cdot\nabla e_{\mathrm{m}}\right)-\frac{p}{T\rho}\left(\frac{\partial \rho}{\partial t}+\boldsymbol{u}\cdot\nabla\rho\right)$$

$$=\frac{1}{T}\left(\frac{\partial e}{\partial t}+\nabla\cdot(e\boldsymbol{u})\right)+\frac{p}{T}\nabla\cdot\boldsymbol{u}$$

$$=-\frac{1}{T}\nabla\cdot\boldsymbol{J}-\frac{1}{T}\overleftrightarrow{\boldsymbol{P}}:\nabla\boldsymbol{u}+\frac{1}{T}\hat{q}_{\mathrm{J}}+\frac{p}{T}\nabla\cdot\boldsymbol{u}$$

$$=-\nabla\cdot\left(\frac{1}{T}\boldsymbol{J}\right)+\boldsymbol{J}\cdot\nabla\frac{1}{T}-\frac{1}{T}\overleftrightarrow{\boldsymbol{\Pi}}:\nabla\boldsymbol{u}+\frac{1}{T}\hat{q}_{\mathrm{J}}\,. \tag{4.3.21}$$

Here use has been made of the balance equations of mass and internal energy (4.3.5, 7), and in the last line the stress tensor $\overleftrightarrow{\boldsymbol{\Pi}}$ has been introduced according to (4.3.9). Equation (4.3.21) can now be written in the form of a balance equation of entropy,

$$\frac{\partial s}{\partial t}+\nabla\cdot(s\boldsymbol{u}+\boldsymbol{J}^{s})=\hat{\sigma}\,. \tag{4.3.22}$$

Here the entropy flux density is composed of the convective entropy flow $s\boldsymbol{u}$, and of the entropy conduction flow

$$\boldsymbol{J}^{s}=\frac{1}{T}\boldsymbol{J}\,. \tag{4.3.23}$$

On the other hand, the entropy production rate per unit time and unit volume is given by

$$\hat{\sigma}=-\frac{1}{T^{2}}\boldsymbol{J}\cdot\nabla T-\frac{1}{T}\overleftrightarrow{\boldsymbol{\Pi}}:\nabla\boldsymbol{u}+\frac{1}{T}\hat{q}_{\mathrm{J}}\,. \tag{4.3.24}$$

However, on account of (4.3.8)

$$\boldsymbol{J}\cdot\nabla T=-\zeta(\nabla T)^{2}-\mu\nabla\rho\cdot\nabla T\,, \tag{4.3.25}$$

and using (4.3.10, 11)

$$\overleftrightarrow{\boldsymbol{\Pi}}:\nabla\boldsymbol{u}=-\tfrac{1}{2}\eta\overset{\circ}{\overleftrightarrow{\boldsymbol{S}}}:\overset{\circ}{\overleftrightarrow{\boldsymbol{S}}}-\eta'(\nabla\cdot\boldsymbol{u})^{2} \tag{4.3.26}$$

since

$$(\partial_i u_j + \partial_j u_i - \tfrac{2}{3}\partial_l u_l \delta_{ij})\,\partial_j u_i$$
$$= \tfrac{1}{2}(\partial_i u_j + \partial_j u_i - \tfrac{2}{3}\partial_l u_l \delta_{ij})(\partial_j u_i + \partial_i u_j - \tfrac{2}{3}\partial_k u_k \delta_{ji}) \ ,$$

where $\partial_i \equiv \partial/\partial x_i$, and where summation over repeated indices is understood. Inserting (4.3.25, 26) into (4.3.24), the entropy production takes the form

$$\hat{\sigma} = \frac{\zeta}{T^2}(\nabla T)^2 + \frac{\mu}{T^2}\nabla p \cdot \nabla T + \frac{\eta}{2T}\overset{\circ}{\overleftrightarrow{S}} : \overset{\circ}{\overleftrightarrow{S}} + \frac{\eta'}{T}(\nabla \cdot \boldsymbol{u})^2 + \frac{1}{T}\hat{q}_{\mathrm{J}} \ . \qquad (4.3.27)$$

Since, according to the second law, the local entropy production is positive definite,

$$\hat{\sigma} \geqslant 0 \ , \qquad (4.3.28)$$

(4.3.27) leads to $\zeta, \eta, \eta' \geqslant 0$ and $\mu = 0$, see (4.3.12); likewise, $\sigma_\parallel, \sigma_\perp \geqslant 0$ in view of (4.1.99, 102). The electrical conductivities $\sigma_\parallel, \sigma_\perp$ should not be confused with the entropy production $\hat{\sigma}$.

In this section we have derived the closed set of hydrodynamic equations using phenomenological arguments. Systematic procedures for closing the chain of moment equations in near-LTE situations have been developed and are described in the literature. These questions, as well as the problem of calculating transport coefficients, are standard topics of the kinetic theory of gases and lie outside the scope of this book [4.1 – 7].

4.4 Fluid Description of a Gas with Radiative Interactions

A material gas allows a fluid description in terms of a closed set of hydrodynamic equations if it is in a state near thermal equilibrium. By analogy, one may expect that a fluid description of a gas in the presence of radiative interactions is likewise possible provided that both matter and radiation are in a state near thermal equilibrium. In this section, we shall show that this is indeed the case. Compared to the case without radiative interactions, the new feature is that in the hydrodynamic equations of the gas now appear the radiation pressure p_{R}, the radiative heat conductivity ζ_{R}, and the radiative shear and bulk viscosities η_{R} and η'_{R}.

The following physical situation will be considered:

(1) In the local rest frame Σ_{o} of the gas (that is, in the frame in which the mean mass velocity vanishes, $\boldsymbol{u}_{\mathrm{o}} = 0$), the source function is a Planck function of temperature T,

$$S_{\mathrm{o}}(\nu_{\mathrm{o}}) = B_{\nu_{\mathrm{o}}}(T) \ , \qquad (4.4.1)$$

where here and in the following a subscript o indicates that the corresponding quantity refers to the frame Σ_o.

(2) In the local rest frame Σ_o, the radiation field is approximately blackbody radiation of the same temperature T,

$$I_o(\nu_o) \simeq B_{\nu_o}(T) \ . \tag{4.4.2}$$

In view of the discussion in Sect. 3.1, (3.1.18 – 23), Eq. (4.4.2) ensures that the scattering part of the source function is a Planck function. With respect to emission-absorption processes, the source function (4.4.1) requires that in the local rest frame Σ_o, the material gas be in a state of local thermodynamic equilibrium LTE of temperature T.

In the following we refer to the state of matter and radiation described by (4.4.1,2) as LTE*. Matter and radiation in LTE* is found, for instance, in the interior of stars. For nonrelativistic gas velocities $\boldsymbol{u}$, to which we shall restrict ourselves in the following, it follows from (4.4.2) that the radiation field in the observer's frame Σ is likewise approximately blackbody radiation because of $I(\nu) = I_o(\nu_o) + O(u/c)$ so that in this case, LTE* corresponds to a situation where thermal matter moves in an almost thermal radiation field.

4.4.1 Preliminary Discussion

Before undertaking more detailed calculations, let us first give some estimates of the order of magnitude of radiative effects on a gas in LTE*. Since, by assumption, the radiation field is approximately Planckian, its energy density e_R is of the order of the energy density e_{BB} of blackbody radiation. The latter follows immediately from (4.2.1, 1.4.45), and hence

$$e_R \sim e_{BB} \equiv a T^4 \ , \quad \text{where} \tag{4.4.3}$$

$$a = \frac{8\pi^5 k^4}{15 h^3 c^3} \tag{4.4.4}$$

is the so-called *radiation constant.* Likewise, the radiation pressure p_R is of the order of the radiation pressure p_{BB} of blackbody radiation. Since, by symmetry, the pressure tensor of a blackbody radiation field is proportional to the unit tensor, $\overleftrightarrow{\boldsymbol{P}}_{BB} = p_{BB}\overleftrightarrow{\boldsymbol{I}}$, one obtains from (4.2.6, 4.3) and using $\mathrm{Tr}\{\overleftrightarrow{\boldsymbol{P}}_{BB}\} = 3\,p_{BB}$

$$p_R \sim p_{BB} \equiv \frac{a}{3} T^4 \ . \tag{4.4.5}$$

We now turn to the transport coefficients, first considering radiative heat conduction in a stationary gas at rest. The conditions for LTE* (4.4.1,2) can here be written

$$S_\nu = B_\nu(T) \ , \tag{4.4.6}$$

$$I_\nu = B_\nu(T) + I'_\nu \ , \tag{4.4.7}$$

where I'_ν is a small correction term ($I'_\nu \ll B_\nu$). With these relations, the equation of radiative transfer (3.1.16b) takes the form

$$\boldsymbol{n} \cdot \nabla I_\nu = \kappa_\nu (B_\nu - I_\nu) = -\kappa_\nu I'_\nu \ , \tag{4.4.8}$$

where κ_ν is the absorption coefficient of the gas. The left-hand side of this equation can now be approximated by

$$\boldsymbol{n} \cdot \nabla I_\nu \simeq \boldsymbol{n} \cdot \nabla B_\nu \equiv \boldsymbol{n} \cdot \frac{\partial B_\nu}{\partial T} \nabla T \ , \tag{4.4.9}$$

so that (4.4.8) yields

$$I'_\nu(\boldsymbol{n}) = -\frac{1}{\kappa_\nu} \frac{\partial B_\nu(T)}{\partial T} \boldsymbol{n} \cdot \nabla T \ . \tag{4.4.10}$$

The corresponding flow of radiant energy follows from (4.2.2),

$$\boldsymbol{f}_\mathrm{R} = \iint \boldsymbol{n} I_\nu(\boldsymbol{n})\, d\nu\, d\Omega = \iint \boldsymbol{n} I'_\nu(\boldsymbol{n})\, d\nu\, d\Omega \ , \tag{4.4.11}$$

as the isotropic Planck function B_ν does not contribute to the integral. Let the temperature gradient be in z direction, so that

$$\nabla T = \frac{dT}{dz} \boldsymbol{e}_z \ , \quad \boldsymbol{n} \cdot \nabla T = \frac{dT}{dz} \cos\vartheta \ ,$$

where $\boldsymbol{e}_z$ denotes the unit vector in z direction, and $\cos\vartheta \equiv \boldsymbol{n} \cdot \boldsymbol{e}_z$. The only nonvanishing component of the radiative energy flux is, of course, the z component which through (4.4.10, 11) is given by

$$\begin{aligned} f_{\mathrm{R}z} &= -\int d\nu \frac{1}{\kappa_\nu} \frac{\partial B_\nu}{\partial T} \int \cos^2\vartheta\, d\Omega \frac{dT}{dz} \\ &= -\frac{4\pi}{3} \int \frac{1}{\kappa_\nu} \frac{\partial B_\nu}{\partial T} d\nu \frac{dT}{dz} \ . \end{aligned} \tag{4.4.12}$$

Hence, the coefficient of radiative heat conductivity in LTE* is

$$\zeta_\mathrm{R} = \frac{4\pi}{3} \int \frac{1}{\kappa_\nu} \frac{\partial B_\nu}{\partial T} d\nu \ . \tag{4.4.13}$$

Radiative transport coefficients are usually expressed in terms of the so-called *Rosseland mean* of the absorption coefficient $\bar{\kappa}$, defined by

$$\int_0^\infty \frac{1}{\kappa_\nu} \frac{\partial B_\nu(T)}{\partial T} d\nu = \frac{1}{\bar{\kappa}} \int_0^\infty \frac{\partial B_\nu(T)}{\partial T} d\nu \ . \tag{4.4.14}$$

Now,

$$\int \frac{\partial B_\nu}{\partial T} d\nu = \frac{\partial}{\partial T} \int B_\nu d\nu = \frac{\partial}{\partial T}\left(\frac{c}{4\pi} a T^4\right) = \frac{ca}{\pi} T^3 , \qquad (4.4.15)$$

where a is the radiation constant (4.4.4), so that (4.4.14) becomes

$$\int \frac{1}{\kappa_\nu} \frac{\partial B_\nu}{\partial T} d\nu = \frac{1}{\bar{\kappa}} \frac{ca}{\pi} T^3 . \qquad (4.4.16)$$

The radiative heat conductivity in terms of the Rosseland mean $\bar{\kappa}$ is therefore

$$\zeta_R = \frac{4acT^3}{3\bar{\kappa}} . \qquad (4.4.17)$$

The preceding calculation shows that in LTE* it is the Rosseland mean $\bar{\kappa}$ that fixes the scale length of the photon gas. Indeed, $\bar{\kappa}^{-1}$ can be interpreted as the effective mean free path of photons in LTE*. To see this more explicitly, we use the following well-known estimate of the order of magnitude of transport coefficients in gases. Let A be a molecular quantity (such as the mean energy or the mean momentum of a particle) whose gradient $dA/dz \neq 0$ gives rise to a transport process. The corresponding flux density ϕ_A is then of the order of

$$\phi_A \simeq n\bar{v}\left[A\left(z-\frac{1}{2}l\right) - A\left(z+\frac{1}{2}l\right)\right] \simeq -n\bar{v}l\frac{dA}{dz} , \qquad (4.4.18)$$

where n is the number density of the particles, $\bar{v}$ their mean thermal speed, and l their mean free path.

Consider now a photon gas, and let A be the mean energy of a photon, so that $nA = \varepsilon$ is the energy density of the radiation field. The corresponding flux density ϕ_A is the radiative heat current which, through (4.4.18), is of the order of

$$f \simeq -\bar{v}l\frac{d\varepsilon}{dz} . \qquad (4.4.19)$$

Now, for photons in LTE*,

$$\varepsilon \sim aT^4 , \quad \bar{v} \sim c , \quad l \sim \frac{1}{\bar{\kappa}} , \qquad (4.4.20)$$

so that

$$f \simeq -\frac{4acT^3}{\bar{\kappa}} \frac{dT}{dz} . \qquad (4.4.21)$$

Hence, the radiative heat conductivity in LTE* is of the order of

$$\zeta_R \sim \frac{acT^3}{\bar{\kappa}} , \tag{4.4.22}$$

in agreement with (4.4.17). At the same time, this agreement corroborates the interpretation of $\bar{\kappa}^{-1}$ as the effective mean free path of photons in LTE*.

To estimate now the order of magnitude of radiative viscosity, we choose the quantity A in (4.4.18) to represent the mean photon momentum $A = -m^* u_x$. Here u_x is the macroscopic velocity of the material gas in the x direction, which has a gradient in the perpendicular z direction, $du_x/dz \neq 0$, and m^* is a fictitious photon mass to be defined presently. Thus, instead of considering the material gas flowing with macroscopic velocity u_x through a stationary thermal radiation field, we use the equivalent picture of the gas being at rest and the photon gas flowing with velocity $-u_x$. The fictitious photon mass m^* is simply defined such that $-m^* u_x$ is the mean photon momentum due to this macroscopic flow of the radiation field. The flux density ϕ_A in (4.4.18) is now the momentum flux density of the photons M, which, apart from the sign, is equal to the force density R that the photon gas exerts on the material gas, $R = -M$. This friction force R is now, on account of (4.4.18), of the order of

$$R \simeq -m^* n \bar{v} l \frac{du_x}{dz} , \tag{4.4.23}$$

leading to a viscosity coefficient

$$\eta \sim \rho^* \bar{v} l , \tag{4.4.24}$$

where $\rho^* = m^* n$ is the fictitious mass density of the photon gas. Now, for photons in LTE*,

$$\rho^* \sim \frac{aT^4}{c^2} , \quad \bar{v} \sim c , \quad l \sim \frac{1}{\bar{\kappa}} , \tag{4.4.25}$$

where in the first expression the relation $m = E/c^2$ between mass and energy has been used. Thus, the radiative viscosity in LTE* is of the order of

$$\eta_R \sim \frac{aT^4}{c\bar{\kappa}} . \tag{4.4.26}$$

More accurate calculations in Sect. 4.4.3 confirm the order-of-magnitude estimates of this section. The correct expressions for radiative transport coefficients in LTE* were first derived by *Thomas* [4.10].

4.4.2 Covariant Form of the Transfer Equation and the Energy-Momentum Tensor of the Radiation Field

The production rates of particle momentum and energy due to radiative interactions $\hat{\pi}_{\text{mat}}$ and $\hat{\varepsilon}_{\text{mat}}$ enter the balance equations of gas momentum and energy (4.1.86, 103). According to (4.3.1, 2, 2.8, 10) these production rates are known if one knows the density of radiant energy e_{R}, the flux density of radiant energy $\boldsymbol{f}_{\text{R}}$, the density of radiant momentum $\boldsymbol{g}_{\text{R}}$, and the flux density of radiant momentum $\overleftrightarrow{\boldsymbol{P}}_{\text{R}}$. For the special case of LTE* to which we restrict our discussion, we anticipate that these quantities enter the hydrodynamic equations of the gas only through the radiation pressure p_{R} and the coefficients of radiative heat conductivity and viscosity ζ_{R}, η_{R}, η'_{R}.

The hydrodynamic equations of the gas are formulated in the laboratory frame Σ. However, the radiation quantities e_{R}, $\boldsymbol{f}_{\text{R}}$, $\boldsymbol{g}_{\text{R}}$, $\overleftrightarrow{\boldsymbol{P}}_{\text{R}}$ are most easily obtained by calculating them in the local rest frame Σ_0 of the gas, and after that transforming them to the frame Σ [4.11]. As is always the case with radiation quantities, these transformations are best carried out using the relativistic transformation laws, even if, at the end, one is interested only in the nonrelativistic limit. (This procedure is also applied in Appendix A.1 to derive the nonrelativistic transformation laws for photon frequency, direction of propagation, and radiation intensity.)

This method is particularly convenient here since, as shown below, the four quantities in question form one single relativistic quantity, the energy-momentum tensor of the radiation field.

When writing relativistic expressions, the following conventions are used. Latin indices $i, j, \ldots$ run over the three space labels 1, 2, 3 and Greek indices $\alpha, \beta, \ldots$ over the four space-time labels 1, 2, 3, 0. The metric chosen is such that $a_i = a^i$ for $i = 1, 2, 3$ and $a_0 = -a^0$. Hence, the contravariant four-vector

$$a^\alpha \equiv (a^1, a^2, a^3, a^0) = (\boldsymbol{a}, a^0) \ , \tag{4.4.27a}$$

corresponds to the covariant four-vector

$$a_\alpha \equiv (a_1, a_2, a_3, a_0) = (\boldsymbol{a}, -a^0) \ , \tag{4.4.27b}$$

where $\boldsymbol{a} \equiv (a^1,\ a^2,\ a^3)$. Adopting the usual convention of summing over repeated indices, the scalar product of two four-vectors is written

$$a^\alpha b_\alpha = \boldsymbol{a} \cdot \boldsymbol{b} - a^0 b^0 \ , \quad \text{where} \tag{4.4.28}$$

$$\boldsymbol{a} \cdot \boldsymbol{b} = a^i b_i \equiv a^i b^i \equiv a_i b_i \tag{4.4.29}$$

is the ordinary scalar product of the vectors $\boldsymbol{a}$ and $\boldsymbol{b}$. In particular, the space-time coordinate four-vector is

$$x^\alpha = (\boldsymbol{r}, ct) \ , \tag{4.4.30}$$

and the energy-momentum four-vector of a photon $(\nu, \boldsymbol{n})$ is

$$p^\alpha = \frac{h\nu}{c}(\boldsymbol{n}, 1) \ ; \quad p^\alpha p_\alpha = 0 \ . \tag{4.4.31}$$

For later use, recall the definition of the four-velocity corresponding to the velocity $\boldsymbol{u}$,

$$U^\alpha = \frac{1}{(1-u^2/c^2)^{1/2}}(\boldsymbol{u}, c) \ ; \quad U^\alpha U_\alpha = -c^2 \ , \tag{4.4.32}$$

where $u^2 \equiv \boldsymbol{u} \cdot \boldsymbol{u}$.

Let us first write the equation of radiative transfer in covariant form. For this purpose, observe that the photon distribution function $\phi = \phi(p^\alpha, x^\alpha)$ is (approximately) a relativistic invariant, see (A.1.23). That ϕ is not an exact relativistic invariant is discussed in greater detail in Appendix A.2. However, assuming relativistic invariance of ϕ [and of E, K, ψ, (4.4.35) below] is usually an excellent approximation that will be adopted here. In particular, this approximation leads in no case to errors when calculating radiation quantities in the nonrelativistic limit as we do below.

On account of (A.1.22), the covariant transfer equation is simply obtained by multiplying the transfer equation (3.1.16) by the factor $c/h^3\nu^2$, yielding

$$p^\alpha \frac{\partial \phi}{\partial x^\alpha} = E - K\phi = K(\psi - \phi) \ . \tag{4.4.33}$$

Here

$$\phi = \frac{c^2}{h^4\nu^3} I_\nu \tag{4.4.34}$$

is the invariant photon distribution function, see (A.1.22), and

$$E = \frac{c}{h^3\nu^2}\varepsilon_\nu \ , \quad K = \frac{h\nu}{c}\kappa_\nu \ , \quad \psi = \frac{E}{K} = \frac{c^2}{h^4\nu^3} S_\nu \tag{4.4.35}$$

are, respectively, the invariant emission coefficient, the invariant absorption coefficient, and the invariant source function, which are in general functions of p^α and x^α.

Next, the energy-momentum four-tensor of the radiation field is defined by

$$P_{\mathrm{R}}^{\alpha\beta} = c \int \frac{p^\alpha p^\beta}{|\boldsymbol{p}|} \phi \, d^3p \ . \tag{4.4.36}$$

That this is indeed a four-tensor follows from the fact that ϕ is invariant, and that $d^3p/|\boldsymbol{p}|$ is invariant through (A.1.7), because the photon energy is $E = cp^0 = c|\boldsymbol{p}|$. Hence $P_{\mathrm{R}}^{\alpha\beta}$ transforms like $p^\alpha p^\beta$, that is, it is a four-tensor. It is symmetric,

$$P_{\mathrm{R}}^{\alpha\beta}=P_{\mathrm{R}}^{\beta\alpha}\ , \tag{4.4.37}$$

and traceless,

$$P_{\mathrm{R}\alpha}^{\alpha}=0 \tag{4.4.38}$$

owing to the relation $p^{\alpha}p_{\alpha}=0$, see (4.4.31). Using (4.4.31, 34) together with

$$d^3p=p^2dp\,d\Omega=\frac{h^3\nu^2}{c^3}d\nu\,d\Omega\ , \tag{4.4.39}$$

one sees readily that the components P_{R}^{ij} of the four-tensor (4.4.36) are identical with the components P_{R}^{ij} of the pressure tensor $\overleftrightarrow{\boldsymbol{P}}_{\mathrm{R}}$ as defined in (4.2.5), and that furthermore [see (4.2.1–3)]

$$P_{\mathrm{R}}^{i0}=P_{\mathrm{R}}^{0i}=cg_{\mathrm{R}}^{i}=\frac{1}{c}f_{\mathrm{R}}^{i}\ ,\quad P_{\mathrm{R}}^{00}=e_{\mathrm{R}}\ . \tag{4.4.40}$$

The energy-momentum four-tensor of the radiation field is thus built up of the four quantities $\overleftrightarrow{\boldsymbol{P}}_{\mathrm{R}}, \boldsymbol{g}_{\mathrm{R}}, \boldsymbol{f}_{\mathrm{R}}, e_{\mathrm{R}}$ in the following way,

$$P_{\mathrm{R}}^{\alpha\beta}=\left(\begin{array}{c|c}\overleftrightarrow{\boldsymbol{P}}_{\mathrm{R}} & c\boldsymbol{g}_{\mathrm{R}}\\ \hline \frac{1}{c}\boldsymbol{f}_{\mathrm{R}} & e_{\mathrm{R}}\end{array}\right)\ . \tag{4.4.41}$$

Note that (4.4.38) is identical with (4.2.6). Finally, the balance equations (4.2.8, 10) can be written in covariant form as

$$\frac{\partial P_{\mathrm{R}}^{\alpha\beta}}{\partial x^{\beta}}=\hat{\chi}_{\mathrm{rad}}^{\alpha},\quad \text{where} \tag{4.4.42}$$

$$\hat{\chi}_{\mathrm{rad}}^{\alpha}=\left(\hat{\boldsymbol{\pi}}_{\mathrm{rad}},\frac{1}{c}\hat{\varepsilon}_{\mathrm{rad}}\right) \tag{4.4.43}$$

is the four-vector of the production rate of radiant energy-momentum.

We now apply a Lorentz transformation to the four-tensor (4.4.36) to derive the transformation formulas of $\overleftrightarrow{\boldsymbol{P}}_{\mathrm{R}}, \boldsymbol{g}_{\mathrm{R}}, \boldsymbol{f}_{\mathrm{R}}, e_{\mathrm{R}}$. Consider the laboratory frame $\Sigma(x,y,z,t)$ and the local rest frame of the gas $\Sigma_{\mathrm{o}}(x_{\mathrm{o}},y_{\mathrm{o}},z_{\mathrm{o}},t_{\mathrm{o}})$ whose velocity measured in Σ is $\boldsymbol{u}$. Let us choose our coordinate systems temporarily such that the velocity $\boldsymbol{u}$ points in the x direction. Writing $x^1=x$, $x^2=y$, $x^3=z$, $x^0=ct$, and restricting ourselves from the outset to the lowest order in u, the Lorentz transformation (A.1.1) takes the form

$$x_{\mathrm{o}}^1=x^1-\frac{u}{c}x^0\ ,\quad x_{\mathrm{o}}^2=x^2\ ,\quad x_{\mathrm{o}}^3=x^3\ ,\quad x_{\mathrm{o}}^0=x^0-\frac{u}{c}x^1\ , \tag{4.4.44}$$

where $x_{\mathrm{o}}^{\alpha}\equiv(x_{\mathrm{o}})^{\alpha}$ is the coordinate four-vector in the frame Σ_{o}. Notice that (4.4.44) is a Lorentz transformation up to the order $\mathrm{O}(u)$, and not a Galileo transformation. As it is a four-tensor, $P_{\mathrm{R}}^{\alpha\beta}$ transforms by definition like $x^{\alpha}x^{\beta}$. For example,

$$x_o^1 x_o^1 = x^1 x^1 - \frac{2u}{c} x^1 x^0 \ , \quad x_o^1 x_o^0 = x^1 x^0 - \frac{u}{c}(x^1 x^1 + x^0 x^0)$$

yields

$$P_{\mathrm{Ro}}^{11} = P_{\mathrm{R}}^{11} - \frac{2u}{c^2} f_{\mathrm{R}}^1 \ , \quad f_{\mathrm{Ro}}^1 = f_{\mathrm{R}}^1 - u(P_{\mathrm{R}}^{11} + e_{\mathrm{R}})$$

using (4.4.40). As is easily checked, the final result can be written in the following form which is independent of the orientation of the coordinate systems with respect to the macroscopic gas velocity $\boldsymbol{u}$,

$$\overleftrightarrow{\boldsymbol{P}}_{\mathrm{Ro}} = \overleftrightarrow{\boldsymbol{P}}_{\mathrm{R}} - \frac{1}{c^2}(\boldsymbol{f}_{\mathrm{R}}\boldsymbol{u} + \boldsymbol{u}\,\boldsymbol{f}_{\mathrm{R}}) \ ,$$

$$\boldsymbol{f}_{\mathrm{Ro}} = \boldsymbol{f}_{\mathrm{R}} - \overleftrightarrow{\boldsymbol{P}}_{\mathrm{R}} \cdot \boldsymbol{u} - e_{\mathrm{R}}\boldsymbol{u} \ , \tag{4.4.45}$$

$$e_{\mathrm{Ro}} = e_{\mathrm{R}} - \frac{2}{c^2} \boldsymbol{f}_{\mathrm{R}} \cdot \boldsymbol{u} \ .$$

These are the relativistic transformation formulas in the lowest order in u that relate the pressure tensor $\overleftrightarrow{\boldsymbol{P}}_{\mathrm{Ro}}$, the energy flux $\boldsymbol{f}_{\mathrm{Ro}}$, and the energy density e_{Ro} in the local rest frame Σ_o to the corresponding quantities in the laboratory frame Σ. The inverse transformation formulas are obtained from (4.4.45) by substituting $\overleftrightarrow{\boldsymbol{P}}_{\mathrm{Ro}} \leftrightarrow \overleftrightarrow{\boldsymbol{P}}_{\mathrm{R}}$, $\boldsymbol{f}_{\mathrm{Ro}} \leftrightarrow \boldsymbol{f}_{\mathrm{R}}$, $e_{\mathrm{Ro}} \leftrightarrow e_{\mathrm{R}}$, $\boldsymbol{u} \to -\boldsymbol{u}$.

We now turn to the covariant formulation of the conditions for LTE*. By definition, in the local rest frame Σ_o, the source function is the Planck function, see (4.4.1),

$$S_o(\nu_o) = B_{\nu_o}(T) \ , \tag{4.4.46}$$

and hence, according to (3.2.6), the absorption coefficient is given by

$$\kappa_o(\nu_o) = a_o(\nu_o)(1 - \mathrm{e}^{-h\nu_o/kT}) \ , \tag{4.4.47}$$

if $a(\nu)$ denotes the absorption coefficient without stimulated emissions, see (3.1.8, 10b). Hence the invariant source function is

$$\psi = \phi_{\mathrm{Bo}} \ , \quad \text{where} \tag{4.4.48}$$

$$\phi_{\mathrm{Bo}} = \frac{c^2}{h^4 \nu_o^3} B_{\nu_o}(T) = \frac{2}{h^3}(\mathrm{e}^{h\nu_o/kT} - 1)^{-1} \tag{4.4.49}$$

is the photon distribution function corresponding to a blackbody radiation field in Σ_o, and the invariant absorption coefficient is

$$K = \frac{h\nu_o}{c} \kappa_o(\nu_o) = a_o(\nu_o) \frac{h\nu_o}{c}(1 - \mathrm{e}^{-h\nu_o/kT}) \ . \tag{4.4.50}$$

The two invariants ϕ_{Bo} and K must now be written in covariant form. To this end, observe that in Σ_{o} the four-velocity (4.4.32) is

$$U_{\mathrm{o}}^{\alpha}=(0,0,0,c) \tag{4.4.51}$$

since $\boldsymbol{u}_{\mathrm{o}}=0$ by definition of the local rest frame. On the other hand, from (4.4.31),

$$p_{\mathrm{o}}^{\alpha}=\frac{h\nu_{\mathrm{o}}}{c}(n_{\mathrm{o}}^{1},n_{\mathrm{o}}^{2},n_{\mathrm{o}}^{3},1)\ , \tag{4.4.52}$$

and thus

$$p_{\mathrm{o}}^{\alpha}U_{\mathrm{o}\alpha}=-h\nu_{\mathrm{o}}\ . \tag{4.4.53}$$

But the scalar product is an invariant, $p_{\mathrm{o}}^{\alpha}U_{\mathrm{o}\alpha}=p^{\alpha}U_{\alpha}$, and therefore

$$h\nu_{\mathrm{o}}=-p^{\alpha}U_{\alpha} \tag{4.4.54}$$

quite generally. Hence,

$$\phi_{\mathrm{Bo}}=\frac{2}{h^{3}}\left[\exp\left(-\frac{p^{\alpha}U_{\alpha}}{kT}\right)-1\right]^{-1}, \tag{4.4.55}$$

$$K=a_{\mathrm{o}}(\nu_{\mathrm{o}})\left(-\frac{1}{c}p^{\alpha}U_{\alpha}\right)\left[1-\exp\left(\frac{p^{\alpha}U_{\alpha}}{kT}\right)\right] \tag{4.4.56}$$

are the desired covariant expressions. Here the temperature T and the absorption coefficient $a_{\mathrm{o}}(\nu_{\mathrm{o}})$ are quantities determined once and for all by measuring them in the frame Σ_{o}, and are thus relativistic invariants. Equations (4.4.48, 55, 56) express the first condition for LTE* in covariant form. In particular, the corresponding covariant transfer equation is here

$$p^{\alpha}\frac{\partial\phi}{\partial x^{\alpha}}=K(\phi_{\mathrm{Bo}}-\phi)\ . \tag{4.4.57}$$

The second condition for LTE*, (4.4.2), can at once be written in covariant form as

$$\phi\simeq\phi_{\mathrm{Bo}}\ . \tag{4.4.58}$$

Inserting this in the left-hand side of the transfer equation (4.4.57) yields the approximate solution

$$\phi=\phi_{\mathrm{Bo}}-\frac{1}{K}p^{\alpha}\frac{\partial\phi_{\mathrm{Bo}}}{\partial x^{\alpha}}\ . \tag{4.4.59}$$

This is, in a first approximation, the radiation field corresponding to LTE*, ϕ_{Bo} and K being given by (4.4.55, 56). The desired quantities $\vec{\vec{P}}_R$, f_R, e_R can be obtained by directly substituting (4.4.59) into the energy-momentum tensor (4.4.36). However, as already mentioned, it is easier to calculate these quantities in the frame Σ_o, and then to transform to the frame Σ.

4.4.3 Radiation Pressure, Radiative Heat Conductivity, Radiative Viscosity

We now proceed to calculate the radiative energy-momentum tensor for LTE* in the local rest frame Σ_o of the gas. This gives explicit expressions for the hydrostatic radiation pressure and for the coefficients of radiative heat conductivity and radiative viscosity.

In the present section, we drop for simplicity all subscripts o referring to the frame Σ_o, and hence, simply,

$$x^\alpha \equiv x_o^\alpha \;, \quad p^\alpha \equiv p_o^\alpha \;, \quad P_R^{\alpha\beta} \equiv P_{Ro}^{\alpha\beta} \;, \quad \nu \equiv \nu_o \;, \quad \text{etc.}$$

According to (4.4.59), the invariant photon distribution function in LTE* is given by

$$\phi = \phi_B + \phi' \;, \tag{4.4.60}$$

where ϕ_B is the blackbody radiation field of temperature T, and

$$\phi' = -\frac{1}{K} p^\alpha \frac{\partial \phi_B}{\partial x^\alpha} \tag{4.4.61}$$

is a small correction term ($\phi' \ll \phi_B$). In view of (4.4.55), this correction term can be different from zero only when the space-time derivatives of the four-velocity U^α or of the temperature T are different from zero. However, in the frame Σ_o,

$$\frac{\partial U^i}{\partial x^\alpha} = \frac{\partial u^i}{\partial x^\alpha} \;, \quad \frac{\partial U^0}{\partial x^\alpha} = 0 \;, \tag{4.4.62}$$

where $\boldsymbol{u} = (u^1, u^2, u^3)$ is the macroscopic mass velocity of the gas (in Σ_o, $\boldsymbol{u} = 0$ by definition), as can be seen by differentiating (4.4.32) and using (4.4.51). Hence, in Σ_o, the correction term ϕ' is different from zero if

$$\frac{\partial u^i}{\partial x^\alpha} \neq 0 \;, \quad \frac{\partial T}{\partial x^\alpha} \neq 0 \;.$$

We now use $h\nu/c = p^0$, (4.4.31), to write the blackbody distribution function (4.4.49) and the invariant absorption coefficient (4.4.50) as functions of p^0,

$$\phi_B = \frac{2}{h^3} (e^{cp^0/kT} - 1)^{-1} \;, \tag{4.4.63}$$

$$K = p^0 \kappa(p^0) \ , \tag{4.4.64}$$

where $\kappa(p^0)$ is given by (4.4.47). On the other hand, to obtain the photon distribution function $\phi' = \phi'(p^\alpha)$ from (4.4.61), one must start from the general covariant blackbody distribution function (4.4.55) which is an implicit function of x^α via $U^\beta = U^\beta(x^\alpha)$ and $T = T(x^\alpha)$. Carrying out the differentiations and taking (4.4.51, 62, 64) into account, this leads to

$$\phi' = -\frac{2}{h^3} \frac{1}{p^0 \kappa(p^0)} \frac{\mathrm{e}^{cp^0/kT}}{(\mathrm{e}^{cp^0/kT}-1)^2} p^\gamma \left(\frac{p^l}{kT} \frac{\partial u^l}{\partial x^\gamma} + \frac{cp^0}{kT^2} \frac{\partial T}{\partial x^\gamma} \right) . \tag{4.4.65}$$

According to the decomposition of the photon distribution function (4.4.60), the radiative energy-momentum tensor (4.4.36) is likewise composed of two parts,

$$P_{\mathrm{R}}^{\alpha\beta} = P_{\mathrm{B}}^{\alpha\beta} + \Pi_{\mathrm{R}}^{\alpha\beta} \ . \tag{4.4.66}$$

Here the energy-momentum tensor of blackbody radiation

$$P_{\mathrm{B}}^{\alpha\beta} = c \int \frac{p^\alpha p^\beta}{|\boldsymbol{p}|} \phi_{\mathrm{B}} d^3p \tag{4.4.67}$$

is readily calculated. Inserting (4.4.63) yields at once the diagonal tensor

$$P_{\mathrm{B}}^{ij} = \frac{a}{3} T^4 \delta^{ij} \ , \quad P_{\mathrm{B}}^{i0} = 0 \ , \quad P_{\mathrm{B}}^{00} = aT^4 \ , \tag{4.4.68}$$

where a is again the radiation constant (4.4.4). [Compare also (4.4.3, 5).]

The second tensor in (4.4.66) is more interesting. Explicitly,

$$\Pi_{\mathrm{R}}^{\alpha\beta} = c \int \frac{p^\alpha p^\beta}{|\boldsymbol{p}|} \phi' d^3p \ . \tag{4.4.69}$$

To evaluate it, we write

$$p^\alpha = p n^\alpha \ , \quad p \equiv p^0 \equiv |\boldsymbol{p}| = \frac{h\nu}{c} \tag{4.4.70}$$

with

$$n^\alpha = (\boldsymbol{n}, 1) \ . \tag{4.4.71}$$

Notice that n^α is not a four-vector, but merely an abbreviation for $(n^1, n^2, n^3, 1)$. Using spherical coordinates p, ϑ, φ with the polar axis pointing in the 3 direction,

$$n^\alpha = (\sin\vartheta\cos\varphi,\ \sin\vartheta\sin\varphi,\ \cos\vartheta,\ 1) \ , \tag{4.4.72}$$

$$d^3p = p^2 dp\, d\Omega \ , \quad d\Omega = \sin\vartheta\, d\vartheta\, d\varphi \ , \tag{4.4.73}$$

and the tensor (4.4.69) takes the form

$$\Pi_{\mathrm{R}}^{\alpha\beta} = c \int_0^\infty p^3 dp \int \phi' n^\alpha n^\beta d\Omega$$

$$= -\frac{2c}{h^3 kT} \int_0^\infty \frac{p^4}{\kappa(p)} \frac{\mathrm{e}^{cp/kT}}{(\mathrm{e}^{cp/kT}-1)^2} dp$$

$$\cdot \left(\frac{\partial u^l}{\partial x^\gamma} \int n^\alpha n^\beta n^\gamma n^l d\Omega + \frac{c}{T} \frac{\partial T}{\partial x^\gamma} \int n^\alpha n^\beta n^\gamma d\Omega \right) , \tag{4.4.74}$$

which is hence the product of an integral over p alone, on the one hand, and the sum of two angular integrals, on the other.

We first consider the p integral, which we write in terms of the dimensionless variable $\xi = cp/kT$ as

$$\frac{2c}{h^3 kT} \int_0^\infty \frac{p^4}{\kappa(p)} \frac{\mathrm{e}^{cp/kT}}{(\mathrm{e}^{cp/kT}-1)^2} dp = \frac{2(kT)^4}{h^3 c^4} \int_0^\infty \frac{\xi^4}{\kappa(\xi)} \frac{\mathrm{e}^\xi}{(\mathrm{e}^\xi - 1)^2} d\xi . \tag{4.4.75}$$

Now observe that the derivative of the Planck function (1.4.45) with respect to T is given by

$$\frac{\partial B_\nu}{\partial T} = \frac{2h^2\nu^4}{c^2 kT^2} \frac{\mathrm{e}^{h\nu/kT}}{(\mathrm{e}^{h\nu/kT}-1)^2} , \tag{4.4.76}$$

so that, using $\xi = cp/kT = h\nu/kT$ again,

$$\int_0^\infty \frac{1}{\kappa_\nu} \frac{\partial B_\nu}{\partial T} d\nu = \frac{2(kT)^4}{h^3 c^2 T} \int_0^\infty \frac{\xi^4}{\kappa(\xi)} \frac{\mathrm{e}^\xi}{(\mathrm{e}^\xi - 1)^2} d\xi , \tag{4.4.77}$$

which is, up to a numerical factor, precisely our p integral. Since here, according to (4.4.16), the left-hand side is related to the Rosseland mean $\bar{\kappa}$ of the absorption coefficient by

$$\int_0^\infty \frac{1}{\kappa_\nu} \frac{\partial B_\nu}{\partial T} d\nu = \frac{acT^3}{\pi\bar{\kappa}} , \tag{4.4.78}$$

then finally

$$\frac{2c}{h^3 kT} \int_0^\infty \frac{p^4}{\kappa(p)} \frac{\mathrm{e}^{cp/kT}}{(\mathrm{e}^{cp/kT}-1)^2} dp = \frac{aT^4}{\pi c\bar{\kappa}} , \tag{4.4.79}$$

where a is the radiation constant (4.4.4), and $\bar{\kappa}$ is the Rosseland mean defined by (4.4.14). In view of the discussion in Sect. 4.4.1, the appearance of the Rosseland mean is no surprise.

We now turn to the angular integrals in (4.4.74),

$$M(\alpha,\beta,\gamma,l)=\int n^{\alpha}n^{\beta}n^{\gamma}n^{l}d\Omega \ , \tag{4.4.80a}$$

$$N(\alpha,\beta,\gamma)=\int n^{\alpha}n^{\beta}n^{\gamma}d\Omega \ . \tag{4.4.80b}$$

Obviously, these two quantities are symmetric in their arguments. Using spherical coordinates (4.4.72, 73), one readily shows that the only non-vanishing values are

$$M(0,0,i,i)=\frac{4\pi}{3} \ , \quad M(i,i,i,i)=\frac{4\pi}{5} \ ,$$
$$M(i,i,j,j)=\frac{4\pi}{15} \quad (i\neq j) \ , \tag{4.4.81a}$$

$$N(0,0,0)=4\pi \ , \quad N(i,i,0)=\frac{4\pi}{3} \ . \tag{4.4.81b}$$

Hence,

$$M(1,1,\gamma,l)\frac{\partial u^l}{\partial x^\gamma}=\frac{4\pi}{15}\left(\frac{\partial u^l}{\partial x^l}+2\frac{\partial u^1}{\partial x^1}\right) \ ,$$
$$M(1,2,\gamma,l)\frac{\partial u^l}{\partial x^\gamma}=\frac{4\pi}{15}\left(\frac{\partial u^2}{\partial x^1}+\frac{\partial u^1}{\partial x^2}\right) \ ,$$
$$M(1,0,\gamma,l)\frac{\partial u^l}{\partial x^\gamma}=\frac{4\pi}{3}\frac{\partial u^1}{\partial x^0} \ ,$$
$$M(0,0,\gamma,l)\frac{\partial u^l}{\partial x^\gamma}=\frac{4\pi}{3}\frac{\partial u^l}{\partial x^l} \ , \tag{4.4.82a}$$

and

$$N(1,1,\gamma)\frac{\partial T}{\partial x^\gamma}=\frac{4\pi}{3}\frac{\partial T}{\partial x^0} \ ,$$
$$N(1,2,\gamma)\frac{\partial T}{\partial x^\gamma}=0 \ ,$$
$$N(1,0,\gamma)\frac{\partial T}{\partial x^\gamma}=\frac{4\pi}{3}\frac{\partial T}{\partial x^1} \ ,$$
$$N(0,0,\gamma)\frac{\partial T}{\partial x^\gamma}=4\pi\frac{\partial T}{\partial x^0} \ . \tag{4.4.82b}$$

Putting everything together finally gives the tensor components:

$$\Pi_{\mathrm{R}}^{ij} = -\frac{4aT^4}{15c\bar{\kappa}}\left[\frac{\partial u^j}{\partial x^i}+\frac{\partial u^i}{\partial x^j}+\delta^{ij}\left(\frac{\partial u^l}{\partial x^l}+\frac{5c}{T}\frac{\partial T}{\partial x^0}\right)\right],$$

$$\Pi_{\mathrm{R}}^{i0} = -\frac{4aT^4}{3c\bar{\kappa}}\left(\frac{\partial u^i}{\partial x^0}+\frac{c}{T}\frac{\partial T}{\partial x^i}\right), \tag{4.4.83}$$

$$\Pi_{\mathrm{R}}^{00} = -\frac{4aT^4}{3c\bar{\kappa}}\left(\frac{\partial u^l}{\partial x^l}+\frac{3c}{T}\frac{\partial T}{\partial x^0}\right).$$

In a last step, using (4.4.40, 41), (4.4.66, 68, 83) yield the radiative pressure tensor $\overleftrightarrow{\boldsymbol{P}}_{\mathrm{R}}$, the radiative energy flux density $\boldsymbol{f}_{\mathrm{R}}$, and the radiative energy density e_{R} in the frame Σ_{o}, which we write in the following convenient form, setting at the same time $\partial/\partial x^0 = (1/c)\,\partial/\partial t$,

$$P_{\mathrm{R}}^{ij} = \frac{a}{3}T^4\delta^{ij} - \frac{4aT^4}{15c\bar{\kappa}}\left(\frac{\partial u^j}{\partial x^i}+\frac{\partial u^i}{\partial x^j}-\frac{2}{3}\frac{\partial u^l}{\partial x^l}\delta^{ij}\right)$$

$$-\frac{4aT^4}{9c\bar{\kappa}}\frac{\partial u^l}{\partial x^l}\delta^{ij} - \frac{4aT^3}{3c\bar{\kappa}}\frac{\partial T}{\partial t}\delta^{ij}, \tag{4.4.84}$$

$$f_{\mathrm{R}}^i = -\frac{4acT^3}{3\bar{\kappa}}\frac{\partial T}{\partial x^i} - \frac{4aT^4}{3c\bar{\kappa}}\frac{\partial u^i}{\partial t}, \tag{4.4.85}$$

$$e_{\mathrm{R}} = aT^4 - \frac{4aT^4}{3c\bar{\kappa}}\frac{\partial u^l}{\partial x^l} - \frac{4aT^3}{c\bar{\kappa}}\frac{\partial T}{\partial t}. \tag{4.4.86}$$

Hence

$$\mathrm{Tr}\{\overleftrightarrow{\boldsymbol{P}}_{\mathrm{R}}\} = e_{\mathrm{R}}, \tag{4.4.87}$$

as required by (4.2.6).

Let us now consider a stationary state ($\partial/\partial t = 0$). Here the radiative pressure tensor (4.4.84) and the radiative energy flux (4.4.85) in Σ_{o} are given by

$$\overleftrightarrow{\boldsymbol{P}}_{\mathrm{R}} = \frac{a}{3}T^4\overleftrightarrow{\boldsymbol{I}} - \frac{4aT^4}{15c\bar{\kappa}}\overset{\circ}{\overleftrightarrow{\boldsymbol{S}}} - \frac{4aT^4}{9c\bar{\kappa}}(\nabla\cdot\boldsymbol{u})\overleftrightarrow{\boldsymbol{I}}, \tag{4.4.88}$$

$$\boldsymbol{f}_{\mathrm{R}} = -\frac{4acT^3}{3\bar{\kappa}}\nabla T, \tag{4.4.89}$$

where $\overleftrightarrow{\boldsymbol{I}}$ denotes the unit tensor, and the traceless tensor $\overset{\circ}{\overleftrightarrow{\boldsymbol{S}}}$ is defined by (4.3.11). The pressure tensor (4.4.88) is clearly of the same form as the pressure tensor of a material gas, (4.3.9,10), and the radiative energy flux of the same form as the heat flux of a material gas, (4.3.8) with $\mu = 0$, see (4.3.12). Thus, comparison with those formulas shows that for LTE* and in the frame Σ_o, the hydrostatic radiation pressure is

$$p_R = \frac{a}{3} T^4 , \tag{4.4.90}$$

as expected from (4.4.5); the coefficient of radiative heat conductivity is

$$\zeta_R = \frac{4acT^3}{3\bar{\kappa}} , \tag{4.4.91}$$

as expected from (4.4.17); and the coefficients of radiative shear and bulk viscosity are

$$\eta_R = \frac{4aT^4}{15c\bar{\kappa}} , \quad \eta'_R = \frac{4aT^4}{9c\bar{\kappa}} \tag{4.4.92}$$

which are in accord with the order-of-magnitude estimate (4.4.26).

4.4.4 Hydrodynamic Equations with Radiative Terms

Having calculated the radiative quantities $\overleftrightarrow{P}_{Ro}, f_{Ro}, e_{Ro}$ in the local rest frame of the gas, we must now transform them to the laboratory frame in which the hydrodynamic equations are formulated. We consider here only nonrelativistic hydrodynamics, referring to the literature for the relativistic case [4.12–15].

First of all observe that the validity of (4.4.59), on which the calculations in Sect. 4.4.3 were based, requires the second term of the right-hand side to be small compared with the first term, that is,

$$\left| \frac{1}{K\phi_{Bo}} p^\alpha \frac{\partial \phi_{Bo}}{\partial x^\alpha} \right| \ll 1 . \tag{4.4.93}$$

Explicitly, from (4.4.55)

$$\frac{1}{\phi_{Bo}} \frac{\partial \phi_{Bo}}{\partial x^\alpha} = \left[1 - \exp\left(\frac{p^\gamma U_\gamma}{kT} \right) \right]^{-1} \left(\frac{p^\gamma}{kT} \frac{\partial U_\gamma}{\partial x^\alpha} - \frac{p^\gamma U_\gamma}{kT^2} \frac{\partial T}{\partial x^\alpha} \right) , \tag{4.4.94}$$

and K is given by (4.4.56). The orders of magnitude of the various quantities may be estimated as follows: $p^\alpha \sim h\nu/c$, $p^\alpha U_\alpha \sim h\nu$, $p^\alpha U_\alpha/kT \sim h\nu/kT \sim 1$, $K \sim (h\nu/c)\bar{\kappa}$, and hence

$$\frac{p^\alpha}{K} \frac{1}{\phi_{Bo}} \frac{\partial \phi_{Bo}}{\partial x^\alpha} \sim \frac{1}{\bar{\kappa}} \left(\frac{1}{c} \frac{\partial U_\gamma}{\partial x^\alpha} + \frac{1}{T} \frac{\partial T}{\partial x^\alpha} \right) ,$$

where $\bar{\kappa}$ is the Rosseland mean of the absorption coefficient. The validity of the inequality (4.4.93) therefore requires that simultaneously

$$\left|\frac{1}{\bar{\kappa}c}\frac{\partial U_\gamma}{\partial x^\alpha}\right| \ll 1 \ , \quad \left|\frac{1}{\bar{\kappa}T}\frac{\partial T}{\partial x^\alpha}\right| \ll 1 \ . \tag{4.4.95}$$

For nonrelativistic gas velocities $\boldsymbol{u} = \boldsymbol{u}(\boldsymbol{r}, t)$, the two coordinate systems $\Sigma(x, y, z, t)$ and $\Sigma_{\rm o}(x_{\rm o}, y_{\rm o}, z_{\rm o}, t_{\rm o})$ are connected through a Galileo transformation, so that $t_{\rm o} = t$, and furthermore

$$\frac{\partial u_{\rm o}^i}{\partial x_{\rm o}^k} = \frac{\partial u^i}{\partial x^k} \ , \quad \frac{\partial u_{\rm o}^i}{\partial t} = \frac{\partial u^i}{\partial t} \ , \quad \frac{\partial T}{\partial x_{\rm o}^k} = \frac{\partial T}{\partial x^k} \ . \tag{4.4.96}$$

On the other hand, in nonrelativistic hydrodynamics

$$\left|\frac{u^i}{c}\right| \ll 1 \ , \quad \left|\frac{1}{c}\frac{\partial M}{\partial t}\right| \ll \left|\frac{\partial M}{\partial x^k}\right| \ , \tag{4.4.97}$$

where the second inequality is valid for all macroscopic quantities M, such as u^i or T. Then the conditions (4.4.95) become

$$\left|\frac{1}{\bar{\kappa}c}\frac{\partial u^i}{\partial x^k}\right| \ll 1 \ , \quad \left|\frac{1}{\bar{\kappa}T}\frac{\partial T}{\partial x^k}\right| \ll 1 \ , \tag{4.4.98}$$

because the inequalities involving time derivatives are automatically fulfilled in view of the second inequality (4.4.97). Conditions (4.4.97, 98) may be written in the following compact form

$$u \sim \varepsilon c \ , \quad \partial_t \sim \varepsilon c \, \partial_x \ , \quad \partial_x u \sim \varepsilon c \bar{\kappa} \ , \quad \partial_x T \sim \varepsilon \bar{\kappa} T \ , \tag{4.4.99}$$

where ε is a small quantity ($\varepsilon \ll 1$). The nonrelativistic limit then corresponds to keeping terms up to the order $\mathrm{O}(\varepsilon)$ and neglecting terms of the order $\mathrm{O}(\varepsilon^2)$.

Taking now the nonrelativistic limit of (4.4.84 – 86), all terms containing time derivatives drop out. In view of (4.4.90 – 92) one may write the nonrelativistic radiative quantities in the form, see (4.4.88, 89),

$$\overleftrightarrow{\boldsymbol{P}}_{\rm Ro} = p_{\rm R}\overleftrightarrow{\boldsymbol{I}} - \eta_{\rm R}\overset{\circ}{\overleftrightarrow{\boldsymbol{S}}} - \eta_{\rm R}'(\nabla \cdot \boldsymbol{u})\overleftrightarrow{\boldsymbol{I}} \ , \tag{4.4.100a}$$

$$\boldsymbol{f}_{\rm Ro} = -\zeta_{\rm R}\nabla T \ , \tag{4.4.100b}$$

$$e_{\rm Ro} = 3(p_{\rm R} - \eta_{\rm R}'\nabla \cdot \boldsymbol{u}) \ , \tag{4.4.100c}$$

where $\overleftrightarrow{\boldsymbol{I}}$ is again the unit tensor, and the traceless tensor $\overset{\circ}{\overleftrightarrow{\boldsymbol{S}}}$ is defined by (4.3.11). Note that we used (4.4.96) by setting $\nabla_{\rm o} \cdot \boldsymbol{u}_{\rm o} = \nabla \cdot \boldsymbol{u}$ and $\nabla_{\rm o} T = \nabla T$, and that

$$e_{\rm Ro} = \mathrm{Tr}\{\overleftrightarrow{\boldsymbol{P}}_{\rm Ro}\} \tag{4.4.101}$$

is fulfilled, see (4.4.87).

To obtain the radiative quantities $\overleftrightarrow{P}_R, f_R, e_R$ in the laboratory frame from (4.4.100), we apply the inverse transformation formulas of (4.4.45), which in the nonrelativistic limit yield

$$\overleftrightarrow{P}_R = \overleftrightarrow{P}_{Ro} , \tag{4.4.102a}$$

$$f_R = f_{Ro} + p_R \overleftrightarrow{I} \cdot u + 3 p_R u = f_{Ro} + 4 p_R u , \tag{4.4.102b}$$

$$e_R = e_{Ro} , \tag{4.4.102c}$$

or explicitly,

$$\overleftrightarrow{P}_R = p_R \overleftrightarrow{I} - \eta_R \overset{\circ}{\overleftrightarrow{S}} - \eta'_R (\nabla \cdot u) \overleftrightarrow{I} , \tag{4.4.103a}$$

$$f_R = -\zeta_R \nabla T + 4 p_R u , \tag{4.4.103b}$$

$$e_R = 3(p_R - \eta'_R \nabla \cdot u) . \tag{4.4.103c}$$

Again,

$$e_R = \mathrm{Tr}\{\overleftrightarrow{P}_R\} . \tag{4.4.104}$$

In the last step one must calculate the production rates of particle momentum and energy due to radiative interactions $\hat{\pi}_{mat}$ and $\hat{\varepsilon}_{mat}$, which enter the momentum and energy balances of the material gas, (4.1.86,103). According to (4.3.1,2.10,2.4)

$$\hat{\pi}_{mat} = -\frac{1}{c^2} \frac{\partial f_R}{\partial t} - \nabla \cdot \overleftrightarrow{P}_R , \tag{4.4.105}$$

and according to (4.3.2,2.8),

$$\hat{\varepsilon}_{mat} = -\frac{\partial e_R}{\partial t} - \nabla \cdot f_R . \tag{4.4.106}$$

In the nonrelativistic limit, these quantities reduce to

$$\hat{\pi}_{mat} = -\nabla \cdot \overleftrightarrow{P}_R , \tag{4.4.107}$$

$$\hat{\varepsilon}_{mat} = -3 \frac{\partial p_R}{\partial t} - \nabla \cdot f_R . \tag{4.4.108}$$

Let us also calculate the quantity $\hat{\varepsilon}_{mat} - u \cdot \hat{\pi}_{mat}$ which enters the balance equation (4.1.103) for the internal energy of the gas. In the nonrelativistic limit, (4.4.107) yields

$$u \cdot \hat{\pi}_{mat} = -u \cdot \nabla p_R , \tag{4.4.109}$$

and hence

$$\hat{\varepsilon}_{mat} - u \cdot \hat{\pi}_{mat} = -3 \frac{\partial p_R}{\partial t} - \nabla \cdot f_R + u \cdot \nabla p_R . \tag{4.4.110}$$

Inserting now (4.4.107) into (4.1.86) leads to the momentum balance

$$\frac{\partial \boldsymbol{g}}{\partial t} + \nabla \cdot (\boldsymbol{g}\boldsymbol{u} + \overleftrightarrow{\boldsymbol{P}} + \overleftrightarrow{\boldsymbol{P}}_{\mathrm{R}}) = \boldsymbol{k} \ . \tag{4.4.111}$$

On the other hand, substituting (4.4.110) into (4.1.103) yields the energy balance

$$\frac{\partial e}{\partial t} + \nabla \cdot (e\boldsymbol{u} + \boldsymbol{J}) = -\overleftrightarrow{\boldsymbol{P}} : \nabla \boldsymbol{u} + \hat{q}_{\mathrm{J}} - 3\frac{\partial p_{\mathrm{R}}}{\partial t} - \nabla \cdot \boldsymbol{f}_{\mathrm{R}} + \boldsymbol{u} \cdot \nabla p_{\mathrm{R}} \ , \tag{4.4.112}$$

which may be rewritten as

$$\frac{\partial}{\partial t}(e + 3p_{\mathrm{R}}) + \nabla \cdot (e\boldsymbol{u} + \boldsymbol{J} + \boldsymbol{f}_{\mathrm{R}} - p_{\mathrm{R}}\boldsymbol{u}) = -\overleftrightarrow{\boldsymbol{P}} : \nabla \boldsymbol{u} - p_{\mathrm{R}} \nabla \cdot \boldsymbol{u} + \hat{q}_{\mathrm{J}} \ . \tag{4.4.113}$$

Equations (4.4.111, 113) are our final result. However, to display more clearly the symmetry of particles and photons in the momentum and energy balance equations, let us again consider the one-component gas of Sect. 4.3 for which, in particular, (4.3.8 – 12) hold. We then define the following quantities that refer to the total system, material gas plus radiation in LTE*:

total pressure tensor

$$\overleftrightarrow{\boldsymbol{P}}_{\mathrm{tot}} = p_{\mathrm{tot}}\overleftrightarrow{\boldsymbol{I}} - \eta_{\mathrm{tot}}\overset{\circ}{\overleftrightarrow{\boldsymbol{S}}} - \eta'_{\mathrm{tot}}(\nabla \cdot \boldsymbol{u})\overleftrightarrow{\boldsymbol{I}} \ , \tag{4.4.114}$$

total heat flux density

$$\boldsymbol{J}_{\mathrm{tot}} = -\zeta_{\mathrm{tot}} \nabla T \ , \tag{4.4.115}$$

total internal energy density

$$e_{\mathrm{tot}} = e + e_{\mathrm{R}} \ , \tag{4.4.116}$$

total hydrostatic pressure

$$p_{\mathrm{tot}} = p + p_{\mathrm{R}} \ , \tag{4.4.117}$$

total transport coefficients

$$\zeta_{\mathrm{tot}} = \zeta + \zeta_{\mathrm{R}} \ , \quad \eta_{\mathrm{tot}} = \eta + \eta_{\mathrm{R}} \ , \quad \eta'_{\mathrm{tot}} = \eta' + \eta'_{\mathrm{R}} \ . \tag{4.4.118}$$

Furthermore, in (4.4.113) one may set

$$3p_{\mathrm{R}} \simeq e_{\mathrm{R}} \ , \tag{4.4.119}$$

$$p_{\mathrm{R}} \nabla \cdot \boldsymbol{u} \simeq \overleftrightarrow{\boldsymbol{P}}_{\mathrm{R}} : \nabla \boldsymbol{u} \ , \tag{4.4.120}$$

because, as easily shown, the neglected terms are smaller than the dominant ones by a factor ε, (2.4.103c, a, 99).

Using (4.4.103, 114 – 120), the balance equations (4.4.111, 113) now take the following form, see (4.3.6, 7),

$$\frac{\partial(\rho \boldsymbol{u})}{\partial t} + \nabla \cdot (\rho \boldsymbol{u}\boldsymbol{u} + \overleftrightarrow{\boldsymbol{P}}_{\text{tot}}) = \boldsymbol{k} \ , \tag{4.4.121}$$

$$\frac{\partial e_{\text{tot}}}{\partial t} + \nabla \cdot (e_{\text{tot}}\boldsymbol{u} + \boldsymbol{J}_{\text{tot}}) = -\overleftrightarrow{\boldsymbol{P}}_{\text{tot}} : \nabla \boldsymbol{u} + \hat{q}_{\text{J}} \ . \tag{4.4.122}$$

Since the newly introduced radiative quantities p_{R}, ζ_{R}, η_{R}, η'_{R} are known functions of the temperature T and the Rosseland mean $\bar{\kappa} = \bar{\kappa}(T)$, (4.4.121, 122) together with the continuity equation (4.3.5) form a closed set of hydrodynamic equations for the total system consisting of gas and radiation in LTE*. However, it should be realized that their applicability is very limited since the assumption of LTE* is restrictive.

5. H Theorem for Gases and Radiation

This chapter deals with the entropy of gases and radiation. Boltzmann's definition of entropy, applicable to gases in equilibrium and nonequilibrium alike, is used in Sect. 5.1 to derive the entropy and the entropy production rate of material gases and radiation. In Sect. 5.2, the H theorem is proved for three distinct reactions, and it is shown that thermal equilibrium, detailed balance, and vanishing entropy production are equivalent to each other for gases and radiation.

5.1 Entropy and Entropy Production

Irreversibility in systems composed of material gases and radiation is the subject of this chapter. Apart from the basic physical interest of this question, the corresponding study of entropy production in such systems provides, in addition, a quite general method of demonstrating the principle of detailed balance. Again, it will be seen that the validity of the principle of detailed balance (1.2.1) depends crucially on the validity of the reciprocity relation (1.2.2).

Irreversible behavior of a macroscopic system is expressed by the statement that the entropy production of the system is positive definite, and that it vanishes only when the system is in thermodynamic equilibrium. This supposes that the notion of entropy, which in classical thermodynamics is defined only for thermal equilibrium, can be extended to nonequilibrium states. Historically, the first example of a physical quantity that could be interpreted as nonequilibrium entropy was Boltzmann's H function of a gas, see (2.2.12). (Boltzmann's letter H is a capital eta, standing for entropy.) For this reason, the law of positive definite entropy production is generally referred to as the H theorem.

5.1.1 General Definitions

To characterize a macroscopic equilibrium or nonequilibrium state of a gas composed of particles and photons, we divide as in Sects. 1.2, 3 the quantum states of the various particle types into groups of neighboring states having almost identical physical properties, so that each group contains a great number of single quantum states. A particular macrostate of the gas is then specified by the set of occupation numbers of all of these groups.

According to ***Boltzmann***, the *entropy* S of an arbitrary equilibrium or nonequilibrium macrostate of a gas is defined by

$$S = k \ln W \ , \tag{5.1.1}$$

where k is Boltzmann's constant, and W is the number of independent microstates compatible with the macrostate considered.

Let us consider one particular type of particles such as excited H(3d) atoms, free electrons, or photons, and let its quantum states be ordered in groups as described above. (Explicit examples are given in Sects. 5.1.2, 3.) Let G_i be the number of quantum states in the group i (its degeneracy), and let N_i be the number of particles belonging to this group (its occupation number). In addition to the assumption $G_i \gg 1$ already made above, we also assume that $N_i \gg 1$. Then the number of independent arrangements of N_i particles within a group of G_i quantum states is for Bose-Einstein statistics given by [5.1 – 3]

$$W_i^{\mathrm{BE}} = \frac{(N_i + G_i - 1)!}{N_i!\,(G_i - 1)!} \simeq \frac{(N_i + G_i)!}{N_i!\,G_i!} \ , \tag{5.1.2}$$

and for Fermi-Dirac statistics by

$$W_i^{\mathrm{FD}} = \frac{G_i!}{N_i!\,(G_i - N_i)!} \ . \tag{5.1.3}$$

Equations (5.1.2, 3) are, respectively, the number of possible arrangements of N_i identical particles and $G_i - 1$ identical walls, and of N_i identical particles and $G_i - N_i$ identical holes. For fermions, one must, of course, require that $G_i \geqslant N_i$, owing to the exclusion principle according to which a quantum state can be occupied by one fermion at most.

The total number of independent arrangements of particles compatible with the given macrostate is now

$$W = \prod_i W_i \ , \tag{5.1.4}$$

where the product runs over all groups of quantum states. With the help of Stirling's formula

$$\ln(A!) \simeq A \ln A - A \ , \tag{5.1.5}$$

which is valid for $A \gg 1$, from (5.1.1 – 4) the following expression for the entropy of the (equilibrium or nonequilibrium) macrostate of the considered particle type arises:

$$S = \alpha k \sum_i G_i \left[\left(1 + \alpha \frac{N_i}{G_i}\right) \ln \left(1 + \alpha \frac{N_i}{G_i}\right) - \alpha \frac{N_i}{G_i} \ln \frac{N_i}{G_i} \right] , \tag{5.1.6}$$

where, see (1.3.9),

$$\alpha = \begin{cases} +1 & \text{for bosons} \;, \\ -1 & \text{for fermions} \;. \end{cases} \tag{5.1.7}$$

The limit of a classical gas corresponds to $N_i/G_i \ll 1$, that is, to low occupation probabilities of single quantum states, so that quantum correlations (and hence also the distinction between bosons and fermions) can be ignored. To lowest order in N_i/G_i, (5.1.6) yields the entropy of a classical (Maxwell-Boltzmann) gas

$$S^{\text{cl}} = -k \sum_i N_i \left(\ln \frac{N_i}{G_i} - 1 \right) \;, \tag{5.1.8}$$

which, as expected, is independent of the parameter α and hence of the quantum statistical nature of the particles. Note that (5.1.8) can be derived from (5.1.1, 4, 5) if instead of (5.1.2 or 3) one uses

$$W_i = \frac{1}{N_i!} G_i^{N_i} \;. \tag{5.1.9}$$

This quantity is just the number of possibilities to distribute N_i distinguishable (i.e., classical) particles over G_i states, apart from the factor $1/N_i!$ which is an ad hoc correction for the fact that in reality particles of a given type are indistinguishable. This is precisely the procedure applied in classical statistical mechanics to avoid Gibbs' paradox, that is, to ensure that for given pressure and temperature, the entropy of a gas is proportional to the particle number. No such ad hoc prescription is necessary if one starts from quantum statistics.

Since particles of different types [e.g., $\mathrm{H}(1s)$, $\mathrm{H}(2s)$, $\mathrm{He}(1\,^1S_0)$, photons] are distinguishable, for the total system

$$W_{\text{tot}} = \prod_\gamma W_\gamma \;, \tag{5.1.10}$$

where the product runs over all particle types present, each W_γ being an expression of the form of (5.1.2, 3 or 9). In view of (5.1.1), it follows from (5.1.10) that the total entropy is the sum of the entropies of all particle types present,

$$S_{\text{tot}} = \sum_\gamma S_\gamma \;. \tag{5.1.11}$$

The *entropy production* $\delta S/\delta t$ is defined as the rate of change of entropy connected with the rate of change of occupation numbers $\delta N_i/\delta t$ due to *collisions* of particles and photons. (External and self-consistent fields do not contribute to entropy production, Sect. 5.1.2.) Thus, following (5.1.6),

$$\frac{\delta S}{\delta t} = k \sum_i \frac{\delta N_i}{\delta t} \ln \left(\frac{G_i}{N_i} + \alpha \right) \;. \tag{5.1.12}$$

In the classical limit $N_i/G_i \ll 1$, this reduces to

$$\frac{\delta S^{\text{cl}}}{\delta t} = -k \sum_i \frac{\delta N_i}{\delta t} \ln \frac{N_i}{G_i} , \tag{5.1.13}$$

which can also be obtained directly from (5.1.8).

The *H theorem* (or *second law*) can be stated in the form

$$\frac{\delta S_{\text{tot}}}{\delta t} = \sum_\gamma \frac{\delta S_\gamma}{\delta t} \geqslant 0 , \tag{5.1.14}$$

expressing that the total entropy production of all particle types, including photons, is positive definite, and vanishes only if the system is in complete thermodynamic equilibrium. The proof of this important theorem is given in Sect. 5.2.

Despite its plausibility, the definition of entropy production as the rate of change of entropy due to collisions needs to be justified by showing that it is equivalent to the usual definition of entropy production through the corresponding local balance equation, see (4.3.22),

$$\frac{\partial s}{\partial t} + \nabla \cdot \boldsymbol{h} = \hat{\sigma} , \tag{5.1.15}$$

where s and $\boldsymbol{h}$ are the entropy density and entropy flux density, respectively, so that $\hat{\sigma}$ is the entropy production per unit time and unit volume. For a system composed of particles and photons, each quantity in (5.1.15) is composed of two parts, one part due to the material gas, and the other due to the radiation field,

$$s_{\text{tot}} = s_{\text{mat}} + s_{\text{rad}} , \quad \boldsymbol{h}_{\text{tot}} = \boldsymbol{h}_{\text{mat}} + \boldsymbol{h}_{\text{rad}} , \quad \hat{\sigma}_{\text{tot}} = \hat{\sigma}_{\text{mat}} + \hat{\sigma}_{\text{rad}} . \tag{5.1.16}$$

These parts are considered separately in Sects. 5.1.2, 3, respectively. In terms of (5.1.15), the second law can be written in the form, see (4.3.28),

$$\hat{\sigma}_{\text{tot}} \geqslant 0 . \tag{5.1.17}$$

Sections 5.1.2, 3 show that the formulations (5.1.14, 17) of the second law are indeed equivalent. The reason why entropy is produced only by collisions (as opposed to interactions with external and self-consistent fields) is easily understood. On the one hand, the entropy density is a functional of the distribution functions [that is, s_{mat} can be expressed in terms of the various distribution functions $F_a(\boldsymbol{v})$, and s_{rad} in terms of the specific intensity $I_\nu(\boldsymbol{n})$], while on the other hand, according to Liouville's theorem, the distribution functions in the absence of collisions are constant along the particle or photon trajectories. Hence, entropy can be produced only by collisions.

5.1.2 Entropy of a Classical Gas

We first consider a gas of one particle type a, e.g., H(3d) atoms, in the classical limit, so that no difference between bosons and fermions arises.

A group of neighboring quantum states of the particles is defined by a small volume ΔV inside the gas, and by a velocity range $(\boldsymbol{v}, d^3v)$. In view of (1.4.7), the number of quantum states in this group is given by

$$G_i = g_a \frac{m_a^3}{h^3} d^3v \, \Delta V \,, \tag{5.1.18}$$

where m_a is the particle mass, and g_a the statistical weight of the particle type, e.g., for a free electron $g_e = 2$. On the other hand, the occupation number of this group is obviously

$$N_i = F_a(\boldsymbol{v}) \, d^3v \, \Delta V \,, \tag{5.1.19}$$

where $F_a(\boldsymbol{v}) \equiv F_a(\boldsymbol{r}, \boldsymbol{v}, t)$ is the distribution function.

Using (5.1.18, 19), (5.1.8) gives the classical entropy density $s = S/\Delta V$,

$$\begin{aligned} s_a &= -k \int F_a \left[\ln \left(\frac{h^3}{g_a m_a^3} F_a \right) - 1 \right] d^3v \\ &= -k \int F_a \ln F_a d^3v + k \left[1 - \ln \left(\frac{h^3}{g_a m_a^3} \right) \right] n_a \,, \end{aligned} \tag{5.1.20}$$

where n_a is the number density, (4.1.6). Since the first term in the second line is, up to the constant factor k, equal to the negative H function as defined by (C.28), it follows that for a given number density, Boltzmann's H function is essentially the negative entropy density of the gas.

With the aim of deriving an entropy balance equation, (5.1.15), we form the time derivative of (5.1.20),

$$\frac{\partial s_a}{\partial t} = -k \int (1 + \ln F_a) \frac{\partial F_a}{\partial t} d^3v + k \left[1 - \ln \left(\frac{h^3}{g_a m_a^3} \right) \right] \frac{\partial n_a}{\partial t} \,, \tag{5.1.21}$$

and substitute into its right-hand side the time derivatives that follow from the kinetic equation (4.1.1) and from the continuity equation (4.1.27),

$$\frac{\partial F_a}{\partial t} = -\boldsymbol{v} \cdot \frac{\partial F_a}{\partial \boldsymbol{r}} - \frac{\boldsymbol{K}_a}{m_a} \cdot \frac{\partial F_a}{\partial \boldsymbol{v}} + \left(\frac{\delta F_a}{\delta t} \right)_{\text{coll}} \,, \tag{5.1.22}$$

$$\frac{\partial n_a}{\partial t} = -\nabla \cdot (n_a \boldsymbol{u}_a) + \hat{\nu}_a \,, \tag{5.1.23}$$

where the mean velocity $\boldsymbol{u}_a$ and the particle production rate $\hat{\nu}_a$ are defined by (4.1.8, 18), respectively. Note that $\nabla \equiv \partial/\partial \boldsymbol{r}$. One checks readily that

$$\int (1+\ln F_a)\boldsymbol{v} \cdot \frac{\partial F_a}{\partial \boldsymbol{r}} d^3v = \nabla \cdot \int \boldsymbol{v} F_a \ln F_a d^3v \ . \tag{5.1.24}$$

On the other hand, partial integration shows that

$$\int (1+\ln F_a)\boldsymbol{K}_a \cdot \frac{\partial F_a}{\partial \boldsymbol{v}} d^3v = 0 \tag{5.1.25}$$

on account of

$$\frac{\partial}{\partial \boldsymbol{v}} \cdot \boldsymbol{K}_a = 0 \ , \tag{5.1.26}$$

which holds for the velocity-independent gravitational and electrostatic forces as well as for the velocity-dependent Lorentz force. Since the force $\boldsymbol{K}_a$ is in general composed of external and self-consistent forces [see (2.1.4, 8)], (5.1.25) implies that neither external nor self-consistent fields contribute to the entropy production.

With (5.1.24, 25) one verifies that (5.1.21) can be cast into the form of a balance equation,

$$\frac{\partial s_a}{\partial t} + \nabla \cdot \boldsymbol{h}_a \equiv \frac{\partial s_a}{\partial t} + \nabla \cdot (s_a \boldsymbol{u}_a + \boldsymbol{J}_a^{\mathrm{S}}) = \hat{\sigma}_a \ . \tag{5.1.27}$$

Here the entropy flux density $\boldsymbol{h}_a$ is composed of the convective flow $s_a \boldsymbol{u}_a$ and the conduction flow

$$\boldsymbol{J}_a^{\mathrm{S}} = -k \int (\boldsymbol{v} - \boldsymbol{u}_a) F_a \ln \left(\frac{h^3}{g_a m_a^3} F_a \right) d^3v \ , \tag{5.1.28}$$

see (4.3.22, 23). The dimensionless number $h^3 F_a / g_a m_a^3 \equiv N_i / G_i \ll 1$ is always used here as argument of the logarithm. The entropy production rate is given by

$$\hat{\sigma}_a = -k \int \left(\frac{\delta F_a}{\delta t} \right)_{\mathrm{coll}} \ln \left(\frac{h^3}{g_a m_a^3} F_a \right) d^3v \ . \tag{5.1.29}$$

Using now the first line of (5.1.20), one verifies at once that

$$\hat{\sigma}_a = \frac{\delta s_a}{\delta t} \ , \tag{5.1.30}$$

where $\delta/\delta t$ stands for the rate of change due to collisions. This shows explicitly that the two definitions of entropy production, (5.1.13, 15), are identical.

The generalization to several particle types is trivial. Evidently, for the total material gas

$$s_{\text{mat}} = \sum_a s_a \ , \quad \boldsymbol{h}_{\text{mat}} = \sum_a \boldsymbol{h}_a \ , \quad \hat{\sigma}_{\text{mat}} = \sum_a \hat{\sigma}_a \ , \tag{5.1.31}$$

where the sums run over all particle types present. (With the notation used in Chap. 4, these sums would be written as $\sum_A \sum_a$, where A denotes the particle species composed of particle types $a_1, a_2 \ldots.$) Obviously, these quantities obey the balance equation (5.1.15).

5.1.3 Entropy of Radiation

We now turn to radiation. Since photons are bosons, the general definition of entropy (5.1.6), with $\alpha = +1$, pertains. For photons, a group of neighboring quantum states is defined by a small volume ΔV inside the radiation field, by a frequency range $(\nu, d\nu)$, and by a range of propagation directions $(\boldsymbol{n}, d\Omega)$. According to (1.4.41), the number of quantum states in this group is

$$G_i = \frac{2\nu^2}{c^3} d\nu \, d\Omega \, \Delta V \ . \tag{5.1.32}$$

On the other hand, according to (1.4.43), the mean occupation number of a quantum state of this group is given by

$$\frac{N_i}{G_i} = \frac{c^2}{2h\nu^3} I_\nu(\boldsymbol{n}) \tag{5.1.33}$$

in terms of the specific intensity $I_\nu(\boldsymbol{n}) \equiv I_\nu(\boldsymbol{n}, \boldsymbol{r}, t)$.

Using (5.1.32, 33), (5.1.6) gives the entropy density of the radiation field $s_{\text{rad}} = S_{\text{rad}}/\Delta V$,

$$s_{\text{rad}} = \iint s_\nu(\boldsymbol{n}) \, d\nu \, d\Omega \quad \text{with} \tag{5.1.34}$$

$$s_\nu(\boldsymbol{n}) = k\frac{2\nu^2}{c^3}\left[\left(1+\frac{c^2}{2h\nu^3}I_\nu(\boldsymbol{n})\right)\ln\left(1+\frac{c^2}{2h\nu^3}I_\nu(\boldsymbol{n})\right)\right.$$
$$\left. -\frac{c^2}{2h\nu^3}I_\nu(\boldsymbol{n})\ln\left(\frac{c^2}{2h\nu^3}I_\nu(\boldsymbol{n})\right)\right] \ . \tag{5.1.35}$$

The local balance equation of radiant entropy is easily obtained from the equation of radiative transfer, see (3.1.16a),

$$\frac{1}{c}\frac{\partial I_\nu}{\partial t} + \boldsymbol{n} \cdot \nabla I_\nu = \varepsilon_\nu - \kappa_\nu I_\nu \ . \tag{5.1.36}$$

Indeed, differentiating (5.1.35) yields

$$\frac{\partial s_\nu}{\partial t} = \frac{1}{cT_\nu} \frac{\partial I_\nu}{\partial t} , \tag{5.1.37a}$$

$$\boldsymbol{n} \cdot \nabla s_\nu = \frac{1}{cT_\nu} \boldsymbol{n} \cdot \nabla I_\nu , \tag{5.1.37b}$$

where T_ν is defined by

$$\frac{1}{T_\nu(\boldsymbol{n})} = \frac{k}{h\nu} \ln\left(1 + \frac{2h\nu^3}{c^2 I_\nu(\boldsymbol{n})}\right) . \tag{5.1.38}$$

Hence, on account of (5.1.37), multiplying (5.1.36) by $1/T_\nu$ yields

$$\frac{\partial s_\nu}{\partial t} + \nabla \cdot \boldsymbol{h}_\nu = \hat{\sigma}_\nu , \tag{5.1.39}$$

where $s_\nu(\boldsymbol{n})$ is given by (5.1.35), and the two other quantities are defined by

$$\boldsymbol{h}_\nu(\boldsymbol{n}) = c\boldsymbol{n}\, s_\nu(\boldsymbol{n}) , \tag{5.1.40}$$

$$\hat{\sigma}_\nu(\boldsymbol{n}) = \frac{1}{T_\nu(\boldsymbol{n})} [\varepsilon_\nu(\boldsymbol{n}) - \kappa_\nu(\boldsymbol{n}) I_\nu(\boldsymbol{n})] . \tag{5.1.41}$$

The local balance equation of radiant entropy is now simply obtained by integrating (5.1.39) over all frequencies and directions:

$$\frac{\partial s_{\text{rad}}}{\partial t} + \nabla \cdot \boldsymbol{h}_{\text{rad}} = \hat{\sigma}_{\text{rad}} . \tag{5.1.42}$$

Here s_{rad} is the total entropy density of the radiation field (5.1.34),

$$\boldsymbol{h}_{\text{rad}} = \iint \boldsymbol{h}_\nu(\boldsymbol{n})\, d\nu\, d\Omega \tag{5.1.43}$$

is the total flux density of radiant entropy, and

$$\hat{\sigma}_{\text{rad}} = \iint \hat{\sigma}_\nu(\boldsymbol{n})\, d\nu\, d\Omega \tag{5.1.44}$$

is the production of radiant entropy per unit time and unit volume. Evidently, (5.1.39) is to be interpreted as the entropy balance equation of photons $(\nu, \boldsymbol{n})$.

For radiant entropy, the equivalence of the two definitions of entropy production (5.1.12, 15) is readily shown. Let $\delta/\delta t$ again denote the rate of change due to collisions (that is, due to emission, absorption, and scattering), then by definition, (5.1.36),

$$\frac{1}{c} \frac{\delta I_\nu}{\delta t} \equiv \varepsilon_\nu - \kappa_\nu I_\nu . \tag{5.1.45}$$

From (5.1.41, 45, 37a),

$$\hat{\sigma}_\nu = \frac{1}{T_\nu}(\varepsilon_\nu - \kappa_\nu I_\nu) = \frac{1}{cT_\nu}\frac{\delta I_\nu}{\delta t} = \frac{\delta s_\nu}{\delta t} , \tag{5.1.46}$$

which upon integration leads to

$$\hat{\sigma}_{\mathrm{rad}} = \frac{\delta s_{\mathrm{rad}}}{\delta t} . \tag{5.1.47}$$

Equations (5.1.46, 47) express the plausible fact that radiant entropy is produced by emission, absorption, and scattering, but not by free streaming of photons.

The following considerations by *Planck* [5.4–6] conclude this section. From (5.1.38), the specific intensity I_ν can be written formally as a Planck function,

$$I_\nu(\boldsymbol{n}) = B_\nu[T_\nu(\boldsymbol{n})] \equiv \frac{2h\nu^3}{c^2}\left[\exp\left(\frac{h\nu}{kT_\nu(\boldsymbol{n})}\right) - 1\right]^{-1} . \tag{5.1.48}$$

This suggests that the quantity T_ν may be interpreted as the temperature of a light beam of specific intensity I_ν. Thermal blackbody radiation then corresponds to a situation where all radiant temperatures are equal, $T_\nu(\boldsymbol{n}) = T$. This interpretation of T_ν as radiant temperature is also supported by the following reasoning. Consider a light beam $I_\nu(\boldsymbol{n})\,d\nu\,d\Omega$ in a given volume ΔV. The corresponding radiant entropy S_ν and radiant energy E_ν are given by, see (5.1.34) and (1.4.38),

$$S_\nu = s_\nu(\boldsymbol{n})\,d\nu\,d\Omega\,\Delta V \quad \text{and} \tag{5.1.49a}$$

$$E_\nu = \frac{1}{c} I_\nu(\boldsymbol{n})\,d\nu\,d\Omega\,\Delta V . \tag{5.1.49b}$$

Now, taking (5.1.35, 38) into account, one notices that these two quantities are related to T_ν by the well-known thermodynamic relation

$$\left(\frac{\partial S_\nu}{\partial E_\nu}\right)_{\Delta V} = \frac{\partial s_\nu}{\partial I_\nu} = \frac{1}{T_\nu} . \tag{5.1.50}$$

This shows that T_ν is in fact a temperature. Finally, in addition to the energy intensity I_ν, one can attribute to a light beam the entropy intensity

$$L_\nu(\boldsymbol{n}) = c s_\nu(\boldsymbol{n}) , \tag{5.1.51}$$

which obeys the entropy transfer equation [5.7]

$$\frac{1}{c}\frac{\partial L_\nu}{\partial t} + \boldsymbol{n} \cdot \nabla L_\nu = \frac{1}{T_\nu}(\varepsilon_\nu - \kappa_\nu I_\nu) \tag{5.1.52}$$

through (5.1.36, 37).

5.2 Proof of the H Theorem

Pauli originally proved the H theorem for gases and radiation [5.8]. Here we specifically investigate three particular types of reactions, namely, elastic collisions of identical particles, elastic collisions of unlike particles, and emission and absorption of photons, which display in sufficient detail the general principles underlying the validity of the H theorem for other reactions as well.

Pauli's approach to the H theorem is closely related to the principle of detailed balance (Chap. 1). Physical processes are viewed as transitions between stationary quantum states, and the reciprocity relation of the transition probabilities (1.2.2) is the key relation in Pauli's proof. In particular, it will be shown that deviation from detailed balance is a necessary and sufficient condition for positive entropy production, or, which amounts to the same, that detailed balance and thermal equilibrium are strictly equivalent. Thus, Pauli's proof of the H theorem constitutes at the same time a proof of the principle of detailed balance, which is perhaps more satisfactory than the considerations presented in Sects. 1.3,4.

First of all let us show that the validity of the H theorem in Pauli's approach can be traced back to the use of transition probabilities for describing the time evolution of a system. This fact emerges from the following simple, but instructive reasoning [5.9]. Consider the entropy of a classical one-component gas (5.1.8),

$$S^{\mathrm{cl}} = kN - k\sum_i G_i z_i \ln z_i = kN - k\sum_\alpha z_\alpha \ln z_\alpha \ , \qquad (5.2.1)$$

where $N=\sum_i N_i$ is the total number of particles, and $z_i = N_i/G_i$ is the mean occupation number of a quantum state. In (5.2.1), the sum over i denotes summation over groups of quantum states, whereas the sum over α denotes summation over single quantum states $|\alpha\rangle$. Of course, $z_\alpha = z_i$ if $|\alpha\rangle$ belongs to the group i. Now let $S\equiv S^{\mathrm{cl}}(t)$ and $S'\equiv S^{\mathrm{cl}}(t+dt)$ be the entropies at times t and $t+dt$, respectively. If the total number of particles N is conserved, then

$$\frac{1}{k}(S'-S) = \sum_\alpha z_\alpha \ln z_\alpha - \sum_\alpha z'_\alpha \ln z'_\alpha \ . \qquad (5.2.2)$$

Here the mean occupation numbers of quantum states at time $t+dt$, z'_α, are given in terms of the mean occupation numbers at time t, z_α, by

$$z'_\alpha = \sum_\beta w_{\alpha\beta} z_\beta \ , \qquad (5.2.3)$$

where $w_{\alpha\beta}$ is the probability that the transition $|\beta\rangle \to |\alpha\rangle$ occurs during the time dt. Clearly,

$$0 \leqslant w_{\alpha\beta} \leqslant 1 \ , \tag{5.2.4a}$$

$$\sum_{\alpha} w_{\alpha\beta} = \sum_{\beta} w_{\alpha\beta} = 1 \ . \tag{5.2.4b}$$

The positive definite character of the entropy production now follows at once because

$$\begin{aligned} \frac{1}{k}(S'-S) &= \sum_{\alpha} z_{\alpha} \ln z_{\alpha} - \sum_{\alpha,\beta} w_{\alpha\beta} z_{\beta} \ln z'_{\alpha} \\ &\geqslant \sum_{\alpha} z_{\alpha} \ln z_{\alpha} - \sum_{\alpha,\beta} w_{\alpha\beta}(z_{\beta} \ln z_{\beta} - z_{\beta} + z'_{\alpha}) = 0 \ , \end{aligned} \tag{5.2.5}$$

where use has been made of (5.2.4) and of the trivial inequality

$$z'_{\alpha} \int_{1}^{z_{\beta}/z'_{\alpha}} \ln x \, dx \geqslant 0 \tag{5.2.6}$$

which upon integration yields

$$z_{\beta} \ln z'_{\alpha} \leqslant z_{\beta} \ln z_{\beta} - z_{\beta} + z'_{\alpha} \ . \tag{5.2.7}$$

We may now ask why it is not possible to derive the inequality $S \geqslant S'$ in the same way, starting from

$$z_{\beta} = \sum_{\alpha} v_{\beta\alpha} z'_{\alpha}$$

obtained by inverting (5.2.3). Since $v_{\alpha\beta}$ is the inverse matrix to $w_{\alpha\beta}$,

$$\sum_{\gamma} v_{\beta\gamma} w_{\gamma\alpha} = \delta_{\beta\alpha} \ .$$

Thus, for $\beta \neq \alpha$,

$$\sum_{\gamma} v_{\beta\gamma} w_{\gamma\alpha} = 0 \ ,$$

so that at least one $v_{\beta\gamma}$ must be negative in view of (5.2.4). Similarly, for $\beta = \alpha$,

$$\sum_{\gamma} v_{\alpha\gamma} w_{\gamma\alpha} = 1 \ ,$$

so that at least one $v_{\alpha\gamma}$ must be greater than unity in view of (5.2.4). In other words, the quantities $v_{\alpha\beta}$ cannot be interpreted as transition probabilities, and (5.2.4), which ensure the validity of (5.2.5), do not apply to them.

We now proceed to the explicit proof of the H theorem. As already mentioned, we consider the following three types of reactions:

(1) Elastic binary collisions of identical particles,

$$A+A \rightleftarrows A+A ,$$

e.g., $H(1s)+H(1s) \rightleftarrows H(1s)+H(1s)$.

(2) Elastic binary collisions of unlike particles,

$$A+B \rightleftarrows A+B ,$$

e.g., $H(1s)+H(2s) \rightleftarrows H(1s)+H(2s)$, or Rayleigh scattering $H(1s)+h\nu \rightleftarrows H(1s)+h\nu$; the proof is easily extended to comprise inelastic collisions, e.g., $H(1s)+e \rightleftarrows H(2p)+e$.

(3) Emissions and absorptions of the type

$$A' \rightleftarrows A+B ,$$

e.g., $H(2p) \rightleftarrows H(1s)+h\nu$.

However, as the method of proving the H theorem is always the same, we do not consider more complex reactions like

$$A+B \rightleftarrows C+D ,$$

e.g., charge exchange $H_2+H^+ \rightleftarrows H_2^+ +H$, or photoionization and radiative recombination, $H+h\nu \rightleftarrows H^+ +e$; or

$$A+B \rightleftarrows A+B+C ,$$

e.g., free-free emission and absorption $H^+ +e \rightleftarrows H^+ +e+h\nu$; or

$$A+B \rightleftarrows C+D+E ,$$

e.g., collisional ionization and three-body recombination $H+e \rightleftarrows H^+ +e+e$, or collisional dissociation and recombination $H_2+H \rightleftarrows H+H+H$, and so on.

In the following we shall start from the general expression for the entropy production due to collisions, (5.1.12), valid for bosons and fermions alike, and hence for the limiting case of classical particles too, which we write in the form

$$\frac{\delta S}{\delta t} = k \sum_i \frac{\delta N_i}{\delta t} \ln \omega_i , \tag{5.2.8}$$

where

$$\omega_i = \frac{G_i}{N_i} + \alpha . \tag{5.2.9}$$

Notice that ω_i is a nonnegative quantity, $\omega_i \geqslant 0$. This is evident for bosons, for which even $\omega_i \geqslant 1$ holds because $\alpha = +1$. However, it is also true for fermions

($\alpha = -1$) since $N_i \leqslant G_i$ owing to the Pauli principle. Moreover, since $\omega_i = 0$, corresponding to a fully occupied "Fermi sea" with $N_i = G_i$, can be excluded from our consideration which is restricted to dilute gases, then always

$$\omega_i > 0 \ . \tag{5.2.10}$$

The reaction rates, which determine the rate of change of the occupation numbers $\delta N_i/\delta t$ appearing in the entropy production (5.2.8), take a particularly convenient form in terms of the quantities ω_i, (5.2.9). For example, taking (1.3.2, 3) into account, one verifies easily that the rate of the reaction

$$A + a \rightarrow B + b$$

can be written as

$$[A\,a \rightarrow B\,b] = w(A\,a \rightarrow B\,b) N_A N_a N_B N_b \omega_B \omega_b \ , \tag{5.2.11}$$

where $w(A\,a \rightarrow B\,b)$ is the corresponding transition probability per unit time. (Recall that as in Sect. 1.3 the symbols A, a, B, b here denote not only the type of the particle, but also its translational state.) The generalization to other processes is straightforward: the reaction rate is the product of the transition probability w, of all particle numbers N involved in the reaction, and of the quantities ω of all final particles.

5.2.1 Elastic Collisions (Identical Particles)

We first discuss the H theorem for elastic binary collisions of identical particles

$$A + A \rightleftarrows A + A \ .$$

According to (5.2.8), the corresponding entropy production rate is

$$\frac{1}{k} \frac{\delta S}{\delta t} = \sum_{\kappa} \frac{\delta N_\kappa}{\delta t} \ln \omega_\kappa \ , \tag{5.2.12}$$

where the letter κ refers to a group of neighboring translational quantum states of the A particles. The rate of change of the occupation numbers due to the considered elastic collisions is evidently given by

$$\frac{\delta N_\kappa}{\delta t} = -\sum_{\lambda} \sum_{(\mu\nu)} [\kappa\lambda \rightarrow \mu\nu] + \sum_{\lambda} \sum_{(\mu\nu)} [\mu\nu \rightarrow \kappa\lambda] \tag{5.2.13}$$

in terms of reaction rates $[i \rightarrow f]$, and where $\sum_{(\mu\nu)}$ denotes the sum over all pairs $(\mu\nu)$ irrespective of the order of μ and ν, that is, $(\mu\nu)$ and $(\nu\mu)$ are considered identical, and are counted only once. Using (5.2.11) for the reaction rate, we write (5.2.13) in terms of transition probabilities $w(i \rightarrow f)$ as

$$\frac{\delta N_\kappa}{\delta t} = -\sum_\lambda \sum_{(\mu\nu)} w(\kappa\lambda\to\mu\nu)N^4\omega_\mu\omega_\nu$$

$$+\sum_\lambda \sum_{(\mu\nu)} w(\mu\nu\to\kappa\lambda)N^4\omega_\kappa\omega_\lambda\,, \tag{5.2.14}$$

where for brevity

$$N^4 \equiv N_\kappa N_\lambda N_\mu N_\nu\,. \tag{5.2.15}$$

It is readily checked that the number of A particles is conserved. Indeed,

$$\frac{\delta}{\delta t}\sum_\kappa N_\kappa = \sum_\kappa \frac{\delta N_\kappa}{\delta t} = -\sum_\kappa\sum_\lambda\sum_{(\mu\nu)} w(\kappa\lambda\to\mu\nu)N^4\omega_\mu\omega_\nu$$

$$+\sum_\kappa\sum_\lambda\sum_{(\mu\nu)} w(\mu\nu\to\kappa\lambda)N^4\omega_\kappa\omega_\lambda$$

$$= -2\sum_{(\kappa\lambda)}\sum_{(\mu\nu)} w(\kappa\lambda\to\mu\nu)N^4\omega_\mu\omega_\nu$$

$$+2\sum_{(\kappa\lambda)}\sum_{(\mu\nu)} w(\mu\nu\to\kappa\lambda)N^4\omega_\kappa\omega_\lambda = 0 \tag{5.2.16}$$

since

$$\sum_\kappa\sum_\lambda = 2\sum_{(\kappa\lambda)} \quad \text{and} \tag{5.2.17}$$

$$w(\kappa\lambda\to\mu\nu) = w(\lambda\kappa\to\mu\nu)\,,\quad w(\mu\nu\to\kappa\lambda) = w(\mu\nu\to\lambda\kappa) \tag{5.2.18}$$

because of the identity of the A particles.

Substituting (5.2.14) into (5.2.12) and using the symmetry properties (5.2.17, 18) leads to the following expression for the entropy production rate:

$$\frac{1}{k}\frac{\delta S}{\delta t} = -\sum_\kappa\sum_\lambda\sum_{(\mu\nu)} w(\kappa\lambda\to\mu\nu)N^4\omega_\mu\omega_\nu \ln\omega_\kappa$$

$$+\sum_\kappa\sum_\lambda\sum_{(\mu\nu)} w(\mu\nu\to\kappa\lambda)N^4\omega_\kappa\omega_\lambda \ln\omega_\kappa$$

$$= -\sum_{(\kappa\lambda)}\sum_{(\mu\nu)} w(\kappa\lambda\to\mu\nu)N^4\omega_\mu\omega_\nu \ln(\omega_\kappa\omega_\lambda)$$

$$+\sum_{(\kappa\lambda)}\sum_{(\mu\nu)} w(\mu\nu\to\kappa\lambda)N^4\omega_\kappa\omega_\lambda \ln(\omega_\kappa\omega_\lambda)$$

$$= -\frac{1}{2}\sum_{(\kappa\lambda)}\sum_{(\mu\nu)} w(\kappa\lambda\to\mu\nu)N^4\omega_\mu\omega_\nu \ln(\omega_\kappa\omega_\lambda)$$

$$-\frac{1}{2}\sum_{(\kappa\lambda)}\sum_{(\mu\nu)} w(\mu\nu\to\kappa\lambda)N^4\omega_\kappa\omega_\lambda \ln(\omega_\mu\omega_\nu)$$

$$+\frac{1}{2}\sum_{(\kappa\lambda)}\sum_{(\mu\nu)} w(\mu\nu\rightarrow\kappa\lambda)N^4\omega_\kappa\omega_\lambda\ln(\omega_\kappa\omega_\lambda)$$

$$+\frac{1}{2}\sum_{(\kappa\lambda)}\sum_{(\mu\nu)} w(\kappa\lambda\rightarrow\mu\nu)N^4\omega_\mu\omega_\nu\ln(\omega_\mu\omega_\nu) \ . \tag{5.2.19}$$

For example, the first term has been transformed here as follows,

$$\sum_\kappa\sum_\lambda(\dots)=2\sum_{(\kappa\lambda)}(\dots)=\sum_{(\kappa\lambda)}(\dots)+\sum_{(\kappa\lambda)}(\kappa\leftrightarrow\lambda) \ ,$$

$$\sum_{(\kappa\lambda)}\sum_{(\mu\nu)}(\dots)=\frac{1}{2}\sum_{(\kappa\lambda)}\sum_{(\mu\nu)}(\dots)+\frac{1}{2}\sum_{(\kappa\lambda)}\sum_{(\mu\nu)}(\kappa\lambda\leftrightarrow\mu\nu) \ ,$$

using an obvious notation.

At this point we employ the reciprocity relation for the transition probabilities (1.2.2), which here reads

$$w(\kappa\lambda\rightarrow\mu\nu)=w(\mu\nu\rightarrow\kappa\lambda) \ . \tag{5.2.20}$$

With the aid of this relation, and taking (5.2.15) into account, we obtain from (5.2.19)

$$\frac{1}{k}\frac{\delta S}{\delta t}=\frac{1}{2}\sum_{(\kappa\lambda)}\sum_{(\mu\nu)} w(\kappa\lambda\rightarrow\mu\nu)N_\kappa N_\lambda N_\mu N_\nu(\omega_\kappa\omega_\lambda-\omega_\mu\omega_\nu)\ln\left(\frac{\omega_\kappa\omega_\lambda}{\omega_\mu\omega_\nu}\right), \tag{5.2.21}$$

which is our final result.

Inspecting (5.2.21) shows that the H theorem is indeed valid. First of all, the entropy production is nonnegative,

$$\frac{\delta S}{\delta t}\geqslant 0 \ , \tag{5.2.22}$$

for in (5.2.21), apart from nonnegative factors w and N, each term of the sum is of the form

$$(x-y)\ln(x/y)\geqslant 0 \tag{5.2.23}$$

where $x, y>0$ [as $\omega>0$, (5.2.10)], which is a nonnegative quantity. Moreover, because the equality sign in (5.2.23) obtains if and only if $x=y$, it follows that the necessary and sufficient condition for a vanishing entropy production $\delta S/\delta t=0$ is the validity of

$$\omega_\kappa\omega_\lambda=\omega_\mu\omega_\nu \tag{5.2.24}$$

for all pairs $(\kappa\lambda)$ and $(\mu\nu)$ of groups of translational quantum states connected by a nonvanishing transition probability $w(\kappa\lambda\rightarrow\mu\nu)>0$.

Starting from (5.2.24), let us now show that for dilute gases the notions of thermal equilibrium, detailed balance, and vanishing entropy production are equivalent.

To show the equivalence of vanishing entropy production and thermodynamic equilibrium, which from a physical point of view is almost trivial, we consider for simplicity an isotropic gas of A particles. It is physically plausible that an anisotropic distribution function $F(\boldsymbol{v}) \equiv F(v, \boldsymbol{n})$ has a smaller entropy density and a higher entropy production than the corresponding isotropic distribution function $F(v) \equiv \int F(v, \boldsymbol{n})\, d\Omega/4\pi$. An explicit proof of this fact can be found in [5.10] for the photon gas (radiation field).

The quantity ω_κ then depends only on the kinetic energy E_κ of the particles, $\omega_\kappa = \omega(E_\kappa)$. On the other hand, comparing (5.2.9) with (1.3.8, 10) shows that ω is related to the quantity χ used in Sect. 1.3 by

$$\omega(E) = \frac{1}{\chi(E)} , \tag{5.2.25}$$

and hence $\chi(E) > 0$ on account of (5.2.10). The condition for vanishing entropy production (5.2.24) may therefore be written alternatively as

$$\chi(E_\kappa)\,\chi(E_\lambda) = \chi(E_\mu)\,\chi(E_\nu) , \tag{5.2.26}$$

where energy conservation requires

$$E_\kappa + E_\lambda = E_\mu + E_\nu . \tag{5.2.27}$$

Writing

$$E_\kappa = E , \quad E_\lambda = E' , \quad E_\mu = E + W , \quad E_\nu = E' - W ,$$

clearly (5.2.26) can be written in the form

$$\frac{\chi(E)}{\chi(E+W)} = \frac{\chi(E'-W)}{\chi(E')} = r(W) , \tag{5.2.28}$$

where the function $r(W)$ is independent of E and E'. Equation (5.2.28) is identical with (1.3.20) whose solution is given by, see (1.3.15),

$$\chi(E) = \zeta \mathrm{e}^{-\beta E} , \tag{5.2.29}$$

which in turn corresponds to the thermal distribution function, see (1.3.17),

$$z(E) = \frac{1}{\zeta^{-1} \mathrm{e}^{\beta E} - \alpha} . \tag{5.2.30}$$

Here $z(E)$ denotes the mean occupation number of a single quantum state of energy E. On the other hand, if the gas is in thermodynamic equilibrium so that (5.2.29, 30) hold, the validity of (5.2.24) and hence of vanishing entropy production follows immediately. This completes the proof that vanishing entropy production is a necessary and sufficient condition for thermal equilibrium.

The equivalence of vanishing entropy production and detailed balance, on the other hand, is easily established. To this end, one simply writes down explicitly the reaction rates, (5.2.11),

$$[\kappa\lambda \to \mu\nu] = w(\kappa\lambda \to \mu\nu) N_\kappa N_\lambda N_\mu N_\nu \omega_\mu \omega_\nu \ , \tag{5.2.31a}$$

$$[\mu\nu \to \kappa\lambda] = w(\kappa\lambda \to \mu\nu) N_\mu N_\nu N_\kappa N_\lambda \omega_\kappa \omega_\lambda \ , \tag{5.2.31b}$$

where in the second equation the reciprocity relation (5.2.20) has been used. Thus, the condition for zero entropy production, (5.2.24), is seen to be equivalent to the equality of the reaction rates

$$[\kappa\lambda \to \mu\nu] = [\mu\nu \to \kappa\lambda] \tag{5.2.32}$$

for all pairs $(\kappa\lambda)$ and $(\mu\nu)$, and therefore to detailed balance as defined in (1.2.1).

The preceding discussion may be summarized in the following diagram.

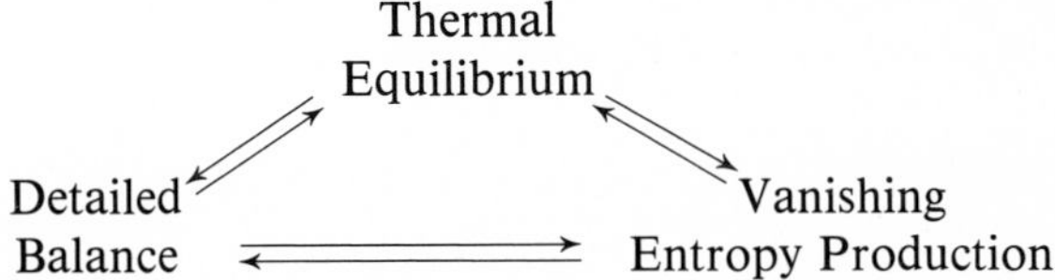

We emphasize again that the principle of detailed balance holds in a dilute gas only up to an accuracy determined by the accuracy of the reciprocity relation $w(i \to f) = w(f \to i)$, (1.2.2). Indeed, we have seen that this latter relation provides the crucial steps in the above demonstrations, (5.2.21, 31).

Let us conclude this section by showing that Boltzmann's H theorem as discussed in Appendix C, (C.27), is contained in the above consideration. Indeed, the time variation of the H function due to elastic collisions according to (C.29) is given by

$$\frac{\delta H}{\delta t} = \int \ln F \frac{\delta F}{\delta t} d^3v \ , \tag{5.2.33}$$

where $F \equiv F(\boldsymbol{r}, \boldsymbol{v}, t)$ is the distribution function of the A particles. On the other hand, the local production rate of entropy is via (5.1.29)

$$\hat{\sigma} = -k \int \frac{\delta F}{\delta t} \ln F \, d^3v \ , \tag{5.2.34}$$

where we have used the fact that

$$\int \frac{\delta F}{\delta t} d^3v = 0 \ , \tag{5.2.35}$$

because elastic collisions conserve the number of A particles. Thus,

$$\frac{\delta H}{\delta t} = -\frac{1}{k}\hat{\sigma} \ , \tag{5.2.36}$$

and $\delta H/\delta t \leqslant 0$ and $\hat{\sigma} \geqslant 0$ are equivalent statements.

5.2.2 Elastic Collisions (Unlike Particles)

Let us now consider elastic binary collisions of particles of two different types,

$$A + B \rightleftarrows A + B \ .$$

As already mentioned, this case includes nonresonant scattering of photons, like Rayleigh or Thomson scattering.

The rate of entropy production is given by, see (5.2.8),

$$\frac{1}{k}\frac{\delta S}{\delta t} = \frac{1}{k}\frac{\delta S_A}{\delta t} + \frac{1}{k}\frac{\delta S_B}{\delta t}$$

$$= \sum_{\kappa} \frac{\delta N_\kappa}{\delta t} \ln \omega_\kappa + \sum_{s} \frac{\delta N_s}{\delta t} \ln \omega_s \ , \tag{5.2.37}$$

where Greek (κ, λ) and Latin (s, t) letters specify the groups of translational quantum states of the A and B particles, respectively. Here the corresponding rates of change of the occupation numbers are

$$\frac{\delta N_\kappa}{\delta t} = -\sum_{\lambda st} [\kappa s \rightarrow \lambda t] + \sum_{\lambda st} [\lambda t \rightarrow \kappa s]$$

$$= -\sum_{\lambda st} w(\kappa s \rightarrow \lambda t) N^4 \omega_\lambda \omega_t + \sum_{\lambda st} w(\lambda t \rightarrow \kappa s) N^4 \omega_\kappa \omega_s \ , \tag{5.2.38a}$$

$$\frac{\delta N_s}{\delta t} = -\sum_{\kappa \lambda t} [\kappa s \rightarrow \lambda t] + \sum_{\kappa \lambda t} [\lambda t \rightarrow \kappa s]$$

$$= -\sum_{\kappa \lambda t} w(\kappa s \rightarrow \lambda t) N^4 \omega_\lambda \omega_t + \sum_{\kappa \lambda t} w(\lambda t \rightarrow \kappa s) N^4 \omega_\kappa \omega_s \tag{5.2.38b}$$

with

$$N^4 \equiv N_\kappa N_\lambda N_s N_t \ . \tag{5.2.39}$$

One verifies immediately that the numbers of A and B particles are both conserved,

$$\frac{\delta}{\delta t}\sum_{\kappa} N_{\kappa}=0\ ,\quad \frac{\delta}{\delta t}\sum_{s} N_{s}=0\ . \tag{5.2.40}$$

Inserting (5.2.38) into (5.2.37) now leads to

$$\begin{aligned}
\frac{1}{k}\frac{\delta S}{\delta t} = & -\sum_{\kappa\lambda st} w(\kappa s\rightarrow\lambda t)N^4\omega_{\lambda}\omega_t \ln\omega_{\kappa}\\
& +\sum_{\kappa\lambda st} w(\lambda t\rightarrow\kappa s)N^4\omega_{\kappa}\omega_s \ln\omega_{\kappa}\\
& -\sum_{\kappa\lambda st} w(\kappa s\rightarrow\lambda t)N^4\omega_{\lambda}\omega_t \ln\omega_{s}\\
& +\sum_{\kappa\lambda st} w(\lambda t\rightarrow\kappa s)N^4\omega_{\kappa}\omega_s \ln\omega_{s}\\
= & -\frac{1}{2}\sum_{\kappa\lambda st} w(\kappa s\rightarrow\lambda t)N^4\omega_{\lambda}\omega_t \ln\omega_{\kappa}\\
& -\frac{1}{2}\sum_{\kappa\lambda st} w(\lambda t\rightarrow\kappa s)N^4\omega_{\kappa}\omega_s \ln\omega_{\lambda}\\
& +\frac{1}{2}\sum_{\kappa\lambda st} w(\lambda t\rightarrow\kappa s)N^4\omega_{\kappa}\omega_s \ln\omega_{\kappa}\\
& +\frac{1}{2}\sum_{\kappa\lambda st} w(\kappa s\rightarrow\lambda t)N^4\omega_{\lambda}\omega_t \ln\omega_{\lambda}\\
& -\frac{1}{2}\sum_{\kappa\lambda st} w(\kappa s\rightarrow\lambda t)N^4\omega_{\lambda}\omega_t \ln\omega_{s}\\
& -\frac{1}{2}\sum_{\kappa\lambda st} w(\lambda t\rightarrow\kappa s)N^4\omega_{\kappa}\omega_s \ln\omega_{t}\\
& +\frac{1}{2}\sum_{\kappa\lambda st} w(\lambda t\rightarrow\kappa s)N^4\omega_{\kappa}\omega_s \ln\omega_{s}\\
& +\frac{1}{2}\sum_{\kappa\lambda st} w(\kappa s\rightarrow\lambda t)N^4\omega_{\lambda}\omega_t \ln\omega_{t}\ ,
\end{aligned} \tag{5.2.41}$$

where we have used the following evident relation

$$\sum_{\kappa\lambda st}(\dots)=\frac{1}{2}\sum_{\kappa\lambda st}(\dots)+\frac{1}{2}\sum_{\kappa\lambda st}(\kappa s\leftrightarrow\lambda t)\ .$$

Making use of the reciprocity relation

$$w(\kappa s \to \lambda t) = w(\lambda t \to \kappa s) \ , \tag{5.2.42}$$

and taking (5.2.39) into account, the rate of entropy production (5.2.41) is

$$\frac{1}{k}\frac{\delta S}{\delta t} = \frac{1}{2}\sum_{\kappa\lambda st} w(\kappa s \to \lambda t) N_\kappa N_\lambda N_s N_t (\omega_\kappa \omega_s - \omega_\lambda \omega_t) \ln\left(\frac{\omega_\kappa \omega_s}{\omega_\lambda \omega_t}\right), \tag{5.2.43}$$

which shows at once that $\delta S/\delta t \geqslant 0$, and $\delta S/\delta t = 0$ if and only if detailed balance obtains.

This demonstration of the H theorem also includes *inelastic binary collisions* of particles of different species, for example,

$$\mathrm{H} + \mathrm{e} \rightleftarrows \mathrm{H}' + \mathrm{e} \ ,$$

where H and H′ denote H atoms in different excitation states. The only change would concern the interpretation of the letter κ, which now refers to a group of quantum states characterized by translational *and* internal quantum numbers of the H atom.

5.2.3 Emission and Absorption

The last example of the H theorem is emission and absorption of particles of species B by particles of species A,

$$A' \rightleftarrows A + B \ .$$

For the applications discussed in this book, the B particles are of course photons, and A and A' denote atoms of the same species in different excitation states.

The rate of entropy production is given by, see (5.2.37),

$$\frac{1}{k}\frac{\delta S}{\delta t} = \frac{1}{k}\frac{\delta S_A}{\delta t} + \frac{1}{k}\frac{\delta S_B}{\delta t}$$

$$= \sum_\kappa \frac{\delta N_\kappa}{\delta t} \ln \omega_\kappa + \sum_s \frac{\delta N_s}{\delta t} \ln \omega_s \ . \tag{5.2.44}$$

Here the Greek letters κ, λ specify the groups of quantum states of the A particles (characterized by translational *and* internal quantum numbers), and the Latin letter s those of the B particles.

The A particles in group κ are both destroyed and created by emissions and absorptions,

$$\frac{\delta N_\kappa}{\delta t} = -\sum_{\lambda s}[\kappa\to\lambda s] - \sum_{\lambda s}[\kappa s\to\lambda] + \sum_{\lambda s}[\lambda\to\kappa s] + \sum_{\lambda s}[\lambda s\to\kappa]$$
$$= -\sum_{\lambda s} w(\kappa\to\lambda s)N^3\omega_\lambda\omega_s - \sum_{\lambda s} w(\kappa s\to\lambda)N^3\omega_\lambda + \sum_{\lambda s} w(\lambda\to\kappa s)N^3\omega_\kappa\omega_s + \sum_{\lambda s} w(\lambda s\to\kappa)N^3\omega_\kappa\ . \tag{5.2.45}$$

On the other hand,

$$\frac{\delta N_s}{\delta t} = -\sum_{\kappa\lambda}[\lambda s\to\kappa] + \sum_{\kappa\lambda}[\kappa\to\lambda s]$$
$$= -\sum_{\kappa\lambda} w(\lambda s\to\kappa)N^3\omega_\kappa + \sum_{\kappa\lambda} w(\kappa\to\lambda s)N^3\omega_\lambda\omega_s\ . \tag{5.2.46}$$

Again, we have used the abbreviation

$$N^3 \equiv N_\kappa N_\lambda N_s\ . \tag{5.2.47}$$

From (5.2.45) follows at once

$$\frac{\delta}{\delta t}\sum_\kappa N_\kappa = 0\ , \tag{5.2.48}$$

that is, the number of A particles is conserved. However, (5.2.46) shows that $(\delta/\delta t)(\sum_s N_s)$ is undetermined, as it should be.

The rate of entropy production may now be written as

$$\frac{1}{k}\frac{\delta S}{\delta t} = -\sum_{\kappa\lambda s} w(\kappa\to\lambda s)N^3\omega_\lambda\omega_s\ln\omega_\kappa$$
$$-\sum_{\kappa\lambda s} w(\kappa s\to\lambda)N^3\omega_\lambda\ln\omega_\kappa$$
$$+\sum_{\kappa\lambda s} w(\lambda\to\kappa s)N^3\omega_\kappa\omega_s\ln\omega_\kappa$$
$$+\sum_{\kappa\lambda s} w(\lambda s\to\kappa)N^3\omega_\kappa\ln\omega_\kappa$$
$$-\sum_{\kappa\lambda s} w(\lambda s\to\kappa)N^3\omega_\kappa\ln\omega_s$$
$$+\sum_{\kappa\lambda s} w(\kappa\to\lambda s)N^3\omega_\lambda\omega_s\ln\omega_s$$

$$= - \sum_{\kappa\lambda s} w(\kappa\to\lambda s) N^3 \omega_\lambda \omega_s \ln \omega_\kappa$$

$$- \sum_{\kappa\lambda s} w(\lambda s\to\kappa) N^3 \omega_\kappa \ln \omega_\lambda$$

$$+ \sum_{\kappa\lambda s} w(\kappa\to\lambda s) N^3 \omega_\lambda \omega_s \ln \omega_\lambda$$

$$+ \sum_{\kappa\lambda s} w(\lambda s\to\kappa) N^3 \omega_\kappa \ln \omega_\kappa$$

$$- \sum_{\kappa\lambda s} w(\lambda s\to\kappa) N^3 \omega_\kappa \ln \omega_s$$

$$+ \sum_{\kappa\lambda s} w(\kappa\to\lambda s) N^3 \omega_\lambda \omega_s \ln \omega_s \ , \tag{5.2.49}$$

where in some expressions we have simply substituted $\kappa\leftrightarrow\lambda$.

We now use the reciprocity relation

$$w(\kappa\to\lambda s) = w(\lambda s\to\kappa) \tag{5.2.50}$$

to rewrite (5.2.49) in the form

$$\frac{1}{k}\frac{\delta S}{\delta t} = \sum_{\kappa\lambda s} w(\kappa\to\lambda s) N_\kappa N_\lambda N_s (\omega_\kappa - \omega_\lambda\omega_s) \ln\left(\frac{\omega_\kappa}{\omega_\lambda\omega_s}\right) . \tag{5.2.51}$$

Hence, again $\delta S/\delta t \geqslant 0$, and $\delta S/\delta t = 0$ if and only if detailed balance obtains.

The following remarks conclude this section. On physical grounds, it must be required that the H theorem be valid independently of the specific interactions between the particles, whereas the Pauli proof applies only to interactions that fulfill the reciprocity relation (1.2.2). We refer to the literature for proofs of the H theorem which do not make use of the reciprocity relation [5.11, 12]. On the other hand, the time evolution of the entropy production of radiative processes has been discussed in [5.13] in analogy to the general time evolution criterion of *Glansdorff* and *Prigogine* for material systems [5.14, 15].

6. Energy Exchange Between Matter and Radiation

This chapter treating the energy exchange between matter and radiation represents the core of this book. Section 6.1 recalls the definitions of LTE and non-LTE, and of optically thin and optically thick gases. Optically thin plasmas are briefly discussed in Sect. 6.2. Optically thick plasmas are then considered for the special case of two-level atoms, where non-LTE line transfer is treated using a self-consistent approach via the radiative transfer equation and the kinetic equation of excited atoms. Section 6.3 derives the basic relations and discusses the physical assumptions underlying various approximations. Numerical results for the standard problem (line transfer by two-level atoms in a homogeneous and plane parallel gas) are presented in Sect. 6.4. Section 6.5 derives atomic and laboratory line profile coefficients of three-level atoms. Finally, Sect. 6.6 discusses non-Maxwellian electron distributions due to unbalanced inelastic collision processes in non-LTE plasmas.

6.1 General Remarks

Chapter 2 dealt with the kinetic equations of particles, and Chap. 3 with the kinetic equation of photons, i.e., the equation of radiative transfer. These equations are coupled to one another, namely, on the one hand, through the photon collision terms of the particle distribution functions (Sect. 2.4), which depend on the radiation intensity, and, on the other hand, through the emission, absorption, and scattering coefficients of the transfer equation (Sects. 3.2, 3), which depend on the particle distribution functions. So, in principle, to know the microscopic state of a system composed of particles and photons, one must solve the set of coupled equations consisting of the kinetic equations of all particle types present, and of the equation of radiative transfer. Clearly, drastic simplifications are required to obtain a set of tractable equations.

The most important simplification, almost universally made in plasma spectroscopy, consists in assuming that all particle distribution functions are nearly Maxwellian distributions. Each particle type a [to recapitulate, the *species* H (atomic hydrogen), say, consists of the particle *types* H(1s), H(2s), H(2p), ...] is then characterized by just five parameters: the number density n_a, the three components of the mean velocity $\boldsymbol{u}_a$, and the internal energy e_a, or equivalently the kinetic temperature T_a, defined through $e_a = n_a(3kT_a/2 + E_a)$, where E_a is the excitation energy of a particle of type a (Sect. 4.1). In order to avoid misunderstandings, it should be stressed that the assumption of nearly Maxwellian distribution functions is not equivalent to the assumption of LTE, which requires additionally that Boltzmann and Saha distributions obtain (see below).

Roughly speaking, this assumption is a reasonable approximation provided that elastic collisions among particles are sufficiently frequent. However, sometimes the assumption of Maxwellian velocity distributions is questionable, for example, if strong electric fields are present, if steep density or temperature gradients occur, if inelastic collision rates are high compared to elastic collision rates, if the gas or plasma is in a turbulent state, and the like. Moreover, Sects. 6.3.4, 4.6 show that the velocity distribution of excited atoms can be non-Maxwellian as a result of line absorption if the radiation intensity varies appreciably within the frequency range of the spectral line. In all such cases, one has to go back to the kinetic equations to determine (or at least to estimate) the particle distribution functions actually present.

The photons' kinetic equation (the radiative transfer equation) must be solved for all photon modes $(\nu, \boldsymbol{n})$ for which the considered system is *optically thick*, that is, for which $\int \kappa_\nu ds \gtrsim 1$, where κ_ν is the phenomenological absorption coefficient (Sect. 3.1), integration being along the photon trajectory. In the opposite limiting case $\int \kappa_\nu ds \ll 1$, the system is called *optically thin* for photons $(\nu, \boldsymbol{n})$. Here the probability for absorption or scattering inside the system is negligible, so that these photons, once created by a spontaneous emission or by scattering from another photon mode, escape from the system freely.

Another important notion, introduced in Sect. 1.1, is that of *local thermodynamic equilibrium* (LTE) defined there as follows: a gas at a given point and at a given time is said to be in LTE if *all* particle distribution functions are (more precisely, deviate only slightly from) thermal distributions (Maxwell, Boltzmann, Saha) corresponding to a unique temperature. By contrast, a gas is said to be in a non-LTE state if at least one of the particle distribution functions is nonthermal, or if there is no unique local temperature.

We emphasize that the notions LTE and non-LTE refer only to the *particles,* but not to the photons. In other words, the intensity of the radiation field is *not* required to be a Planck function for LTE to obtain. Note, in particular, that the radiation field near the boundary of a radiating gas (a gas discharge or a stellar atmosphere, say) is necessarily nonthermal, simply because it is anisotropic, regardless of the spectral distribution of the emitted radiation.

Roughly speaking, LTE is expected to hold in rather dense gases where collision processes are more frequent than radiative processes. On the other hand, in gases of low or moderate densities, the main reason why non-LTE states occur is often simply the escape of photons from the gas, that is, the simple fact that hot gases and plasmas emit radiation. To understand this crucial point, consider the emission of a spectral line from a gas. In the radiating outer layer of the gas, the emission rate exceeds the absorption rate, otherwise no overall emission could take place. As the absorptions do not balance the emissions, it follows from the principle of detailed balance that the two atomic levels corresponding to the considered spectral line are not populated according to a Boltzmann distribution. If transitions between these levels due to excitation and de-excitation collisions are much more frequent than radiative transitions, the deviation from a thermal Boltzmann distribution is only small, that is, LTE obtains. If not, a non-LTE state results where the occupation number of the upper atomic level is sensibly lower than the corresponding Boltzmann value.

The interaction of photons with particles via emission, absorption, and scattering leads to an exchange of energy and momentum between radiation field and material gas. The momentum exchange, however, may often be neglected, as the ratio of the photon momentum to the mean thermal momentum of the particles is usually very small. Accordingly, in this chapter we shall ignore momentum exchange between particles and photons, postponing a discussion of some physical effects related to it to Chap. 7. However, it must be realized that the above argument concerns only the overall momentum balance of a system composed of particles and photons. For example, it may well be that the recoil of atoms due to emission and absorption of photons is significant in the formation of a narrow spectral line even if the associated momentum transfer is negligible when computing the pressure balance of the gas.

6.2 Optically Thin Plasmas

We first consider optically thin plasmas. Compared with the enormous amount of work done in this field, this discussion is rather short. For a more detailed discussion and for a comprehensive list of references refer to [6.1].

6.2.1 Physical Reactions

In optically thin gases, absorptions and stimulated emissions of photons are negligible. Figure 6.1 shows a schematic energy flow diagram for a very simple example of an optically thin system: inside the gas, energy is transferred between atoms and free electrons through excitation and de-excitation collisions, and from the atoms to the radiation field through spontaneously emitted photons which escape freely from the gas. In a stationary state, this energy loss in the form of emitted radiation must be counterbalanced by transferring to the electrons a corresponding amount of energy from another energy source (from a chemical reaction, say).

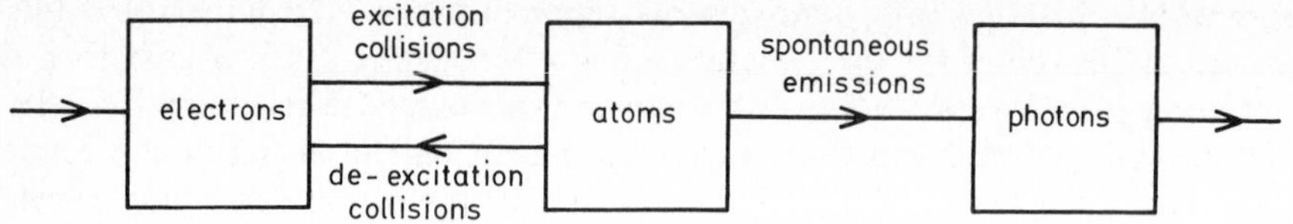

Fig. 6.1. Schematic energy flow diagram of an optically thin plasma

Here and in the following sections we always assume that all inelastic collision processes are due to free electrons, although some exceptions are mentioned below and in Sect. 6.6. Furthermore, with the exception of Sect. 6.6, we also assume that the velocity distribution of the electrons is Maxwellian, so that a local electron temperature T_e is defined.

Provided that a macroscopic description by number densities, mean velocities, and kinetic temperatures is sufficient (Sect. 6.1), the behavior of the considered gas is governed by the coupled set of hydrodynamic equations of all particle types present. In these equations not only the relevant transport coefficients occur, but also the collisional and radiative production rates of particle number, momentum, and energy of all particle types. Obviously, a proper treatment of such a system of equations is impossible. Roughly speaking, two alternative approaches are found in the literature. One approach concentrates on the dynamics of the system, treating the spectroscopic aspects (radiative energy losses, ionization and recombination rates, etc.) only approximately. Calculations of the transport of impurity ions in high-temperature (thermonuclear) plasmas are very sophisticated examples of this type of approach, see [6.2] and the literature cited therein. The other concentrates on the spectroscopy of the system without discussing the hydrodynamics in detail. Here we consider only the second approach.

One thus supposes the hydrodynamical state of the gas – density, chemical composition, kinetic temperatures, macroscopic velocities – to be known. The problem, then, is to determine the spectroscopic state of the gas – the various ionization degrees and the occupation numbers of the bound levels of all atomic and ionic species – taking into account all relevant collisional and radiative processes. Once the various occupation numbers are known, the radiation emitted by an optically thin gas can be calculated straightforwardly, for the only radiative processes taking place in the gas are spontaneous bound-bound, free-bound, and free-free emissions, and every spontaneously emitted photon escapes from the gas.

Hence the only equations to be considered are the balance equations for the particle numbers of all particle types i present, see (4.1.27),

$$\frac{\partial n_i}{\partial t} + \nabla \cdot \boldsymbol{\Gamma}_i = \hat{\nu}_i \; , \tag{6.2.1}$$

where $\boldsymbol{\Gamma}_i \equiv n_i \boldsymbol{u}_i$ is the flux density of the particles. The balance equations (6.2.1) are coupled to one another, for the production rate $\hat{\nu}_i$ depends on the number densities n_j of all particle types j involved in collisional or radiative processes that create or destroy particles of type i. However, if only inelastic collision processes due to free electrons are important, the balance equations corresponding to an atomic species and its various ionization states, e.g., He, He^+, He^{2+}, are decoupled from those of all the other atomic species, apart from the indirect effect of their contributions on the total electron density through ionization and recombination processes.

Let us consider the net production rate of the ion A^{z+} in the bound level i, $A^{z+}(i)$ (Fig. 6.2). In an optically thin plasma, the following reactions must be considered: collisional excitations and de-excitations, collisional ionizations and three-body recombinations, spontaneous radiative bound-bound transitions, and spontaneous radiative recombinations.

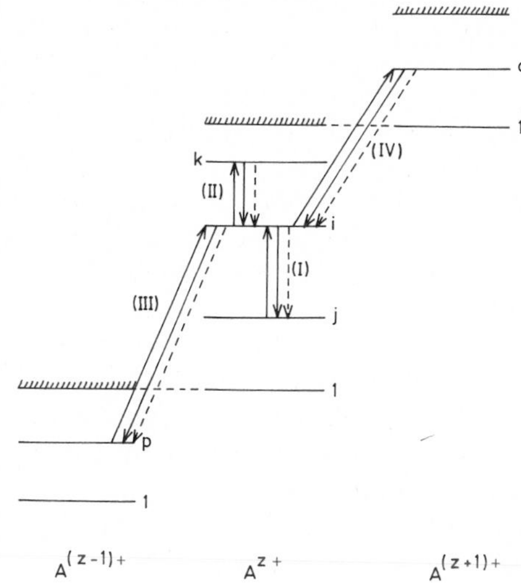

Fig. 6.2. Collisional and radiative processes involving level i of the ion A^{z+}. The ground states of the various ions are labeled 1. Solid (dashed) arrows indicate collisional (radiative) processes

(I) Reactions involving lower lying levels $j(<i)$ of the same ion A^{z+}:

$$A^{z+}(i)+e \rightleftarrows A^{z+}(j)+e \ , \tag{Ia}$$

$$A^{z+}(i) \rightarrow A^{z+}(j)+h\nu \ . \tag{Ib}$$

(II) Reactions involving higher lying levels $k\ (>i)$ of the same ion A^{z+}:

$$A^{z+}(k)+e \rightleftarrows A^{z+}(i)+e \ , \tag{IIa}$$

$$A^{z+}(k) \rightarrow A^{z+}(i)+h\nu \ . \tag{IIb}$$

(III) Reactions involving the ion $A^{(z-1)+}$ of lower ionization degree:

$$A^{z+}(i)+e+e \rightleftarrows A^{(z-1)+}(p)+e \ , \tag{IIIa}$$

$$A^{z+}(i)+e \rightarrow A^{(z-1)+}(p)+h\nu \ . \tag{IIIb}$$

(IV) Reactions involving the ion $A^{(z+1)+}$ of higher ionization degree:

$$A^{(z+1)+}(q) + \mathrm{e} + \mathrm{e} \rightleftarrows A^{z+}(i) + \mathrm{e} \ , \tag{IVa}$$

$$A^{(z+1)+}(q) + \mathrm{e} \rightarrow A^{z+}(i) + h\nu \ . \tag{IVb}$$

It should be observed that negative ions may occur in the ionization sequence, e.g., H^-, H, H^+.

For ionization and recombination processes, in general only those involving the ground state of the higher ionized ion need to be considered, i.e., $i = 1$ in reactions (III), and $q = 1$ in reactions (IV), whereas the corresponding reactions involving excited states of the higher ionized ion are usually negligible, with the possible exception of metastable levels.

In an optically thin plasma where all inelastic collision processes are due to free electrons, the net production rate of particles $A^{z+}(i)$ is hence given by

$$\begin{aligned} \hat{\nu}_i = & \sum_{j<i} [-n_i(A_{ij} + n_\mathrm{e} C_{ij}) + n_j n_\mathrm{e} C_{ji}] \\ & + \sum_{k>i} [n_k(A_{ki} + n_\mathrm{e} C_{ki}) - n_i n_\mathrm{e} C_{ik}] \\ & + \sum_p [-n_i(n_\mathrm{e} R_{ip} + n_\mathrm{e}^2 N_{ip}) + n_p n_\mathrm{e} M_{pi}] \\ & + \sum_q [n_q(n_\mathrm{e} R_{qi} + n_\mathrm{e}^2 N_{qi}) - n_i n_\mathrm{e} M_{iq}] \ . \end{aligned} \tag{6.2.2}$$

For the rate coefficients, the same symbols as in Chap. 2 are used, namely C for collisional excitation and de-excitation, M for collisional ionization, N for three-body recombination, and R for radiative recombination, while A is again the Einstein coefficient for spontaneous bound-bound transitions.

Since by assumption the velocity distribution of the electrons is Maxwellian, the rate coefficients for collisional de-excitation can be expressed in terms of the corresponding rate coefficient for collisional excitation, see (2.3.10), e.g.,

$$C_{ij} = \frac{g_j}{g_i} \mathrm{e}^{E_{ij}/kT_\mathrm{e}} C_{ji} \ , \tag{6.2.3}$$

and the rate coefficient for three-body recombination in terms of the corresponding rate coefficient for collisional ionization, see (2.3.16), e.g.,

$$N_{ip} = \frac{g_p}{2g_i} \lambda_\mathrm{e}^3 \mathrm{e}^{E_{ip}/kT_\mathrm{e}} M_{pi} \ . \tag{6.2.4}$$

So far we have not yet discussed the processes related to dielectronic recombination, which involve doubly excited atomic levels. For simplicity, we consider only the most important case where the recombining ion $A^{(z+1)+}$ is in its ground state $A^{(z+1)+}(1)$ (Fig. 6.3 and Sects. 1.6.3 and 2.3.5). The basic processes for dielectronic recombination are the following:

(V) Radiationless electron capture and autoionization:

$$A^{(z+1)+}(1)+\mathrm{e} \rightleftarrows A^{(z+)**}(s,i) \ . \tag{V}$$

If the core electron is sufficiently highly excited, autoionization may also proceed to an excited state of the ion $A^{(z+1)+}$,

$$A^{(z+)**}(s,i) \rightarrow A^{(z+1)+}(r)+\mathrm{e} \ ,$$

with $1<r<s$ (Fig. 6.3).

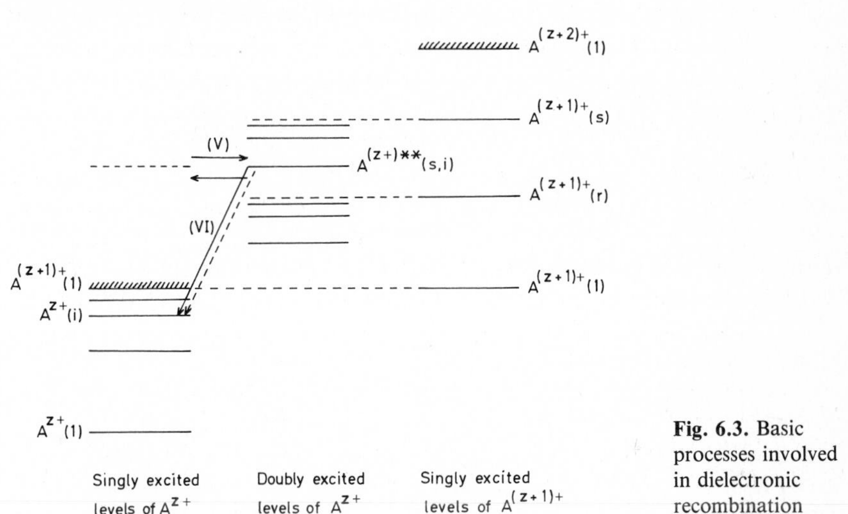

Fig. 6.3. Basic processes involved in dielectronic recombination

(VI) Stabilization of the doubly excited level:

$$A^{(z+)**}(s,i)+\mathrm{e} \rightarrow A^{z+}(i)+\mathrm{e} \ , \tag{VIa}$$

$$A^{(z+)**}(s,i) \rightarrow A^{z+}(i)+h\nu \ . \tag{VIb}$$

Here $A^{**}(s,i)$ denotes a doubly excited atom with s and i standing for appropriate quantum numbers of the core electron and the captured electron, respectively.

In general, collisional stabilization (VI a) and its inverse excitation process are negligible. Then the rate equations for the doubly excited levels are effectively decoupled from those of the singly excited levels.

In the limit of low density, the dielectronic recombination rate in an optically thin plasma is determined by reactions (V, VI b) and radiative cascade (Ib). For a discussion of the net rates of reactions (V, VIb), see Sect. 2.3.5.

At higher densities, the rate of dielectronic recombination is influenced by collision processes. Besides ionization collisions of type (IVa) for singly and doubly excited atoms, the following inelastic collision processes are important:

$$A^{(z+)**}(s, nl) + P \rightleftarrows A^{(z+)**}(s, n'l') + P \ , \qquad \text{(VIIa)}$$

$$A^{z+}(nl) + P \rightleftarrows A^{z+}(n'l') + P \ . \qquad \text{(VIIb)}$$

Here the principal and azimuthal quantum numbers (nl) of the captured electron have been indicated explicitly. Recall that for dielectronic recombination, very highly excited levels (with $n \gg 1$) are relevant. For large principal quantum numbers n, equipartition collisions that leave the principal quantum number unaffected while changing the azimuthal quantum number by unity ($n' = n$, $l' = l \pm 1$), and excitation and de-excitation collisions that change the principal quantum number by unity ($n' = n \pm 1$) are among the most important reactions of type (VII). In these cases, collisions with heavy particles [usually hydrogen or helium nuclei, i.e., $P = H^+$ or He^{2+} in (VIIa, b)] may be important or even predominant compared with electronic collisions. Roughly speaking, the rate of dielectronic recombination is increased by reactions (VIIa) since doubly excited levels with large l become populated which do not or only slowly autoionize, whereas it is decreased by reactions (VIIb) which, for large n, usually lift the captured electron to more highly excited levels, eventually leading to ionization.

It should be emphasized that under special circumstances, still other processes may contribute to the particle production rates. Without going into details, one such is *Penning ionization,* where collision with an atom B in a metastable level m of sufficiently high excitation energy ionizes an atom A which is usually in its ground state,

$$A(1) + B(m) \rightarrow A^+(1) + B(1) + e \ .$$

Another important class of collisions to be mentioned are *charge-exchange collisions,*

$$A^{z+} + B^{z'+} \rightarrow A^{(z+1)+} + B^{(z'-1)+} \ ,$$

which may involve atoms in ground or excited states. This class comprises elastic collisions, e.g.,

$$H^+(\text{fast}) + H(\text{slow}) \rightarrow H^+(\text{slow}) + H(\text{fast}) \ ,$$

and inelastic collisions, e.g.,

$$H(1) + He^+(1) \rightarrow H^+ + He(1) \quad \text{or}$$

$$H(n) + He^{2+} \rightarrow H^+ + He^+(n') \ .$$

6.2.2 Some Illustrative Results

Before quoting some numerical results, we first introduce the often used *b coefficient* corresponding to a bound atomic level, which provides a convenient measure of the departure from LTE. The b coefficient b_i of the atomic level i is defined as the ratio of the actual number density n_i to the Saha density,

$$b_i = \frac{n_i}{n_i^S} \; . \tag{6.2.5}$$

Here n_i^S denotes the number density that follows from Saha's equation, using as temperature the local electron temperature T_e, and using the *actual* values of the local electron density n_e and the local ion density n_1^+, that is, the density of ions in the ground state of the next higher stage of ionization. Explicitly, on account of (1.4.25),

$$b_i = \frac{n_i}{n_1^+ n_e} \frac{2g_1^+}{g_i} \lambda_e^{-3} \exp[-(E_1^+ - E_i)/kT_e] \; , \tag{6.2.6}$$

where λ_e is the thermal de Broglie wavelength of the electrons (1.4.24), and the other symbols have their usual meaning. Hence

$$\frac{n_1^+ n_e}{n_i} = \frac{1}{b_i} \frac{2g_1^+}{g_i} \lambda_e^{-3} \exp[-(E_1^+ - E_i)/kT_e] = \frac{1}{b_i} \left(\frac{n_1^+ n_e}{n_i} \right)^S \; , \tag{6.2.7}$$

and furthermore

$$\frac{n_j}{n_i} = \frac{b_j}{b_i} \frac{g_j}{g_i} \exp[-(E_j - E_i)/kT_e] = \frac{b_j}{b_i} \left(\frac{n_j}{n_i} \right)^B \; , \tag{6.2.8}$$

where $(n_1^+ n_e/n_i)^S$ and $(n_j/n_i)^B$ denote, respectively, the Saha and Boltzmann distributions corresponding to temperature T_e, (1.4.25, 19). In LTE, $b_i = 1$ for all i. Hence, if the set of all b coefficients is known, all deviations from LTE can be read off at once. Note that $b_i = 1$ means that the density n_i is the Saha density, whereas $b_j/b_i = 1$ only means that the ratio n_j/n_i is the Boltzmann ratio, while n_j and n_i themselves need not be Saha densities.

Let us now turn to some examples of optically thin plasmas, and let us first consider stationary ($\partial n_i/\partial t = 0$) and homogeneous ($\nabla n_i = 0$) plasmas without macroscopic motion ($u_i = 0$). Then the balance equations for the particle numbers (6.2.1) reduce to the simple form

$$\hat{\nu}_i = 0 \; . \tag{6.2.9}$$

Tables 6.1 – 4 list some b coefficients of atomic hydrogen and helium for various electron densities and temperatures. It can be seen that LTE (that is, $b_i \simeq 1$ for all i) is attained only at rather high electron densities ($n_e \gtrsim 10^{18}\,\mathrm{cm}^{-3}$).

Table 6.1. Stationary, homogeneous, and optically thin hydrogen plasma. The b coefficient of the ground state b_1 and the ratio of the proton density to the total density of H atoms $\chi = n_+/\sum_i n_i$ for various values of the electron density n_e and the electron temperature T_e. For a pure hydrogen plasma $n_+ = n_e$. The electron distribution function is assumed to be Maxwellian [6.3, 4]

n_e [cm^{-3}]		T_e [K]		
		8000	16000	32000
10^8	b_1	2×10^9	2×10^9	2×10^9
	χ	2×10^{-5}	1.2	4×10^2
10^{10}	b_1	2×10^7	2×10^7	2×10^7
	χ	2×10^{-5}	1.2	4×10^2
10^{12}	b_1	2×10^5	2×10^5	2×10^5
	χ	2.5×10^{-5}	1.2	3×10^2
10^{14}	b_1	10^3	1.5×10^3	1.5×10^3
	χ	4×10^{-5}	1.8	2×10^2
10^{16}	b_1	8	10	10
	χ	5×10^{-5}	2.5	6×10^1
10^{18}	b_1	1.1	1.1	1.1
	χ	4×10^{-5}	2.2	1.5×10^1

Table 6.2. Stationary, homogeneous, and optically thin hydrogen plasma at electron temperature $T_e = 8000$ K. The b coefficients b_n of levels with principal quantum number n for various values of the electron density n_e. Equipartition over the quantum states with the same principal quantum number has been assumed. The electron distribution function is assumed to be Maxwellian [6.4]

n_e [cm^{-3}]	b_1	b_2	b_3	b_4	b_5	b_{10}
10^8	2×10^9	2.8	1.8	1.3	1.1	0.9
10^{10}	2×10^7	2.8	1.8	1.3	1.1	1
10^{12}	2×10^5	2.6	1.7	1.2	1	1
10^{14}	10^3	1.6	1.1	1	1	1
10^{16}	8	1	1	1	1	1
10^{18}	1.1	1	1	1	1	1

Table 6.3. Stationary, homogeneous, and optically thin helium plasma. The b coefficient of the ground state $b_1 \equiv b(1^1S)$ for various values of the electron density n_e and the electron temperature T_e. The electron distribution function is assumed to be Maxwellian [6.5] (and private communication)

n_e [cm^{-3}]	T_e [K]			
	12000	16000	20000	48000
10^{10}	5×10^6	3×10^7	5×10^7	10^8
10^{12}	2×10^4	3×10^5	4×10^5	8×10^5
10^{14}	5×10^2	2×10^3	3×10^3	8×10^3

Table 6.4. Stationary, homogeneous, and optically thin helium plasma at electron temperature $T_e = 16000$ K. The b coefficients of some singlet and triplet levels for various values of electron density n_e. Equipartition over the quantum states with the same principal quantum number n has been assumed for $n \geqslant 3$. The electron distribution function is assumed to be Maxwellian [6.5] (and private communication)

n_e [cm^{-3}]	Singlet levels					
	$b(1^1S)$	$b(2^1S)$	$b(2^1P)$	$b(3)$	$b(4)$	$b(5)$
10^{10}	3×10^7	3.5×10^4	0.4	0.2	0.3	0.5
10^{12}	3×10^5	4×10^2	0.35	0.25	0.4	0.8
10^{14}	2×10^3	6	0.2	0.6	0.9	1

n_e [cm^{-3}]	Triplet levels					
	–	$b(2^3S)$	$b(2^3P)$	$b(3)$	$b(4)$	$b(5)$
10^{10}		6×10^6	1.2×10^3	1.1	0.8	0.75
10^{12}		4×10^4	8×10^2	1.7	1	1
10^{14}		5×10^1	3×10^1	1.4	1	1

At lower electron densities, the plasmas are in non-LTE states where, above all, the ground state is strongly overpopulated compared to LTE ($b_1 \gg 1$). In other words, excitation and ionization are lower in non-LTE than in LTE plasmas. In non-LTE plasmas, the ground state coefficient b_1 is, over large ranges of the electron temperature T_e, roughly proportional to the electron density n_e and rather independent of T_e. Exactly the opposite behavior shows the ionization degree, specified by the ratio of total ion density to total atom density

$$\chi = \sum_i n_i^+ / \sum_i n_i \ .$$

Table 6.1 shows that for atomic hydrogen, χ depends strongly on T_e, but only weakly on n_e. Deviation of the ionization degree from its thermal Saha value is determined by the ground state coefficient b_1, because the number of excited atoms is generally negligible compared with the number of atoms in the ground state. Hence $\chi \simeq n_1^+/n_1 = (n_1^+/n_1)^S/b_1$, the Saha ratio $(n_1^+/n_1)^S \equiv (n_1^+ n_e/n_1)^S/n_e = f(T_e)/n_e$ being a function of electron temperature and density. Finally, we remark that in non-LTE situations, metastable levels (e.g., the levels 2^1S and 2^3S of atomic helium, Table 6.4) have rather large b coefficients, as expected.

Dielectronic recombination, which plays no role for the cases listed in Tables 6.1 – 4, becomes important for nonhydrogenic atoms at high temperatures and low densities. As an example, Fig. 6.4 shows a plot of b coefficients of atomic helium for a very low electron density ($n_e = 10^4\,cm^{-3}$). It can be seen that at high enough electron temperatures, dielectronic recombination

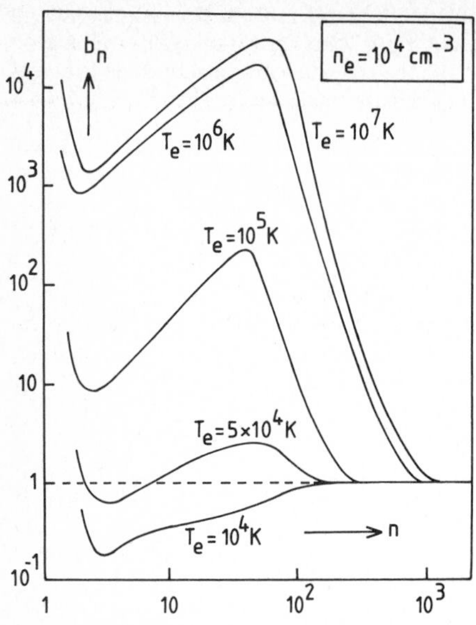

Fig. 6.4. The b coefficient b_n as a function of principal quantum number n for a stationary, homogeneous, and optically thin helium plasma with electron density $n_e = 10^4\ \mathrm{cm}^{-3}$ and various electron temperatures T_e. The values of b_n for $n \lesssim 5$ are only rough estimates as the detailed structure of these levels has not been taken into account [6.6]

leads to strong overpopulation ($b_i \gg 1$) of levels with large principle quantum numbers.

In the general nonstationary and inhomogeneous case, the complete balance equations (6.2.1) [as opposed to their special form (6.2.9)] must be considered. This set of coupled differential equations can be greatly simplified in most cases of practical interest. Indeed, it turns out that the relaxation time of excited levels ($i > 1$) is usually much shorter than the relaxation time of the ground level ($i = 1$). (This statement may not pertain to metastable levels; then the procedure outlined below must be generalized [6.7].) The physical reason for this fact is the short lifetime of excited levels compared with the lifetime of the ground state. Consequently, the populations of the excited levels quickly reach quasi-equilibrium with the instantaneous ground state population. Likewise, the diffusion lengths of excited particles during their lifetime are usually so short that variations of physical quantities like n_1, n_1^+, n_e, T_e are negligible over such distances. In other words, the dependence of the populations of the excited levels on space and time is only implicit through the space- and time-dependent ground state population. The complete balance equations (6.2.1) may therefore be approximated by the following, much simpler set of equations [6.8]:

$$\frac{\partial n_1}{\partial t} + \nabla \cdot \boldsymbol{\Gamma}_1 = \hat{\nu}_1 \ , \tag{6.2.10a}$$

$$\hat{\nu}_i = 0 \quad (i > 1) \ . \tag{6.2.10b}$$

Equation (6.2.10b) forms a system of linear equations for the number densities n_i ($i = 2, 3, \ldots$), with inhomogeneous terms containing the ground state population n_1 and the ion density n_1^+, which may be considered as undetermined parameters. The solution of (6.2.10b) can hence be written as

$$n_i = r_i(n_e, T_e)\, n_1 + s_i(n_e, T_e)\, n_1^+ \quad (i>1)\,, \tag{6.2.11a}$$

or equivalently in terms of b coefficients

$$b_i = \rho_i(n_e, T_e)\, b_1 + \sigma_i(n_e, T_e) \quad (i>1)\ . \tag{6.2.11b}$$

If the considered species is the ion A^{z+}, more explicitly $n_i \equiv n_i^z$, $n_i^+ \equiv n_i^{z+1}$, etc. In (6.2.11), it has been assumed that only those ionization and recombination processes contribute that involve the ground state of the ion of the next higher stage of ionization. Hence, besides n_1^z, only n_1^{z+1} occurs on the right-hand sides, but not the densities n_i^{z+1} with $i>1$ of the ion $A^{(z+1)+}$ nor the densities n_i^{z-1} of the ion $A^{(z-1)+}$.

The equation for the ground state density n_1^z of the ion A^{z+} can now be written in the following form

$$\begin{aligned}\frac{\partial n_1^z}{\partial t} + \nabla \cdot \boldsymbol{\Gamma}_1^z = {} & n_1^{z+1} n_e \alpha^z(n_e, T_e) - n_1^z n_e S^z(n_e, T_e) \\ & - n_1^z n_e \alpha^{z-1}(n_e, T_e) + n_1^{z-1} n_e S^{z-1}(n_e, T_e)\ ,\end{aligned} \tag{6.2.12}$$

where the first two terms on the right-hand side describe recombination and ionization processes involving the ion $A^{(z+1)+}$, and the last two terms recombination and ionization processes involving the ion $A^{(z-1)+}$. Equation (6.2.12) *defines* the quantities α^z, S^z, α^{z-1}, S^{z-1}. The quantities α^z and S^z are called, respectively, the *collisional-radiative recombination coefficient* and the *collisional-radiative ionization coefficient* of the ion A^{z+}. They depend on the populations of the excited levels of the ion A^{z+}, and can thus be readily expressed in terms of the quantities r_i^z, s_i^z $(i = 2, 3, \ldots)$ defined by (6.2.11a). The other two coefficients α^{z-1} and S^{z-1} occurring in (6.2.12) must, of course, be determined from (6.2.10b) corresponding to the ion $A^{(z-1)+}$.

For neutral atoms ($z = 0$) and neglecting the formation of negative ions, only the first two terms on the right-hand side of (6.2.12) pertain, and only two coefficients $\alpha \equiv \alpha^0$ and $S \equiv S^0$ are defined. As an example, Table 6.5 lists some values of α and S for atomic hydrogen. On the other hand, Fig. 6.5 plots the recombination coefficient α of neutral helium, displaying clearly the effect of dielectronic recombination at high temperatures and low electron densities.

In contrast to most astrophysical situations where diffusion is negligible so that homogeneous models can be used, convective particle transport due to density gradients is important in many laboratory plasmas, changing appreciably the level populations in comparison with a homogeneous model. For example, the steady-state balance equation for neutral atoms in the ground state is, see (6.2.12),

$$\nabla \cdot \boldsymbol{\Gamma}_1 = n_1^+ n_e \alpha - n_1 n_e S\ , \tag{6.2.13}$$

Table 6.5. The collisional-radiative coefficients of recombination α [$cm^3 s^{-1}$] and ionization S [$cm^3 s^{-1}$] of an optically thin hydrogen plasma for various values of the electron density n_e and the electron temperature T_e. The electron distribution function is assumed to be Maxwellian [6.8]

n_e [cm^{-3}]		T_e [K] 8000	16000	32000
10^8	α	5×10^{-13}	3×10^{-13}	2×10^{-13}
	S	10^{-17}	4×10^{-13}	8×10^{-11}
10^{10}	α	6×10^{-13}	3×10^{-13}	2×10^{-13}
	S	1.5×10^{-17}	4×10^{-13}	9×10^{-11}
10^{12}	α	10^{-12}	4×10^{-13}	2×10^{-13}
	S	3×10^{-17}	7×10^{-13}	10^{-10}
10^{14}	α	5×10^{-12}	10^{-12}	3×10^{-13}
	S	2×10^{-16}	2.5×10^{-12}	3×10^{-10}
10^{16}	α	8×10^{-11}	5×10^{-12}	7×10^{-13}
	S	6×10^{-15}	2×10^{-11}	10^{-9}
10^{18}	α	2.5×10^{-9}	10^{-10}	10^{-11}
	S	10^{-14}	2×10^{-11}	10^{-9}

whereas the number density $\bar{n}_1$ in the corresponding homogeneous plasma is given by

$$\bar{n}_1 = \frac{\alpha}{S} n_1^+ \ . \tag{6.2.14}$$

Thus

$$n_1 = \bar{n}_1 - \frac{1}{n_e S} \nabla \cdot \boldsymbol{\Gamma}_1 \ , \tag{6.2.15a}$$

or after dividing by the Saha density n_1^S,

$$b_1 = \bar{b}_1 - \frac{1}{n_1^S n_e S} \nabla \cdot \boldsymbol{\Gamma}_1 \ . \tag{6.2.15b}$$

Furthermore, from (6.2.11b) and using (6.2.15b),

$$b_i = \bar{b}_i - \frac{\rho_i}{n_1^S n_e S} \nabla \cdot \boldsymbol{\Gamma}_1 \quad (i>1) \ . \tag{6.2.16}$$

For instance, in a stationary gas discharge, neutral atoms in the ground state diffuse from the cold layer near the wall into the plasma, where they become excited and ionized. Thus $\nabla \cdot \boldsymbol{\Gamma}_1 < 0$ and hence $b_1 > \bar{b}_1$, so that departure from LTE is increased compared with the homogeneous case.

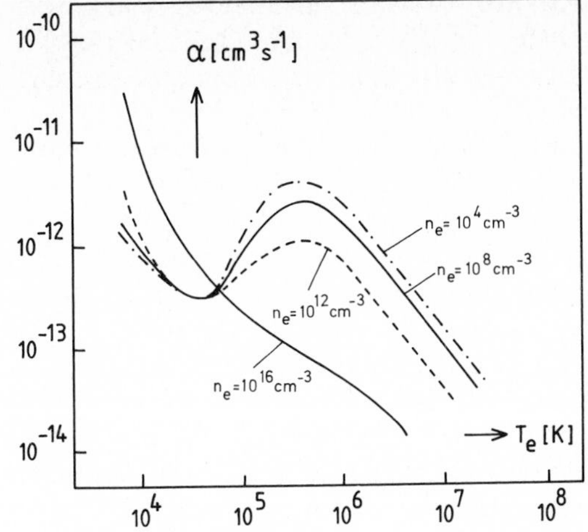

Fig. 6.5. Collisional-radiative recombination coefficient α of an optically thin helium plasma as a function of electron temperature T_e for various electron densities n_e. For low n_e, the increase of α above $T_e \simeq 10^5$ K is due to dielectronic recombination [6.6]

The spectroscopic state of optically thin plasmas shows simple behavior in two limiting cases. In the limit of high electron densities ($n_e \gtrsim 10^{17}\,cm^{-3}$, say), LTE is approached. On the other hand, at low electron densities ($n_e \lesssim 10^{11}\,cm^{-3}$, say), the so-called *corona model* is valid (named after the solar corona to which it was first applied). According to this model, the ionization equilibrium is determined by the balance between collisional ionization from the ground state and radiative recombination into all levels, supposing that the electrons in the excited levels rapidly cascade down to the ground state under spontaneous emission of photons. Hence, instead of (6.2.12), in the corona model the following balance equation is appropriate:

$$\frac{\partial n_1^z}{\partial t} + \nabla \cdot \boldsymbol{\Gamma}_1^z = n_1^{z+1} n_e (\alpha_{rad}^z + \alpha_{diel}^z) - n_1^z n_e M_{11}^z$$

$$- n_1^z n_e (\alpha_{rad}^{z-1} + \alpha_{diel}^{z-1}) + n_1^{z-1} n_e M_{11}^{z-1} \ . \qquad (6.2.17)$$

Here the recombination coefficient has been decomposed for clarity into coefficients for ordinary radiative and dielectronic recombination, respectively, and M_{11}^z denotes the rate coefficient for ionization collisions, see (2.3.14a), that lead from the ground state of the ion A^{z+} to the ground state of the ion $A^{(z+1)+}$,

$$A^{z+}(1) + e \rightarrow A^{(z+1)+}(1) + e + e \ .$$

Let us disregard dielectronic recombination for the moment. Since $\alpha_{rad}^z(T_e)$ and $M_{11}^z(T_e)$ are independent of the electron density n_e, it follows from (6.2.17) that the ionization equilibrium of a stationary and homogeneous plasma ($\partial n/\partial t = 0$, $\nabla \cdot \boldsymbol{\Gamma} = 0$) depends only on the electron temperature T_e, because n_e drops out of (6.2.17): this is the characteristic feature of the corona limit. At very low electron densities, dielectronic recombination does not change this picture because α_{diel}^{z} here is also a function of T_e only. At higher densities, however, the coefficient α_{diel}^{z} is no longer independent of the density due to inelastic collisions of type (VII) (Sect. 6.2.1) and due to collisional ionization of highly excited levels populated mainly through dielectronic recombination. Thus, at such densities, dielectronic recombination introduces a certain dependence on density into the corona equilibrium.

Once the ground state densities are evaluated from (6.2.17), the corona model assumes that the populations of the excited levels are determined by the balance between collisional excitation from the ground state and radiative transitions to all lower levels,

$$n_1 n_e C_{1i} = n_i \sum_{j<i} A_{ij} \quad (i>1) \ . \tag{6.2.18}$$

However, this approximation breaks down for highly excited levels populated by dielectronic recombination. Furthermore, the existence of metastable levels may limit the validity of (6.2.17, 18) of the simple corona model.

6.3 Non-LTE Line Transfer by Two-Level Atoms (I): Basic Relations

Optically thin gases as discussed in the preceding section are rather trivial systems: once a photon has been created by a spontaneous emission, it escapes from the gas without interacting with material particles. We now turn to optically thick systems in which interaction of photons with material particles via absorption, stimulated emission, or scattering does take place.

This section and Sect. 6.4 treat in some detail the radiative transfer in a spectral line due to two-level atoms, that is, atoms that have a ground state and only one excited level, so that they can emit and absorb only one spectral line. This simple model has the great merit of allowing a detailed study of many important questions of radiative line transfer [6.9–14].

From a physical point of view, one special case of this problem is trivial. If elastic and inelastic collisions among particles are very frequent, LTE obtains where the atomic levels are populated according to a Boltzmann distribution, and where the velocity distributions of excited and nonexcited atoms are Maxwellian. Thus, the state of the material gas is determined by collision processes alone, and is not affected by the radiation field. In more technical terms (Sect. 3.2.1), the line source function here is the Planck function corresponding to the local (atomic and electronic) temperature of the gas, and the line absorption coefficient is a known quantity (assuming, of course, that the atomic line broadening profile is known). Then the radiative transfer equation can in principle be integrated straightforwardly to yield the radiation intensity at each point inside and outside the considered gas.

By contrast, in a non-LTE situation, the state of the material gas (level populations, velocity distributions) depends generally on the radiation field. In particular, the atomic level populations cannot be assumed to be given by a Boltzmann distribution, rather, they must be determined simultaneously with the radiation field in the spectral line. Clearly, LTE line transfer is a special case of general non-LTE line transfer.

Figure 6.6 shows a schematic energy flow diagram of the optically thick systems to be considered: free electrons exchange energy with the two-level atoms via excitation and de-excitation collisions, and the two-level atoms in turn exchange energy with the radiation field via emission and absorption of photons. Energy is lost from the system as photons escaping through the boundaries. In a steady state, energy conservation then requires a corresponding energy flow into the electron gas.

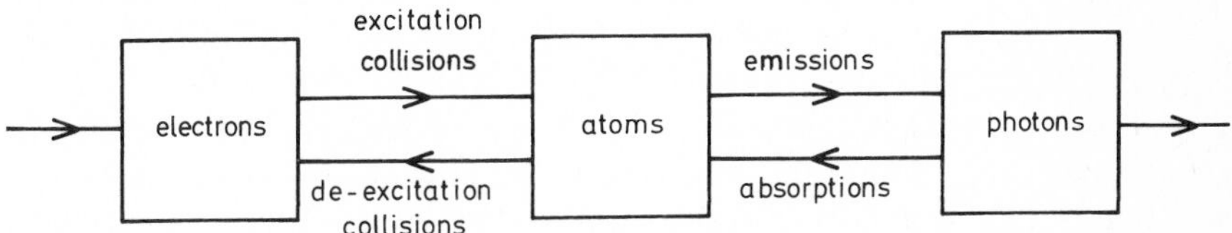

Fig. 6.6. Energy flow diagram of an optically thick plasma

In this section all equations relevant to non-LTE line transfer by two-level atoms are derived, while Sect. 6.4 discusses some results. We consider only systems in a stationary state (or in a quasi-stationary state with only an implicit dependence on time via densities, temperatures, and radiation intensities).

6.3.1 Kinetic Equation of Photons

The time-independent transfer equation for the specific intensity I_ν in the considered spectral line can be written as, see (3.1.16b),

$$\boldsymbol{n} \cdot \nabla I_\nu(\boldsymbol{n}) = \kappa_\nu(\boldsymbol{n})[S_\nu(\boldsymbol{n}) - I_\nu(\boldsymbol{n})] \quad . \tag{6.3.1}$$

Here the line absorption coefficient κ_ν and the line source function S_ν of the two-level atoms are given by, see (3.2.15, 16),

$$\kappa_\nu(\boldsymbol{n}) = \frac{h\nu_0}{4\pi}[n_1 B_{12}\varphi_\nu(\boldsymbol{n}) - n_2 B_{21}\psi_\nu(\boldsymbol{n})] \quad , \tag{6.3.2}$$

$$S_\nu(\boldsymbol{n}) = \frac{n_2 A_{21}\psi_\nu(\boldsymbol{n})}{n_1 B_{12}\varphi_\nu(\boldsymbol{n}) - n_2 B_{21}\psi_\nu(\boldsymbol{n})} \quad . \tag{6.3.3}$$

The line profile coefficients for absorption, φ_ν, and for emission, ψ_ν, are given in terms of the corresponding atomic profiles α_{12} and η_{21} by, see (3.2.10, 13),

$$\varphi_\nu(\boldsymbol{n}) = \int f_1(\boldsymbol{v})\,\alpha_{12}(\xi)\,d^3v \;, \tag{6.3.4}$$

$$\psi_\nu(\boldsymbol{n}) = \int f_2(\boldsymbol{v})\,\eta_{21}(\xi,\boldsymbol{n})\,d^3v \;, \tag{6.3.5}$$

with

$$\xi \equiv \xi(\nu,\boldsymbol{n};\boldsymbol{v}) = \nu - \frac{\nu_0}{c}\boldsymbol{n}\cdot\boldsymbol{v} \tag{6.3.6}$$

being the frequency of the photon $(\nu,\boldsymbol{n})$ in the rest frame of an atom of velocity $\boldsymbol{v}$. The profile coefficients φ_ν and ψ_ν are discussed further in Sect. 6.3.4.

Recall that in the above relations, n_i and $f_i(\boldsymbol{v})$ are the number density and the velocity distribution of two-level atoms in level $i=1$ or 2, respectively, A_{ij} and B_{ij} are the usual Einstein coefficients, and ν_0 is the frequency of the line center in the atomic rest frame. Apart from atomic constants, all quantities are functions of position $\boldsymbol{r}$.

As seen from (6.3.2 – 5), the absorption coefficient and the source function depend of course on the distribution functions $F_1(\boldsymbol{v}) \equiv n_1 f_1(\boldsymbol{v})$ and $F_2(\boldsymbol{v}) \equiv n_2 f_2(\boldsymbol{v})$ of nonexcited and excited two-level atoms. Distribution functions of other particle species do not intervene because it has been tacitly assumed in (6.3.2, 3) that other radiation processes (e.g., Thomson scattering by free electrons) contribute negligibly to the line absorption coefficient and the line source function compared with the absorption and emission processes of the two-level atoms themselves.

6.3.2 Kinetic Equations of Two-Level Atoms

In a stationary state and in the absence of (external or self-consistent) force fields, the distribution functions of two-level atoms in the ground state $i=1$ and in the excited state $i=2$ obey the kinetic equations, see (2.1.7, 11),

$$\boldsymbol{v}\cdot\nabla F_i = \left(\frac{\delta F_i}{\delta t}\right)_{\rm rad} + \left(\frac{\delta F_i}{\delta t}\right)_{\rm inel} + \left(\frac{\delta F_i}{\delta t}\right)_{\rm el} \;, \tag{6.3.7}$$

where $i=1$ or 2, and where the collision terms on the right-hand side are due to radiative interactions, inelastic collisions, and elastic collisions, respectively.

According to (2.4.10), the radiative collision terms of (6.3.7) are given by

$$\left(\frac{\delta F_2}{\delta t}\right)_{\rm rad} = -\left(\frac{\delta F_1}{\delta t}\right)_{\rm rad}$$

$$= B_{12} I_{12}(\boldsymbol{v}) F_1(\boldsymbol{v}) - [A_{21} + B_{21} I_{21}(\boldsymbol{v})]\, F_2(\boldsymbol{v}) \tag{6.3.8}$$

where, see (2.4.11, 12),

$$I_{12}(\boldsymbol{v}) = \iint I_\nu(\boldsymbol{n})\,\alpha_{12}(\xi)\,d\nu\,\frac{d\Omega}{4\pi}\;, \tag{6.3.9}$$

$$I_{21}(\boldsymbol{v}) = \iint I_\nu(\boldsymbol{n})\,\eta_{21}(\xi,\boldsymbol{n})\,d\nu\,\frac{d\Omega}{4\pi}\;. \tag{6.3.10}$$

In (6.3.8), the first term on the right-hand side is a production rate due to absorptions, and the second term is a production rate due to emissions. Recoil due to photon emission and absorption has been neglected.

Note that the quantities I_{12} and I_{21}, (6.3.9, 10), are analogous to the profiles φ_ν and ψ_ν, (6.3.4, 5), the (unnormalized) photon distribution function I_ν replacing the (normalized) velocity distributions $f_i(\boldsymbol{v})$.

On the other hand, supposing that inelastic collisions of two-level atoms only with free electrons are important, the inelastic collision terms of (6.3.7) are, according to (2.3.6), given by

$$\left(\frac{\delta F_2}{\delta t}\right)_{\text{inel}} = -\left(\frac{\delta F_1}{\delta t}\right)_{\text{inel}} = n_e C_{12} F_1(\boldsymbol{v}) - n_e C_{21} F_2(\boldsymbol{v})\;, \tag{6.3.11}$$

where n_e is the electron density, and C_{12}, C_{21} are the rate coefficients for electronic excitation and de-excitation collisions, assumed to be independent of the velocity $\boldsymbol{v}$ of the colliding two-level atom, (2.3.7). Again, velocity changes of two-level atoms due to inelastic collisions with electrons have been neglected.

Finally, the elastic collision terms of (6.3.7) are in general sums of Boltzmann or Fokker-Planck collision terms not specified here (Sect. 2.2). Note, however, that for many purposes it may be appropriate to approximate the elastic collision terms of the two-level atoms by relaxation terms of the form

$$\left(\frac{\delta F_i}{\delta t}\right)_{\text{el}} = \gamma_i n_i [f^{\mathrm{M}}(\boldsymbol{v}) - f_i(\boldsymbol{v})] \tag{6.3.12}$$

which describe the approach of the distribution function $F_i = n_i f_i$ to a local Maxwell distribution $n_i f^{\mathrm{M}}$ as a result of elastic collisions occurring with constant (i.e., $\boldsymbol{v}$-independent) collision frequency γ_i. For a gas in motion, the normalized Maxwell distribution in (6.3.12) has to be taken as, see (4.3.4),

$$f^{\mathrm{M}}(\boldsymbol{v}) = \left(\frac{m}{2\pi kT}\right)^{3/2} \exp\left(-\frac{m}{2kT}(\boldsymbol{v}-\boldsymbol{u})^2\right)\;, \tag{6.3.13}$$

where m is the mass of a two-level atom, and $\boldsymbol{u}$ is the local macroscopic velocity of the gas.

The distribution functions $F_1(\boldsymbol{v})$ and $F_2(\boldsymbol{v})$ of nonexcited and excited two-level atoms thus depend on the photon distribution function I_ν through the radiative collision terms (6.3.8), on the electron distribution function $F_e = n_e f_e$ through the inelastic collision terms (6.3.11), and on all particle distribution functions that contribute to the elastic collision terms in (6.3.7).

6.3.3 Approximations

In studying spectral line formation by two-level atoms one has to deal with at least four distinct distribution functions: the distribution functions F_1 and F_2 of nonexcited and excited two-level atoms, the electron distribution function F_e, and the photon distribution function I_ν. In the simplest model, no other particle species are considered, electric charge neutrality being provided by an appropriate background of positive ions.

The four distribution functions just mentioned must in principle be determined as solutions of the four corresponding kinetic equations, which form a set of coupled equations owing to mutual interactions. As this is a formidable task, various approximations are introduced.

(1) The electron distribution function $F_e(\boldsymbol{v}) = n_e f_e(\boldsymbol{v})$ is usually supposed to be known at each point of the gas. More specifically, in most models it is assumed to be a Maxwellian distribution $f_e = f^M$ corresponding to a local electron temperature T_e. However, Sect. 6.6 discusses models that do not make this assumption.

(2) Likewise, the distribution function of nonexcited two-level atoms $F_1(\boldsymbol{v}) = n_1 f_1(\boldsymbol{v})$ is assumed to be a given function at each point of the gas. Because of the various excitation and de-excitation processes $1 \leftrightarrow 2$ giving rise to transitions between the two atomic states, this assumption is a self-consistent approximation only if

$$F_2(\boldsymbol{v}) \ll F_1(\boldsymbol{v}) \tag{6.3.14}$$

is valid for all atomic velocities $\boldsymbol{v}$, and hence, a fortiori, if

$$n_2 \ll n_1 \ . \tag{6.3.15}$$

In view of (6.3.2, 3), (6.3.15) implies that *stimulated emissions can be neglected* in comparison with absorptions. Stimulated emissions are considered in Sect. 6.3.6.

If the inequalities (6.3.14, 15) hold, the absorption coefficient (6.3.2) becomes

$$\kappa_\nu(\boldsymbol{n}) = \frac{h\nu_0}{4\pi} n_1 B_{12} \varphi_\nu(\boldsymbol{n}) \tag{6.3.16}$$

which is a known quantity since n_1 and $f_1(\boldsymbol{v})$, and hence $\varphi_\nu(\boldsymbol{n})$, see (6.3.4), are known by assumption. On the other hand, the source function (6.3.3) reduces to

$$S_\nu(\boldsymbol{n}) = \frac{n_2 A_{21} \psi_\nu(\boldsymbol{n})}{n_1 B_{12} \varphi_\nu(\boldsymbol{n})} \ , \tag{6.3.17}$$

which depends on n_2, and on $f_2(\boldsymbol{v})$ via $\psi_\nu(\boldsymbol{n})$, (6.3.5).

If the distribution functions $F_e(\boldsymbol{v})$ and $F_1(\boldsymbol{v})$ are specified, only two distribution functions $F_2(\boldsymbol{v})$ and $I_\nu(\boldsymbol{n})$ remain to be determined. Further approximations discussed in the following concern the distribution function of excited two-level atoms $F_2(\boldsymbol{v})$.

(3) The approximation of *complete redistribution* is very often used in non-LTE line transfer calculations. Here, *by assumption,* one sets the emission profile equal to the absorption profile,

$$\psi_\nu(\boldsymbol{n}) = \varphi_\nu(\boldsymbol{n}) \ . \tag{6.3.18}$$

Comparison with (6.3.4, 5) shows that (6.3.18) is valid provided that the two atomic profiles are equal,

$$\eta_{21}(\xi, \boldsymbol{n}) = \alpha_{12}(\xi) \ , \tag{6.3.19}$$

and that the two velocity distributions are equal,

$$f_2(\boldsymbol{v}) = f_1(\boldsymbol{v}) \ . \tag{6.3.20}$$

In typical non-LTE situations, usually neither (6.3.19) nor (6.3.20) holds (Sect. 6.3.4).

In the approximation of complete redistribution (6.3.18), the source function (6.3.17) reduces to

$$S = \frac{n_2 A_{21}}{n_1 B_{12}} \ , \tag{6.3.21}$$

which is constant in the spectral line (i.e., independent of ν and $\boldsymbol{n}$), and depends only on the ratio n_2/n_1 of the atomic densities.

To determine the ratio n_2/n_1, one considers approximations to the kinetic equation of the excited two-level atoms which, according to (6.3.7, 8, 11), can be written as

$$\begin{aligned} \boldsymbol{v} \cdot \nabla F_2(\boldsymbol{v}) = {} & [B_{12} I_{12}(\boldsymbol{v}) + n_e C_{12}] F_1(\boldsymbol{v}) \\ & - (A_{21} + n_e C_{21}) F_2(\boldsymbol{v}) + \gamma_2 n_2 [f^M(\boldsymbol{v}) - f_2(\boldsymbol{v})] \ , \end{aligned} \tag{6.3.22}$$

where stimulated emissions have been neglected, and where the approximate elastic collision term (6.3.12) has been used.

(4) In the *static approximation,* streaming of excited atoms is neglected,

$$\boldsymbol{v} \cdot \nabla F_2 = 0 \ . \tag{6.3.23}$$

Integration of (6.3.22) over all velocities now yields

$$0=(B_{12}J_{12}+n_e C_{12})\,n_1-(A_{21}+n_e C_{21})\,n_2 \tag{6.3.24a}$$

or

$$\frac{n_2}{n_1}=\frac{B_{12}J_{12}+n_e C_{12}}{A_{21}+n_e C_{21}}\,, \tag{6.3.24b}$$

as the elastic collision term drops out. In (6.3.24),

$$J_{12}=\int f_1(\boldsymbol{v})\,I_{12}(\boldsymbol{v})\,d^3v=\iint I_\nu(\boldsymbol{n})\,\varphi_\nu(\boldsymbol{n})\,d\nu\frac{d\Omega}{4\pi}\,, \tag{6.3.25}$$

see (2.4.14).

Inserting (6.3.24b) in (6.3.21), using

$$\frac{A_{21}}{B_{12}}=\frac{g_1}{g_2}\,\frac{2h\nu_0^3}{c^2}\,, \tag{6.3.26}$$

which follows from (1.6.31, 32), and the detailed balance relation, see (2.3.10),

$$\frac{C_{21}}{C_{12}}=\frac{g_1}{g_2}\,\mathrm{e}^{h\nu_0/kT_e} \tag{6.3.27}$$

valid in a Maxwellian electron gas of temperature T_e, where g_1, g_2 are the statistical weights of the atomic levels, the line source function for complete redistribution can be written as

$$S=(1-\varepsilon)\,J_{12}+\varepsilon B^{\mathrm{W}}\,. \tag{6.3.28}$$

Here the dimensionless parameter ε, $0<\varepsilon<1$, is defined by

$$\varepsilon=\frac{\varepsilon_0}{1+\varepsilon_0}\,,\quad \text{where} \tag{6.3.29}$$

$$\varepsilon_0=\frac{n_e C_{21}}{A_{21}} \tag{6.3.30}$$

is the ratio of collisional de-excitations to spontaneous emissions, and, see (1.6.33),

$$B^{\mathrm{W}}\equiv B^{\mathrm{W}}_{\nu_0}(T_e)=\frac{2h\nu_0^3}{c^2}\,\mathrm{e}^{-h\nu_0/kT_e} \tag{6.3.31}$$

is the Planck-Wien function of electron temperature T_e at the frequency ν_0 of the spectral line. In non-LTE situations, $\varepsilon_0\ll 1$ and $\varepsilon\ll 1$, while in collision dominated gases, $\varepsilon_0\gg 1$ and $\varepsilon\simeq 1$.

The line source function (6.3.28), corresponding to complete redistribution in the static approximation, is frequently used in non-LTE line transfer calculations.

(5) In the *diffusion approximation,* the streaming of excited atoms is taken into account in form of a diffusion flow. Integrating the kinetic equation (6.3.22) over all $\boldsymbol{v}$ gives the continuity equation, see (4.1.27, 6.2.1),

$$\nabla \cdot \boldsymbol{\Gamma}_2 = \hat{\nu}_2 \tag{6.3.32}$$

where, see (4.1.8),

$$\boldsymbol{\Gamma}_2 = \int \boldsymbol{v} F_2(\boldsymbol{v})\, d^3v \tag{6.3.33}$$

is the flux density of excited atoms, and, see (4.1.18),

$$\hat{\nu}_2 = (B_{12}J_{12} + n_e C_{12})\, n_1 - (A_{21} + n_e C_{21})\, n_2 \tag{6.3.34}$$

is the local production rate of excited atoms due to radiative interactions and inelastic collisions. In the diffusion approximation, one sets

$$\boldsymbol{\Gamma}_2 = -D_2 \nabla n_2 \ , \tag{6.3.35}$$

where D_2 is the diffusion constant of the excited atoms.

The diffusion constant D_2 can be estimated using the method outlined in Sect. 4.4.1. In (4.4.18),

$$\phi_A \simeq -n\bar{v}l \frac{dA}{dz} \ ,$$

the physical quantity whose gradient gives rise to the considered transport process of diffusion is the ratio $A = n_2/n$, where $n = n_1 + n_2$ is the total density of two-level atoms, while the corresponding flux is $\phi_A = \Gamma_2$. Thus,

$$\Gamma_2 \simeq -\bar{v}l \frac{dn_2}{dz} \tag{6.3.36}$$

and hence, by comparison with (6.3.35),

$$D_2 \simeq \bar{v}l \ . \tag{6.3.37}$$

As the mean thermal velocity is $\bar{v} \sim (kT/m)^{1/2}$, and the mean free path is $l \sim \bar{v}/\gamma_2$, one obtains

$$D_2 = \frac{kT}{m\gamma_2} \ , \tag{6.3.38}$$

where m is again the mass of a two-level atom, and T is the temperature of the gas.

To derive (6.3.38) more formally, one starts from the kinetic equation for F_2 (without inelastic processes)

$$\boldsymbol{v} \cdot \nabla F_2 = \gamma_2 n_2 [f^{\mathrm{M}}(v) - f_2(\boldsymbol{v})] \tag{6.3.39}$$

where $f^{\mathrm{M}}(v)$ is a normalized Maxwell distribution of temperature T. [The Maxwell distribution (6.3.13) containing a macroscopic velocity $\boldsymbol{u}$ is denoted by $f^{\mathrm{M}}(\boldsymbol{v})$, and the isotropic Maxwell distribution with $\boldsymbol{u}=0$ by $f^{\mathrm{M}}(v)$.] Since in a diffusion regime the distribution function deviates only slightly from a Maxwell distribution, one solves (6.3.39) for $F_2 = n_2 f_2$ by setting in the streaming term $F_2 \simeq n_2 f^{\mathrm{M}}$. Hence

$$F_2(\boldsymbol{v}) = n_2 f^{\mathrm{M}}(v) - \frac{1}{\gamma_2} \boldsymbol{v} f^{\mathrm{M}}(v) \cdot \nabla n_2 \tag{6.3.40}$$

and therefore

$$\Gamma_2 = \int \boldsymbol{v} F_2(\boldsymbol{v}) d^3 v = -\frac{kT}{m\gamma_2} \nabla n_2 \tag{6.3.41}$$

where we have used

$$\int \boldsymbol{v}\boldsymbol{v} f^{\mathrm{M}}(v) d^3 v = \frac{kT}{m} \overleftrightarrow{\boldsymbol{I}} \tag{6.3.42}$$

with $\overleftrightarrow{\boldsymbol{I}}$ denoting the unit tensor. Equation (6.3.41) again leads to the diffusion constant (6.3.38).

Returning to the diffusion approximation, (6.3.32, 34, 35) give

$$-D_2 \nabla^2 n_2 = n_1 (B_{12} J_{12} + n_e C_{12}) - n_2 (A_{21} + n_e C_{21}) \ . \tag{6.3.43}$$

As a final step,

$$\frac{1}{n_1} \nabla^2 n_2 \simeq \nabla^2 \left(\frac{n_2}{n_1} \right) , \tag{6.3.44}$$

which expresses the fact that gradients of n_1 must be much smaller than the gradients of n_2 due to non-LTE line transfer, otherwise the diffusion approximation discussed here does not make sense. Substituting the source function (6.3.21) into (6.3.43) and taking (6.3.44) into account, we finally obtain the *diffusion equation* of the source function for complete redistribution

$$S - \Delta_2 \nabla^2 S = (1-\varepsilon) J_{12} + \varepsilon B^{\mathrm{W}} \ , \tag{6.3.45}$$

where

$$\Delta_2 = \frac{D_2/A_{21}}{1+\varepsilon_0} \ . \tag{6.3.46}$$

If $D_2 \to 0$, then $\Delta_2 \to 0$, and (6.3.45) reduces to (6.3.28) as it should.

From a physical point of view it is evident that the diffusion approximation (6.3.45) [or its limiting case, the static approximation (6.3.28)] describes non-LTE line transfer correctly only if the elastic collision frequency of the excited atoms is sufficiently high. However, in most cases of physical interest, elastic collisions of excited atoms are, on the contrary, negligible.

(6) For completeness, we mention the *LTE approximation* where the atomic levels are supposed to be populated according to the Boltzmann distribution,

$$\frac{n_2}{n_1} = \frac{g_2}{g_1} e^{-h\nu_0/kT_e} , \tag{6.3.47}$$

and complete redistribution holds. Substituting (6.3.47) in (6.3.21) yields

$$S = B^W . \tag{6.3.48}$$

Physically, this approximation is valid if inelastic collisions are very frequent, $\varepsilon_0 = n_e C_{21}/A_{21} \gg 1$. Indeed, for $\varepsilon_0 \to \infty$, both (6.3.28, 45) reduce to (6.3.48).

(7) Lastly, the formulation of the line source function in terms of *redistribution functions* (Sect. 6.3.5) uses the complete source function (6.3.17) [as opposed to the source function (6.3.21) corresponding to complete redistribution], and it makes implicit use of the solution of the kinetic equation (6.3.22) in the static approximation (6.3.23). Hence, this approximation of the source function neglects only streaming of the excited atoms.

6.3.4 Line Profile Coefficients

The two line profile coefficients φ_ν and ψ_ν of a two-level atom are according to (6.3.4, 5) the averages of the corresponding atomic profile coefficients over the appropriate atomic velocity distributions. As shown below, the atomic absorption profile α_{12} can be considered as a given quantity. If the velocity distribution of nonexcited atoms $f_1(\boldsymbol{v})$ is known (Sect. 6.3.3), the laboratory absorption profile φ_ν is known, too. By contrast, both the atomic emission profile η_{21} and the velocity distribution of excited atoms $f_2(\boldsymbol{v})$, which together form the laboratory emission profile ψ_ν, are in general dependent on the radiation field. Consequently, in a line transfer problem, I_ν and ψ_ν must be determined self-consistently.

The reason why the emission profile ψ_ν is such an intricate quantity is that in a process consisting of absorption followed by re-emission, the re-emitted photon is generally correlated with the previously absorbed one. In other words, absorption followed by re-emission is a non-Markovian process. These photon correlations are provided by the atomic process itself, on the one hand, and by the velocity of the atom, on the other.

We therefore distinguish between two limiting cases. We speak of *correlated re-emission* if during the lifetime of the excited atomic state no elastic collisions take place that give rise to velocity changes of the atom. By contrast, we speak of *uncorrelated re-emission* if elastic collisions with an excited atom occur that change its velocity [6.15].

For correlated re-emission, the constancy of the velocity of the atom provides, via the Doppler effect, a correlation between the frequencies of the absorbed and re-emitted photons, quite independently of whether or not there is an additional correlation between the photon frequencies in the atomic rest frame. In the case of uncorrelated re-emission, the absorbed and re-emitted photons are uncorrelated because, on the one hand, the velocities of the atom at the times of absorption and re-emission are uncorrelated as a result of elastic collisions, while, on the other hand, one may assume that the frequencies of the two photons in the atomic rest frame are likewise uncorrelated (complete redistribution in the atomic rest frame) owing to the strong interaction of the excited atom with the elastically colliding particles.

The limiting case of correlated re-emission, which applies in a good approximation to many cases of physical interest, is more important and will be adopted in the following. (It is likewise adopted in Appendix B, which provides an introductory discussion of atomic profile coefficients.) The effect of elastic velocity-changing collisions with excited atoms is readily incorporated into the final expressions of the emission profiles η_{21} and ψ_ν, as shown below.

More precisely, we assume that the velocity of an atom is constant during an excitation – de-excitation process $1 \to 2 \to 1$,

$$\boldsymbol{v} = \text{const} \ . \tag{6.3.49}$$

This assumption requires: (1) no velocity-changing elastic collisions with the excited atoms occur (correlated re-emission), (2) the recoil of an atom due to absorption and emission of photons, and due to inelastic collisions (with electrons) is negligible, and (3) the velocity of the atom is not affected by the interaction with perturber particles giving rise to collision broadening of the atomic levels. We ignore the fact that collisions generally have both elastic and inelastic components so that, strictly speaking, the possibility of velocity-changing inelastic collisions with excited atoms should also be considered.

As a consequence of (6.3.49), as far as the atomic profile coefficients are concerned, the ensemble of atoms within a small volume element of the gas with velocities in the range $(\boldsymbol{v}, d^3v)$ can be treated independently of all other atoms with velocities outside this range.

In an absorption – re-emission process, $(\nu', \boldsymbol{n}')$ denotes the absorbed photon, and $(\nu, \boldsymbol{n})$ the re-emitted photon. In the usual approximation which transforms only the photon frequency [but not the photon direction nor the specific intensity, see (2.4.3)], these photons are described in the atomic rest frame by $(\xi', \boldsymbol{n}')$ and $(\xi, \boldsymbol{n})$, respectively, where

$$\xi = \nu - \frac{\nu_0}{c} \boldsymbol{n} \cdot \boldsymbol{v} \ , \quad \xi' = \nu' - \frac{\nu_0}{c} \boldsymbol{n}' \cdot \boldsymbol{v} \tag{6.3.50}$$

with the *same* atomic velocity $\boldsymbol{v}$ on account of (6.3.49).

We now come to the notion of *natural population* of an atomic level, which is crucial for understanding the atomic profile coefficients α_{12} and η_{21}. An atomic level is said to be naturally populated if the probability of emitting (absorbing) a well-defined photon $(\xi,\boldsymbol{n})$ in a transition to a lower (higher) level, when averaged over an ensemble of identical atoms, is independent of the previous history of the ensemble, that is, of the process by which the level has been populated. It should be realized that any property of an atom, such as the absorption or emission probability, always refers to the quantum mechanical average over an ensemble of identical atoms.

An atomic level is naturally populated by processes involving particles or photons whose energy spectrum is constant over the width of the level, and by spontaneous radiative processes. Thus, excitation and de-excitation collisions and spontaneous emissions lead to natural population, as do all ionization and recombination processes (collisional and photoionization, three-body and radiative recombination). However, absorptions and stimulated emissions of photons due to bound-bound transitions populate the final atomic level naturally only if the intensity of the radiation field is constant over the frequency range of the spectral line, otherwise these processes lead to deviations from natural population. It may be helpful to recall that in the Weisskopf-Woolley model (Appendix B), which views a broadened atomic level as a continuous distribution of sublevels, natural population corresponds to uniform occupation of the sublevels of an atomic level.

The concept of ensemble averaging explicitly used in this definition of natural population is important in order to avoid possible misinterpretations connected with spontaneous emissions by multilevel atoms. Consider two consecutive spontaneous emissions of photons by a multilevel atom due to the transitions $3\to2\to1$. In general, the photon $(\xi,\boldsymbol{n})$ emitted in the transition $2\to1$ is correlated with the photon $(\xi',\boldsymbol{n}')$ emitted in the previous transition $3\to2$. Nevertheless, as discussed in Appendix B.6, spontaneous emissions $3\to2$ lead to natural population of the atomic level 2. Thus, the fact that spontaneous emissions $3\to2$ populate level 2 naturally does not exclude the possibility of correlations between the photons $(\xi',\boldsymbol{n}')$ and $(\xi,\boldsymbol{n})$ involved in consecutive radiative transitions $3\to2\to1$. It means only that the spectrum of spontaneously emitted photons $(\xi,\boldsymbol{n})$ is identical with that of an ensemble of atoms whose levels 2 are populated through inelastic collisions.

Let us now define atomic quantities $r_{12}(\xi)$, $r_{21}(\xi)$, and $r_{121}(\xi',\boldsymbol{n}';\xi,\boldsymbol{n})$. In view of the more complicated case of multilevel atoms discussed in Sect. 6.5, we deliberately use a seemingly complicated notation. The more conventional notation (Appendices B.1, 2, 5), which writes α_{12} and r instead of r_{12} and r_{121}, and which does not introduce r_{21}, obscures the central role played by the natural population of the atomic levels in their definitions. Note that the quantities r_{12}, r_{21}, r_{121} are generalized atomic redistribution functions, as defined in Appendix B.6. We define:

$r_{12}(\xi)\,d\xi\,d\Omega/4\pi$ is the probability of absorbing a photon $(\xi, d\xi; \boldsymbol{n}, d\Omega)$ in an isotropic white radiation field if the atom is in the *naturally populated* level

1*. (Here and in the following i^* indicates a naturally populated atomic level i.)

$r_{21}(\xi)\,d\xi\,d\Omega/4\pi$ is the probability of emitting a photon $(\xi, d\xi; \boldsymbol{n}, d\Omega)$ spontaneously if the atom is in the *naturally populated* level 2*.

$r_{121}(\xi',\boldsymbol{n}';\xi,\boldsymbol{n})\,d\xi'\,d\Omega'\,d\xi\,d\Omega/(4\pi)^2$ is the probability of absorbing a photon $(\xi', d\xi'; \boldsymbol{n}', d\Omega')$ in an isotropic white radiation field, and of re-emitting a photon $(\xi, d\xi; \boldsymbol{n}, d\Omega)$ spontaneously if the atom is initially in the *naturally populated* level 1*.

As far as photon directions are concerned, $r_{12}(\xi)$ and $r_{21}(\xi)$ are supposed to be isotropic, and $r_{121}(\xi',\boldsymbol{n}';\xi,\boldsymbol{n})$ to depend only on the angle between $\boldsymbol{n}'$ and $\boldsymbol{n}$. For our purposes, the quantities r_{12}, r_{21}, r_{121} are considered as known quantities which can be calculated by quantum mechanics, taking the interactions with the surrounding gas (line-broadening collisions, electric microfields) into account.

The normalizations are, see (B.6.1, 2.3),

$$\int r_{12}(\xi)\,d\xi = \int r_{21}(\xi)\,d\xi = 1 \;, \tag{6.3.51a}$$

$$\iiiint r_{121}(\xi',\boldsymbol{n}';\xi,\boldsymbol{n})\,d\xi'\frac{d\Omega'}{4\pi}\,d\xi\frac{d\Omega}{4\pi} = 1 \;. \tag{6.3.51b}$$

Furthermore, see (B.6.2, 2.4),

$$\iint r_{121}(\xi',\boldsymbol{n}';\xi,\boldsymbol{n})\,d\xi\frac{d\Omega}{4\pi} = r_{12}(\xi') \;, \tag{6.3.52a}$$

$$\iint r_{121}(\xi',\boldsymbol{n}';\xi,\boldsymbol{n})\,d\xi'\frac{d\Omega'}{4\pi} = r_{21}(\xi) \;. \tag{6.3.52b}$$

Lastly, there are the symmetry relations, see (B.6.41, 2.15),

$$r_{12}(\xi) = r_{21}(\xi) \;, \tag{6.3.53a}$$

$$r_{121}(\xi',\boldsymbol{n}';\xi,\boldsymbol{n}) = r_{121}(\xi,\boldsymbol{n};\xi',\boldsymbol{n}') \;. \tag{6.3.53b}$$

It is helpful to recall that in some cases the *atomic redistribution function* can be written as (Appendices B.2, 5)

$$r_{121}(\xi',\boldsymbol{n}';\xi,\boldsymbol{n}) = \gamma(\boldsymbol{n}',\boldsymbol{n})\,r_{12}(\xi')\,[\beta\delta(\xi-\xi') + (1-\beta)\,r_{21}(\xi)] \;, \tag{6.3.54}$$

with $0 \leqslant \beta \leqslant 1$, where $\gamma(\boldsymbol{n}',\boldsymbol{n})$ is a normalized phase function, (B.2.9 – 11). The redistribution function (6.3.54) describes the re-emission of a radiation-excited atom as a superposition of coherent re-emission and completely redistributed re-emission.

We now proceed to the explicit expressions for the atomic profile coefficients α_{12} and η_{21} [6.16]. Since in the absence of stimulated emissions the ground state of an atom is naturally populated, it follows at once that the atomic absorption profile is given by

$$\alpha_{12}(\xi) = r_{12}(\xi) \ . \tag{6.3.55}$$

On the other hand, to derive the emission profile η_{21}, one must treat the absorptions, which may lead to deviations from natural population of the excited level and which are denoted by $1^* \Rightarrow 2$, separately from all the other processes which populate the excited level naturally, denoted collectively by $\rightarrow 2^*$. Thus,

$$\eta_{21}(\xi,\boldsymbol{n}) = \text{prob}\{1^* \Rightarrow 2\} j_{121}(\xi,\boldsymbol{n}) + \text{prob}\{\rightarrow 2^*\} r_{21}(\xi) \tag{6.3.56}$$

where prob$\{\ldots\}$ denote the relative probabilities of the corresponding processes. Here the quantity j_{121} is the normalized emission profile of radiation-excited atoms for correlated re-emission, given by [cf. (B.6.30); see also (B.2.22) where $\rho_{21} \equiv j_{121}$]

$$j_{121}(\xi,\boldsymbol{n}) = \frac{1}{I_{12}^*} \iint I_{\nu'}(\boldsymbol{n}') r_{121}(\xi',\boldsymbol{n}';\xi,\boldsymbol{n}) d\nu' \frac{d\Omega'}{4\pi} \ , \tag{6.3.57}$$

$$\iint j_{121}(\xi,\boldsymbol{n}) d\xi \frac{d\Omega}{4\pi} = 1 \ . \tag{6.3.58}$$

In (6.3.57),

$$I_{12}^* = \iint I_\nu(\boldsymbol{n}) r_{12}(\xi) d\nu \frac{d\Omega}{4\pi} \ , \tag{6.3.59}$$

which is equal to the quantity I_{12},(6.3.9), on account of (6.3.55), see (B.6.16). One readily verifies that in an isotropic white radiation field, $I_\nu(\boldsymbol{n}) = I = \text{const}$, radiation-excited atoms re-emit according to complete redistribution in the atomic rest frame, $j_{121}(\xi,\boldsymbol{n}) = r_{21}(\xi)$.

It remains to determine the two probabilities prob$\{\ldots\}$ in (6.3.56). Here velocity-changing elastic collisions with excited two-level atoms are easily taken into account in the framework of a strong collision model (on which also our definition of the limiting case of uncorrelated re-emission is based) according to which elastic collisions give rise to a large momentum transfer between the colliding particles, so that it is reasonable to assume that they also lead to natural population of the excited level 2.

Writing the elastic collision term of the excited two-level atoms in terms of a creation and a destruction term,

$$\left(\frac{\delta F_2}{\delta t}\right)_{\text{el}} = \Gamma_2^+(\boldsymbol{v}) - \Gamma_2^-(\boldsymbol{v}) \ , \tag{6.3.60}$$

then

$$\mathrm{prob}\{1^* \Rightarrow 2\} = \frac{1}{N} B_{12} I_{12}^* F_1(\boldsymbol{v}) \ , \qquad (6.3.61a)$$

$$\mathrm{prob}\{\rightarrow 2^*\} = \frac{1}{N} [n_e C_{12} F_1(\boldsymbol{v}) + \Gamma_2^+(\boldsymbol{v})] \ , \qquad (6.3.61b)$$

where

$$N = (B_{12} I_{12}^* + n_e C_{12}) F_1(\boldsymbol{v}) + \Gamma_2^+(\boldsymbol{v}) \qquad (6.3.61c)$$

so that

$$\mathrm{prob}\{1^* \Rightarrow 2\} + \mathrm{prob}\{\rightarrow 2^*\} = 1 \ . \qquad (6.3.62)$$

Equations (6.3.56, 57, 61) determine the atomic emission profile $\eta_{21}(\xi, \boldsymbol{n})$ of two-level atoms of velocity $\boldsymbol{v}$ immersed in the radiation field $I_\nu(\boldsymbol{n})$. For correlated re-emission, $\Gamma_2^+ = 0$, they reduce to (B.2.26, 6.32), the distribution function of nonexcited two-level atoms $F_1(\boldsymbol{v})$ dropping out.

Note that apart from the implicit dependence on $\boldsymbol{v}$ through the rest frame frequency $\xi(\boldsymbol{v})$, all quantities introduced may also depend explicitly on the atomic velocity $\boldsymbol{v}$ in the laboratory frame. The explicit $\boldsymbol{v}$ dependence of the atomic quantities r_{12}, r_{21}, r_{121} and hence of α_{12} is usually neglected, whereas in general

$$I_{12}^* \equiv I_{12}^*(\boldsymbol{v}) \ , \quad j_{121} \equiv j_{121}(\xi, \boldsymbol{n}; \boldsymbol{v}) \ , \quad \eta_{21} \equiv \eta_{21}(\xi, \boldsymbol{n}; \boldsymbol{v}) \ .$$

An explicit $\boldsymbol{v}$ dependence $r_{12}(\xi; \boldsymbol{v})$ arises if the line broadening depends on the relative velocity between atom and perturbers.

Apart from the assumption of a (quasi-)stationary state, this discussion has also tacitly assumed a quasi-homogeneous state so that consecutive excitations and de-excitations take place under the same physical conditions. To make this point clearer, consider a light beam in z direction passing through an atomic beam in x direction. If the photon frequency corresponds to the resonance transition of the atoms, one observes resonance fluorescence whose intensity varies along the x axis according to

$$I(x) \propto \exp(-A_{21} x / v) \ ,$$

where v is the velocity of the atoms, owing to the exponential decay of the excited atoms through spontaneous emissions. This example illustrates the fact that the atomic emission profile η_{21} is, in principle, a nonlocal quantity because successive excitations and de-excitations of atoms with nonvanishing velocity take place at different points in space. More specifically, for correlated re-emission ($\Gamma_2^+ = 0$), the nonlocal character of η_{21} is taken into account through

$$\eta_{21}(\xi, \boldsymbol{n}; \boldsymbol{r}) = \int_0^\infty \eta_{21}^{\mathrm{loc}}(\xi, \boldsymbol{n}; \boldsymbol{r} - \boldsymbol{v} t) \, \mathrm{e}^{-A_{21} t} A_{21} \, dt \ , \qquad (6.3.63)$$

where η_{21}^{loc} denotes the local version of η_{21}, (6.3.56, 61), because $\exp(-A_{21}t) \cdot A_{21}dt$ is the probability that a spontaneous emission $2 \rightarrow 1$ occurs during a time interval dt at time t after the previous excitation $1 \rightarrow 2$. The nonlocal character of η_{21} is usually negligible, but it may play a role in the presence of very steep gradients as provided, for example, by shock waves or walls.

The laboratory profile coefficients φ_ν and ψ_ν are obtained from the atomic profile coefficients through (6.3.4, 5). Here the velocity distribution $f_1(\boldsymbol{v})$ is usually assumed to be a Maxwell distribution. On the other hand, the velocity distribution $f_2(\boldsymbol{v})$ must, in principle, be determined as a solution of the kinetic equation of the excited atoms (Sect. 6.4).

The velocity distribution of excited atoms may differ from a Maxwellian for a variety of reasons.

(a) From a macroscopic point of view, owing to the escape of photons from the gas, non-LTE line transfer gives rise to density gradients of excited atoms near the boundary of the gas, which in turn produce a convective flow of excited atoms so as to reduce these gradients.

(b) Boundary layers with non-Maxwellian velocity distributions of excited atoms can be formed as a result of particular boundary conditions; for example, a wall that de-excites the atoms hitting it gives rise to a macroscopic flow velocity of excited atoms toward the wall as no excited atoms flow back into the gas.

(c) For non-LTE line transfer, the radiation intensity near the boundary of the gas is anisotropic and in general a strongly varying function of frequency, so that, via the Doppler effect, non-Maxwellian velocity distributions of excited atoms may result from selective radiative excitation of atoms in certain velocity ranges.

To give a simple illustration of this last point, consider a Maxwellian gas of two-level atoms with sharp levels, so that they can absorb only photons of frequency $\xi = \nu_0$, and let this gas be irradiated with a monochromatic light beam of this frequency ν_0. If the light beam propagates in the z direction, then only atoms with vanishing velocity component $v_z = 0$ absorb photons from the beam, because atoms with $v_z \neq 0$ see frequencies $\xi \neq \nu_0$ they cannot absorb. Hence, if recoil is neglected, the excited atoms have the non-Maxwellian velocity distribution

$$f_2(v_x, v_y, v_z) = f^{\text{M}}(v_x)\, f^{\text{M}}(v_y)\, \delta(v_z) \quad ,$$

where $f^{\text{M}}(v_i)$ denotes a one-dimensional Maxwell distribution and $\delta(v_i)$ is the one-dimensional delta function.

Let us pursue the example of two-level atoms with sharp levels a little further by considering isotropic radiation, neglecting collisions and the streaming of excited atoms. In this situation, $\eta_{21}(\xi)$ and $f_2(v)$ are isotropic quantities. We introduce the dimensionless frequency x and the dimensionless velocity y through

$$x = \frac{\nu - \nu_0}{\Delta\nu_D}\ , \quad y = \frac{1}{w}\boldsymbol{v}\ , \tag{6.3.64}$$

where

$$w = \left(\frac{2kT}{m}\right)^{1/2} \tag{6.3.65}$$

is the thermal velocity of the atoms, and

$$\Delta\nu_D = \frac{\nu_0}{c} w \tag{6.3.66}$$

is the Doppler width of the spectral line.

For a Maxwellian velocity distribution $f_1(v)$ and zero linewidth in the atomic rest frame, the absorption profile is a pure Doppler profile,

$$\varphi(x) = \int f^M(y)\,\delta(x - \boldsymbol{n}\cdot\boldsymbol{y})\,d^3y = \frac{1}{\pi^{1/2}} e^{-x^2}\ , \tag{6.3.67}$$

where

$$f^M(y) = \frac{1}{\pi^{3/2}} e^{-y^2}\ , \quad \int f^M(y)\,d^3y = 1 \tag{6.3.68}$$

is the normalized Maxwell distribution. On the other hand, the emission profile is given by

$$\psi(x) = \int f_2(y)\,\delta(x - \boldsymbol{n}\cdot\boldsymbol{y})\,d^3y\ . \tag{6.3.69}$$

Here the velocity distribution $f_2(y)$ is obtained from (6.3.22, 24b), setting there $\boldsymbol{v}\cdot\nabla F_2 = C_{12} = \gamma_2 = 0$, yielding

$$f_2(y) = \frac{I_{12}(y)}{J_{12}} f^M(y)\ , \tag{6.3.70}$$

where

$$I_{12}(y) = \int_{-\infty}^{\infty} dx\, I(x) \int_{(4\pi)} \frac{d\Omega}{4\pi}\,\delta(x - \boldsymbol{n}\cdot\boldsymbol{y}) = \frac{1}{2y}\int_{-y}^{y} I(x)\,dx \tag{6.3.71}$$

and

$$J_{12} = \int_{-\infty}^{\infty} I(x)\,\varphi(x)\,dx = \frac{1}{\pi^{1/2}}\int_{-\infty}^{\infty} I(x)\,e^{-x^2}dx\ . \tag{6.3.72}$$

Equations (6.3.70–72) show that $f_2(y) \neq f^M(y)$ except in the case of white radiation $I(x) = I = \text{const}$. Above we have used

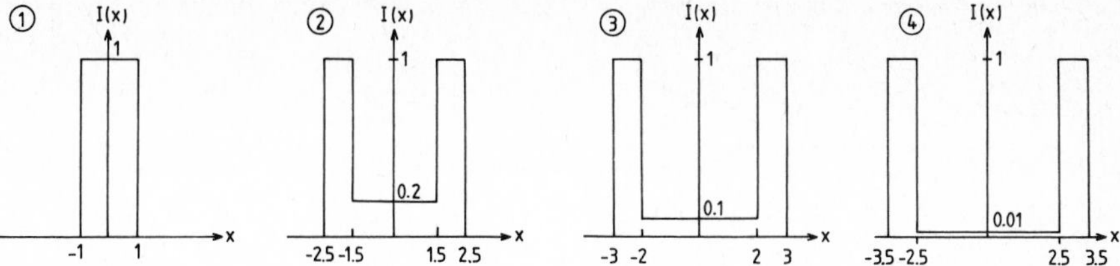

Fig. 6.7. Isotropic radiation intensities $I(x)$ vs dimensionless frequency x, used to calculate the emission profiles $\psi(x)$ and the velocity distributions $f_2(y)$ in Figs. 6.8, 9. The single-humped intensity profile 1 corresponds roughly to a spectral line in a gas of moderate optical thickness, the double-humped intensity profiles 2 – 4 to spectral lines near the boundary of non-LTE gases of large optical thickness

$$\int_{(4\pi)} \delta(x-\boldsymbol{n}\cdot y)\,d\Omega = \frac{2\pi}{y}\,\Theta(y-|x|) \;, \tag{6.3.73}$$

where

$$\Theta(z) = \begin{cases} 1 & \text{if } z>0 \\ 0 & \text{if } z<0 \end{cases} \tag{6.3.74}$$

is the step function. Substituting (6.3.70) into (6.3.69) now leads to

$$\psi(x) = \frac{2}{\pi^{1/2} J_{12}} \int_{|x|}^{\infty} I_{12}(y)\, y e^{-y^2} dy \;. \tag{6.3.75}$$

Figures 6.8, 9 show the ratios $\psi(x)/\varphi(x)$ and $f_2(y)/f^{\mathrm{M}}(y)$, respectively, for the four simple radiation intensities plotted in Fig. 6.7. The intensity profile "1" corresponds roughly to a spectral line in a gas of moderate optical thickness, while the intensity profiles "2" – "4" correspond to spectral lines near the boundary of non-LTE gases of large optical thickness (Sect. 6.4). The functions $\psi(x)$ and $f_2(y)$ do not depend critically on deviations from isotropy, nor are they very sensitive with respect to the exact shape of $I(x)$ as long as its maxima and minima are kept fixed.

Figures 6.8, 9 may give some feeling for the domain of validity of the approximation of complete redistribution $\psi(x) = \varphi(x)$, which here is equivalent to the equality $f_2(y) = f^{\mathrm{M}}(y)$ since $\alpha_{12} = \eta_{21}$. As can be seen, this approximation is usually quite good in the line core, rather independently of the form of the intensity profile, but it may completely fail in the line wings.

6.3.5 Redistribution Functions

We now discuss the line source function of a two-level atom in terms of so-called redistribution functions [6.10, 13, 16]. As we shall see, this formulation

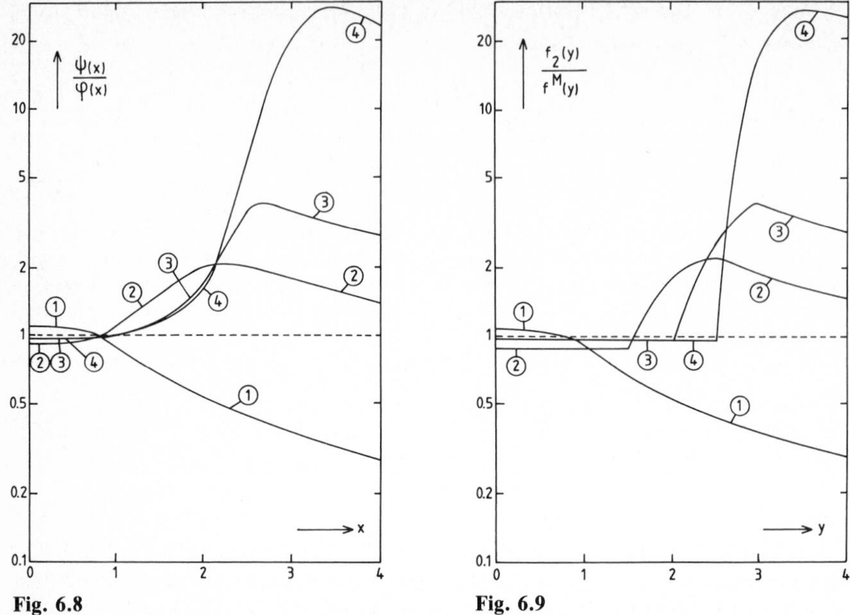

Fig. 6.8. Ratio of emission to absorption profile $\psi(x)/\varphi(x)$ of radiation-excited atoms vs dimensionless frequency x for the limiting case of pure Doppler broadening. All profiles are symmetric about the line center $x=0$. Labels 1–4 refer to the radiation intensities 1–4 shown in Fig. 6.7. Complete redistribution corresponds to $\psi(x)/\varphi(x)\equiv 1$

Fig. 6.9. Ratio of the velocity distributions of radiation-excited and nonexcited atoms $f_2(y)/f^{\mathrm{M}}(y)$ vs dimensionless velocity y for the limiting case of pure Doppler broadening. All velocity distributions are isotropic. Labels 1–4 refer to the radiation intensities 1–4 in Fig. 6.7. A Maxwell distribution of excited atoms corresponds to $f_2(y)/f^{\mathrm{M}}(y)\equiv 1$

of the source function neglects only the streaming of excited atoms, and therefore should provide an excellent approximation in many cases of physical interest.

As usual, we assume a Maxwellian velocity distribution of nonexcited atoms,

$$f_1(\boldsymbol{v}) = f^{\mathrm{M}}(\boldsymbol{v}) \;, \tag{6.3.76}$$

with $f^{\mathrm{M}}(\boldsymbol{v})$ given by (6.3.13). On account of (6.3.55), the absorption profile is then

$$\varphi_\nu(\boldsymbol{n}) = \int d^3v\, f^{\mathrm{M}}(\boldsymbol{v})\, r_{12}(\xi) \tag{6.3.77}$$

which is anisotropic if the local macroscopic velocity of the gas is different from zero, $\boldsymbol{u}\neq 0$, (6.3.13).

The emission profile is given by

$$\psi_\nu(\boldsymbol{n}) = \int d^3v\, f_2(\boldsymbol{v})\, \eta_{21}(\xi,\boldsymbol{n};\boldsymbol{v}) \;. \tag{6.3.78}$$

According to (6.3.56, 61) and taking (6.3.76) into account, the atomic emission profile can be written as

$$\eta_{21}(\xi, \boldsymbol{n}; \boldsymbol{v}) = \frac{1}{L(\boldsymbol{v})}\left[B_{12}I_{12}^{*}(\boldsymbol{v})j_{121}(\xi, \boldsymbol{n}; \boldsymbol{v}) + \left(n_e C_{12} + \gamma_2 \frac{n_2}{n_1}\right) r_{21}(\xi)\right] \tag{6.3.79}$$

with

$$L(\boldsymbol{v}) = B_{12}I_{12}^{*}(\boldsymbol{v}) + n_e C_{12} + \gamma_2 \frac{n_2}{n_1} \tag{6.3.80}$$

if the elastic collision term of the excited atoms is approximated by (6.3.12) so that, by comparison with (6.3.60),

$$\Gamma_2^{+}(\boldsymbol{v}) = \gamma_2 n_2 f^{\mathrm{M}}(\boldsymbol{v}) \ . \tag{6.3.81}$$

Hence, as can be seen from (6.3.78, 79), it is the ratio $f_2(\boldsymbol{v})/L(\boldsymbol{v})$ that appears in the emission profile ψ_ν. It turns out that this ratio can be obtained directly from the kinetic equation of the excited atoms if the streaming term is neglected. Indeed, in the *static approximation* (6.3.23), and using (6.3.76), the kinetic equation (6.3.22) takes the form

$$(1 + \gamma_2\tau_2) f_2(\boldsymbol{v}) = \left(\frac{n_1}{\tau_2^{-1} n_2}[B_{12}I_{12}^{*}(\boldsymbol{v}) + n_e C_{12}] + \gamma_2\tau_2\right) f^{\mathrm{M}}(\boldsymbol{v}) \ , \tag{6.3.82}$$

with τ_2 denoting the mean lifetime of an excited atom, defined through

$$\tau_2^{-1} = A_{21} + n_e C_{21} \ , \tag{6.3.83}$$

while, according to (6.3.24b),

$$\frac{\tau_2^{-1} n_2}{n_1} = B_{12}J_{12}^{*} + n_e C_{12} \ , \tag{6.3.84}$$

with $J_{12}^{*} \equiv J_{12}$ in order to conform with our present notation which was chosen to facilitate comparison with multilevel atoms, (6.3.106a). Notice in passing that (6.3.82) shows explicitly that $f_2 \neq f^{\mathrm{M}}$ whenever $I_{12}^{*}(\boldsymbol{v})$ varies as a function of $\boldsymbol{v}$. In view of the definition (6.3.80) of $L(\boldsymbol{v})$, (6.3.82) yields immediately the desired ratio

$$\frac{f_2(\boldsymbol{v})}{L(\boldsymbol{v})} = \frac{1}{1 + \gamma_2\tau_2} \, \frac{n_1}{\tau_2^{-1} n_2} f^{\mathrm{M}}(\boldsymbol{v}) \tag{6.3.85}$$

which, together with (6.3.78, 79), leads to the emission profile

$$\psi_\nu(\boldsymbol{n}) = \frac{1}{1+\gamma_2\tau_2}\left(\frac{n_1}{\tau_2^{-1}n_2}\int d^3v\, f^{\mathrm{M}}(\boldsymbol{v})[B_{12}I_{12}^*(\boldsymbol{v})\, j_{121}(\xi,\boldsymbol{n};\boldsymbol{v}) + n_{\mathrm{e}}C_{12}r_{21}(\xi)] + \gamma_2\tau_2\int d^3v\, f^{\mathrm{M}}(\boldsymbol{v})\, r_{21}(\xi)\right). \qquad (6.3.86)$$

Because of $r_{21}(\xi) = r_{12}(\xi)$, (6.3.53a), the integrals over r_{21} in (6.3.86) yield the absorption profile φ_ν, (6.3.77). On the other hand, recalling from (6.3.57) that

$$I_{12}^*(\boldsymbol{v})\, j_{121}(\xi,\boldsymbol{n};\boldsymbol{v}) = \iint I_{\nu'}(\boldsymbol{n}')\, r_{121}(\xi',\boldsymbol{n}',\xi,\boldsymbol{n})\, d\nu' \frac{d\Omega'}{4\pi}, \qquad (6.3.87)$$

in the first term on the right-hand side of (6.3.86) the quantity

$$\begin{aligned} K_{121}^*(\nu,\boldsymbol{n}) &= \int d^3v\, f_1(\boldsymbol{v})\, I_{12}^*(\boldsymbol{v})\, j_{121}(\xi,\boldsymbol{n};\boldsymbol{v}) \\ &= \iint I_{\nu'}(\boldsymbol{n}')\, R_{121}(\nu',\boldsymbol{n}';\nu,\boldsymbol{n})\, d\nu' \frac{d\Omega'}{4\pi} \end{aligned} \qquad (6.3.88)$$

appears, where we have defined the *redistribution function* in the laboratory frame as the average of the atomic redistribution function r_{121} over the velocity distribution of nonexcited atoms,

$$R_{121}(\nu',\boldsymbol{n}';\nu,\boldsymbol{n}) = \int d^3v\, f_1(\boldsymbol{v})\, r_{121}(\xi',\boldsymbol{n}';\xi,\boldsymbol{n}) . \qquad (6.3.89)$$

Here $f_1(\boldsymbol{v}) = f^{\mathrm{M}}(\boldsymbol{v})$, (6.3.76). Recall that K_{121}^* and R_{121} refer to correlated reemission with ξ, ξ' given by (6.3.50).

The following relations are readily derived using (6.3.51, 52):

$$\iiiint R_{121}(\nu',\boldsymbol{n}';\nu,\boldsymbol{n})\, d\nu' \frac{d\Omega'}{4\pi} d\nu \frac{d\Omega}{4\pi} = 1 , \qquad (6.3.90)$$

$$\iint R_{121}(\nu',\boldsymbol{n}';\nu,\boldsymbol{n})\, d\nu \frac{d\Omega}{4\pi} = \varphi_{\nu'}(\boldsymbol{n}') , \qquad (6.3.91)$$

$$\iint K_{121}^*(\nu,\boldsymbol{n})\, d\nu \frac{d\Omega}{4\pi} = J_{12}^* . \qquad (6.3.92)$$

Using (6.3.84, 88), we finally obtain from (6.3.86) the emission profile in terms of the redistribution function, valid in the static approximation,

$$\psi_\nu(\boldsymbol{n}) = \frac{1}{1+\gamma_2\tau_2}\left(\frac{B_{12}K_{121}^*(\nu,\boldsymbol{n}) + n_{\mathrm{e}}C_{12}\varphi_\nu(\boldsymbol{n})}{B_{12}J_{12}^* + n_{\mathrm{e}}C_{12}} + \gamma_2\tau_2\varphi_\nu(\boldsymbol{n})\right) . \qquad (6.3.93)$$

The emission profile (6.3.93) is easily interpreted by observing that $(1+\gamma_2\tau_2)^{-1}$ and $\gamma_2\tau_2(1+\gamma_2\tau_2)^{-1}$ are the probabilities for correlated and uncorrelated re-emission, respectively, and that the emission profile corresponding to correlated re-emission is a superposition of the contributions from radiation-excited and collision-excited atoms, while the emission profile corresponding to uncorrelated re-emission is simply given by the absorption profile φ_ν.

From now on we adopt the simpler notation

$$K_\nu(\boldsymbol{n}) \equiv K^*_{121}(\nu,\boldsymbol{n}) \ , \quad R \equiv R_{121} \ , \quad J_{12} \equiv J^*_{12} \ .$$

Let us rewrite the emission profile (6.3.93) in another form to gain some additional insight into the notion of complete redistribution. Obviously,

$$\psi_\nu(\boldsymbol{n}) = \varphi_\nu(\boldsymbol{n}) + [\psi_\nu(\boldsymbol{n}) - \varphi_\nu(\boldsymbol{n})] \ . \tag{6.3.94}$$

Now, from (6.3.86),

$$\begin{aligned}\psi_\nu(\boldsymbol{n}) - \varphi_\nu(\boldsymbol{n}) &= \frac{1}{1+\gamma_2\tau_2} \frac{n_1}{\tau_2^{-1} n_2} \left[B_{12} K_\nu(\boldsymbol{n}) + n_e C_{12} \varphi_\nu(\boldsymbol{n}) - \frac{\tau_2^{-1} n_2}{n_1} \varphi_\nu(\boldsymbol{n}) \right] \\ &= \frac{1}{1+\gamma_2\tau_2} \frac{n_1}{\tau_2^{-1} n_2} B_{12} [K_\nu(\boldsymbol{n}) - J_{12} \varphi_\nu(\boldsymbol{n})] \ ,\end{aligned} \tag{6.3.95}$$

using (6.3.84) in the last step. Hence, combining (6.3.94, 95),

$$\psi_\nu(\boldsymbol{n}) = \varphi_\nu(\boldsymbol{n}) + \frac{1}{1+\gamma_2\tau_2} \frac{n_1}{\tau_2^{-1} n_2} B_{12} [K_\nu(\boldsymbol{n}) - J_{12} \varphi_\nu(\boldsymbol{n})] \ . \tag{6.3.96}$$

On the other hand, according to (6.3.25, 88), the term in brackets on the right-hand side of (6.3.96) may be written as

$$K_\nu(\boldsymbol{n}) - J_{12} \varphi_\nu(\boldsymbol{n}) = \iint I_{\nu'}(\boldsymbol{n}') [R(\nu',\boldsymbol{n}';\nu,\boldsymbol{n}) - \varphi_{\nu'}(\boldsymbol{n}') \varphi_\nu(\boldsymbol{n})] \, d\nu' \frac{d\Omega'}{4\pi} \ , \tag{6.3.97}$$

which shows that in the static approximation, *complete redistribution* $\psi_\nu = \varphi_\nu$ is equivalent to

$$R(\nu',\boldsymbol{n}';\nu,\boldsymbol{n}) = \varphi_{\nu'}(\boldsymbol{n}') \varphi_\nu(\boldsymbol{n}) \ . \tag{6.3.98}$$

Lastly, we derive the line source function in the static approximation. To this end, we start from the general expression (6.3.17)

$$S_\nu(\boldsymbol{n}) = \frac{n_2 A_{21}}{n_1 B_{12}} \frac{\psi_\nu(\boldsymbol{n})}{\varphi_\nu(\boldsymbol{n})} \ . \tag{6.3.99}$$

From (6.3.21, 28, 29, 30)

$$\frac{n_2 A_{21}}{n_1 B_{12}} = \frac{J_{12} + \varepsilon_0 B^{\mathrm{W}}}{1+\varepsilon_0} \ , \tag{6.3.100}$$

and from (6.3.93)

$$\frac{\psi_\nu(\boldsymbol{n})}{\varphi_\nu(\boldsymbol{n})} = \frac{1}{1+\gamma_2\tau_2}\left(\frac{K_\nu(\boldsymbol{n})/\varphi_\nu(\boldsymbol{n})+\varepsilon_0 B^{\mathrm{W}}}{J_{12}+\varepsilon_0 B^{\mathrm{W}}}+\gamma_2\tau_2\right) , \tag{6.3.101}$$

using

$$\frac{n_{\mathrm{e}}C_{12}}{B_{12}} = \varepsilon_0 B^{\mathrm{W}} \tag{6.3.102}$$

which follows from (6.3.26, 27, 30, 31). Introducing again the quantity ε defined by (6.3.29) gives the line source function in the static approximation

$$S_\nu(\boldsymbol{n}) = \frac{1}{1+\gamma_2\tau_2}\,[S_\nu^0(\boldsymbol{n})+\gamma_2\tau_2 S] , \tag{6.3.103}$$

where

$$\begin{aligned} S_\nu^0(\boldsymbol{n}) &= (1-\varepsilon)\frac{K_\nu(\boldsymbol{n})}{\varphi_\nu(\boldsymbol{n})}+\varepsilon B^{\mathrm{W}} \\ &= (1-\varepsilon)\frac{1}{\varphi_\nu(\boldsymbol{n})}\iint I_{\nu'}(\boldsymbol{n}')R(\nu',\boldsymbol{n}';\nu,\boldsymbol{n})\,d\nu'\,\frac{d\Omega'}{4\pi}+\varepsilon B^{\mathrm{W}} \end{aligned} \tag{6.3.104}$$

is the source function corresponding to correlated re-emission which, in general, is anisotropic and frequency-dependent within the spectral line, while, see (6.3.28),

$$\begin{aligned} S &= (1-\varepsilon)\,J_{12}+\varepsilon B^{\mathrm{W}} \\ &= (1-\varepsilon)\iint I_{\nu'}(\boldsymbol{n}')\,\varphi_{\nu'}(\boldsymbol{n}')\,d\nu'\,\frac{d\Omega'}{4\pi}+\varepsilon B^{\mathrm{W}} \end{aligned} \tag{6.3.105}$$

is the isotropic and frequency-independent source function corresponding to complete redistribution. For the redistribution function (6.3.98), (6.3.104) reduces to (6.3.105), as it should.

For explicit forms of the redistribution function $R(\nu',\boldsymbol{n}';\nu,\boldsymbol{n})$, we refer to the literature [6.13, 17, 18]. The functions quoted there suppose a static Maxwell distribution $f_1 = f^{\mathrm{M}}$ (with $\boldsymbol{u} = 0$), and they apply to correlated re-emission ($\boldsymbol{v}$ = const) in the absence of elastic collisions ($\gamma_2 = 0$). Their use in the line source function supposes the validity of the static approximation ($\boldsymbol{v}\cdot\nabla F_2 = 0$) and the absence of stimulated emissions.

6.3.6 Stimulated Emission

Stimulated emissions, which we have neglected up to now, turn out to complicate the problem of non-LTE line transfer considerably, as seen in the following [6.15, 19–22]. When discussing stimulated emission we suppose, as elsewhere in this book, that the radiation intensity is still so low that all induced radiation processes (absorptions and stimulated emissions) are correctly described by first-order perturbation theory. In particular, the interaction of atoms with high-intensity laser light is excluded from our consideration. Rather, our discussion is intended to show the *kinetic* implications of the presence of stimulated emissions in the context of non-LTE line transfer. Recall that in lowest order of the radiation intensity, the atomic emission profile η_{21}, and hence the laboratory emission profile ψ_ν too, apply to both spontaneous and stimulated emissions, which is a simple consequence of (1.3.3a), see Appendix B.1.

Stimulated emissions give rise to two fundamental complications in the problem of non-LTE line transfer by two-level atoms:

(1) Stimulated emissions may lead to deviations from natural population of the atomic ground state (Sect. 6.3.4 and Appendix B.3). This means that in the presence of stimulated emissions, the atomic absorption profile α_{12} may differ from the generalized redistribution function r_{12}, thereby becoming a function of the radiation field.

(2) In the presence of stimulated emissions, the distribution function of nonexcited atoms $F_1(\boldsymbol{v})$ cannot be specified a priori even when the total distribution function of two-level atoms $F(\boldsymbol{v}) = F_1(\boldsymbol{v}) + F_2(\boldsymbol{v})$ is known, because the validity of $F_2(\boldsymbol{v}) \ll F_1(\boldsymbol{v})$ for all velocities $\boldsymbol{v}$, (6.3.14), can no longer be assumed. Consequently, even if one adopts $\alpha_{12}(\xi) = r_{12}(\xi)$ as an approximation, the absorption profile φ_ν cannot be specified a priori, rather it has to be determined self-consistently together with the radiation field.

In this connection let us recall that absorptions and stimulated emissions also contribute to the broadening of the spectral line via the corresponding shortening of the lifetime of the atomic levels (Appendices B.1, 3).

From all of these remarks it is apparent that a proper treatment of stimulated emissions in line transfer calculations requires rather complex iteration procedures in order to obtain self-consistent solutions.

When discussing kinetic effects due to stimulated emissions, it is important to distinguish between the quantities $I_{12}(\boldsymbol{v})$ and $I_{21}(\boldsymbol{v})$ which are defined in terms of the atomic profiles α_{12} and η_{21}, (2.4.11, 12), on the one hand, and the quantity $I^*_{12}(\boldsymbol{v}) = I^*_{21}(\boldsymbol{v})$ which is defined in terms of the generalized redistribution function $r_{12} = r_{21}$, (6.3.59), on the other. Likewise, the quantities J_{12} and J_{21}, (2.4.14, 15), must be distinguished from J^*_{12} and J^*_{21} defined by

$$J^*_{12} = \int d^3v\, f_1(\boldsymbol{v}) I^*_{12}(\boldsymbol{v}) \;, \tag{6.3.106a}$$

$$J^*_{21} = \int d^3v\, f_2(\boldsymbol{v}) I^*_{21}(\boldsymbol{v}) \;. \tag{6.3.106b}$$

Note that $I^*_{12} = I^*_{21}$ always, but $J^*_{12} = J^*_{21}$ only if $f_1(\boldsymbol{v}) = f_2(\boldsymbol{v})$.

To get some further insight into the kinetic effects of stimulated emissions, let us now simply *assume* that (a) the atomic ground state is naturally populated, and (b) the distribution function $F_1(\boldsymbol{v}) = n_1 f_1(\boldsymbol{v})$ is known.

Assumption (a) is justified if the broadening of the ground state is negligible, or if elastic collisions of nonexcited atoms take place during the lifetime of the ground state. Assumption (b) may be viewed either as an approximation or as being an intermediate result of an iteration procedure which determines F_1 self-consistently together with F_2 and I_ν.

It follows from assumptions (a) and (b) that the absorption profile φ_ν, (6.3.4), is known, because $\alpha_{12} = r_{12}$ on account of (a), while $f_1(\boldsymbol{v})$ is known on account of (b). However, the line absorption coefficient κ_ν, (6.3.2), is *not* known as neither n_2 nor ψ_ν can be specified. In other words, even if one makes the simplifying assumptions (a) and (b), in addition to the line source function S_ν one has to determine also the line absorption coefficient κ_ν in a self-consistent way by solving the coupled kinetic equations for I_ν and F_2, respectively.

To have something specific at hand, let us now assume that

(1) the ground state is naturally populated, $1 = 1^*$;

(2) the distribution function F_1 is Maxwellian, $F_1 = n_1 f^{\mathrm{M}}$;

(3) the static approximation is valid, $\boldsymbol{v} \cdot \nabla F_2 = 0$;

(4) elastic collisions with excited atoms are described by (6.3.12).

In other words, we consider the same case as in Sect. 6.3.5 except that now stimulated emissions are taken into consideration.

As in Sect. 6.3.5, the absorption and emission profiles φ_ν and ψ_ν are given by (6.3.77) and (6.3.78 – 80), respectively. However, the velocity distribution $f_2(\boldsymbol{v})$ now follows from (6.3.22, 23) to be

$$f_2(\boldsymbol{v}) = \frac{n_1}{n_2} \frac{B_{12} I_{12}^*(\boldsymbol{v}) + n_e C_{12} + \gamma_2 \dfrac{n_2}{n_1}}{A_{21} + B_{21} I_{21}(\boldsymbol{v}) + n_e C_{21} + \gamma_2} f^{\mathrm{M}}(\boldsymbol{v}) \; . \tag{6.3.107}$$

Integrating this equation over all velocities $\boldsymbol{v}$ yields

$$\frac{n_2}{n_1} = \frac{B_{12} J_{12}^* + n_e C_{12}}{A_{21} + B_{21} J_{21} + n_e C_{21}} \; . \tag{6.3.108}$$

Here J_{12}^* is given by (6.3.25), the asterisk indicating that $1 = 1^*$, whereas, see (2.4.15),

$$J_{21} = \int f_2(\boldsymbol{v}) I_{21}(\boldsymbol{v}) d^3v = \iint I_\nu(\boldsymbol{n}) \psi_\nu(\boldsymbol{n}) d\nu \frac{d\Omega}{4\pi} \; . \tag{6.3.109}$$

Note that $I_{21}(\boldsymbol{v})$ in (6.3.107) is a function of the atomic emission profile η_{21}, and J_{21} in (6.3.108) is a function of the laboratory emission profile ψ_ν.

In view of (6.3.80), (6.3.107) can be written as

$$\frac{f_2(\boldsymbol{v})}{L(\boldsymbol{v})}=\frac{n_1}{n_2}\,\frac{f^{\mathrm{M}}(\boldsymbol{v})}{A_{21}+B_{21}I_{21}(\boldsymbol{v})+n_{\mathrm{e}}C_{21}+\gamma_2}\;, \tag{6.3.110}$$

which upon substitution in (6.3.78,79), and using (6.3.108), yields the emission profile

$$\begin{aligned}\psi_\nu(\boldsymbol{n})=&\frac{A_{21}+B_{21}J_{21}+n_{\mathrm{e}}C_{21}}{B_{12}J^{*}_{12}+n_{\mathrm{e}}C_{12}}\int d^3v\,f^{\mathrm{M}}(\boldsymbol{v})\\&\cdot\frac{B_{12}I^{*}_{12}(\boldsymbol{v})\,j_{121}(\xi,\boldsymbol{n};\boldsymbol{v})+n_{\mathrm{e}}C_{12}r_{21}(\xi)}{A_{21}+B_{21}I_{21}(\boldsymbol{v})+n_{\mathrm{e}}C_{21}+\gamma_2}\\&+\gamma_2\int d^3v\,f^{\mathrm{M}}(\boldsymbol{v})\frac{r_{21}(\xi)}{A_{21}+B_{21}I_{21}(\boldsymbol{v})+n_{\mathrm{e}}C_{21}+\gamma_2}\;.\end{aligned} \tag{6.3.111}$$

Comparing (6.3.111) with (6.3.86), the following two facts emerge:

(1) Since $I_{21}(\boldsymbol{v})$ and J_{21}, which appear on the right-hand side of (6.3.111), contain η_{21} and ψ_ν, respectively, (6.3.111) is actually an implicit equation rather than an explicit expression for the emission profile ψ_ν.

(2) The presence of the $\boldsymbol{v}$-dependent quantity $I_{21}(\boldsymbol{v})$ in the denominators on the right-hand side of (6.3.111) prevents the $\boldsymbol{v}$ integrations from yielding K^{*}_{121} and φ_ν, which appear in the analogous emission profile without stimulated emissions, (6.3.93).

To arrive at an emission profile analogous to (6.3.93), two approximations have to be made. First, averaging over the Maxwellian of numerators and denominators must be performed separately,

$$\int d^3v\,f^{\mathrm{M}}(\boldsymbol{v})\frac{N(\boldsymbol{v})}{D(\boldsymbol{v})}\simeq\frac{\int d^3v\,f^{\mathrm{M}}(\boldsymbol{v})N(\boldsymbol{v})}{\int d^3v\,f^{\mathrm{M}}(\boldsymbol{v})D(\boldsymbol{v})}\;, \tag{6.3.112}$$

and second, one has to set approximately

$$\int d^3v\,f^{\mathrm{M}}(\boldsymbol{v})I_{21}(\boldsymbol{v})\simeq J_{21}\;. \tag{6.3.113}$$

With these two approximations, (6.3.111) takes the form

$$\psi_\nu(\boldsymbol{n})=\frac{1}{1+\gamma_2\hat{\tau}_2}\left(\frac{B_{12}K^{*}_{121}(\nu,\boldsymbol{n})+n_{\mathrm{e}}C_{12}\varphi_\nu(\boldsymbol{n})}{B_{12}J^{*}_{12}+n_{\mathrm{e}}C_{12}}+\gamma_2\hat{\tau}_2\varphi_\nu(\boldsymbol{n})\right)\;, \tag{6.3.114}$$

corresponding to the emission profile (6.3.93). However, the mean lifetime of the excited level $\hat{\tau}_2$, defined by

$$\hat{\tau}_2^{-1}=A_{21}+B_{21}J_{21}+n_{\mathrm{e}}C_{21}\;, \tag{6.3.115}$$

still depends on ψ_ν through J_{21}.

From now on we treat only the collisionless case of correlated re-emission ($\gamma_2 = 0$). Writing again simply

$$K_\nu(\boldsymbol{n}) \equiv K^*_{121} \ , \quad J_{12} \equiv J^*_{12} \ ,$$

the emission profile (6.3.114) is given by

$$\psi_\nu(\boldsymbol{n}) = \frac{B_{12}K_\nu(\boldsymbol{n}) + n_e C_{12}\varphi_\nu(\boldsymbol{n})}{B_{12}J_{12} + n_e C_{12}} \tag{6.3.116}$$

which coincides with (6.3.93) for $\gamma_2 = 0$. Recall, however, that in the presence of stimulated emissions, (6.3.116) was derived using the approximations (6.3.112, 113). Evidently, in (6.3.116) the velocity distribution $f_1(\boldsymbol{v})$ need no longer be assumed to be Maxwellian,

$$f_1(\boldsymbol{v}) \neq f^{\mathrm{M}}(\boldsymbol{v}) \ .$$

Adopting for the following the emission profile (6.3.116), we now derive the line absorption coefficient and the line source function in the presence of stimulated emissions. From (6.3.2, 3)

$$\kappa_\nu = \frac{h\nu_0}{4\pi} n_1 B_{12}\varphi_\nu \left(1 - \frac{n_2 B_{21}}{n_1 B_{12}} \frac{\psi_\nu}{\varphi_\nu}\right) \ , \tag{6.3.117}$$

$$S_\nu = \frac{n_2 A_{21}}{n_1 B_{12}} \frac{\psi_\nu}{\varphi_\nu} \frac{1}{1 - \dfrac{n_2 B_{21}}{n_1 B_{12}} \dfrac{\psi_\nu}{\varphi_\nu}} \ . \tag{6.3.118}$$

Here, n_2/n_1 is given by (6.3.108), while ψ_ν/φ_ν follows from (6.3.116).

By analogy with (6.3.29, 30) we define the dimensionless quantities

$$\varepsilon_0' = \varepsilon_0(1 - \mathrm{e}^{-h\nu_0/kT_e}) = \frac{n_e C_{21}}{A_{21}}(1 - \mathrm{e}^{-h\nu_0/kT_e}) \ , \tag{6.3.119}$$

$$\varepsilon' = \frac{\varepsilon_0'}{1 + \varepsilon_0'} \ , \tag{6.3.120}$$

and we introduce the Planck function, (1.4.45),

$$B \equiv B_{\nu_0}(T_e) = \frac{2h\nu_0^3}{c^2} \frac{1}{\mathrm{e}^{h\nu_0/kT_e} - 1} \ . \tag{6.3.121}$$

Using (1.6.31, 32) and (6.3.27), one readily verifies that, see (6.3.102),

$$\frac{n_e C_{12}}{B_{12}} = \varepsilon_0' B \ . \tag{6.3.122}$$

In terms of these quantities, from (6.3.108, 116)

$$\frac{n_2 A_{21}}{n_1 B_{12}} = \frac{J_{12} + \varepsilon_0' B}{1 + \varepsilon_0' + \dfrac{B_{21}}{A_{21}}(J_{21} + \varepsilon_0' B)} , \tag{6.3.123}$$

$$\frac{n_2 B_{21}}{n_1 B_{12}} = \frac{B_{21}}{A_{21}} \frac{J_{12} + \varepsilon_0' B}{1 + \varepsilon_0' + \dfrac{B_{21}}{A_{21}}(J_{21} + \varepsilon_0' B)} , \tag{6.3.124}$$

$$\frac{\psi_\nu}{\varphi_\nu} = \frac{K_\nu/\varphi_\nu + \varepsilon_0' B}{J_{12} + \varepsilon_0' B} . \tag{6.3.125}$$

Substituting now (6.3.123 – 125) into (6.3.117, 118) and using (6.3.120) finally gives

$$\kappa_\nu(\boldsymbol{n}) = \frac{h\nu_0}{4\pi} n_1 B_{12} \varphi_\nu(\boldsymbol{n}) \frac{1 + (1-\varepsilon')\dfrac{B_{21}}{A_{21}}\left(J_{21} - \dfrac{K_\nu(\boldsymbol{n})}{\varphi_\nu(\boldsymbol{n})}\right)}{1 + \dfrac{B_{21}}{A_{21}}[(1-\varepsilon')J_{21} + \varepsilon' B]} , \tag{6.3.126}$$

$$S_\nu(\boldsymbol{n}) = \frac{(1-\varepsilon')\dfrac{K_\nu(\boldsymbol{n})}{\varphi_\nu(\boldsymbol{n})} + \varepsilon' B}{1 + (1-\varepsilon')\dfrac{B_{21}}{A_{21}}\left(J_{21} - \dfrac{K_\nu(\boldsymbol{n})}{\varphi_\nu(\boldsymbol{n})}\right)} . \tag{6.3.127}$$

Recall that, according to (6.3.88), K_ν is given by

$$K_\nu(\boldsymbol{n}) = \iint I_{\nu'}(\boldsymbol{n}') R(\nu', \boldsymbol{n}'; \nu, \boldsymbol{n}) d\nu' \frac{d\Omega'}{4\pi} \tag{6.3.128}$$

in terms of the redistribution function R. On the other hand,

$$\begin{aligned} J_{21} &= \iint I_\nu(\boldsymbol{n}) \psi_\nu(\boldsymbol{n}) d\nu \frac{d\Omega}{4\pi} \\ &= \iint I_\nu(\boldsymbol{n}) \frac{K_\nu(\boldsymbol{n}) + \varepsilon_0' B \varphi_\nu(\boldsymbol{n})}{J_{12} + \varepsilon_0' B} d\nu \frac{d\Omega}{4\pi} \\ &= \frac{M + \varepsilon_0' B J_{12}}{J_{12} + \varepsilon_0' B} = \frac{(1-\varepsilon')M + \varepsilon' B J_{12}}{(1-\varepsilon')J_{12} + \varepsilon' B} , \end{aligned} \tag{6.3.129}$$

where

$$M = \iiiint I_{\nu'}(\boldsymbol{n}') I_\nu(\boldsymbol{n}) R(\nu', \boldsymbol{n}'; \nu, \boldsymbol{n}) d\nu' \frac{d\Omega'}{4\pi} d\nu \frac{d\Omega}{4\pi} . \tag{6.3.130}$$

In the approximation of *complete redistribution,* $\psi_\nu = \varphi_\nu$, and $R = \varphi_{\nu'}\varphi_\nu$ in the static approximation, see (6.3.98), and hence

$$J_{12} = J_{21} \ , \quad K_\nu = J_{12}\varphi_\nu \ , \quad M = J_{12}^2 \ ,$$

so that (6.3.126, 127) reduce to

$$\kappa_\nu(\boldsymbol{n}) = \frac{h\nu_0}{4\pi} n_1 B_{12}\varphi_\nu(\boldsymbol{n}) \frac{1}{1+\dfrac{B_{21}}{A_{21}}[(1-\varepsilon')J_{12}+\varepsilon' B]} \ , \tag{6.3.131}$$

$$S = (1-\varepsilon')J_{12}+\varepsilon' B \ . \tag{6.3.132}$$

Here the isotropic and frequency-independent source function (6.3.132) corresponds exactly to the source function (6.3.28), while the absorption coefficient (6.3.131) should be compared with (6.3.16).

In conclusion, we want to stress again that all expressions for κ_ν and S_ν derived above depend on the number density n_1 (entering κ_ν) and on the velocity distribution $f_1(\boldsymbol{v})$ (entering φ_ν, K_ν, and R). However, the distribution function $F_1(\boldsymbol{v}) = n_1 f_1(\boldsymbol{v})$ cannot be specified a priori in situations where stimulated emissions are important, rather, it must in principle be obtained through an iterative procedure.

6.4 Non-LTE Line Transfer by Two-Level Atoms (II): Results

This section presents numerical results of non-LTE line transfer by two-level atoms for some simple cases collectively referred to as the standard problem of radiative line transfer, adopting a term coined by *Hummer* and *Rybicki* [6.10]. The standard problem originally defined by these authors corresponds to the special case of complete redistribution in the static approximation, Sect. 6.4.4.

6.4.1 The Standard Problem

The standard problem of non-LTE line transfer refers to radiative transfer in a spectral line due to two-level atoms taking place in a homogeneous and plane parallel gas layer. More precisely, we consider a stationary, homogeneous, and isothermal gas composed of two-level atoms and free electrons, the negative charge of the electron gas being neutralized by an otherwise unspecified background of positive ions. The gas is self-excited with no radiation incident on it from outside, so that the energy of the emitted spectral line is entirely provided by excitation collisions of electrons with two-level atoms. The electron temperature is supposed to be small compared with the atomic excitation energy ($kT_e \ll h\nu_0$) so that (6.3.14, 15) are valid, and the distribution function $F_1(\boldsymbol{v})$ of nonexcited atoms is not affected by the occurrence of excitation and de-excitation processes within the gas. Furthermore, in this low-temperature limit, stimulated emissions can be ignored (Sect. 6.3.3).

Specifically, the distribution function of nonexcited two-level atoms is supposed to be Maxwellian

$$F_1(\boldsymbol{v}) = n_1 f^{\mathrm{M}}(v) \ , \tag{6.4.1}$$

where

$$f^{\mathrm{M}}(v) = \left(\frac{m}{2\pi kT}\right)^{3/2} \exp\left(-\frac{mv^2}{2kT}\right) \tag{6.4.2}$$

is the static Maxwell distribution, (6.3.13) with $\boldsymbol{u} = 0$. Moreover, the kinetic temperature of the atoms is assumed to be equal to the electron temperature,

$$T = T_e \ . \tag{6.4.3}$$

Our assumption of a homogeneous gas means that the densities n_1 and n_e as well as the kinetic temperature T are constant within the gas. By contrast, owing to the presence of boundaries, the specific intensity $I_\nu(\boldsymbol{n},\boldsymbol{r})$ of the radiation field and the distribution function $F_2(\boldsymbol{r},\boldsymbol{v})$ of excited two-level atoms are in general functions of position $\boldsymbol{r}$. These functions must be determined from their respective kinetic equations.

The radiative transfer equation is (6.3.1) with κ_ν and $S_\nu(\boldsymbol{n},\boldsymbol{r})$ given by (6.3.16, 17), the line profile coefficients φ_ν and $\psi_\nu(\boldsymbol{n},\boldsymbol{r})$ being defined by (6.3.4, 5). Note that in the standard problem, κ_ν and φ_ν are isotropic quantities and independent of $\boldsymbol{r}$.

On the other hand, the kinetic equation for the excited atoms is (6.3.22), where $I_{12}(\boldsymbol{r},\boldsymbol{v})$ is defined by (6.3.9). For the standard problem, the quantities C_{12}, C_{21}, and γ_2 are independent of $\boldsymbol{r}$.

Recall that in (6.3.22) the simple elastic collision term (6.3.12) has been used. A better approximation of the elastic collision term is provided by the following expression which describes the effect of elastic collisions of excited atoms with a thermal gas of nonexcited atoms if excited and nonexcited atoms are treated as hard spheres:

$$\left(\frac{\delta F_2}{\delta t}\right)_{\mathrm{el}} = \gamma_2 \left\{ \frac{1}{\pi^{3/2} w^2} \int \frac{1}{u} \exp\left(-\left[\frac{\boldsymbol{v}\cdot\boldsymbol{u}}{wu}\right]^2\right) F_2(\boldsymbol{v}+\boldsymbol{u})\, d^3u \right.$$
$$\left. - \left[\left(\frac{v}{w}+\frac{w}{2v}\right)\mathrm{erf}\left(\frac{v}{w}\right)+\frac{1}{\pi^{1/2}}\exp\left(-\frac{v^2}{w^2}\right)\right] F_2(\boldsymbol{v}) \right\} . \tag{6.4.4}$$

Here

$$\operatorname{erf} x = \frac{2}{\pi^{1/2}} \int_0^x e^{-t^2} dt \tag{6.4.5}$$

is the error function, $w = (2kT/m)^{1/2}$ is the thermal velocity of the atoms, see (6.3.65), and the collision frequency γ_2 is related to the total cross section Q by $\gamma_2 = n_1 Q w$. In the hard sphere approximation adopted in collision term (6.4.4), Q and hence γ_2 are independent of the atomic velocity $\boldsymbol{v}$. Elastic collisions of excited atoms with excited atoms and with free electrons have been ignored in (6.4.4). The elastic collision term (6.4.4) is derived in Appendix F.

6.4.2 Dimensionless Quantities

In order to see which combinations of physical quantities characterize the standard problem, the kinetic equations of photons and excited atoms are written in dimensionless form.

Starting from the thermal velocity of the two-level atoms, (6.3.65),

$$w = \left(\frac{2kT}{m}\right)^{1/2} \tag{6.4.6}$$

and the Doppler width of the considered spectral line, (6.3.66),

$$\Delta\nu_D = \frac{\nu_0}{c} w , \tag{6.4.7}$$

we introduce the dimensionless photon frequency, (6.3.64),

$$x = \frac{\nu - \nu_0}{\Delta\nu_D} \tag{6.4.8}$$

and the dimensionless atomic velocity, (6.3.64),

$$\boldsymbol{y} = \frac{1}{w} \boldsymbol{v} . \tag{6.4.9}$$

We define dimensionless velocity distributions by

$$f(\boldsymbol{y}) = w^3 f(\boldsymbol{v}) , \quad \int f(\boldsymbol{y}) d^3y = 1 . \tag{6.4.10}$$

(In some cases the same symbol denotes a physical quantity and its dimensionless analogue, which should not cause any confusion.) In particular, the three-dimensional Maxwell distribution takes the form

$$f^{\mathrm{M}}(\boldsymbol{y}) = \frac{1}{\pi^{3/2}} e^{-y^2} , \quad \int f^{\mathrm{M}}(\boldsymbol{y}) d^3y = 1 , \tag{6.4.11}$$

where $y^2 = \boldsymbol{y} \cdot \boldsymbol{y}$.

Turning to the line profile coefficients, it is convenient to use atomic profile coefficients α_{12}^0 and η_{21}^0 defined such that the center of the spectral line is the origin of the frequency axis, in agreement with definition (6.4.8) of the frequency x. Observing that $\xi - \nu_0 = \Delta\nu_D(x - \boldsymbol{n}\cdot\boldsymbol{y})$ according to (6.3.6), we therefore define the dimensionless atomic line profile coefficients as

$$\alpha_{12}^0(x - \boldsymbol{n}\cdot\boldsymbol{y}) = \Delta\nu_D\,\alpha_{12}(\xi)\ , \quad \eta_{21}^0(x - \boldsymbol{n}\cdot\boldsymbol{y}, \boldsymbol{n}) = \Delta\nu_D\,\eta_{21}(\xi, \boldsymbol{n})\ , \tag{6.4.12}$$

which are normalized according to

$$\int \alpha_{12}^0(x)\,dx = \iint \eta_{21}^0(x, \boldsymbol{n})\,dx\,\frac{d\Omega}{4\pi} = 1\ . \tag{6.4.13}$$

The limiting case of zero linewidth in the atomic rest frame corresponds to

$$\alpha_{12}^0(x) = \delta(x)\ , \tag{6.4.14}$$

while the atomic Lorentz profile is given by

$$\alpha_{12}^0(x) = \frac{a}{\pi}\,\frac{1}{x^2 + a^2}\ , \tag{6.4.15}$$

where $a = \gamma/4\pi\Delta\nu_D$, γ being the frequency of phase-changing collisions.

Likewise, the dimensionless laboratory line profile coefficients are defined by

$$\varphi_x = \Delta\nu_D\,\varphi_\nu\ , \quad \psi_x(\boldsymbol{n}) = \Delta\nu_D\,\psi_\nu(\boldsymbol{n})\ , \tag{6.4.16}$$

which are normalized according to

$$\int \varphi_x dx = \iint \psi_x(\boldsymbol{n})\,dx\,\frac{d\Omega}{4\pi} = 1\ . \tag{6.4.17}$$

Here we have used the fact that in the standard problem, the absorption profile φ_x is an isotropic quantity. Explicitly, from (6.3.4, 5) and using $f_1 = f^{\mathrm{M}}$, (6.4.1),

$$\varphi_x = \int f^{\mathrm{M}}(y)\,\alpha_{12}^0(x - \boldsymbol{n}\cdot\boldsymbol{y})\,d^3y\ , \tag{6.4.18}$$

$$\psi_x(\boldsymbol{n}) = \int f_2(y)\,\eta_{21}^0(x - \boldsymbol{n}\cdot\boldsymbol{y}, \boldsymbol{n})\,d^3y\ . \tag{6.4.19}$$

In particular, zero atomic linewidth (6.4.14) yields the Doppler absorption profile

$$\varphi_x = \frac{1}{\pi^{1/2}}\,\mathrm{e}^{-x^2}\ , \tag{6.4.20}$$

and the atomic Lorentz profile (6.4.15) yields the Voigt profile

$$\varphi_x = \frac{a}{\pi^{3/2}} \int_{-\infty}^{\infty} \frac{e^{-s^2}}{(s-x)^2+a^2} ds \ , \tag{6.4.21}$$

where again $a = \gamma/4\pi\Delta\nu_D$. The limit $a \to \infty$ of (6.4.21) corresponds to the laboratory Lorentz profile

$$\varphi_\omega = \frac{1}{\pi} \frac{1}{\omega^2+1} \tag{6.4.22}$$

which, however, must be expressed in terms of the dimensionless frequency

$$\omega = \frac{\nu - \nu_0}{\Delta\nu_L} \ , \quad \Delta\nu_L = \frac{\gamma}{4\pi} \ , \tag{6.4.23}$$

where $\Delta\nu_L$ is the Lorentz width of the spectral line, rather than in terms of the frequency x, since in the limit $a \to \infty$ the Doppler width $\Delta\nu_D$ is no longer defined. Notice that also for the Voigt profile (6.4.21), x is a good measure of the frequency within the spectral line only as long as $a \ll 1$.

The number density of excited two-level atoms is expressed by the dimensionless quantity

$$\tilde{n}_2 = \frac{n_2}{n_2^B} \ , \quad \text{where} \tag{6.4.24}$$

$$n_2^B = n_1 \frac{g_2}{g_1} e^{-h\nu_0/kT} \tag{6.4.25}$$

is the corresponding Boltzmann density.

To make radiative quantities dimensionless, we refer them to the Wien function, see (6.3.31),

$$B^W \equiv B_{\nu_0}^W(T) = \frac{2h\nu_0^3}{c^2} e^{-h\nu_0/kT} \ . \tag{6.4.26}$$

Thus, the dimensionless specific intensity is defined by

$$I_x(\boldsymbol{n}) = \frac{1}{B^W} I_\nu(\boldsymbol{n}) \ , \tag{6.4.27}$$

the dimensionless source function (6.3.17) by

$$S_x(\boldsymbol{n}) = \frac{1}{B^W} S_\nu(\boldsymbol{n}) = \tilde{n}_2 \frac{\psi_x(\boldsymbol{n})}{\varphi_x} \ , \tag{6.4.28}$$

and the dimensionless quantity I_{12}, (6.3.9), by

$$I_{12}(\boldsymbol{y}) = \frac{1}{B^W} I_{12}(\boldsymbol{v}) = \iint I_x(\boldsymbol{n})\, \alpha_{12}^0(x - \boldsymbol{n}\cdot\boldsymbol{y})\, dx \frac{d\Omega}{4\pi} \ . \tag{6.4.29}$$

Lastly, we use the mean absorption coefficient in the spectral line, see (6.3.16),

$$\kappa = \frac{h\nu_0}{4\pi}\frac{n_1 B_{12}}{\Delta\nu_D} \tag{6.4.30}$$

to define the dimensionless optical depth variable

$$\tau = -\kappa z , \tag{6.4.31}$$

where the z axis is the symmetry axis of the plane parallel gas layer (Fig. 6.10).

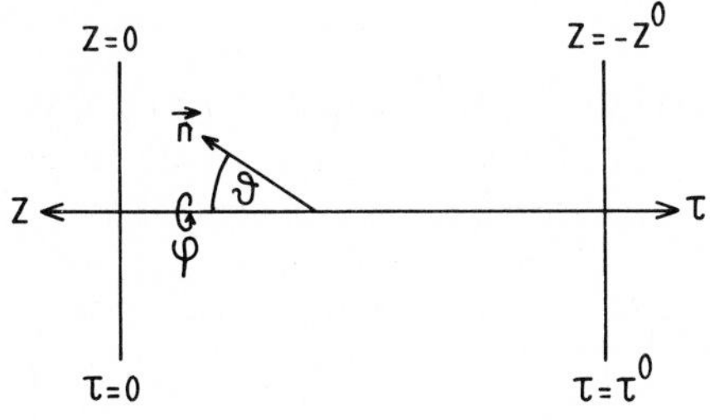

Fig. 6.10. Geometrical quantities of the standard problem. The plane parallel gas layer extends to infinity in the two directions perpendicular to the z axis, and it can be finite ($z^0 < \infty$) or semi-infinite ($z^0 = \infty$) in z direction

At any point inside the gas the direction of motion of a particle (photon or atom) is specified by the polar angle ϑ with respect to the positive z axis, and by the azimuthal angle φ about the z axis. Then, by symmetry, the distribution functions of the standard problem are independent of φ. Instead of ϑ we use

$$\mu = \cos\vartheta , \tag{6.4.32}$$

as usual. In particular,

$$\boldsymbol{n}\cdot\nabla = -\kappa\mu\frac{\partial}{\partial\tau} , \tag{6.4.33}$$

where $\boldsymbol{n}$ is the unit vector that specifies the direction of motion of the considered particle (photon or atom).

Using the absorption coefficient (6.3.16) and the source function (6.3.17) corresponding to the low-temperature limit where stimulated emissions are neglected, the transfer equation (6.3.1) can be written in dimensionless form as

$$\mu\frac{\partial}{\partial\tau}I_x(\mu;\tau) = \varphi_x[I_x(\mu;\tau) - S_x(\mu;\tau)] \tag{6.4.34}$$

which needs no further interpretation.

On the other hand, to write the kinetic equation (6.3.22) in dimensionless form, the following dimensionless quantities are first defined:

$$\varepsilon_0 = \frac{n_e C_{21}}{A_{21}} , \tag{6.4.35}$$

$$\zeta_0 = \frac{\gamma_2}{A_{21}} , \tag{6.4.36}$$

$$\eta_0 = \frac{\kappa w}{A_{21}} . \tag{6.4.37}$$

Here ε_0, see (6.3.30), and ζ_0 are, respectively, the number of inelastic and elastic collisions per excited atom during the radiative lifetime $1/A_{21}$ of an excited atom, and η_0 is the optical length traveled by an excited atom with thermal speed w during the radiative lifetime $1/A_{21}$. Using these definitions, and taking into account the detailed balance relations

$$n_1 C_{12} = n_2^{\mathrm{B}} C_{21} , \tag{6.4.38}$$

$$n_1 B_{12} B^{\mathrm{W}} = n_2^{\mathrm{B}} A_{21} , \tag{6.4.39}$$

the kinetic equation for the excited two-level atoms becomes

$$\begin{aligned} \eta y \mu \frac{\partial}{\partial \tau} [\tilde{n}_2(\tau) f_2(y,\mu;\tau)] = {} & \tilde{n}_2(\tau) f_2(y,\mu;\tau) \\ & - [\varepsilon + (1-\varepsilon) I_{12}(y,\mu;\tau)] f^{\mathrm{M}}(y) \\ & - \zeta \tilde{n}_2(\tau) [f^{\mathrm{M}}(y) - f_2(y,\mu;\tau)] , \end{aligned} \tag{6.4.40}$$

where

$$\varepsilon = \frac{\varepsilon_0}{1+\varepsilon_0} , \tag{6.4.41}$$

$$\zeta = \frac{\zeta_0}{1+\varepsilon_0} , \tag{6.4.42}$$

$$\eta = \frac{\eta_0}{1+\varepsilon_0} . \tag{6.4.43}$$

In (6.4.40), the first term on the right-hand side describes the destruction of excited atoms of velocity $\boldsymbol{v} = w\boldsymbol{y}$ by the combined effect of de-excitation collisions and spontaneous emissions. The second term describes their creation by excitation collisions and absorptions, respectively. The third term describes the effect of elastic collisions according to the simple elastic collision term (6.3.12).

The quantity ε was already introduced in (6.3.29). Recall that $\varepsilon = 1$ for LTE, while $\varepsilon \ll 1$ in typical non-LTE situations. On the other hand, as seen from (6.4.40), η is some measure of the importance of streaming of excited atoms, and ζ is some measure of the importance of elastic collisions of the excited atoms. Note that $\eta = l_{\mathrm{exc}}/l_{\mathrm{ph}}$ and $\zeta = l_{\mathrm{exc}}/l_{\mathrm{el}}$, where $l_{\mathrm{el}} = w/\gamma_2$ is the usual mean free path of excited atoms due to elastic collisions, $l_{\rm exc} = w/(A_{21} + n_e C_{21})$ is the mean free path of an excited atom in the absence of elastic collisions, and $l_{\rm ph} = 1/\kappa$ is the mean free path of a photon.

In (6.4.34, 40) we have explicitly indicated the dependence of the various quantities on the independent variables whose respective domains of definition are $-\infty < x < \infty$, $0 \leqslant y < \infty$, $-1 \leqslant \mu \leqslant 1$, $0 \leqslant \tau \leqslant \tau^0$. Note that in the standard problem ε, ζ, η are constants, with $0 \leqslant \varepsilon \leqslant 1$, $0 \leqslant \zeta$, $\eta < \infty$.

6.4.3 Boundary Conditions

The kinetic equations (6.4.34, 40) must be supplemented by boundary conditions for the distribution functions I_x and $\tilde{n}_2 f_2$. For the standard problem, the natural boundary conditions are those that specify the distribution functions "incident" on the boundary from outside. These boundary conditions are most easily formulated by introducing distribution functions I_x^+, I_x^-, f_2^+, f_2^- such that

$$I_x^+(\mu) = I_x(\mu) \ , \quad I_x^-(\mu) = I_x(-\mu) \ , \quad (0 \leqslant \mu \leqslant 1) \ , \tag{6.4.44}$$

$$f_2^+(y,\mu) = f_2(y,\mu) \ , \quad f_2^-(y,\mu) = f_2(y,-\mu) \ , \quad (0 \leqslant \mu \leqslant 1) \ . \tag{6.4.45}$$

Thus, these functions are defined in the interval $0 \leqslant \mu \leqslant 1$ in contrast to the functions I_x and f_2 which are defined in $-1 \leqslant \mu \leqslant 1$. In particular, the intensities I_x^+ and I_x^- obey the transfer equations

$$\mu \frac{\partial I_x^+}{\partial \tau} = \varphi_x (I_x^+ - S_x) \ , \quad -\mu \frac{\partial I_x^-}{\partial \tau} = \varphi_x (I_x^- - S_x) \ , \tag{6.4.46}$$

respectively.

Therefore, the natural boundary conditions for the standard problem specify at $\tau = 0$ (see Fig. 6.10)

$$I_x^-(\mu;0) \quad \text{and} \quad \tilde{n}_2(0) f_2^-(y,\mu;0) \ ,$$

and at $\tau = \tau^0$

$$I_x^+(\mu;\tau^0) \quad \text{and} \quad \tilde{n}_2(\tau^0) f_2^+(y,\mu;\tau^0) \ .$$

As mentioned in Sect. 6.4.1, it is assumed in the standard problem as formulated here that no external radiation is incident on the gas layer considered. If, in addition, the boundaries are transparent for the radiation incident from inside, the boundary conditions for the radiation field are

$$I_x^-(\mu;0) = 0 \ , \quad I_x^+(\mu;\tau^0) = 0 \ . \tag{6.4.47}$$

For the excited atoms, we consider two types of boundaries. For reflecting boundaries at $\tau = 0$ and $\tau = \tau^0$,

$$f_2^-(y,\mu;0) = f_2^+(y,\mu;0) \ , \quad f_2^+(y,\mu;\tau^0) = f_2^-(y,\mu;\tau^0) \ , \tag{6.4.48}$$

and for destroying boundaries at $\tau = 0$ and $\tau = \tau^0$,

$$f_2^-(y,\mu;0) = 0 \ , \quad f_2^+(y,\mu;\tau^0) = 0 \ . \tag{6.4.49}$$

Destruction of an excited atom at a boundary may be achieved, for example, by its de-excitation at the boundary wall, while the analogous destruction of a photon, (6.4.47), simply means its escape through a transparent boundary. Strictly speaking, the standard problem with destroying boundary conditions for photons and/or atoms violates energy and/or mass conservation. [Appropriate sources are tacitly assumed in the following.]

For a semi-infinite gas layer ($\tau^0 = \infty$), the boundary conditions for $\tau = \tau^0$ in (6.4.47 – 49) must be replaced by

$$\lim_{\tau\to\infty} [I_x^+(\mu;\tau)\mathrm{e}^{-\tau/\mu}] = 0 \ , \tag{6.4.50}$$

and

$$\lim_{\tau\to\infty} [\tilde{n}_2(\tau) f_2^+(y,\mu;\tau)\mathrm{e}^{-\tau/\mu}] = 0 \ , \tag{6.4.51}$$

respectively [6.23].

In conclusion, any particular case of the standard problem is specified by three intrinsic parameters ε, ζ, η, by one geometrical parameter τ^0, and by the boundary conditions for I_x and $\tilde{n}_2 f_2$.

6.4.4 Complete Redistribution: Static Approximation

This section presents some numerical results of the standard problem for the important special case of complete redistribution in the static approximation.

Recall that in the approximation of complete redistribution, the line profile coefficients for emission and absorption are assumed to be equal, see (6.3.18),

$$\psi_x = \varphi_x \ . \tag{6.4.52}$$

This assumption is equivalent to postulating the equality of the atomic profile coefficients for emission and absorption, $\eta_{21}^0 = \alpha_{12}^0$, see (6.3.19), and the equality of the atomic velocity distributions $f_2 = f_1$, or, since $f_1 = f^{\mathrm{M}}$ in the standard problem,

$$f_2(y) = f^{\mathrm{M}}(y) \ . \tag{6.4.53}$$

The line source function corresponding to complete redistribution is isotropic and frequency-independent, for according to (6.3.21, 4.28)

$$S = \tilde{n}_2 \ . \tag{6.4.54}$$

On the other hand, the static approximation ignores the streaming of excited atoms, (6.3.23), which in our present notation means

$$\eta = 0 \ . \tag{6.4.55}$$

Consequently, integration of the kinetic equation (6.4.40) over all velocities yields for any ζ, see (6.3.28),

$$S = \tilde{n}_2 = (1-\varepsilon) J_{12} + \varepsilon \ , \tag{6.4.56}$$

where (6.4.54) has been used. Here, see (6.3.25),

$$J_{12} = \int f^{\mathrm{M}}(y) I_{12}(y) d^3y = \iint I_x(\boldsymbol{n}) \varphi_x dx \frac{d\Omega}{4\pi} \ . \tag{6.4.57}$$

Hence, in the approximation of the standard problem considered, one must solve the transfer equation (6.4.34) whose source function is given by (6.4.56). One possibility for solving these two coupled equations for I_x and S consists in eliminating the radiation intensity so as to obtain an equation for the source function alone. To this end, one formally integrates the transfer equations (6.4.46), taking the boundary conditions (6.4.47) into account, yielding

$$I_x^+(\mu;\tau) = \int_\tau^{\tau_0} S(\tau') \exp\left(-\frac{\varphi_x}{\mu}(\tau'-\tau)\right) \frac{\varphi_x}{\mu} d\tau' \ , \tag{6.4.58a}$$

$$I_x^-(\mu;\tau) = \int_0^{\tau} S(\tau') \exp\left(-\frac{\varphi_x}{\mu}(\tau-\tau')\right) \frac{\varphi_x}{\mu} d\tau' \ . \tag{6.4.58b}$$

The quantity J_{12}, (6.4.57), can now be written as

$$\begin{aligned} J_{12}(\tau) &= \frac{1}{2} \int_{-\infty}^{\infty} dx\, \varphi_x \int_0^1 [I_x^+(\mu;\tau) + I_x^-(\mu;\tau)]\, d\mu \\ &= \frac{1}{2} \int_{-\infty}^{\infty} dx\, \varphi_x^2 \int_0^1 \frac{d\mu}{\mu} \int_0^{\tau_0} S(\tau') \exp\left(-\frac{\varphi_x}{\mu}|\tau-\tau'|\right) d\tau' \\ &= \int_0^{\tau_0} S(\tau') K(|\tau-\tau'|) d\tau' \ . \end{aligned} \tag{6.4.59}$$

Here the kernel function is given by

$$K(t) = \frac{1}{2} \int_{-\infty}^{\infty} \varphi_x^2 E_1(\varphi_x t)\, dx \quad (t \geqslant 0) \tag{6.4.60}$$

with $E_1(t)$ denoting the first exponential integral

$$E_1(t) = \int_1^\infty \frac{\mathrm{e}^{-ts}}{s} ds = \int_t^\infty \frac{\mathrm{e}^{-s}}{s} ds \ . \tag{6.4.61}$$

It is normalized according to

$$\int_0^\infty K(t)\,dt = \frac{1}{2} \,. \tag{6.4.62}$$

From (6.4.56, 59) one derives for the source function the integral equation

$$S(\tau) = (1-\varepsilon)\int_0^{\tau^0} S(\tau')K(|\tau-\tau'|)\,d\tau' + \varepsilon \tag{6.4.63}$$

whose kernel is defined by (6.4.60).

Once the source function has been determined as a solution of (6.4.63), the standard problem in the considered approximation is solved. Indeed, the radiation field can be calculated from (6.4.58), while the distribution function $F_2 = \tilde{n}_2 f_2$ follows at once from (6.4.53, 54).

The most significant features of the non-LTE line transfer problem considered are best recognized by considering the limiting case of a semi-infinite gas layer. In Fig. 6.11, the spatial distribution of excited two-level atoms is plotted for various values of the parameter ε for pure Doppler broadening, (6.4.20), while Fig. 6.12 compares the spatial distributions of excited atoms corresponding to Doppler, Voigt, and Lorentz profiles (6.4.20 – 22) for a fixed value of ε.

Inspection of Figs. (6.11, 12) shows that in a semi-infinite atmosphere, the number density of excited atoms for any line profile φ_x takes at the surface $\tau = 0$ the value

$$\tilde{n}_2(0) = \varepsilon^{1/2} \,. \tag{6.4.64}$$

Consequently, a non-LTE gas with $\varepsilon \ll 1$ emits much less radiation than the corresponding LTE gas with $\varepsilon = 1$, even in the extreme case of a semi-infinite layer.

In non-LTE situations with $\varepsilon \ll 1$, the Boltzmann distribution $\tilde{n}_2 = 1$ for the excited atoms is attained only at quite large optical depths, as can be seen from Figs. 6.11, 12. The *thermalization length* Λ is the order of magnitude of that optical depth where the density of excited atoms approaches the Boltzmann distribution,

$$\tilde{n}_2(\Lambda) \sim 1 \,. \tag{6.4.65}$$

Then, according to Figs. 6.11, 12, the thermalization lengths for the Doppler, Voigt, and Lorentz profiles (6.4.20 – 22) are of the order of

$$\Lambda \sim \frac{1}{\varepsilon} \qquad \text{(Doppler, } a = 0) \,, \tag{6.4.66a}$$

$$\Lambda \sim \frac{a}{\varepsilon^2} \qquad \text{(Voigt, } \varepsilon < a < 1) \,, \tag{6.4.66b}$$

$$\Lambda \sim \frac{1}{\varepsilon^2} \qquad \text{(Lorentz, } a = \infty) \,, \tag{6.4.66c}$$

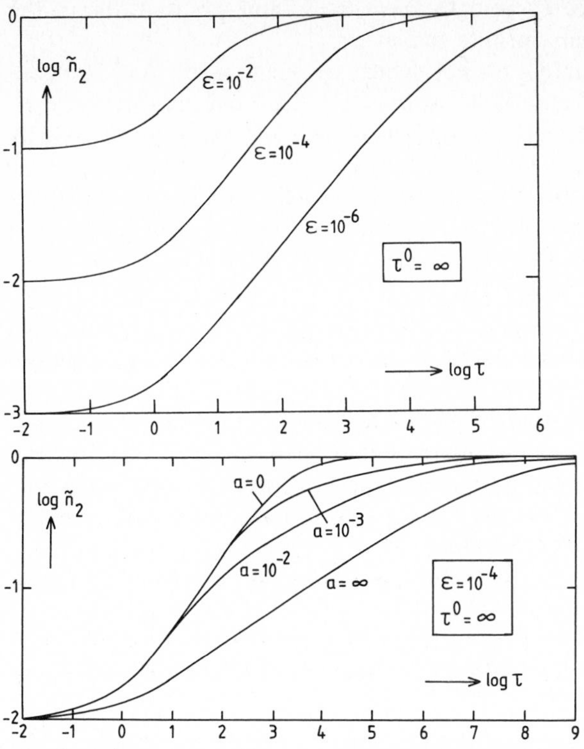

Fig. 6.11. Complete redistribution, static approximation. Density of excited two-level atoms $\tilde{n}_2(\tau)$ vs optical depth τ in a semi-infinite gas layer for various values of ε, for the limiting case of pure Doppler broadening [6.24]

Fig. 6.12. Complete redistribution, static approximation. Density of excited two-level atoms $\tilde{n}_2(\tau)$ vs optical depth τ in a semi-infinite gas layer with $\varepsilon = 10^{-4}$. The curve labeled $a = 0$ corresponds to a Doppler profile, the curves labeled $a = 10^{-3}$ and $a = 10^{-2}$ to Voigt profiles, and that labeled $a = \infty$ to a Lorentz profile, where a is the parameter characterizing the Voigt profile (6.4.21) [6.24]

where $a = \gamma/4\pi\Delta\nu_D$ is the dimensionless parameter characterizing the Voigt profile (6.4.21).

The noteworthy point of (6.4.66) is the strong dependence of the thermalization length on the line profile, in contrast to the surface value $\tilde{n}_2(0)$ which is independent of φ_x, see (6.4.64). As can be seen from Fig. 6.12 and (6.4.66), for a given value of ε the thermalization length increases with increasing values of the parameter a, which means that the stronger the line wings, the larger the thermalization length. This result is corroborated by the (somewhat unphysical) case of purely coherent scattering $\varphi_x = \delta(x)$, which yields the smallest thermalization length $\Lambda \sim \varepsilon^{-1/2}$. All of these results show the importance of photon escape from the atmosphere through the line wings.

As a result, if the optical thickness of a gas is smaller than the thermalization length, $\tau^0 < \Lambda$, the density of excited atoms attains the Boltzmann distribution nowhere inside the gas, and one speaks of an *effectively thin* gas. By contrast, when $\tau^0 > \Lambda$, the density of excited atoms attains $\tilde{n}_2 \sim 1$ inside the gas, which is then referred to as *effectively thick,* and which in the region $0<\tau<\Lambda$ behaves like a semi-infinite gas layer.

The results contained in (6.4.64, 66), which are due to *Avrett* and *Hummer* [6.24], belong to the most important results of non-LTE plasma spectroscopy.

Figures 6.11, 12 show that non-LTE line transfer gives rise to density gradients of excited atoms near the plasma boundary. As a result of these density gradients, the emitted spectral line shows self-reversal. This fact can easily be understood by considering, for example, the emergent intensity in the direction perpendicular ($\vartheta=0$, $\mu=1$) to the boundary surface at $\tau=0$. According to (6.4.58a) it is given by

$$I_x^{\text{em}} \equiv I_x(1;0) = \int_0^{\tau^0} S(\tau)\exp(-\varphi_x\tau)\,\varphi_x d\tau \ . \tag{6.4.67}$$

Hence, owing to the rapid decrease of the exponential function for $\tau>1/\varphi_x$, the emergent intensity I_x^{em} is essentially some mean value of the function $\varphi_x S(\tau)$ over the interval $0<\tau<1/\varphi_x$, resulting in a spectral line with a self-reversed core (Fig. 6.13). In the case of a semi-infinite gas, the emitted spectral line has the shape of an absorption line.

Two comments conclude this section. First, in the approximation considered, only one intrinsic parameter ε appears instead of the three parameters ε, ζ, η of the complete standard problem. Indeed, we have set $\eta=0$, (6.4.55), whereas ζ drops out when integrating the kinetic equation (6.4.40) to obtain the moment equation (6.4.56).

Second, it is easily seen that the approximation considered is inconsistent. Indeed, using (6.4.54–56), the kinetic equation (6.4.40) can be solved for f_2, yielding, see (6.3.82),

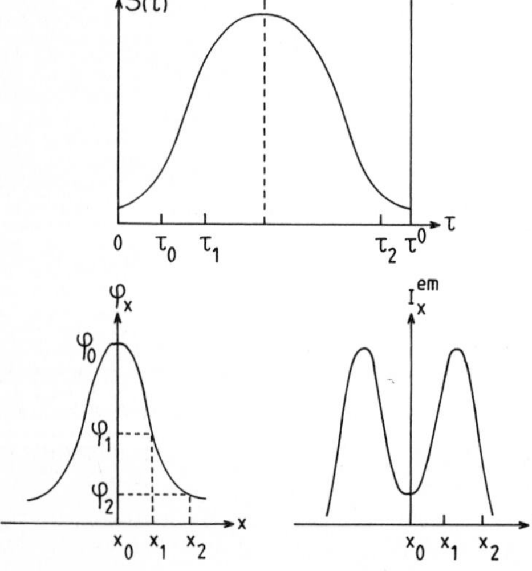

Fig. 6.13. Schema to explain the self-reversal of spectral lines emitted by optically thick non-LTE gases. The values $\varphi_i = \varphi(x_i)$ of the line profile coefficients, and the optical depths $\tau_i = 1/\varphi_i$ correspond to the arbitrarily chosen frequencies x_i ($i=0,1,2$). See text for further explanation

$$f_2(y) = \frac{1}{1+\zeta}\left(\frac{(1-\varepsilon)I_{12}(y)+\varepsilon}{(1-\varepsilon)J_{12}+\varepsilon}+\zeta\right)f^{\mathrm{M}}(y) \ . \tag{6.4.68}$$

Hence $f_2(y) \neq f^{\mathrm{M}}(y)$ in general, in contradiction to (6.4.53). Using some idealized radiation intensities, this point has been discussed in Sect. 6.3.4 for the collisionless case $\zeta = 0$, see (6.3.70) and Fig. 6.9. On the other hand, in the collision-dominated case $\zeta \gg 1$, (6.4.68) leads to $f_2 = f^{\mathrm{M}}$ independently of the radiation field, that is, of $I_{12}(y)$, as expected on physical grounds.

6.4.5 Complete Redistribution: Diffusion Approximation

We now study the standard problem for complete redistribution in the diffusion approximation, which comprises the static approximation considered in the preceding section as a special case. Here the convective transport of excited atoms, caused by density gradients due to non-LTE line transfer, is taken into account in the form of a diffusion current [6.25, 26].

Again we assume complete redistribution $\psi_x = \varphi_x$, (6.4.52), and hence $S = \tilde{n}_2$, (6.4.54). It should be observed, however, that $f_2 = f^{\mathrm{M}}$, (6.4.53), which is implicitly contained in (6.4.52), is in principle inconsistent with the diffusion approximation since, by assumption, the velocity distribution f_2 must be anisotropic to account for a macroscopic flow velocity. On the other hand, because of $n_1 \gg n_2$, the particle flow of nonexcited atoms in the opposite direction is negligible.

According to Sect. 6.3.3, the diffusion approximation leads to the diffusion equation for the line source function, (6.3.45), which for a plane-parallel geometry and in our present notation reads

$$S - \delta\frac{d^2S}{d\tau^2} = (1-\varepsilon)J_{12}+\varepsilon \ , \tag{6.4.69}$$

where, see (6.3.46),

$$\delta = \frac{\eta^2}{2\zeta} \ , \tag{6.4.70}$$

and where J_{12} is given by (6.4.57).

Following the same procedure as in Sect. 6.4.4, one now derives for the source function the integro-differential equation

$$S(\tau) - \delta\frac{d^2S}{d\tau^2} = (1-\varepsilon)\int_0^{\tau^0} S(\tau')K(|\tau-\tau'|)d\tau' + \varepsilon \ , \tag{6.4.71}$$

where the kernel function is defined by (6.4.60).

In contrast to the integral equation (6.4.63), the integrodifferential equation (6.4.71) must be supplemented by boundary conditions for $S(\tau)$. Since destroying boundary conditions (6.4.49) give rise, near the boundary, to flow velocities of the order of the thermal velocity of the atoms (Sect. 6.4.6) incompatible with the assumption of a (slow) diffusion current, the natural boundary conditions for the present problem are reflecting ones, (6.4.48). Here the particle flow at the boundaries vanishes, which in the diffusion approximation is equivalent to the density gradients of the excited atoms vanishing there, on account of (6.3.35). Therefore, in view of (6.4.54), we adopt as boundary conditions

$$\left.\frac{dS}{d\tau}\right|_{\tau=0} = \left.\frac{dS}{d\tau}\right|_{\tau=\tau^0} = 0 \ . \tag{6.4.72}$$

Recall that the boundary conditions for the radiation field, (6.4.47), are implicitly contained in the right-hand side of (6.4.71).

Some numerical results for a Doppler broadened spectral line are presented in Fig. 6.14. As expected, diffusion tends to smooth out the density gradients of the excited atoms. The spatial distribution of excited atoms is affected by diffusion primarily in the vicinity of the boundaries. Here higher values of the parameter δ lead to larger values of the density $\tilde{n}_2$, exhibiting the increasing convective transport of excited atoms from deeper layers into the boundary region. In particular, the surface value $\tilde{n}_2(0)$ in the diffusion approximation is always larger than that in the static approximation, whereas the thermalization length Λ, (6.4.65), is not affected by diffusion, at least for not too high values of δ.

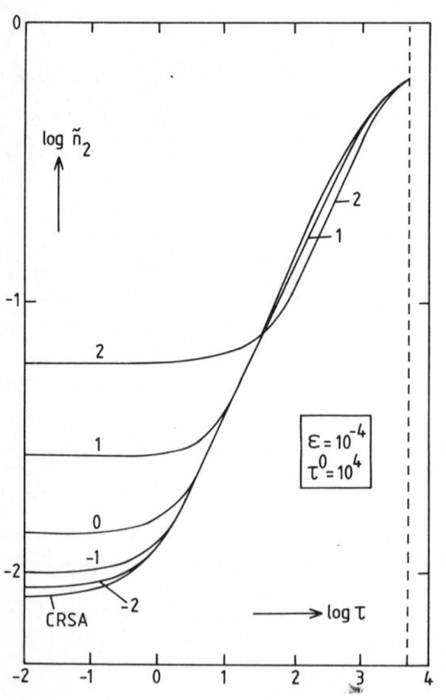

Fig. 6.14. Complete redistribution, diffusion approximation. Density of excited two-level atoms $\tilde{n}_2(\tau)$ vs optical depth τ in a gas with $\varepsilon = 10^{-4}$ and of optical thickness $\tau^0 = 10^4$ for pure Doppler broadening. Curve labeled n ($n = 2, 1, 0, -1, -2$) corresponds to $\delta = 10^n$, that labeled CRSA corresponds to the static approximation $\delta = 0$. The gas layer is symmetric about the optical depth $\tau = 5 \times 10^3$ (dashed line) [6.26]

In contrast to the static approximation (Sect. 6.4.4), the diffusion approximation is characterized by two intrinsic parameters ε and δ. In the collision-dominated limiting case $\zeta\to\infty$, one has $\delta\to 0$ on account of (6.4.70), and the diffusion approximation reduces to the static approximation, as it should.

That the particle flow can be described in terms of a diffusion current supposes that the mean free path of the excited atoms with respect to elastic collisions, $l_{\rm el} = w/\gamma_2$, is much smaller than the characteristic scale length of the density gradients, which is of the order of the photon mean free path $l_{\rm ph} = 1/\kappa$,

$$\frac{l_{\rm el}}{l_{\rm ph}} = \frac{\kappa w}{\gamma_2} = \frac{\eta}{\zeta} \ll 1 \ .$$

Because of (6.4.70), this condition implies for the model considered the further condition

$$\delta \ll \zeta \ .$$

Hence, appreciable diffusion effects, which occur if $\delta = \eta^2/2\zeta \gtrsim 1$, require quite large values of the parameters η and ζ, with $1 \lesssim \delta \ll \eta \ll \zeta$. Therefore, the diffusion approximation is of no great practical importance. It is nevertheless a useful model for gaining insight into the physics of non-LTE line transfer.

6.4.6 Exact Solutions

We finally present some exact solutions to the standard problem obtained by solving simultaneously the radiative transfer equation and the kinetic equation of excited atoms (6.4.34, 40) [6.27, 28].

Specifically, let us consider the limiting case of zero linewidth in the atomic rest frame where the atomic absorption profile is a delta function (6.4.14), and the laboratory absorption profile is the Doppler profile (6.4.20). Likewise, assuming isotropic re-emission, (B.2.11a), the atomic emission profile is also a delta function,

$$\eta^0_{21}(x, \boldsymbol{n}) = \delta(x) \ , \tag{6.4.73}$$

so that the laboratory emission profile according to (6.4.19) is given by

$$\psi_x(\boldsymbol{n}) = \int f_2(\boldsymbol{y})\,\delta(x - \boldsymbol{n}\cdot\boldsymbol{y})\,d^3y \ . \tag{6.4.74}$$

In this special case, deviations from complete redistribution are entirely due to deviations of the velocity distribution of excited atoms $f_2(\boldsymbol{v})$ from the Maxwell distribution $f^{\rm M}(\boldsymbol{v})$, as explicitly seen by comparing (6.4.74) with (6.4.14, 18).

As a specific example, we consider here a gas characterized by

$$\varepsilon = 10^{-4} \ , \quad \tau^0 = 2\times 10^4 \ .$$

In addition, we assume correlated re-emission (Sect. 6.3.4) where no elastic velocity-changing collisions of excited atoms occur, $\gamma_2 = 0$, so that on account of (6.4.36, 42)

$$\zeta = 0 \ .$$

As regards the parameter η, we shall consider the cases

$$\eta = 10^{-2}, 1, 10^2$$

corresponding to various degrees of streaming of the excited atoms. We assume the usual boundary conditions for the photons (6.4.47), and we consider reflecting and destroying boundary conditions for the excited atoms, (6.4.48, 49).

As pointed out in Sect. 6.3.5, approximating the line source function by redistribution functions consists in neglecting the streaming of the excited atoms. Therefore, in the limit $\eta \to 0$, the exact solutions of the standard problem go over into the approximate solutions obtained by using redistribution functions. For the special case considered here (zero linewidth in the atomic rest frame, isotropic re-emission), the redistribution function (6.3.89), written in dimensionless form, is given by

$$R(x', \boldsymbol{n}'; x, \boldsymbol{n}) = \int f^{\mathrm{M}}(\boldsymbol{y})\, \delta(x' - \boldsymbol{n}' \cdot \boldsymbol{y})\, \delta(x - \boldsymbol{n} \cdot \boldsymbol{y})\, d^3y \ . \tag{6.4.75}$$

To evaluate it, we use coordinates $\boldsymbol{y} = (y_1, y_2, y_3)$ such that the 1 axis points in the direction of the unit vector $\boldsymbol{n}$, and that the unit vector $\boldsymbol{n}'$ lies in the 1,2 plane. In other words, $\boldsymbol{n} \cdot \boldsymbol{y} = y_1$ and $\boldsymbol{n}' \cdot \boldsymbol{y} = y_1 \cos\alpha + y_2 \sin\alpha$, where α is the scattering angle defined through $\boldsymbol{n} \cdot \boldsymbol{n}' = \cos\alpha$. Inserting now in (6.4.75) the Maxwell distribution (6.4.11), and carrying out the integration over y_3, one obtains

$$\begin{aligned} R(x', \boldsymbol{n}'; x, \boldsymbol{n}) &= \frac{1}{\pi} \int_{-\infty}^{\infty} dy_2 \mathrm{e}^{-y_2^2} \int_{-\infty}^{\infty} dy_1 \mathrm{e}^{-y_1^2} \\ &\quad \cdot \delta(x' - y_1 \cos\alpha - y_2 \sin\alpha)\, \delta(x - y_1) \\ &= \frac{1}{\pi} \mathrm{e}^{-x^2} \int_{-\infty}^{\infty} dy_2 \mathrm{e}^{-y_2^2} \delta(x' - x\cos\alpha - y_2 \sin\alpha) \\ &= \frac{\mathrm{e}^{-x^2}}{\pi \sin\alpha} \exp\left(-\left[\frac{x' - x\cos\alpha}{\sin\alpha}\right]^2\right) \ . \end{aligned} \tag{6.4.76}$$

For correlated re-emission, as considered in this section, the line source function is given by (6.3.104), which on account of (6.4.20, 76) here takes the form

$$S_x(\boldsymbol{n}) = (1-\varepsilon) \int_{-\infty}^{\infty} dx' \int \frac{d\Omega'}{4\pi} I_{x'}(\boldsymbol{n}') \cdot \frac{1}{\pi^{1/2}\sin\alpha} \exp\left(-\left[\frac{x'-x\cos\alpha}{\sin\alpha}\right]^2\right) + \varepsilon , \qquad (6.4.77)$$

where $\cos\alpha \equiv \boldsymbol{n}' \cdot \boldsymbol{n}$. In the approximation of redistribution functions, solutions to the standard problem are obtained as solutions of the transfer equation (6.4.34) with the boundary conditions (6.4.47) using the absorption profile (6.4.20) and the (frequency-dependent and anisotropic) source function (6.4.77). Note that neither the parameter η nor any boundary conditions for the excited atoms appear in this approximation.

The exact solutions of the standard problem corresponding to $\eta = 10^{-2}$ with reflecting boundary conditions for the excited atoms, presented below, resemble closely the approximate solutions corresponding to the source function (6.4.77) [6.29].

Let us return to the exact solutions of the standard problem. In Figs. 6.15 – 17, the particle density $\tilde{n}_2$, the particle flux density Γ_2, and the mean velocity u_2 of the gas of excited two-level atoms are plotted as functions of the optical depth. In the interval $0 < \tau < 10^4$ shown in these figures, the particle flux and the macroscopic gas velocity are in the direction of the positive z axis (that is, in the direction of the negative τ axis, see Fig. 6.10), and their absolute values are given by

$$\Gamma_2 = \tilde{n}_2 u_2 , \qquad (6.4.78)$$

$$u_2 = 2\pi \int_0^{\infty} dy\, y^3 \int_{-1}^{1} \mu f_2(y,\mu)\, d\mu . \qquad (6.4.79)$$

Figure 6.15 shows that for $\eta \lesssim 1$, the spatial distribution of excited atoms at optical depths $\tau \gtrsim 1$ does not differ appreciably from that corresponding to complete redistribution in the static approximation. Consequently, the relation $\Lambda \sim \varepsilon^{-1}$, (6.4.66a), for the thermalization length of a Doppler broadened spectral line remains valid for $\eta \lesssim 1$. On the other hand, streaming with $\eta \gg 1$ affects the gas of excited atoms up to much deeper layers, as expected.

In the surface region, $0 < \tau \lesssim 1$, the influence of the boundary on the gas of excited atoms is appreciable even for rather small values of the streaming parameter η. As an example, in Fig. 6.15 compare the two cases $\eta = 10^{-2}$ for destroying and reflecting boundaries, respectively. For a given value of η, the optical depth up to which the boundary is felt by the gas of excited atoms can be recognized from Figs. 6.16, 17 as that optical depth where the two curves corresponding to destroying and reflecting boundaries begin to merge. For a more thorough discussion of the boundary layers of excited atoms, refer to [6.28].

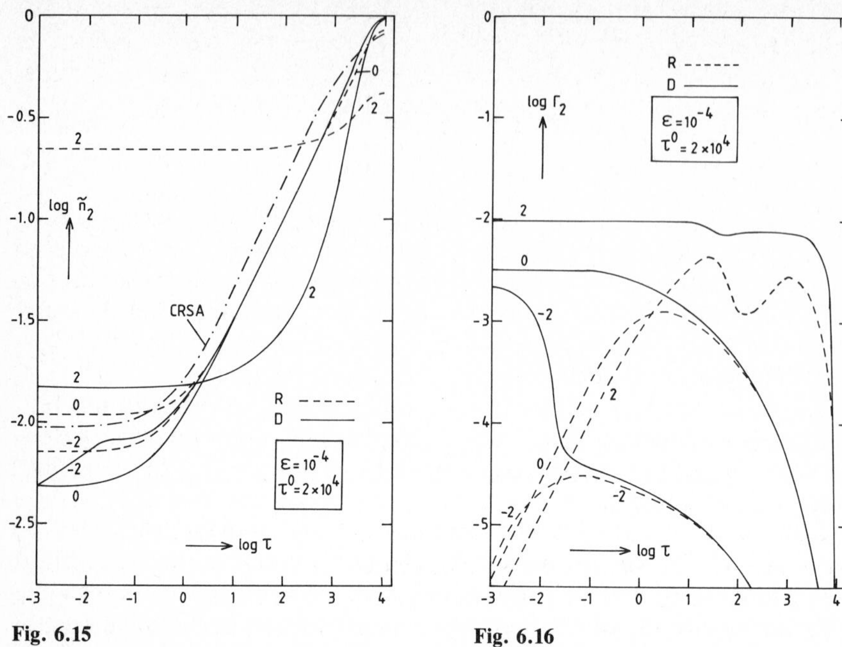

Fig. 6.15 **Fig. 6.16**

Fig. 6.15. Density of excited atoms $\tilde{n}_2(\tau)$ vs optical depth τ in a gas with $\varepsilon = 10^{-4}$, $\zeta = 0$, and of optical thickness $\tau^0 = 2 \times 10^4$, for pure Doppler broadening. Curves labeled n ($n = 2, 0, -2$) correspond to $\eta = 10^n$, and dashed (solid) curves correspond to reflecting (destroying) boundaries for the excited atoms. Curve CRSA corresponds to complete redistribution in the static approximation. The gas layer is symmetric about the optical depth $\tau = 10^4$

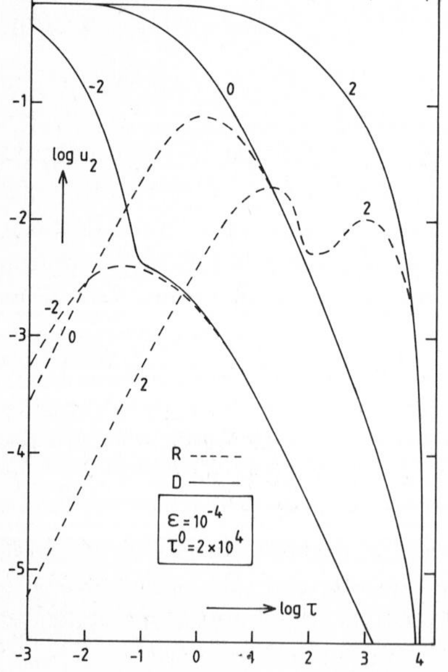

Fig. 6.16. Flux density of excited atoms $\Gamma_2(\tau)$ vs optical depth τ in a gas with $\varepsilon = 10^{-4}$, $\zeta = 0$, and of optical thickness $\tau^0 = 2 \times 10^4$, for pure Doppler broadening. Curves labeled n ($n = 2, 0, -2$) correspond to $\eta = 10^n$, and dashed (solid) curves correspond to reflecting (destroying) boundaries for the excited atoms. The gas layer is symmetric about the optical depth $\tau = 10^4$

◀ **Fig. 6.17.** Macroscopic velocity of excited atoms $u_2(\tau)$ vs optical depth τ in a gas with $\varepsilon = 10^{-4}$, $\zeta = 0$, and of optical thickness $\tau^0 = 2 \times 10^4$, for pure Doppler broadening. Curves labeled n ($n = 2, 0, -2$) correspond to $\eta = 10^n$, and dashed (solid) curves correspond to reflecting (destroying) boundaries for the excited atoms. The gas layer is symmetric about the optical depth $\tau = 10^4$

As can be seen from Figs. 6.16, 17, a destroying boundary gives rise to a considerable flow of excited atoms into the boundary. Near a destroying boundary, the macroscopic velocity of the gas of excited atoms reaches values of the order of $u_2 \sim 1$, that is, of the order of the thermal speed of the atoms $w \sim (kT/m)^{1/2}$, a fact that is easily understood.

Flow velocities $u_2 \neq 0$ are due to anisotropies of the velocity distribution $f_2(y)$, and therefore indicate deviations from the isotropic Maxwell distribution $f^{\mathrm{M}}(y)$. Another quite different measure of the non-Maxwellian character of the velocity distribution of the excited atoms is provided by the ratio $\bar{f}_2(y)/f^{\mathrm{M}}(y)$, where

$$\bar{f}_2(y) = \frac{1}{2} \int_{-1}^{1} f_2(y, \mu)\, d\mu \tag{6.4.80}$$

is the isotropic angle-averaged velocity distribution. In Fig. 6.18, this ratio is plotted as a function of the absolute value of the particle velocity y for various optical depths. Near the boundary and for not too high values of the streaming parameter η, the high-velocity tail of the velocity distribution $\bar{f}_2$ is overpopulated compared with the Maxwellian, but at very high velocities it drops rapidly to zero. Except for these very high velocities, the distribution function $\bar{f}_2$ approaches the Maxwell distribution when going to larger optical depths. On the other hand, for $\eta \gg 1$, the isotropic distribution function $\bar{f}_2$ does not differ very much from a Maxwellian, but it should be remembered that the true velocity distributions $f_2(y)$ are anisotropic, as displayed by the occurrence of nonvanishing particle flows (Figs. 6.16, 17).

Turning now to the radiation field, Fig. 6.19 shows the mean intensity $\bar{I}_x$ as a function of frequency at various optical depths, where

$$\bar{I}_x = \frac{1}{2} \int_{-1}^{1} I_x(\mu)\, d\mu \; . \tag{6.4.81}$$

More precisely, we have plotted the symmetrized mean intensity

$$\bar{\bar{I}}_x = \tfrac{1}{2}(\bar{I}_x + \bar{I}_{-x}) \quad (x \geqslant 0) \tag{6.4.82}$$

which is the mean value of the corresponding intensities of the blue ($x \geq 0$) and red ($x < 0$) parts of the spectral line. The symmetrized mean intensity $\bar{\bar{I}}_x$ is the photon analogue to the angle-averaged distribution function $\tilde{n}_2 \bar{f}_2(y)$ of the excited atoms. For not too high values of η, the blue-red asymmetry of the spectral line is small ($\bar{\bar{I}}_x \simeq \bar{I}_x \simeq \bar{I}_{-x}$), whereas for $\eta \gg 1$ this asymmetry can be appreciable, in particular near a destroying boundary (Fig. 6.20). In the cases shown in Fig. 6.19 (except for $\eta = 10^2$ with reflecting boundaries), the radiation intensity at small optical depths has the double-humped shape of an absorption line, while it becomes isotropic and white in the frequency range $-3 \lesssim x \lesssim 3$ in which the system is optically thick, $\tau^0 \varphi_x \gtrsim 1$, when going to the center of the plane parallel slab. Since the gas considered ($\varepsilon = 10^{-4}$,

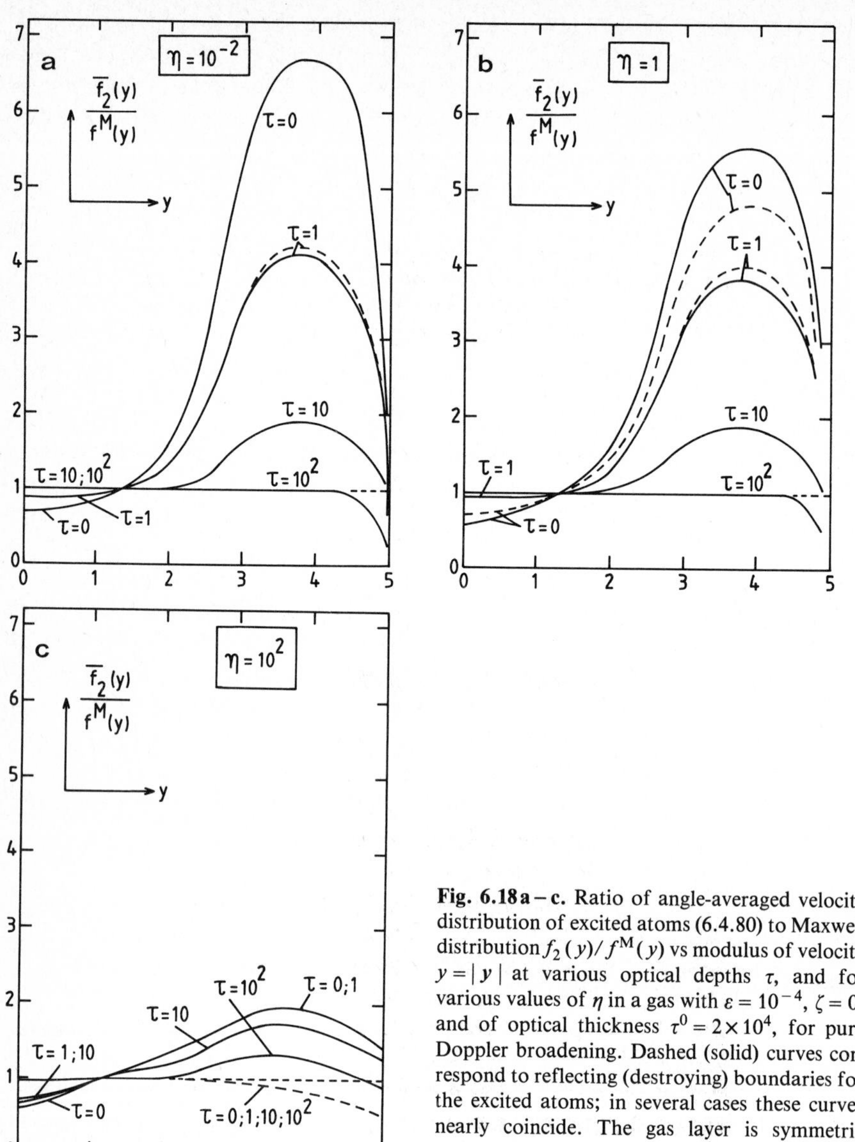

Fig. 6.18a–c. Ratio of angle-averaged velocity distribution of excited atoms (6.4.80) to Maxwell distribution $f_2(y)/f^M(y)$ vs modulus of velocity $y = |\boldsymbol{y}|$ at various optical depths τ, and for various values of η in a gas with $\varepsilon = 10^{-4}$, $\zeta = 0$, and of optical thickness $\tau^0 = 2 \times 10^4$, for pure Doppler broadening. Dashed (solid) curves correspond to reflecting (destroying) boundaries for the excited atoms; in several cases these curves nearly coincide. The gas layer is symmetric about the optical depth $\tau = 10^4$

$\tau^0 = 2 \times 10^4$) is almost effectively thick as defined in Sect. 6.4.4 [because $\tau^0 \simeq \Lambda$, where $\Lambda \sim \varepsilon^{-1}$, see (6.4.66a)], the number density of excited atoms approaches the Boltzmann value $\tilde{n}_2 = 1$ at the center of the slab (Fig. 6.15), and the radiation intensity in the optically thick line core there almost attains the Planck-Wien value $I_x = 1$ (Fig. 6.19).

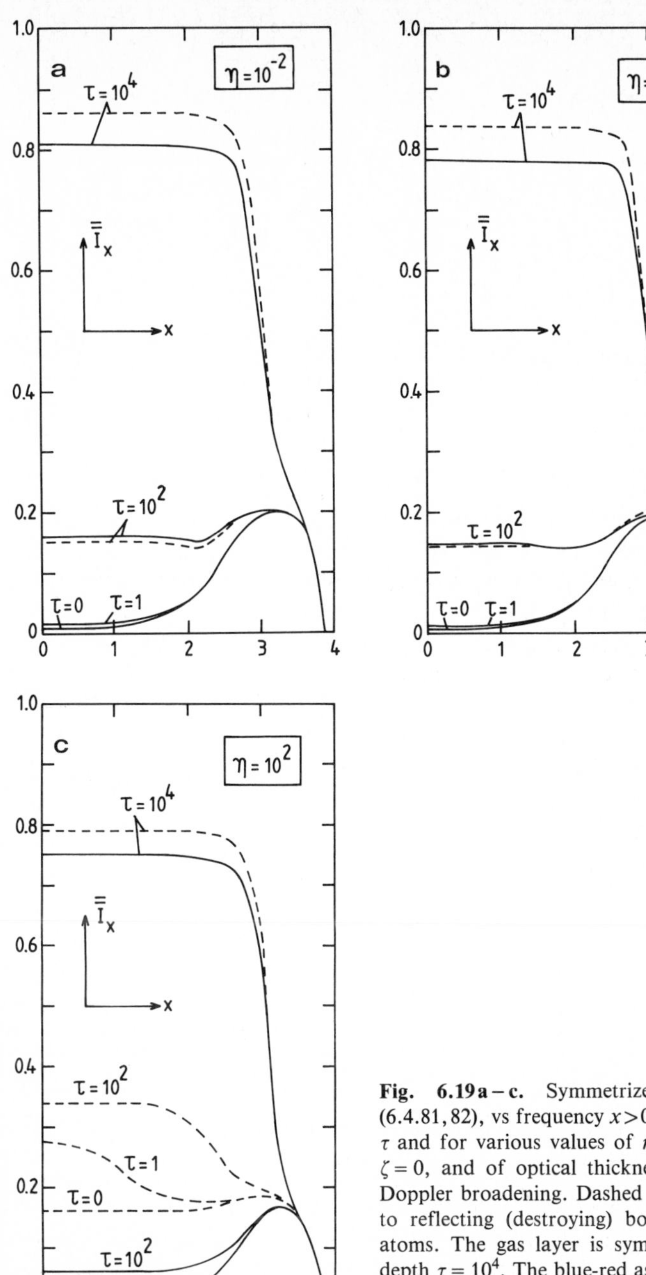

Fig. 6.19a–c. Symmetrized mean intensity $\bar{\bar{I}}_x$, (6.4.81, 82), vs frequency $x>0$ at various optical depths τ and for various values of η in a gas with $\varepsilon=10^{-4}$, $\zeta=0$, and of optical thickness $\tau^0=2\times10^4$, for pure Doppler broadening. Dashed (solid) curves correspond to reflecting (destroying) boundaries for the excited atoms. The gas layer is symmetric about the optical depth $\tau=10^4$. The blue-red asymmetry of the mean intensity $\bar{I}_x$ is small for $\eta=10^{-2}$ and $\eta=1$; for $\eta=10^2$, see Fig. 6.20

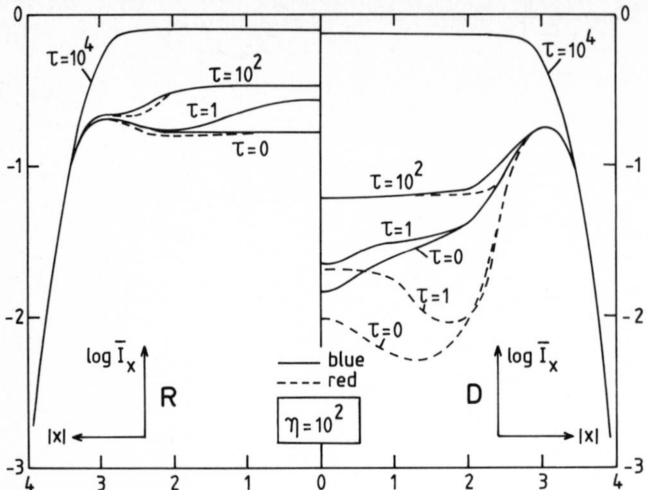

Fig. 6.20. Mean intensity $\bar{I}_x$, (6.4.81), vs modulus of frequency $|x|$ at various optical depths τ for $\eta = 10^2$ (all other parameters as in Fig. 6.19). Solid [dashed] curves correspond to the blue ($x>0$) [red ($x<0$)] part of the spectral line. The left (right) figure corresponds to reflecting (destroying) boundaries for the excited atoms

As a measure of the radiation emitted by the gas layer, Fig. 6.21 plots the quantity $H_x(0)$ as a function of frequency, where $H_x(\tau)$ is generally defined by

$$H_x(\tau) = \frac{1}{2} \int_{-1}^{1} \mu I_x(\mu;\tau)\, d\mu \ , \tag{6.4.83}$$

which at the boundary $\tau = 0$ takes the value

$$H_x(0) = \frac{1}{2} \int_{0}^{1} \mu I_x^+(\mu;0)\, d\mu \tag{6.4.84}$$

since $I_x^-(\mu;0) = 0$ according to the boundary condition (6.4.47). The vector $\boldsymbol{H}_\nu = \int \boldsymbol{n} I_\nu(\boldsymbol{n})\, d\Omega/4\pi$ equals $1/4\pi$ times the monochromatic radiant energy flux $\boldsymbol{f}_\nu = \int \boldsymbol{n} I_\nu(\boldsymbol{n})\, d\Omega$ which, in turn, is related to the total radiant energy flux by $\boldsymbol{f}_\mathrm{R} = \int \boldsymbol{f}_\nu d\nu$, see (4.2.2). Notice that a blackbody radiator $I_x^+(\mu;0) = 1$ yields the constant $H_x(0) = 1/4$. As can be seen from Fig. 6.21, the spectral lines emitted show the characteristic non-LTE self-reversal as explained in Fig. 6.13, except for $\eta = 10^2$ with reflecting boundary conditions, because then the density of excited atoms $\tilde{n}_2(\tau)$ is nearly constant within the gas (Fig. 6.15). On the other hand, Fig. 6.21 shows that for $\eta \gg 1$ the boundary conditions for the excited atoms have a tremendous effect on the shape of the emitted spectral line, as is to be expected on physical grounds.

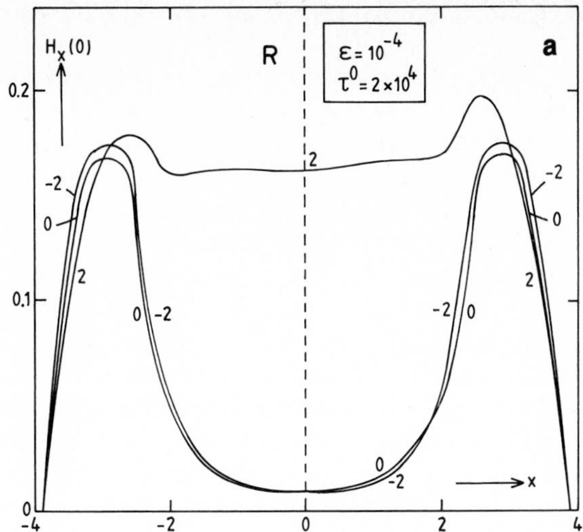

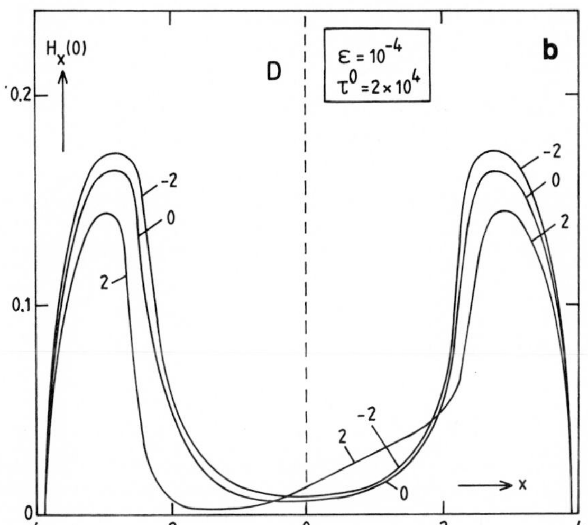

Fig. 6.21a, b. Radiation flux $H_x(0)$, (6.4.84), vs frequency x for a gas with $\varepsilon = 10^{-4}$, $\zeta = 0$, and of optical thickness $\tau^0 = 2\times10^4$, for pure Doppler broadening. Curves labeled n ($n = 2$, 0, -2) correspond to $\eta = 10^n$. Figure a(b) corresponds to reflecting (destroying) boundaries for the excited atoms. For comparison, a blackbody radiator emits $H_x(0) = 0.25$

All characteristic features of the distribution function $\tilde{n}_2 f_2(y)$ have their counterpart in analogous features of the radiation intensity $I_x(\boldsymbol{n})$. This parallelism between the distribution functions of excited atoms and photons is not surprising since in an optically thick non-LTE plasma with $\varepsilon \ll 1$, most of the excited atoms are radiation-excited atoms (as opposed to collision-excited ones), while on the other hand, in a self-excited gas as considered here, all photons have been emitted by excited atoms within the gas (as opposed to photons entering the gas from outside). In this context, it may be helpful to reconsider the simple model discussed in Sect. 6.3.4, see (6.3.64 – 75) and Figs. 6.7 – 9.

To conclude this section, we discuss the various limiting cases arising in the standard model. There are three dimensionless parameters ε, ζ, η (6.4.35 – 37, 41 – 43) which characterize, respectively, inelastic collisions, elastic collisions, and streaming of the excited atoms. The limiting case $\varepsilon_0 \gg 1$, $\varepsilon \simeq 1$ corresponds to a Boltzmann distribution for n_2/n_1 and hence to LTE because, in this case, there is also a Maxwellian velocity distribution $f_2 = f^M$. In contrast, for $\varepsilon \ll 1$ the gas is in a non-LTE state. Note that in the standard model, density gradients and thus streaming of the excited atoms owing to radiative transfer arise only in non-LTE situations.

Now consider an optically thick non-LTE gas with $\varepsilon \ll 1$. The relation $\zeta \ll 1$ expresses the fact that elastic velocity-changing collisions of excited atoms can be ignored (kinetic regime), so that radiation-excited atoms re-emit according to correlated re-emission (Sect. 6.3.4). If here $\eta \ll 1$, streaming of excited atoms is negligible, and non-LTE line transfer is described by static redistribution functions (Sect. 6.3.5). In the opposite limiting case $\eta \gg 1$, free streaming of excited atoms is important and must be described by a kinetic equation (Sect. 6.4.6).

Conversely, for $\zeta \gg 1$ elastic velocity-changing collisions of excited atoms are frequent (diffusion regime), and radiation-excited atoms re-emit according to uncorrelated re-emission (Sect. 6.3.4). If here $\eta \ll 1$, static complete redistribution (Sect. 6.4.4) applies. On the other hand, for $\eta \gg 1$ the dimensionless diffusion constant $\delta = \eta^2/2\zeta$, (6.4.70), intervenes. For $\delta \ll 1$, static complete redistribution (Sect. 6.4.4) applies again, whereas for $\delta \gg 1$ there is diffusion flow of excited atoms (Sect. 6.4.5).

The following scheme summarizes this discussion:

$\varepsilon \simeq 1$: LTE
$\varepsilon \ll 1$: non-LTE
    $\zeta \ll 1$: kinetic regime
        $\eta \ll 1$: static partial redistribution
        $\eta \gg 1$: kinetic flow
    $\zeta \gg 1$: diffusion regime
        $\eta \ll 1$: static complete redistribution
        $\eta \gg 1$: $\delta = \eta^2/2\zeta$
            $\delta \ll 1$: static complete redistribution
            $\delta \gg 1$: diffusion flow.

6.5 Multilevel Atoms

Non-LTE radiative transfer becomes much more complicated if one goes beyond the two-level atom by considering additional bound levels and the continuum. First of all, there is the simple fact that the number of independent physical parameters (such as Einstein coefficients, line profile coefficients, optical thicknesses) increases rapidly with an increasing number of atomic levels. Consequently, systematic investigations along the lines outlined in the preceding sections for two-level atoms have not been carried out. To a large extent, our present knowledge about the physics of non-LTE radiative transfer by multilevel atoms consists of a collection of numerical results referring to various individual cases [6.9,13].

However, independently of the question of the mere number of physical parameters, there are special new problems that arise in the context of non-LTE radiative transfer by multilevel atoms as compared to the case of a two-level atom. Let us illustrate them by considering a three-level atom that has three bound levels 1, 2, 3 with energies $E_1<E_2<E_3$, and no continuum. Consider, in particular, radiative transfer in the spectral line due to the atomic transition $2\leftrightarrow 3$. In the absence of stimulated emissions, the corresponding line emission and absorption coefficients are, according to (3.2.9,10,12,13), given by

$$\varepsilon_\nu^{32}(\boldsymbol{n}) = \frac{h\nu_{23}^0}{4\pi} n_3 A_{32} \int f_3(\boldsymbol{v})\,\eta_{32}(\xi,\boldsymbol{n})\,d^3v \ ,$$

$$\kappa_\nu^{23}(\boldsymbol{n}) = \frac{h\nu_{23}^0}{4\pi} n_2 B_{23} \int f_2(\boldsymbol{v})\,\alpha_{23}(\xi)\,d^3v \ ,$$

where ξ denotes the frequency of the photon $(\nu,\boldsymbol{n})$ as measured in the rest frame of the three-level atom, see (3.2.8). Hence, the emission and the absorption coefficient depend explicitly on the distribution functions $F_3(\boldsymbol{v}) = n_3 f_3(\boldsymbol{v})$ and $F_2(\boldsymbol{v}) = n_2 f_2(\boldsymbol{v})$, respectively, which in general are affected by the radiation field in the spectral line considered, thereby giving rise to the self-consistency problem discussed above. However, for the three-level atom, these distribution functions are also affected by the radiative transfer in the two other spectral lines due to the atomic transitions $1\leftrightarrow 2$ and $1\leftrightarrow 3$. This means that in a non-LTE situation, there is a coupling (interlocking) of the radiative transfer in all spectral lines (and in the continuum). For complete redistribution where $f_1(\boldsymbol{v}) = f_2(\boldsymbol{v}) = f_3(\boldsymbol{v})$, this interlocking is brought about by the number densities n_2 and n_3, while in the general case also the velocity distributions $f_2(\boldsymbol{v})$ and $f_3(\boldsymbol{v})$ contribute to it. In this context it should be noticed that for the subordinate line $2\leftrightarrow 3$ not only the emission coefficient ε_ν^{32} is a priori unknown (as is familiar from the case of a two-level atom), but also the absorption coefficient κ_ν^{23}, so that here *both* quantities have to be determined through an iterative procedure. In the absence of complete redistribution, not even the absorption profile φ_ν^{23} can be specified a priori since the velocity distribution f_2 is not known.

While the physical basis of the interlocking of the spectral lines of a multilevel atom via the number densities n_i and the velocity distributions $f_i(\boldsymbol{v})$ is easy to understand, it is not so obvious that even the *atomic* profile coefficients α_{ij} and η_{ji} contribute to this interlocking. In other words, the atomic profiles α_{ij} and η_{ji} of the transition $i \leftrightarrow j$ of a multilevel atom depend in general not only on the radiation field in its own spectral line (as is also the case for a two-level atom), but also on the radiation fields in the other spectral lines of the atom. This is treated in Sect. 6.5.1 where explicit expressions of the atomic profiles of a three-level atom are discussed in terms of generalized atomic redistribution functions, while Sect. 6.5.2 treats the corresponding velocity-averaged profile coefficients φ_ν and ψ_ν for the important special case where particle streaming is negligible.

6.5.1 Atomic Line Profile Coefficients of a Three-Level Atom

In this section we consider again an atom with three bound levels 1, 2, 3 (of energies $E_1 < E_2 < E_3$) and no continuum. In determining its profile coefficients for absorption and emission, α_{ij} and η_{ij}, we encounter the difficulty that photons involved in consecutive radiative transitions of the atom are in general correlated. A well-known example is provided by resonance fluorescence (Sect. 6.3.4 and Appendix B.2) where the probability of re-emitting a photon of frequency ν in the atomic transition $2 \to 1$ depends on the frequency ν' of the photon previously absorbed in the transition $1 \to 2$. Or, to take another example, the probability of absorbing a photon of frequency ν in transition $2 \to 3$ depends on the frequency ν' of the photon previously absorbed in the transition $1 \to 2$. In other words, consecutive radiative transitions of an atom are non-Markovian processes.

To derive the formal expressions of the atomic profile coefficients, the concept of *natural population* of an atomic level turns out to be crucial. Indeed, as pointed out in Sect. 6.3.4, the emission and absorption probabilities of an ensemble of atoms in a naturally populated level are independent of the previous history of the atoms. Consequently, as far as atomic line profiles are concerned, the non-Markovian character of radiative processes needs to be considered only to the point where a naturally populated level is encountered.

Recall that spontaneous emissions and inelastic collisions (as well as all ionization and recombination processes) lead to natural population of the final atomic level, and these processes are collectively denoted by

$$\to i^* ,$$

with i^* indicating a naturally populated atomic level i. By contrast, absorptions from lower levels j are denoted by

$$j \Rightarrow i \quad (j < i)$$

because they give rise to deviations from natural population of level i if the intensity of the absorbed spectral line varies as a function of frequency.

Before proceeding, let us make the following comments.

(1) Like absorptions, stimulated emissions can also give rise to deviations from natural population of the final atomic level; however, in this section, we

discuss only the low-temperature limit $kT \ll h\nu$ so that stimulated emissions in all considered spectral lines can be ignored.

(2) In the absence of stimulated emissions, the ground state of an atom is always naturally populated, and is denoted 1* in the following.

(3) As pointed out in Sect. 6.3.4 and Appendix B.6, the fact that spontaneous emissions populate the final atomic level naturally does not exclude the existence of correlations between the spontaneously emitted photon and the photons involved in subsequent radiative transitions of the atom.

Let us now turn to the atomic line profile coefficients of a three-level atom, and consider, as a first example, the absorption profile α_{23}. According to the above discussion, one has to consider here all processes that start from a naturally populated initial level and terminate with the absorption $2 \Rightarrow 3$. There are only two such processes, namely,

$$\rightarrow 2^* \Rightarrow 3 \quad \text{and} \quad 1^* \Rightarrow 2 \Rightarrow 3 \ ,$$

and they contribute to the absorption profile α_{23} according to their relative probabilities of occurrence, prob$\{\rightarrow 2^*\}$ and prob$\{1^* \Rightarrow 2\}$. Or, to determine the emission profile η_{31} one must take into account all processes that start from a naturally populated initial level and terminate with the spontaneous emission $3 \rightarrow 1$, namely,

$$\begin{aligned} \rightarrow 3^* \rightarrow 1 \ , \\ 1^* \Rightarrow 3 \rightarrow 1 \ , \\ \rightarrow 2^* \Rightarrow 3 \rightarrow 1 \ , \\ 1^* \Rightarrow 2 \Rightarrow 3 \rightarrow 1 \ , \end{aligned}$$

which contribute to η_{31} according to their relative probabilities prob$\{\rightarrow 3^*\}$, prob$\{1^* \Rightarrow 3\}$, prob$\{\rightarrow 2^* \Rightarrow 3\}$, and prob$\{1^* \Rightarrow 2 \Rightarrow 3\}$.

Quite generally, in the absence of stimulated emissions, determining the line profile coefficients of an atom with a finite number of bound levels requires knowing only a *finite number* of distinct types of non-Markovian radiative processes. Indeed, for an n-level atom, non-Markovian radiative processes relevant to profile coefficients comprise n radiative transitions at most, the longest reaction chains being provided by the $n-1$ consecutive absorptions $1^* \Rightarrow 2 \Rightarrow \cdots \Rightarrow n$ followed by a spontaneous emission $n \rightarrow i$ into a lower level $i < n$.

For a given atom, one therefore defines so-called generalized atomic redistribution functions which describe the photon correlations in each sequence of consecutive radiative transitions of the atom that starts from a naturally populated initial level. These functions are straightforward generalizations of the usual redistribution function describing the correlation between absorbed and re-emitted photons in the case of resonance fluorescence. Like ordinary redistribution functions, generalized redistribution functions must be determined by quantum mechanical calculations, taking into account the effects of perturber particles and fields on the atom under consideration. In this book, we consider atomic redistribution functions as known quantities.

Before giving the exact definition of generalized atomic redistribution functions, we introduce the following simplified notation to be used in Sect. 6.5. From now on, the unit vectors $\boldsymbol{n}$ referring to photon direction are suppressed, so e.g.,

$$r_{121}(\xi', \xi) \equiv r_{121}(\xi', \boldsymbol{n}'; \xi, \boldsymbol{n}) \ , \quad I(\nu) \equiv I_\nu(\boldsymbol{n}) \ ,$$

where ξ denotes again the photon frequency in the atomic rest frame, (6.3.6). In particular,

$$d\xi \equiv d\xi \frac{d\Omega}{4\pi} \ , \quad d\nu \equiv d\nu \frac{d\Omega}{4\pi}$$

and

$$\int d\xi \equiv \iint d\xi \frac{d\Omega}{4\pi} = \iint d\nu \frac{d\Omega}{4\pi} \equiv \int d\nu \ ,$$

where the integrations are understood to be over all frequencies and all directions. Recall that according to approximation (2.4.3), the photon $(\nu, \boldsymbol{n})$ in the laboratory frame corresponds to the photon $(\xi, \boldsymbol{n})$ in the atomic rest frame and vice versa, for we do not transform the photon direction $\boldsymbol{n}$.

Suppose the energy levels of an atom are arranged according to increasing energies, $E_1 < E_2 < E_3 \dots$. As in Appendix B.6, a photon involved in the radiative transition $i \to j$ of the atom is denoted by $\xi_{ij} \equiv (\xi_{ij}, \boldsymbol{n}_{ij})$ and $\nu_{ij} \equiv (\nu_{ij}, \boldsymbol{n}_{ij})$ in the atomic and the laboratory frame, respectively, so that ξ_{ij} and ν_{ij} are absorbed photons if $i < j$, and emitted photons if $i > j$ (Fig. B.4). (In Appendix B, all quantities refer to the atomic rest frame, and hence ν there denotes the frequency in the atomic rest frame.)

For the reader's convenience the *generalized atomic redistribution functions* r_{ij}, r_{ijk}, ... as given in Appendix B.6 are also defined here. Imagine the atom immersed in an isotropic white radiation field (the radiation intensity thus being constant within the frequency range of each spectral line), and consider consecutive radiative transitions $i^* \to j$, $i^* \to j \to k$, ..., of the atom that start from a naturally populated initial level i^*. Then:

$r_{ij}(\xi_{ij})\, d\xi_{ij}$ is the probability that a photon in the range $(\xi_{ij}, d\xi_{ij})$ is involved in the radiative transition $i^* \to j$;

$r_{ijk}(\xi_{ij}, \xi_{jk})\, d\xi_{ij}\, d\xi_{jk}$ is the probability that two photons in the ranges $(\xi_{ij}, d\xi_{ij})$ and $(\xi_{jk}, d\xi_{jk})$, respectively, are involved in the consecutive radiative transitions $i^* \to j \to k$, and so on.

The functions r_{ij}, r_{ijk}, ... are normalized, (B.6.1), and they obey integral relations, (B.6.2), and symmetry relations, (B.6.41).

Generalized atomic redistribution functions are defined in the rest frame of the atom. Therefore, if elastic collisions take place that change the velocity of the atom, the relations between the atomic frequencies ξ_{ij}, ξ_{jk}, ... and the corresponding laboratory frequencies ν_{ij}, ν_{jk}, ... according to (6.3.6) may involve different atomic velocities $\boldsymbol{v}, \boldsymbol{v}'$, ... For example, in the reaction $1^* \Rightarrow 2 \Rightarrow 3$,

$$\xi_{12} = \nu_{12} - \frac{\nu_{12}^0}{c} \boldsymbol{n}_{12} \cdot \boldsymbol{v} \ , \quad \xi_{23} = \nu_{23} - \frac{\nu_{23}^0}{c} \boldsymbol{n}_{23} \cdot \boldsymbol{v}' \ ,$$

with $\boldsymbol{v}' \neq \boldsymbol{v}$ if the velocity of the atom was changed by an elastic collision while it was in the excited level 2. However, below we need only redistribution functions in which all ξ_{ij} correspond to the same atomic velocity $\boldsymbol{v}$. Furthermore, it should be noted that the redistribution functions may depend explicitly on the atomic velocity, $r_{ij\ldots}(\xi_{ij}, \ldots ; \boldsymbol{v})$, for instance, if the line broadening depends on the relative velocity between the atom and the perturbers.

In close analogy to Sect. 6.3.4, we now suppose that in the absence of elastic collisions, an atom has a constant velocity during the whole time it is excited,

$$\boldsymbol{v} = \text{const} \ , \tag{6.5.1}$$

no matter whether it remains in just one excited level or makes transitions between several excited levels before returning to the ground state. The various assumptions underlying (6.5.1) are discussed after (6.3.49) in Sect. 6.3.4. On the other hand, elastic velocity-changing collisions with excited atoms are assumed to be strong collisions in the sense that they give rise to natural population of the corresponding atomic level. Therefore, in a small volume element of the gas, the ensemble of atoms (being in all possible excitation states) with velocity $\boldsymbol{v}$ in the laboratory frame can be considered independently of all the other atoms having velocities different from $\boldsymbol{v}$. Elastic collisions are the only (positive or negative) source for this ensemble, acting at the same time as a source of natural population of excited atomic levels.

We are now ready to write down the atomic absorption and emission profiles of a three-level atom [6.16]. Omitting all independent variables one has, see (B.6.8 – 13),

$$\alpha_{12} = r_{12} \ , \tag{6.5.2}$$

$$\alpha_{13} = r_{13} \ , \tag{6.5.3}$$

$$\alpha_{23} = \text{prob}\{\rightarrow 2^*\} r_{23} + \text{prob}\{1^* \Rightarrow 2\} j_{123} \ , \tag{6.5.4}$$

$$\eta_{21} = \text{prob}\{\rightarrow 2^*\} r_{21} + \text{prob}\{1^* \Rightarrow 2\} j_{121} \ , \tag{6.5.5}$$

$$\eta_{31} = \text{prob}\{\rightarrow 3^*\} r_{31} + \text{prob}\{1^* \Rightarrow 3\} j_{131}$$
$$+ \text{prob}\{\rightarrow 2^* \Rightarrow 3\} j_{231} + \text{prob}\{1^* \Rightarrow 2 \Rightarrow 3\} j_{1231} \ , \quad (6.5.6)$$

$$\eta_{32} = \text{prob}\{\rightarrow 3^*\} r_{32} + \text{prob}\{1^* \Rightarrow 3\} j_{132}$$
$$+ \text{prob}\{\rightarrow 2^* \Rightarrow 3\} j_{232} + \text{prob}\{1^* \Rightarrow 2 \Rightarrow 3\} j_{1232} \ . \quad (6.5.7)$$

Here we have introduced the normalized quantities, see (B.6.14, 15),

$$j_{ijk}(\xi_{jk}; \boldsymbol{v}) = \frac{1}{I_{ij}^*(\boldsymbol{v})} \int I(\nu_{ij}) r_{ijk}(\xi_{ij}, \xi_{jk}) d\xi_{ij} \ , \quad (6.5.8)$$

$$j_{ijkl}(\xi_{kl}; \boldsymbol{v}) = \frac{1}{I_{ijk}^*(\boldsymbol{v})} \iint I(\nu_{ij}) I(\nu_{jk}) r_{ijkl}(\xi_{ij}, \xi_{jk}, \xi_{kl}) d\xi_{ij} d\xi_{jk} \ , \quad (6.5.9)$$

where

$$I_{ij}^*(\boldsymbol{v}) = \int I(\nu_{ij}) r_{ij}(\xi_{ij}) d\xi_{ij} \ , \quad (6.5.10)$$

$$I_{ijk}^*(\boldsymbol{v}) = \iint I(\nu_{ij}) I(\nu_{jk}) r_{ijk}(\xi_{ij}, \xi_{jk}) d\xi_{ij} d\xi_{jk} \ . \quad (6.5.11)$$

Equations (6.5.2 – 7) are readily interpreted along the lines outlined above (also in Appendix B.6).

On the other hand, the various probabilities prob{...} appearing in (6.5.2 – 7) are given by, see (B.6.18 – 26),

$$\text{prob}\{\rightarrow 2^*\} = \frac{1}{N_2} [S_{12} F_1 + (A_{32} + S_{32}) F_3 + \Gamma_2^+] \ , \quad (6.5.12)$$

$$\text{prob}\{1^* \Rightarrow 2\} = \frac{1}{N_2} B_{12} I_{12}^* F_1 \ , \quad (6.5.13)$$

$$\text{prob}\{\rightarrow 3^*\} = \frac{1}{N_3} (S_{13} F_1 + S_{23} F_2 + \Gamma_3^+) \ , \quad (6.5.14)$$

$$\text{prob}\{1^* \Rightarrow 3\} = \frac{1}{N_3} B_{13} I_{13}^* F_1 \ , \quad (6.5.15)$$

$$\text{prob}\{\rightarrow 2^* \Rightarrow 3\} = \frac{1}{N_3} B_{23} I_{23} F_2 \frac{1}{N_2} [S_{12} F_1 + (A_{32} + S_{32}) F_3 + \Gamma_2^+] \ , \quad (6.5.16)$$

$$\text{prob}\{1^* \Rightarrow 2 \Rightarrow 3\} = \frac{1}{N_3} B_{23} I_{23} F_2 \frac{1}{N_2} B_{12} I_{12}^* F_1 \ , \quad (6.5.17)$$

with

$$N_2 = (B_{12}I_{12}^* + S_{12})F_1 + (A_{32} + S_{32})F_3 + \Gamma_2^+ \ , \tag{6.5.18}$$

$$N_3 = (B_{13}I_{13}^* + S_{13})F_1 + (B_{23}I_{23} + S_{23})F_2 + \Gamma_3^+ \ . \tag{6.5.19}$$

Here the inelastic collision rates are

$$S_{ij} \equiv n_e C_{ij} \ ,$$

F_i stands for the distribution function

$$F_i \equiv F_i(\boldsymbol{v}) \ ,$$

and Γ_i^+ is the creation part of the elastic collision term, see (6.3.60),

$$\left(\frac{\delta F_i}{\delta t}\right)_{\rm el} = \Gamma_i^+ - \Gamma_i^- \ . \tag{6.5.20}$$

Furthermore, see (B.6.16a, 35a),

$$I_{ij}(\boldsymbol{v}) = \int I(\nu_{ij})\,\alpha_{ij}(\xi_{ij})\,d\xi_{ij} \quad (i<j) \ , \tag{6.5.21a}$$

$$I_{ij}(\boldsymbol{v}) = \int I(\nu_{ij})\,\eta_{ij}(\xi_{ij})\,d\xi_{ij} \quad (i>j) \ . \tag{6.5.21b}$$

In the absence of stimulated emissions, the second quantity does not appear in the probabilities prob{...}, but has been defined here for completeness. Observe that I_{ij}, (6.5.21), contain the true atomic profile coefficients α_{ij} and η_{ij}, in contrast to I_{ij}^*, (6.5.10), which is formed using atomic redistribution functions r_{ij}.

In (6.5.12 – 19), the absorptions $1^* \Rightarrow 2$ and $1^* \Rightarrow 3$ are described by I_{12}^* and I_{13}^* because the ground state is naturally populated, see (6.5.2, 3). By contrast, absorptions $2 \Rightarrow 3$ are described in (6.5.16, 17, 19) by I_{23} rather than I_{23}^* since level 2 may show deviations from natural population due to previous absorptions $1^* \Rightarrow 2$. Hence, owing to the dependence of I_{23} on the profile coefficient α_{23}, (6.5.2 – 7) together with (6.5.12 – 19) are, strictly speaking, only an implicit system of equations for the profiles α_{ij} and η_{ij}. However, by setting approximately in (6.5.16, 17, 19)

$$I_{23} \simeq I_{23}^* \ , \tag{6.5.22}$$

one obtains explicit expressions of the atomic profile coefficients α_{ij} and η_{ij} in terms of the particle and photon distribution functions $F_i(\boldsymbol{v})$ and $I_\nu(\boldsymbol{n})$.

Due to the non-Markovian character of resonance fluorescence, the atomic emission profile η_{21} of a two-level atom depends on the radiation intensity in the resonance line, see (6.3.56, 57). However, the atomic line profiles of multilevel atoms are far more complicated quantities as they depend in general not only on the radiation field in the corresponding spectral line, but explicitly or implicitly on the radiation field in all of the other spectral lines and on the distribution functions of the atoms in the various excitation states as well.

Going beyond a three-level atom by considering additional bound levels complicates the matter, but does not give rise to any new features. On the other hand, because recombination (ionization) processes involving ions of the next higher (lower) stage of ionization lead to natural population of the bound level considered, they can be included in the above formalism simply by appropriate changes in the probabilities prob$\{\ldots\}$ wherever processes $\rightarrow i^*$ occur.

6.5.2 Laboratory Line Profile Coefficients of a Three-Level Atom

The laboratory emission and absorption line profile coefficients ψ_ν and φ_ν appearing in the radiative transfer equation are the mean values of the corresponding atomic profiles, averaged over the velocity distributions of atoms in the respective initial levels of the atomic transitions considered. In principle, these velocity distributions must be determined as solutions of the corresponding kinetic equations (Sect. 6.4.6). In many cases, however, one may make the following two simplifying assumptions. First, the velocity distribution of atoms in the ground state is Maxwellian, and second, the streaming of excited atoms is negligible owing to their short lifetime. Adopting these two assumptions, this section treats the line profiles of a three-level atom [6.16].

First of all, we introduce for later use various velocity-averaged quantities. Velocity-averaged generalized redistribution functions are defined by

$$\varphi_{ij}^*(\nu_{ij}) = \int d^3v\, f_i(\boldsymbol{v})\, r_{ij}(\xi_{ij}) \ , \tag{6.5.23a}$$

$$R_{ijk}(\nu_{ij}, \nu_{jk}) = \int d^3v\, f_i(\boldsymbol{v})\, r_{ijk}(\xi_{ij}, \xi_{jk}) \ , \tag{6.5.23b}$$

$$R_{ijkl}(\nu_{ij}, \nu_{jk}, \nu_{kl}) = \int d^3v\, f_i(\boldsymbol{v})\, r_{ijkl}(\xi_{ij}, \xi_{jk}, \xi_{kl}) \ , \tag{6.5.23c}$$

etc., although a more consistent notation would be R_{ij}^*, R_{ijk}^*, R_{ijkl}^*, ... instead of φ_{ij}^*, R_{ijk}, R_{ijkl}, ... Here $f_i(\boldsymbol{v})$ is the normalized velocity distribution of the atoms in level i. Provided that level i is naturally populated, φ_{ij}^* is identical with the absorption profile φ_{ij} if $i<j$, and it is identical with the emission profile ψ_{ij} if $i>j$. Note that $r_{ij}=r_{ji}$ always, (B.6.41a), but $\varphi_{ij}^*=\varphi_{ji}^*$ only if $f_i(\boldsymbol{v})=f_j(\boldsymbol{v})$. On the other hand, R_{iji} with $i<j$ is the ordinary redistribution function for the atomic transition $i\rightarrow j\rightarrow i$, see (6.3.89); recall that it supposes a naturally populated lower level i. The functions (6.5.23) are normalized on account of (2.1.3) and (B.6.1),

$$\int \varphi_{ij}^*(\nu_{ij})\, d\nu_{ij} = 1 \ , \tag{6.5.24a}$$

$$\iint R_{ijk}(\nu_{ij}, \nu_{jk})\, d\nu_{ij}\, d\nu_{jk} = 1 \ , \tag{6.5.24b}$$

$$\iiint R_{ijkl}(\nu_{ij}, \nu_{jk}, \nu_{kl})\, d\nu_{ij}\, d\nu_{jk}\, d\nu_{kl} = 1 \ , \tag{6.5.24c}$$

and the following relations hold on account of (B.6.2):

$$\int R_{ijk}(\nu_{ij}, \nu_{jk})\, d\nu_{jk} = \varphi^*_{ij}(\nu_{ij}) \ , \tag{6.5.25a}$$

$$\int R_{ijkl}(\nu_{ij}, \nu_{jk}, \nu_{kl})\, d\nu_{kl} = R_{ijk}(\nu_{ij}, \nu_{jk}) \ . \tag{6.5.25b}$$

Furthermore, we shall need the velocity averages of the quantities (6.5.8 – 11), so we define

$$J^*_{ij} = \int d^3v\, f_i(\boldsymbol{v}) I^*_{ij}(\boldsymbol{v}) = \int I(\nu_{ij})\, \varphi^*_{ij}(\nu_{ij})\, d\nu_{ij} \ , \tag{6.5.26a}$$

$$J^*_{ijk} = \int d^3v\, f_i(\boldsymbol{v}) I^*_{ijk}(\boldsymbol{v}) = \iint I(\nu_{ij}) I(\nu_{jk}) R_{ijk}(\nu_{ij}, \nu_{jk})\, d\nu_{ij} d\nu_{jk} \ , \tag{6.5.26b}$$

$$\begin{aligned} K^*_{ijk}(\nu_{jk}) &= \int d^3v\, f_i(\boldsymbol{v}) I^*_{ij}(\boldsymbol{v})\, j_{ijk}(\xi_{jk}; \boldsymbol{v}) \\ &= \int I(\nu_{ij}) R_{ijk}(\nu_{ij}, \nu_{jk})\, d\nu_{ij} \ , \end{aligned} \tag{6.5.27a}$$

$$\begin{aligned} K^*_{ijkl}(\nu_{kl}) &= \int d^3v\, f_i(\boldsymbol{v}) I^*_{ijk}(\boldsymbol{v})\, j_{ijkl}(\xi_{kl}; \boldsymbol{v}) \\ &= \iint I(\nu_{ij}) I(\nu_{jk}) R_{ijkl}(\nu_{ij}, \nu_{jk}, \nu_{kl})\, d\nu_{ij}\, d\nu_{jk} \ . \end{aligned} \tag{6.5.27b}$$

Through (6.5.25),

$$\int K^*_{ijk}(\nu_{jk})\, d\nu_{jk} = J^*_{ij} \ , \tag{6.5.28a}$$

$$\int K^*_{ijkl}(\nu_{kl})\, d\nu_{kl} = J^*_{ijk} \ . \tag{6.5.28b}$$

After these preliminaries we now turn to the velocity distributions of excited three-level atoms. For clarity, the three simplifying assumptions are repeated.

(1) The distribution function of three-level atoms in the ground state 1 is Maxwellian, $F_1(\boldsymbol{v}) = n_1 f^{\mathrm{M}}(\boldsymbol{v})$, where $f^{\mathrm{M}}(\boldsymbol{v})$ is given by (6.3.13).

(2) The static approximation is adopted, which neglects streaming of excited atoms, so that $\boldsymbol{v} \cdot \nabla F_i = 0$ in the kinetic equations of the excited three-level atoms $i = 2, 3$, see (6.3.23).

(3) The elastic collision terms of the excited three-level atoms are approximated by simple relaxation terms, (6.3.12).

With these approximations, the kinetic equations for three-level atoms in the excited levels 2 and 3 can be written as, see (6.3.7 – 12),

$$\begin{aligned} &\gamma_2 n_2 (f_2 - f^{\mathrm{M}}) + n_2 f_2 (A_{21} + S_{21} + B_{23} I^*_{23} + S_{23}) \\ &\quad = n_3 f_3 (A_{32} + S_{32}) + n_1 f^{\mathrm{M}} (B_{12} I^*_{12} + S_{12}) \ , \end{aligned} \tag{6.5.29}$$

$$\begin{aligned} &\gamma_3 n_3 (f_3 - f^{\mathrm{M}}) + n_3 f_3 (A_{31} + S_{31} + A_{32} + S_{32}) \\ &\quad = n_2 f_2 (B_{23} I^*_{23} + S_{23}) + n_1 f^{\mathrm{M}} (B_{13} I^*_{13} + S_{13}) \ , \end{aligned} \tag{6.5.30}$$

where $f \equiv f(\boldsymbol{v})$, $I^*_{ij} \equiv I^*_{ij}(\boldsymbol{v})$, $S_{ij} \equiv n_e C_{ij}$, and where we have set $I_{23} \simeq I^*_{23}$, (6.5.22).

Integrating (6.5.29, 30) over all velocities, and observing that the elastic collision frequencies γ_i and the inelastic collision rates S_{ij} are assumed to be independent of the atomic velocity $\boldsymbol{v}$, one obtains

$$n_2\tau_2^{-1} = n_3(A_{32}+S_{32}) + n_1(B_{12}J_{12}^*+S_{12}) \ , \tag{6.5.31}$$

$$n_3\tau_3^{-1} = n_2(B_{23}J_{23}^*+S_{23}) + n_1(B_{13}J_{13}^*+S_{13}) \ , \tag{6.5.32}$$

where τ_2 and τ_3 denote the mean lifetimes of the excited levels, so that

$$\tau_2^{-1} = A_{21}+S_{21}+B_{23}J_{23}^*+S_{23} \ , \tag{6.5.33}$$

$$\tau_3^{-1} = A_{31}+S_{31}+A_{32}+S_{32} \ . \tag{6.5.34}$$

As stimulated emissions are neglected, τ_3 is independent of the radiation field, in contrast to τ_2.

Setting now in (6.5.29)

$$A_{21}+S_{21}+B_{23}I_{23}^*+S_{23} \simeq \tau_2^{-1} \ , \tag{6.5.35}$$

(6.5.29, 30) take similar forms,

$$(1+\gamma_2\tau_2)f_2 = \left[\frac{n_1}{n_2\tau_2^{-1}}(B_{12}I_{12}^*+S_{12}) + \gamma_2\tau_2\right]f^{\mathrm{M}} + \frac{n_3}{n_2\tau_2^{-1}}(A_{32}+S_{32})f_3 \ , \tag{6.5.36}$$

$$(1+\gamma_3\tau_3)f_3 = \left[\frac{n_1}{n_3\tau_3^{-1}}(B_{13}I_{13}^*+S_{13}) + \gamma_3\tau_3\right]f^{\mathrm{M}} + \frac{n_2}{n_3\tau_3^{-1}}(B_{23}I_{23}^*+S_{23})f_2 \ . \tag{6.5.37}$$

Note that integrating (6.5.36) over all velocities leads again to (6.5.31), which shows that (6.5.35) is a consistent approximation.

Equations (6.5.36, 37) together with (6.5.31, 32) determine in the adopted approximation the velocity distributions of excited three-level atoms, $f_2(\boldsymbol{v})$ and $f_3(\boldsymbol{v})$, in terms of the specific intensity $I(\nu)$ which enters I_{ij}^* and J_{ij}^*.

We now proceed to determine the absorption and emission profiles of a three-level atom. Explicitly,

$$\varphi_{ij} = \int d^3v\, f_i\alpha_{ij} \ , \quad \psi_{ji} = \int d^3v\, f_j\eta_{ji} \ , \tag{6.5.38}$$

where the atomic profiles α_{ij} and η_{ji} are given by (6.5.2–7, 12–19), while the velocity distributions f_2 and f_3 follow from (6.5.36, 37, 31, 32). However, as the expressions for φ_{ij} and ψ_{ji} thus obtained are unwieldy and not very useful, we adopt a procedure analogous to that used in Sect. 6.3.5 to derive in the following the line profile coefficients in a form that makes the various causes of deviations from complete redistribution more transparent [6.16].

First of all, since the ground state 1 of the three-level atom is naturally populated so that $\alpha_{1i} = r_{1i}$, see (6.5.2, 3), and since the velocity distribution f_1 is Maxwellian, then immediately

$$\varphi_{12} = \varphi_{12}^{\mathrm{M}*} \,, \quad \varphi_{13} = \varphi_{13}^{\mathrm{M}*} \,, \quad \text{where} \tag{6.5.39}$$

$$\varphi_{ij}^{\mathrm{M}*}(\nu_{ij}) = \int d^3v\, f^{\mathrm{M}}(\boldsymbol{v})\, r_{ij}(\xi_{ij}) \tag{6.5.40}$$

denotes φ_{ij}^{*}, (6.5.23a), for the special case of a Maxwellian velocity distribution $f_i = f^{\mathrm{M}}$.

We next consider the absorption profile $\varphi_{23} = \int d^3v\, f_2 \alpha_{23}$ and the emission profile $\psi_{21} = \int d^3v\, f_2 \eta_{21}$, which according to (6.5.4, 5) have the same structure. Using (6.5.12, 13), the quantity f_2/N_2 appears in the integrands of φ_{23} and ψ_{21}, where N_2 is defined by (6.5.18). On the other hand, the same quantity f_2/N_2 appears in the kinetic equation (6.5.36) which may be written as

$$(1 + \gamma_2 \tau_2)\, f_2 = \frac{N_2}{n_2 \tau_2^{-1}} \,, \tag{6.5.41}$$

using $F_i = n_i f_i$ in (6.5.18), and recalling that

$$\Gamma_i^{+} = \gamma_i n_i f^{\mathrm{M}} \tag{6.5.42}$$

according to (6.5.20, 3.12), see (6.3.81).

Thus, starting from (6.5.41) and using (6.5.4, 12, 13) with $F_i = n_i f_i$ and (6.5.42), the absorption profile φ_{23} takes the form

$$\begin{aligned}
\varphi_{23} &= \frac{1}{1+\gamma_2\tau_2}\,\frac{1}{n_2\tau_2^{-1}} \int d^3v\, [n_1 f^{\mathrm{M}} (B_{12} I_{12}^{*} j_{123} + S_{12} r_{23}) \\
&\quad + n_3 f_3 (A_{32} + S_{32})\, r_{23} + \gamma_2 n_2 f^{\mathrm{M}} r_{23}] \\
&= \frac{1}{1+\gamma_2\tau_2} \left[\frac{n_1}{n_2\tau_2^{-1}} (B_{12} K_{123}^{*} + S_{12} \varphi_{23}^{\mathrm{M}*}) \right. \\
&\quad \left. + \frac{n_3}{n_2\tau_2^{-1}} (A_{32} + S_{32})\, \varphi_{23}^{3*} + \gamma_2 \tau_2 \varphi_{23}^{\mathrm{M}*} \right] ,
\end{aligned} \tag{6.5.43}$$

where in the second expression use has been made of the definitions (6.5.27a, 40), and where φ_{ij}^{k*} with $k \neq i$ is defined by

$$\varphi_{ij}^{k*}(\nu_{ij}) = \int d^3v\, f_k(\boldsymbol{v})\, r_{ij}(\xi_{ij}) \;. \tag{6.5.44}$$

Proceeding analogously to (6.3.94 – 96), it is easily shown that φ_{23} can be written as

$$\varphi_{23} = \varphi_{23}^{M*} + \frac{1}{1+\gamma_2\tau_2}\,\frac{1}{n_2\tau_2^{-1}}\,[n_1 B_{12}(K_{123}^* - J_{12}^*\varphi_{23}^{M*}) + n_3(A_{32}+S_{32})(\varphi_{23}^{3*} - \varphi_{23}^{M*})] \ . \tag{6.5.45}$$

On the other hand, by simply substituting $r_{23} \to r_{21} = r_{12}$ and $j_{123} \to j_{121}$, one obtains the emission profile ψ_{21} as

$$\psi_{21} = \varphi_{12}^{M*} + \frac{1}{1+\gamma_2\tau_2}\,\frac{1}{n_2\tau_2^{-1}}\,[n_1 B_{12}(K_{121}^* - J_{12}^*\varphi_{12}^{M*}) + n_3(A_{32}+S_{32})(\varphi_{12}^{3*} - \varphi_{12}^{M*})] \ . \tag{6.5.46}$$

Equations (6.5.45, 46) are written in a form that exhibits the causes of deviations from complete redistribution. Here *complete redistribution* is defined by

$$\varphi_{ij} = \varphi_{ij}^{M*} = \varphi_{ji}^{M*} = \psi_{ji} \ . \tag{6.5.47}$$

Equations (6.5.38, 40) show that this requires

$$\alpha_{ij} = r_{ij} = r_{ji} = \eta_{ji} \ , \tag{6.5.48}$$

on the one hand, and

$$f^M \equiv f_1 = f_2 = f_3 \ , \tag{6.5.49}$$

on the other.

Thus, in (6.5.45, 46), the terms proportional to $\varphi^{3*} - \varphi^{M*}$ describe deviations from complete redistribution due to a non-Maxwellian velocity distribution $f_3 \neq f^M$. On the other hand, if all velocity distributions are Maxwellian, (6.5.49), so that $\varphi_{12}^* = \varphi_{12}^{M*}$, $\varphi_{23}^* = \varphi_{23}^{M*}$, $\varphi_{ij}^{3*} = \varphi_{ij}^{M*}$, (6.5.45, 46) show that complete redistribution applies if

$$K_{ijk}^* = J_{ij}^* \, \varphi_{jk}^* \ , \tag{6.5.50}$$

which in view of (6.5.26a, 27a) requires in turn that the redistribution functions factorize as, see (6.3.98),

$$R_{ijk} = \varphi_{ij}^* \, \varphi_{jk}^* \ . \tag{6.5.51}$$

Note that the two correction terms on the right-hand sides of (6.5.45, 46) are proportional to n_1/n_2 and n_3/n_2, respectively. Since normally $n_1/n_2 \gg 1$ and $n_3/n_2 \lesssim 1$, the second term is usually negligible compared to the first.

Lastly, we consider the emission profiles $\psi_{31} = \int d^3v\, f_3\eta_{31}$ and $\psi_{32} = \int d^3v f_3\eta_{32}$ which have the same structure according to (6.5.6, 7). Taking into account (6.5.12 – 17, 22, 42), and using $r_{31} = r_{13}$, see (B.6.41a), the emission profile ψ_{31} can be written as

$$\psi_{31} = \int d^3v \frac{f_3}{N_3} \{(n_1 f^{\mathrm{M}} S_{13} + n_2 f_2 S_{23} + \gamma_3 n_3 f^{\mathrm{M}}) r_{13} + n_1 f^{\mathrm{M}} B_{13} I_{13}^* j_{131}$$

$$+ n_2 \frac{f_2}{N_2} B_{23} I_{23}^* [(N_2 - n_1 f^{\mathrm{M}} B_{12} I_{12}^*)\, j_{231} + n_1 f^{\mathrm{M}} B_{12} I_{12}^* j_{1231}]\} \,, \tag{6.5.52}$$

where N_2 and N_3 are given by (6.5.18, 19). Now, the ratio f_2/N_2 follows from the kinetic equation (6.5.41), while the ratio f_3/N_3 follows from the kinetic equation (6.5.37) which may be written in the form

$$(1 + \gamma_3 \tau_3)\, f_3 = \frac{N_3}{n_3 \tau_3^{-1}} \,. \tag{6.5.53}$$

Thus, (6.5.52) takes the form

$$\psi_{31} = \frac{1}{1 + \gamma_3 \tau_3} \frac{1}{n_3 \tau_3^{-1}} \left(n_1 B_{13} K_{131}^* + n_1 S_{13} \varphi_{13}^{\mathrm{M}*} + n_2 S_{23} \varphi_{13}^{2*} \right.$$

$$\left. + \gamma_3 n_3 \varphi_{13}^{\mathrm{M}*} + \frac{1}{1 + \gamma_2 \tau_2} \frac{B_{23}}{\tau_2^{-1}} Q \right) \tag{6.5.54}$$

with Q given by

$$Q = \int d^3v\, I_{23}^* [n_1 f^{\mathrm{M}} B_{12} I_{12}^* j_{1231} + (N_2 - n_1 f^{\mathrm{M}} B_{12} I_{12}^*)\, j_{231}]$$

$$= n_1 B_{12} \int d^3v\, f^{\mathrm{M}} I_{12}^* I_{23}^* (j_{1231} - j_{231}) + (1 + \gamma_2 \tau_2)\, n_2 \tau_2^{-1} K_{231}^* \,, \tag{6.5.55}$$

where (6.5.41) has been used in the last step.

In (6.5.55), the two integrals over j_{1231} and j_{231}, respectively, cannot be expressed in terms of the quantities introduced in (6.5.26, 27). However, provided that $I_{ij}^*(\boldsymbol{v})$ and $j_{\dots ij}(\xi_{ij}; \boldsymbol{v})$ are sufficiently slowly varying functions of $\boldsymbol{v}$, one may in a first approximation average the various terms of the integrands separately. [Compare the analogous procedure applied in Sect. 6.3.6, (6.3.111 – 114).] Supposing in addition that the velocity distribution f_2 does not differ too strongly from a Maxwellian, then

$$\int d^3v\, f^{\mathrm{M}} I_{12}^* I_{23}^* j_{231} \simeq \int d^3v f^{\mathrm{M}} I_{12}^* \int d^3v\, f^{\mathrm{M}} I_{23}^* j_{231}$$

$$\simeq \int d^3v\, f^{\mathrm{M}} I_{12}^* \int d^3v\, f_2 I_{23}^* j_{231} = J_{12}^* K_{231}^* \,, \tag{6.5.56}$$

and similarly

$$\int d^3v\, f^{\mathrm{M}} I_{12}^* I_{23}^* j_{1231} = \int d^3v\, f^{\mathrm{M}} \frac{I_{12}^* I_{23}^*}{I_{123}^*} I_{123}^* j_{1231}$$

$$\simeq \frac{J_{12}^* J_{23}^*}{J_{123}^*} K_{1231}^* \; . \tag{6.5.57}$$

Thus, in this approximation,

$$Q \simeq n_1 B_{12} \left(\frac{J_{12}^* J_{23}^*}{J_{123}^*} K_{1231}^* - K_{231}^* \right) + (1 + \gamma_2 \tau_2)\, n_2 \tau_2^{-1} K_{231}^* \; . \tag{6.5.58}$$

Inserting (6.5.58) into (6.5.49), one obtains after a little algebra

$$\psi_{31} = \varphi_{13}^{\mathrm{M}*} + \frac{1}{1+\gamma_3\tau_3} \frac{1}{n_3\tau_3^{-1}} \Bigg[n_1 B_{13} (K_{131}^* - J_{13}^* \varphi_{13}^{\mathrm{M}*})$$

$$+ n_2 B_{23} (K_{231}^* - J_{23}^* \varphi_{13}^{\mathrm{M}*}) + n_2 S_{23} (\varphi_{13}^{2*} - \varphi_{13}^{\mathrm{M}*})$$

$$+ \frac{1}{1+\gamma_2\tau_2} \frac{1}{n_2\tau_2^{-1}} n_1 B_{12} J_{12}^* n_2 B_{23} J_{23}^* \left(\frac{K_{1231}^*}{J_{123}^*} - \frac{K_{231}^*}{J_{23}^*} \right) \Bigg] \; . \tag{6.5.59}$$

In complete analogy,

$$\psi_{32} = \varphi_{23}^{\mathrm{M}*} + \frac{1}{1+\gamma_3\tau_3} \frac{1}{n_3\tau_3^{-1}} \Bigg[n_1 B_{13} (K_{132}^* - J_{13}^* \varphi_{23}^{\mathrm{M}*})$$

$$+ n_2 B_{23} (K_{232}^* - J_{23}^* \varphi_{23}^{\mathrm{M}*}) + n_2 S_{23} (\varphi_{23}^* - \varphi_{23}^{\mathrm{M}*})$$

$$+ \frac{1}{1+\gamma_2\tau_2} \frac{1}{n_2\tau_2^{-1}} n_1 B_{12} J_{12}^* n_2 B_{23} J_{23}^* \left(\frac{K_{1232}^*}{J_{123}^*} - \frac{K_{232}^*}{J_{23}^*} \right) \Bigg] \; . \tag{6.5.60}$$

The emission profiles (6.5.59, 60) are again written in a form that exhibits the physical effects responsible for deviations from complete redistribution. The terms $K^* - J^* \varphi^*$ and $\varphi^* - \varphi^{\mathrm{M}*}$ have already been interpreted above in this section. On the other hand, the significance of the last terms in (6.5.59, 60) becomes clear by observing that if all redistribution functions (6.5.23) factorize according to, see (6.5.51),

$$R_{ijk} = \varphi_{ij}^* \varphi_{jk}^* \; , \quad R_{ijkl} = \varphi_{ij}^* \varphi_{jk}^* \varphi_{kl}^* \; , \tag{6.5.61}$$

then, see (6.5.50),

$$J_{ijk}^* = J_{ij}^* J_{jk}^* \; , \quad K_{ijk}^* = J_{ij}^* \varphi_{jk}^* \; , \quad K_{ijkl}^* = J_{ij}^* J_{jk}^* \varphi_{kl}^* \; , \tag{6.5.62}$$

and hence

$$\frac{K^*_{ijkl}}{J^*_{ijk}} - \frac{K^*_{jkl}}{J^*_{jk}} = 0 \ . \tag{6.5.63}$$

In conclusion, the various absorption and emission profiles of a three-level atom are in the considered approximation given by (6.5.39,45,46,59,60). However, it must be realized that the expressions for the profile coefficients φ_{23}, ψ_{21}, ψ_{31}, ψ_{32} given above depend not only on the radiation intensity $I(\nu)$, but also on the number densities n_2 and n_3, and on the velocity distributions $f_2(\boldsymbol{v})$ and $f_3(\boldsymbol{v})$. Nevertheless, these forms of the line profile coefficients may be the most useful ones for estimating the accuracy of the approximation of complete redistribution, which is usually adopted in line transfer calculations involving multilevel atoms.

6.6 Non-Maxwellian Electron Distribution Functions

Free electrons are of special importance in plasma spectroscopy because in most situations they determine the rates of the inelastic collision processes (excitation, de-excitation, ionization, recombination). As a result, the spectroscopic state of a plasma depends critically on electron density and electron temperature.

Until now it has been assumed in this chapter that the electron distribution function is Maxwellian, so that an electron temperature is defined in every small volume of the considered gas. But this assumption, almost universally made in plasma spectroscopy, is far from evident in general, because many plasma processes are known to be connected with non-Maxwellian electron distribution functions, such as plasma turbulence or the occurrence of runaway electrons. In this section, however, we want to discuss still another physical effect that may lead to non-Maxwellian electron distributions, especially at low ionization degrees, and which in principle is universal in contrast to special plasma effects.

The physical origin of this effect is easily understood by looking at Figs. 6.1,6: in a plasma that emits line radiation, say, the rate of excitation collisions must exceed the rate of de-excitation collisions to provide the radiant energy emitted. But according to the principle of detailed balance, the existence of unbalanced electronic collision processes is necessarily connected with the existence of a nonthermal (i.e., non-Maxwellian) distribution function of the electrons.

Let us repeat the same argument in other words. The principle of detailed balance ensures that in thermal equilibrium, loss of electrons out of the high-energy tail of the distribution function due to excitation and ionization collisions with atoms is exactly counterbalanced by production of such electrons due to inverse de-excitation and three-body recombination collisions. If, however, the gas is in a non-LTE state, the excited atomic levels and the continuum are underpopulated relative to the respective Boltzmann and Saha values, so that this balancing is perturbed, resulting in a net flow of electrons from the high-energy tail into the low-energy body of the distribution function. In a stationary state, this net flow is compensated by elastic electron collisions which tend to fill up the tail of the distribution function to its Maxwellian values. Departures of the atomic occupation numbers from their thermal values thus result in departures of the electron distribution from a Maxwell distribution. Therefore, to determine the spectroscopic state of a plasma, a self-consistent procedure must be applied, for the occupation numbers of the atomic levels and the continuum depend on the electron distribution function via the collision rates, and the electron distribution function in turn depends on the occupation numbers through the mechanism just outlined [6.30–35].

To gain some insight into this problem, we take the example of a partially ionized hydrogen plasma at temperatures $T \lesssim 10^4$ K ($kT \lesssim 1$ eV). Here the lowest resonance transition $H(1) \rightarrow H(2)$ corresponds to an excitation energy $E_{21} = 10.2$ eV, with a cross section for excitation by electron impact of the order of $Q_{12} \sim 10^{-16}\,\mathrm{cm}^2$. On the other hand, for suprathermal electrons with kinetic energies $E \gtrsim 3\,kT$, the effective cross section for elastic collisions with electrons in the energy range $0 < E < 3\,kT$, which form the (nearly Maxwellian) bulk of the distribution function, is of the order of $\bar{Q}_{ee} \sim 2\pi(e^2/E)^2 \ln \Lambda$ in cgs units, where Λ is the plasma parameter, (2.2.6). With $\ln \Lambda \sim 10$, an electron with $E = 10.2$ eV has an elastic cross section $\bar{Q}_{ee} \sim 10^{-14}\,\mathrm{cm}^2$. Hence, in a pure hydrogen plasma (where the electron density is equal to the proton density, $n_e = n_+$), for the high-energy tail of the electron distribution function to be affected by inelastic collisions, one requires ionization degrees $n_+/n_H \lesssim 0.01$ such that $n_H Q_{12} \gtrsim n_e \bar{Q}_{ee}$, recalling that the density of H atoms in the ground state is approximately equal to the total density of H atoms, $n_1 \simeq n_H$, and that the relative velocity of the colliding particles is in both cases approximately the same, given by the velocity of the high-energy electron.

Now, an interesting situation arises in a partially ionized hydrogen plasma whose ionization state is determined solely by collisional ionization (as opposed to photoionization). Imagine that in an iterative procedure one calculates the atomic level populations and the ionization degree using a Maxwellian electron distribution as a first approximation. For electron densities $n_e \lesssim 10^{17}\,\mathrm{cm}^{-3}$, there are departures from LTE (Tables 6.1,2). But, as outlined above in this section, non-LTE populations of atomic levels give rise to a non-Maxwellian high-energy tail of the electron distribution. Recalculating now the spectroscopic state of the hydrogen plasma with the aid of this new electron distribution which has less high-energy electrons than the Maxwell distribution, there is a lower degree of ionization than before because there are less ionization collisions. This means that also the departure from LTE increases since the electron density has decreased. Thus, deviations of the electron distribution $f_e(E)$ from a Maxwellian increase with decreasing ratio n_+/n_H, while on the other hand n_+/n_H decreases with increasing deviations of the high-energy tail of $f_e(E)$ from a Maxwellian. One is hence faced with a feedback effect which, below a critical density, is intrinsically unstable.

To be more explicit, we write the kinetic equations governing an optically thin hydrogen plasma that is stationary, homogeneous, and force-free. To make the equations as lucid as possible, we consider model H atoms having only two bound levels and a continuum (with excitation and ionization energies $E_{21} = 10.2$ eV and $E_{+1} = 13.6$ eV, respectively), and we shall use the simplest approximations possible.

In the situation considered, the kinetic equation for the (isotropic) electron distribution function $F_e(E) = n_e f_e(E)$ reduces according to (2.1.7,11) to

$$\left(\frac{\delta f_e}{\delta t}\right)_{\text{coll}} \equiv \left(\frac{\delta f_e}{\delta t}\right)_{\text{el}} + \left(\frac{\delta f_e}{\delta t}\right)_{\text{inel}} = 0 \ , \tag{6.6.1}$$

which expresses the stationarity of the distribution function under the simultaneous influence of elastic and inelastic collisions. In a first approximation, one may assume that for kinetic energies $E \leqslant E_{21}$, the distribution function $f_e(E)$ is Maxwellian corresponding to the temperature T. This assumption slightly violates the normalization of f_e, the relative error being at most of the order of $\exp(-E_{21}/kT)$.

On the other hand, for energies $E > E_{21}$, the elastic collision term due to electron-electron collisions is according to (2.2.8) given by

$$\left(\frac{\delta f_e}{\delta t}\right)_{\text{el}}^{\text{ee}} = \frac{2^{3/2}\pi n_e e^4}{m_e^{1/2}} \ln \Lambda \frac{d}{dE}\left[\frac{f_e(E)}{E^{1/2}} + kT\frac{d}{dE}\left(\frac{f_e(E)}{E^{1/2}}\right)\right] \tag{6.6.2}$$

in terms of kinetic energy $E = m_e v^2/2$ rather than velocity v, where T is the temperature of the low-energy body of the distribution f_e. In view of the possible ionization instability at lower densities, mentioned above, elastic collisions of suprathermal electrons with H atoms should also be considered. The corresponding collision term is, according to (2.2.2), given by

$$\left(\frac{\delta f_e}{\delta t}\right)_{\text{el}}^{\text{eH}} = \frac{2^{3/2} m_e^{1/2} n_1}{m_H} \frac{d}{dE}\left\{E^2 Q_{\text{eH}}(E)\left[\frac{f_e(E)}{E^{1/2}} + kT\frac{d}{dE}\left(\frac{f_e(E)}{E^{1/2}}\right)\right]\right\} , \tag{6.6.3}$$

where m_H is the mass of an H atom, Q_{eH} is the momentum-transfer cross section, and n_1 is the number density of H atoms in ground state 1. Elastic collisions with protons and with excited H atoms are neglected in (6.6.2,3), and the same temperature T is assumed to apply to both the atoms and the electrons. Since collisions among suprathermal electrons are negligible, the elastic collision term $(\delta f_e/\delta t)_{\text{el}}$, which is the sum of collision terms (6.6.2,3), is linear in $f_e(E)$. In the approximation used, the boundary conditions are $f_e(E_{21}) = f_e^M(E_{21})$ and $f_e(\infty) = 0$. Collision terms (6.6.2, 3) vanish identically if $f_e = f_e^M$, as is readily verified.

We now turn to the inelastic collision term of the electron distribution function. For hydrogen, it is sufficient to consider suprathermal electrons in the range $E_{21} < E < 2E_{21}$. The inelastic collision term for $f_e(E)$ due to excitation and de-excitation collisions is given by (2.3.22), where the third and fourth terms are negligible compared with the second and first terms, respectively (Fig. 6.22). Hence, for $E > E_{21}$

$$\left(\frac{\delta f_e}{\delta t}\right)_{\text{inel}}^{12} = -n_1\left(\frac{2E}{m_e}\right)^{1/2} Q_{12}(E) f_e(E) + n_2\left(\frac{2E'}{m_e}\right)^{1/2} Q_{21}(E') f_e^M(E') , \tag{6.6.4}$$

where $E' = E - E_{21}$, and where in the last term we have set $f_e(E') = f_e^M(E')$ according to our approximation, since $E' < E_{21}$. Using the Klein-Rosseland relation (1.6.9) together with the explicit forms of the Maxwell and Boltzmann distributions (1.4.14, 19), collision term (6.6.4) can be written as

$$\begin{aligned}\left(\frac{\delta f_e}{\delta t}\right)_{\text{inel}}^{12} &= -n_1\left(\frac{2E}{m_e}\right)^{1/2} Q_{12}(E)\left[f_e(E) - \frac{n_2/n_1}{(n_2/n_1)^B} f_e^M(E)\right] \\ &= -n_1\left(\frac{2E}{m_e}\right)^{1/2} Q_{12}(E)\left[f_e(E) - \frac{b_2}{b_1} f_e^M(E)\right] ,\end{aligned} \tag{6.6.5}$$

where in the second line the b coefficients (6.2.5) have been introduced according to (6.2.8). Collision term (6.6.5) vanishes identically if $n_2/n_1 = (n_2/n_1)^B$ and $f_e = f_e^M$, as required by the principle of detailed balance.

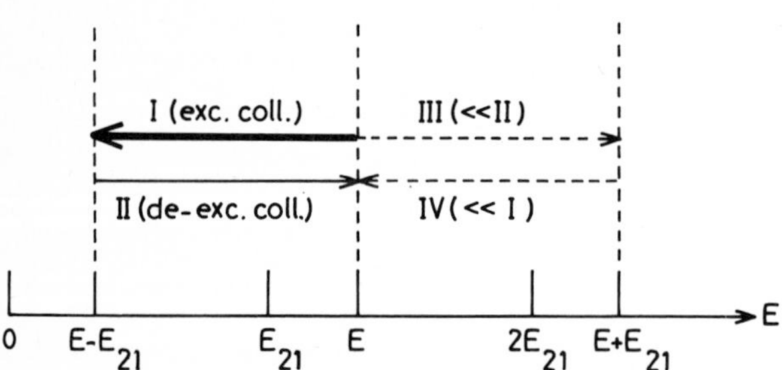

Fig. 6.22. Flow of electron energy due to excitation and de-excitation collisions. The electron energy E chosen lies in the interval $E_{21} < E < 2E_{21}$, where E_{21} is the atomic excitation energy. Processes I–IV correspond to terms 1–4 on rhs of (2.3.22). Processes III and IV are usually negligible compared with II and I, respectively. In a non-LTE plasma, excitation collisions I are more frequent than de-excitation collisions II, resulting in a net flow of electrons from the high-energy tail into the low-energy body of the distribution function

On the other hand, considering the inelastic collision term for suprathermal electrons due to ionization collisions and three-body recombinations, only free-bound transitions involving the ground state 1 need be considered because of the low ionization energy of the excited level, $E_{+2} = 3.4$ eV. Using (2.3.27) and neglecting there again the third and fourth terms, for energies $E > E_{+1}$

$$\left(\frac{\delta f_e}{\delta t}\right)_{\text{inel}}^{1+} = -n_1 \left(\frac{2E}{m_e}\right)^{1/2} Q_{1+}(E)\, f_e(E)$$

$$+ n_+ n_e \iint \left(\frac{2E'}{m_e}\right)^{1/2} \left(\frac{2E''}{m_e}\right)^{1/2} \Omega_{+1}(E', E''; E)$$

$$\cdot f_e^M(E')\, f_e^M(E'')\, dE' dE'' \;, \tag{6.6.6}$$

where $E' + E'' = E - E_{+1}$. Here the distribution functions in the integral describing three-body recombination have been approximated by Maxwell distributions, as primarily electrons with energies below 10.2 eV are involved in these processes. Then, using the explicit form of the Maxwell distribution (1.4.14) and the energy relation $E' + E'' = E - E_{+1}$, the term $(E'E'')^{1/2} f_e^M(E') \cdot f_e^M(E'')$ can be taken out of the integral. Using the Fowler relation (1.6.20), the Saha distribution (1.4.25), and employing (1.6.22), then (6.6.6) can be written in the form

$$\left(\frac{\delta f_e}{\delta t}\right)_{\text{inel}}^{1+} = -n_1 \left(\frac{2E}{m_e}\right)^{1/2} Q_{1+}(E) \left[f_e(E) - \frac{n_+ n_e / n_1}{(n_+ n_e / n_1)^S} f_e^M(E) \right]$$

$$= -n_1 \left(\frac{2E}{m_e}\right)^{1/2} Q_{1+}(E) \left[f_e(E) - \frac{1}{b_1} f_e^M(E) \right] \;, \tag{6.6.7}$$

where in the second line the coefficient b_1 has been introduced according to (6.2.7). Collision term (6.6.7) vanishes identically if $n_+ n_e / n_1 = (n_+ n_e / n_1)^S$ and $f_e = f_e^M$, as required by the principle of detailed balance. The total inelastic collision term for suprathermal electrons $(\delta f_e / \delta t)_{\text{inel}}$ is the sum of (6.6.5, 7).

The kinetic equations of the hydrogen atoms are here simply the balance equations (6.2.2, 9) for the two atomic level populations,

$$\begin{aligned} n_1 (n_e C_{12} + n_e M_{1+}) &= n_2 (A_{21} + n_e C_{21}^M) + n_+ (n_e R_{+1}^M + n_e^2 N_{+1}^M) \;, \\ n_2 (A_{21} + n_e C_{21}^M + n_e M_{2+}^M) &= n_1 n_e C_{12} + n_+ (n_e R_{+2}^M + n_e^2 N_{+2}^M) \;, \end{aligned} \tag{6.6.8}$$

where $n_+ = n_e$ in a pure hydrogen plasma. The balance equation for the protons is implicitly contained in (6.6.8). Here only inelastic collision processes due to free electrons have been taken into account. Because mainly low-energy (thermal) electrons are involved, the Maxwell distribution $f_e = f_e^M$ may be used to calculate the rate coefficients of collisional de-excitation C_{21}, (2.3.8b), of collisional ionization M_{2+}, (2.3.14a), of three-body recombination N_{+1} and N_{+2}, (2.3.14b), and of radiative recombination R_{+1} and R_{+2}, (2.4.23), as indicated by the superscript M in (6.6.8). By contrast, the true electron distribution $f_e(E)$ must be used for calculating the rate coefficients of collisional excitation C_{12}, (2.3.8a), and of collisional ionization M_{1+}, (2.14a). Because of the unbalanced processes of spontaneous emission (A_{21}) and spontaneous radiative recombination (R_{+1}, R_{+2}), (6.6.8) is not satisfied when substituting there Boltzmann, Saha, and Maxwell distributions for n_1, n_2, n_+, and $f_e(E)$, which expresses the often mentioned fact that an optically thin plasma is necessarily in a non-LTE state.

Equations (6.6.5, 7) show that the high-energy tail of the electron distribution function depends on the heavy-particle densities n_1, n_2, n_+, while on the other hand these densities depend on the high-energy tail of f_e through the rate coefficients C_{12} and M_{1+}. Therefore, (6.6.1 – 3, 5, 7, 8) must be solved self-consistently.

Some numerical results are plotted in Fig. 6.23, obtained from a model that takes more bound levels into account, and that uses a better approximation of the elastic collision term, allowing for deviations of $f_e(E)$ from a Maxwellian also at energies $E < E_{21}$. One observes that the high-energy tail of f_e is strongly depleted as a result of the feedback mechanism discussed above. It should be noted that the distribution functions depicted in Fig. 6.23 correspond to given electron and proton densities $n_e = n_+$. The self-consistently

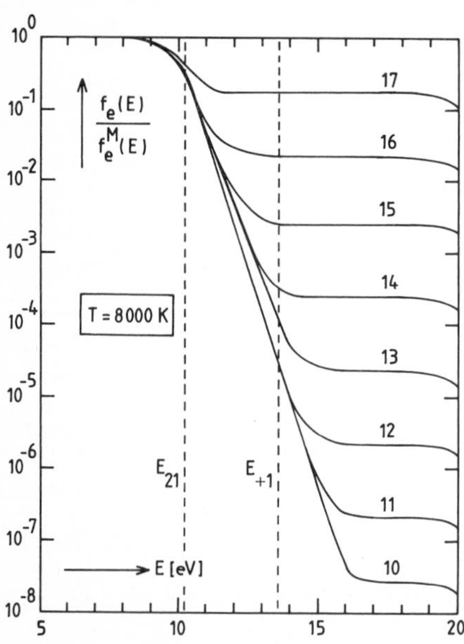

Fig. 6.23. Ratio of electron distribution to Maxwell distribution $f_e(E)/f_e^M(E)$ vs electron energy E in an optically thin hydrogen plasma at temperature $T = 8000$ K. Curve N ($N = 10 - 17$) corresponds to electron density $n_e = 10^N$ cm^{-3}; $E_{21} = 10.2$ eV and $E_{+1} = 13.6$ eV are respectively the excitation energy of the first resonance transition and the ionization energy of an H atom [6.32]

calculated densities of hydrogen atoms turn out to be high, and always to lie in the neighborhood of about $n_H \sim 10^{20}\,cm^{-3}$, while the electron density varies over many orders of magnitude.

Therefore, to investigate the ionization equilibrium of atomic hydrogen at temperatures $T \lesssim 10^4$ K, one has to specify the total density of heavy particles, $n = n_H + n_+$, rather than the electron density n_e. Figure 6.24 plots the ratio $n_1/n_+ \simeq n_H/n_+$ as a function of temperature T for two different total densities n. Consider first the curves labeled by e where, as earlier in this section, inelastic collision processes due only to free electrons have been taken into account. Note again the huge differences between self-consistent and Maxwellian electron distributions. However, at lower densities ($n = 10^{12}\,cm^{-3}$), the ionization equilibrium shows an instability where within a certain temperature interval three different degrees of ionization correspond to one temperature. At higher densities ($n = 10^{16}\,cm^{-3}$), this instability disappears. In any case,

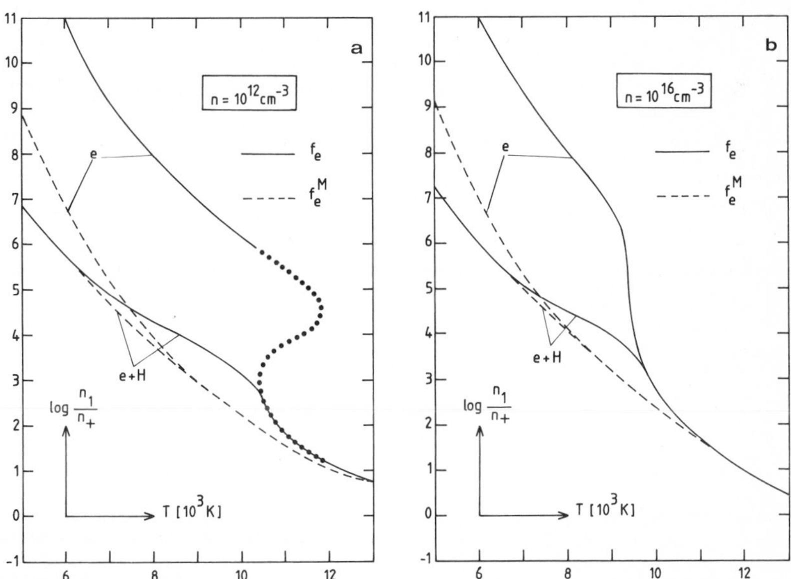

Fig. 6.24 a, b. Ionization equilibrium of an optically thin hydrogen plasma as a function of temperature T, where $n = n_H + n_+$ = total density of heavy particles; n_H = density of H atoms; n_+ = proton density (= electron density n_e); $n_1 (\simeq n_H)$ = density of H atoms in the ground state. Solid curves: self-consistent electron distributions; dashed curves: Maxwellian. To calculate self-consistent electron distributions, elastic collisions of suprathermal electrons both with electrons and with H atoms are taken into account. Curves e(e + H) take into account inelastic collisions of heavy particles only with electrons (both with electrons and with H atoms). At low densities ($n = 10^{12}\,cm^{-3}$, Fig. 6.24a), the ionization equilibrium without inelastic collisions of heavy particles with H atoms shows an instability where 3 values of n_1/n_+ correspond to a given temperature T (dotted part of curve). At higher densities ($n = 10^{16}\,cm^{-3}$, Fig. 6.24b), this instability does not occur. Likewise, this instability does not occur when inelastic collisions of heavy particles with H atoms are taken into account [6.34]

the calculated ionization degrees are so low that inelastic collision processes with H atoms in the ground state H(1) are expected to play a role. This is indeed the case, as shown in Fig. 6.24 by the curves labeled by e+H where, in addition to inelastic electron collisions, also the inelastic collision processes

$$\mathrm{H}(n)+\mathrm{H}(1)\rightleftarrows\mathrm{H}(n')+\mathrm{H}(1) \ ,$$

$$\mathrm{H}(n)+\mathrm{H}(1)\rightleftarrows\mathrm{H}^{+}+\mathrm{e}+\mathrm{H}(1)$$

have been taken into account. The values of n_1/n_+ indicated in Fig. 6.24 should not be taken too literally because of the great uncertainty regarding the cross sections for inelastic H-H collisions. Nevertheless, it is clear that owing to the feedback mechanism that operates on the high-energy tail of the electron distribution function, the ionization equilibrium of optically thin hydrogen below $T\simeq 10000$ K is strongly influenced by inelastic atom-atom collisions. In any case, the tail of the electron distribution deviates from a Maxwellian whenever $n_+/n_\mathrm{H}\lesssim 10^{-2}$.

In optically thick plasmas, the effect discussed is less pronounced because the populations of the excited atomic levels will be closer to the Boltzmann values than in an optically thin plasma, due to the absorption of photons in resonance lines. As an example, Fig. 6.25 shows the results of calculations for an optically thick cesium plasma which confirm this argument. Here self-absorption in the lowest resonance line is taken into account via the radiation escape factor Λ_{21}, that is, the Einstein coefficient A_{21} is replaced by $\Lambda_{21}A_{21}$ so that for $\Lambda_{21}=10^{-3}$, say, only 10^{-3} of all photons emitted in transitions $2\rightarrow 1$ escape from the plasma. Figure 6.25 also shows very nicely that collisional excitation in the lowest resonance transition causes the electron distribution to drop not only at energy E_{21}, but also at $2E_{21}$, $3E_{21}$, etc.

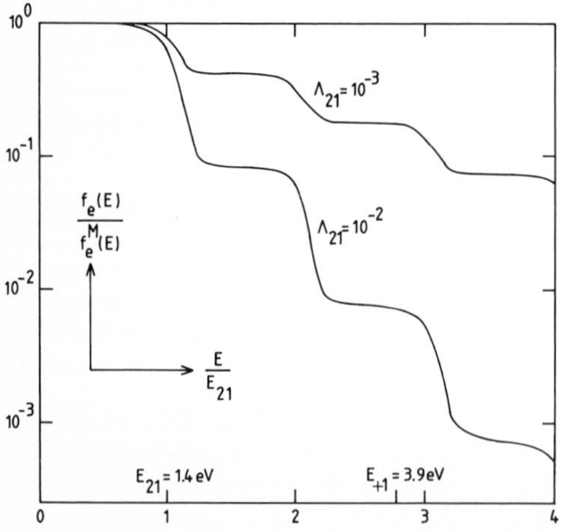

Fig. 6.25. Ratio of electron distribution to Maxwell distribution $f_\mathrm{e}(E)/f_\mathrm{e}^\mathrm{M}(E)$ vs electron energy E in an optically thick cesium plasma with gas temperature $T_0=1000$ K and electron temperature $T_\mathrm{e}=2000$ K. An electron density $n_\mathrm{e}=0.1\,n_\mathrm{e}^\mathrm{Saha}(T_\mathrm{e})$ has been assumed. $E_{21}=1.4$ eV and $E_{+1}=3.9$ eV are respectively the energy of the first resonance transition and the ionization energy of a Cs atom. Photon absorption in the optically thick system has been taken into account through the radiation escape factor of the first resonance line Λ_{21} (see text) [6.36]

More accurate calculations of self-consistent electron distributions in optically thick systems have been carried out by considering non-LTE line transfer in the lowest resonance line rather than merely introducing a simple radiation escape factor [6.34, 37]. In the simplest case of a two-level atom with continuum and assuming complete redistribution, one is faced with a system of coupled equations that comprise the radiative transfer equation, the kinetic equation of the electrons, and the balance equations of the atomic levels. Owing to the great complexity of this problem, the calculations carried out thus far still use unsatisfactory approximations. At present, non-LTE line transfer with a self-consistent electron distribution function is an unsolved problem.

7. Momentum Exchange Between Matter and Radiation

This chapter considers some processes for which momentum exchange between matter and radiation is essential. Section 7.2 treats the approach to thermal equilibrium of a gas through interaction with a blackbody radiation field. Section 7.3 presents a discussion of radiative forces. And the final Sect. 7.4 derives the basic relations for relativistic Compton scattering of photons by free electrons.

7.1 General Remarks

In many cases the momentum exchange between radiation field and material gas may be ignored, since the mean photon momentum is usually much smaller than the mean thermal momentum of a particle. However, some physical effects are due to this momentum exchange between matter and radiation, such as radiation pressure and radiative viscosity (Chap. 4). These effects are a result of particle recoil due to emission, absorption, or scattering of photons. Evidently, the emission of a photon $(\nu, \boldsymbol{n})$ by a particle of mass m gives rise to a velocity change of this particle

$$\boldsymbol{v} \rightarrow \boldsymbol{v}' = \boldsymbol{v} - \frac{h\nu}{mc}\boldsymbol{n} \ , \tag{7.1.1a}$$

while the corresponding velocity change due to the absorption of a photon $(\nu, \boldsymbol{n})$ is obviously

$$\boldsymbol{v} \rightarrow \boldsymbol{v}'' = \boldsymbol{v} + \frac{h\nu}{mc}\boldsymbol{n} \ . \tag{7.1.1b}$$

These recoil effects, neglected in Chap. 6, may sometimes play a role in the formation of spectral lines via the corresponding Doppler shifts of the photons involved [7.1, 2]. Since the changes are obvious in the expressions for line emission and absorption profiles, redistribution functions, etc., which result when (7.1.1) is taken into account, they are not written here.

The subsequent sections discuss some topics where momentum exchange between particles and photons plays an essential role.

7.2 Approach to Thermal Equilibrium Through Interaction with Blackbody Radiation

Let us consider the following situation. A box filled with blackbody radiation of temperature T contains atoms which initially are in the ground state and at rest. All particle interactions except radiative ones are neglected. Since from a thermodynamical point of view the blackbody radiation acts as a heat bath of temperature T, the gas must approach complete thermodynamic equilibrium through radiative interactions. How long are the relaxation times for the excitation degrees of freedom to reach a Boltzmann distribution, for the ionization degrees of freedom to reach a Saha distribution, and for the kinetic degrees of freedom of atoms, ions, and electrons to reach Maxwell distributions, where all distributions correspond to the temperature T of the blackbody radiation field?

The first to consider a related question was *Einstein* [7.3], who showed that atoms undergo Brownian motion when emitting and absorbing line radiation in a thermal radiation field. More precisely, assuming Brownian motion of the atoms, he showed that a Maxwellian velocity distribution of the atoms is stationary in a blackbody radiation field if line emission and absorption is described by the well-known coefficients A_{21}, B_{21}, B_{12} introduced in Sect. 1.6.4. This investigation was later completed by *Milne* [7.4] and *Fowler* [7.5], who showed that Einstein's radiation laws not only preserve the equilibrium velocity distribution of the atoms, but also lead to an actual approach to the equilibrium distribution starting from an arbitrary initial distribution. Hereby the Brownian motion of atoms in a blackbody radiation field was really proved for bound-bound transitions.

These results suggest the following simple procedure for calculating the relaxation times of kinetic degrees of freedom. One first shows that particles interacting with thermal radiation actually undergo Brownian motion, and then applies the relaxation time as given by the theory of Brownian motion. Because of the spherical symmetry of a blackbody radiation field, it is sufficient to consider one arbitrary fixed direction, and hence one-dimensional Brownian motion.

The starting point of the theory of Brownian motion is Langevin's equation, (D.1.2),

$$\dot{v} = -\zeta v + A(t) \ , \tag{7.2.1}$$

which decomposes the total force of a thermal medium on a test particle of velocity v into a systematic friction force and a rapidly fluctuating random force. One assumes that the coefficient of dynamical friction ζ and the fluctuating acceleration $A(t)$ are independent of the particle velocity v, and that the time average of $A(t)$ vanishes. Denoting by $\langle \dots \rangle$ the average over an ensemble of equivalent test particles, Langevin's equation (7.2.1) leads to the following relations for the rate of change of the velocity and of its square, see (D.1.25,26),

$$\left\langle \frac{\Delta v}{\Delta t} \right\rangle = -\zeta v \ , \tag{7.2.2}$$

$$\left\langle \frac{\Delta v^2}{\Delta t} \right\rangle = \frac{2kT}{m} \zeta \ , \tag{7.2.3}$$

where m is the mass of the test particle and T the temperature of the surrounding medium, thus, in our case, of the blackbody radiation. Equation (7.2.2) obviously describes the slowing down of the test particles due to dynamical friction, whereas (7.2.3) describes their thermal velocity fluctuations, which are a consequence of the equipartition law. These two relations are characteristic of Brownian motion.

The time evolution of the normalized velocity distribution function of the test particles is then given by

$$f(v,t) = \left(\frac{m}{2\pi k T(1-e^{-2\zeta t})} \right)^{1/2} \exp\left(-\frac{mv^2}{2kT(1-e^{-2\zeta t})} \right) \ , \tag{7.2.4}$$

which follows from (D.1.12, 22) for the special initial condition $f(v,0) = \delta(v)$ where all particles are at rest at time $t = 0$. Since all direct interactions between the particles are neglected, $f(v,t)$ is more exactly interpreted as the time evolution of an ensemble of test particles rather than that of an actual gas. From (7.2.4) it follows that the relaxation time of the kinetic degrees of freedom to reach a Maxwell distribution is given by

$$\vartheta_{\text{kin}} = \frac{1}{2\zeta} \ . \tag{7.2.5}$$

The procedure applied in this section is hence the following. We calculate $\langle \Delta v/\Delta t \rangle$ for a particle of velocity v, and $\langle \Delta v^2/\Delta t \rangle$ for a particle at rest (recalling that according to Langevin's equation the fluctuating force is supposed to be independent of v, while, on the other hand, its average must be zero). If (7.2.2, 3) are satisfied, Brownian motion obtains, and the corresponding relaxation time of the kinetic degrees of freedom is given by (7.2.5).

The physical reason for the occurrence of dynamical friction of a particle moving in a thermal radiation field is different for bound-bound radiation of atoms and Thomson scattering of free electrons, on the one hand, and for free-bound and free-free radiation of atoms and ions, on the other hand. In the first case, friction is produced by direct momentum transfer from the photons to the test particle, since in the particle's rest frame the radiation field is anisotropic so that more photons meet the particle from in front than from behind. In the second case, the contribution of the photons to the momentum transfer is negligible compared to that of the electrons involved in those radiative interactions, so that here the anisotropy of the electron velocity distribution in the particle's rest frame gives rise to dynamical friction. Likewise, thermal fluctuations are maintained by direct momentum exchange with the photon gas in the first case, and by momentum exchange with the electron gas in the second.

To have a test particle motion in a thermal medium, one must require for free-bound and free-free transitions of atoms and ions that in addition to the radiation field, also the electron gas be thermal, that is, that the velocity distribution of the electrons be Maxwellian. In such a medium composed of photons and free electrons, the actual approach to thermal equilibrium of heavy test particles (atoms or ions) is mainly determined by elastic collisions with free electrons. By "switching off" these elastic collisions, fictitious relaxation times due to radiative free-bound and free-free transitions can be calculated [7.6]. However, this section restricts the discussion to line radiation of atoms and Thomson scattering of free electrons, considering only nonrelativistic particle velocities. On the other hand, *Pauli* [7.7] has shown that the thermal (Maxwellian) velocity distribution of relativistic electrons is stationary when interacting with blackbody radiation via Compton scattering.

From the fact that energy and momentum exchange due to radiative interactions with a thermal medium leads to Brownian motion of the particles considered, one can draw the important conclusion that the corresponding radiative collision terms of the kinetic equation are of a Fokker-Planck type (Appendix D.2).

In the nonrelativistic limit, the transformation of radiative quantities from the laboratory frame to the rest frame of a particle of velocity $\boldsymbol{v}$ is governed by the following formulas, Appendix A. A photon of frequency ν and direction $\boldsymbol{n}$ in the laboratory frame has in the particle's rest frame the frequency, (A.1.27),

$$\nu^{\circ} = \nu\left(1 - \frac{1}{c}\boldsymbol{n}\cdot\boldsymbol{v}\right) \tag{7.2.6}$$

and direction, (A.1.28),

$$\boldsymbol{n}^{\circ} = \boldsymbol{n} - \frac{1}{c}[\boldsymbol{v} - (\boldsymbol{n}\cdot\boldsymbol{v})\boldsymbol{n}] \ , \tag{7.2.7}$$

where $\boldsymbol{n}$ and $\boldsymbol{n}^{\circ}$ are unit vectors, and where here and in the following quantities in the particle's rest frame are denoted by the superscript $^{\circ}$. By assumption, in the laboratory frame, the radiation field is isotropic with a blackbody intensity

$$I_{\nu} = B_{\nu}(T) \equiv \frac{2h\nu^3}{c^2}\,\frac{1}{e^{h\nu/kT}-1} \ . \tag{7.2.8}$$

In the special case of isotropic radiation, the transformation formula of the specific intensity (A.1.36) takes the form

$$I^{\circ} = I + \frac{1}{c}(\boldsymbol{n} \cdot \boldsymbol{v})\left(\nu \frac{\partial I}{\partial \nu} - 3I\right) , \tag{7.2.9}$$

where we have used $\cos \vartheta = \boldsymbol{n} \cdot \boldsymbol{v}/v$. Thus, the transformation of the blackbody radiation field (7.2.8) to a coordinate frame moving with velocity $\boldsymbol{v}$ leads to an anisotropic radiation field of specific intensity

$$I^{\circ}(\nu^{\circ}, \boldsymbol{n}^{\circ}) = B_{\nu^{\circ}}(T)\left(1 - \frac{1}{c}\frac{h\nu^{\circ}/kT}{1 - \mathrm{e}^{-h\nu^{\circ}/kT}}\boldsymbol{n}^{\circ} \cdot \boldsymbol{v}\right) . \tag{7.2.10}$$

Note that we have set $\boldsymbol{n} \cdot \boldsymbol{v}/c = \boldsymbol{n}^{\circ} \cdot \boldsymbol{v}/c$ because $(v/c)\cos\vartheta = (v/c)\cos\vartheta^{\circ} + \mathrm{O}(v^2/c^2)$, following (A.1.33).

7.2.1 Line Radiation of Two-Level Atoms

As our first example we consider two-level atoms with lower and upper bound levels 1 and 2, immersed in blackbody radiation. As usual, we neglect the variation of blackbody intensity over the width of the spectral line, and ascribe to all photons in the spectral line considered the same energy $h\nu_0$ and the same modulus of momentum $h\nu_0/c$ both in the laboratory and the atom's rest frame, where $\nu_0 \equiv \nu_0^{\circ}$ denotes the frequency of the line center as measured in the atomic rest frame.

Let us first consider the excitation degree of freedom of the two-level atoms. In terms of the Einstein coefficients A_{21}, B_{21}, B_{12} (Sect. 1.6.4), the time evolution of the occupation probabilities $w_1(t)$ and $w_2(t)$ of the two atomic levels is governed by the differential equation

$$\dot{w}_2(t) = B_{12}B_{\nu_0}(T)\, w_1(t) - [A_{21} + B_{21}B_{\nu_0}(T)]\, w_2(t) \tag{7.2.11}$$

with

$$w_1(t) + w_2(t) = 1 . \tag{7.2.12}$$

The initial condition is

$$w_2(0) = 0 \tag{7.2.13}$$

since, by assumption, all atoms are in the ground state at time $t = 0$. Writing for brevity

$$\xi \equiv B_{12}B_{\nu_0}(T) , \quad \eta \equiv A_{21} + (B_{21} + B_{12})B_{\nu_0}(T) , \tag{7.2.14}$$

the solution of (7.2.11, 13) is given by

$$w_2(t) = \frac{\xi}{\eta}(1 - \mathrm{e}^{-\eta t}) . \tag{7.2.15}$$

Hence, for $t \to \infty$, the occupation probabilities approach the Boltzmann distribution

$$w_1^{\mathrm{B}} = \frac{1}{1+\dfrac{g_2}{g_1}\mathrm{e}^{-h\nu_0/kT}} , \quad w_2^{\mathrm{B}} = \frac{\dfrac{g_2}{g_1}\mathrm{e}^{-h\nu_0/kT}}{1+\dfrac{g_2}{g_1}\mathrm{e}^{-h\nu_0/kT}} \tag{7.2.16}$$

with the relaxation time $\vartheta_{\mathrm{exc}} = 1/\eta$, or

$$\vartheta_{\mathrm{exc}} = \frac{1}{A_{21}} \frac{1-\mathrm{e}^{-h\nu_0/kT}}{1+\dfrac{g_2}{g_1}\mathrm{e}^{-h\nu_0/kT}} , \tag{7.2.17}$$

where g_1 and g_2 are the statistical weights of the atomic levels. In (7.2.16, 17) we have used (1.6.31, 32) as well as the explicit expression of the Planck function (7.2.8). As expected, at low temperatures ($h\nu_0/kT > 1$) the relaxation time is given by the reciprocal transition probability for spontaneous emission $1/A_{21}$, while at higher temperatures ($h\nu_0/kT < 1$) it becomes smaller owing to the contribution of stimulated emissions.

Turning now to the kinetic degrees of freedom, we may assume that the atomic bound levels are already populated according to the Boltzmann distribution (7.2.16), because the relaxation time of the kinetic degrees of freedom is expected to be very much longer than that of the excitation degrees of freedom. Following the general procedure as explained above, we shall demonstrate the Brownian motion of the two-level atoms by calculating $\langle \Delta v/\Delta t\rangle$ for an atom of velocity $\boldsymbol{v}$ and $\langle \Delta v^2/\Delta t\rangle$ for an atom at rest.

The absorption of a photon $(\nu_0, \boldsymbol{n}^{\circ})$ by an atom of velocity $\boldsymbol{v}$ gives rise to a velocity change whose component in the direction of $\boldsymbol{v}$ is given by $(h\nu_0/mc)$ $(\boldsymbol{n}^{\circ}\cdot \boldsymbol{v}/v)$, m being the mass of the two-level atom. The mean velocity change per unit time due to absorptions is hence

$$\left\langle \frac{\Delta v}{\Delta t} \right\rangle_1 = \int B_{12} I^{\circ}(\nu_0, \boldsymbol{n}^{\circ}) \frac{h\nu_0}{mc} \frac{\boldsymbol{n}^{\circ}\cdot \boldsymbol{v}}{v} \frac{d\Omega^{\circ}}{4\pi} , \tag{7.2.18}$$

where I° is the radiation intensity in the rest frame of the atom, and the integration is over all photon directions $\boldsymbol{n}^{\circ}$. Using (7.2.10) and expressing B_{12} by A_{21} through (1.6.31, 32) yields

$$\left\langle \frac{\Delta v}{\Delta t} \right\rangle_1 = -v \frac{A_{21}}{3} \frac{(h\nu_0)^2}{mc^2 kT} \frac{\dfrac{g_2}{g_1}\mathrm{e}^{-h\nu_0/kT}}{(1-\mathrm{e}^{-h\nu_0/kT})^2} . \tag{7.2.19}$$

For the emission of a photon $(\nu_0, \boldsymbol{n}^{\circ})$, the corresponding velocity change simply takes a minus sign, so that the mean velocity change per unit time due to emissions is

$$\left\langle \frac{\Delta v}{\Delta t} \right\rangle_2 = \int [A_{21} + B_{21} I^{\circ}(\nu_0, \boldsymbol{n}^{\circ})] \left(-\frac{h\nu_0}{mc}\right) \frac{\boldsymbol{n}^{\circ}\cdot \boldsymbol{v}}{v} \frac{d\Omega^{\circ}}{4\pi} . \tag{7.2.20}$$

By symmetry, the spontaneous emissions do not contribute, whereas the stimulated emissions give rise to a mean velocity change

$$\left\langle \frac{\Delta v}{\Delta t} \right\rangle_2 = +v \frac{A_{21}}{3} \frac{(h\nu_0)^2}{mc^2kT} \frac{\mathrm{e}^{-h\nu_0/kT}}{(1-\mathrm{e}^{-h\nu_0/kT})^2}, \tag{7.2.21}$$

where again B_{21} has been expressed by A_{21}. The total velocity change per unit time of a two-level atom is obviously

$$\left\langle \frac{\Delta v}{\Delta t} \right\rangle = w_1^{\mathrm{B}} \left\langle \frac{\Delta v}{\Delta t} \right\rangle_1 + w_2^{\mathrm{B}} \left\langle \frac{\Delta v}{\Delta t} \right\rangle_2, \tag{7.2.22}$$

where w_1^{B}, w_2^{B} are the occupation probabilities (7.2.16) corresponding to a Boltzmann distribution. Note that in writing (7.2.22) we have assumed that the Boltzmann ratio $w_2^{\mathrm{B}}/w_1^{\mathrm{B}}$ applies to *every* velocity v, which amounts to the physically reasonable assumption that the velocity distributions of unexcited and excited atoms are equal, $f_1(v)=f_2(v)$. Taking now (7.2.16, 19, 21, 22) into account, we finally obtain

$$\left\langle \frac{\Delta v}{\Delta t} \right\rangle = -\zeta v \tag{7.2.23}$$

with the coefficient of dynamical friction

$$\zeta = \frac{A_{21}}{3} \frac{(h\nu_0)^2}{mc^2kT} \frac{1}{\left(1+\frac{g_1}{g_2}\mathrm{e}^{h\nu_0/kT}\right)(1-\mathrm{e}^{-h\nu_0/kT})}. \tag{7.2.24}$$

We now calculate the quantity $\langle \Delta v^2/\Delta t \rangle$ for an atom at rest. Here the atom's rest frame coincides with the laboratory frame. The absorption or the emission of a photon $(\nu_0, \boldsymbol{n})$ gives rise to an atomic velocity whose component in an arbitrary but fixed direction is $\pm(h\nu_0/mc)\cos\vartheta$, where ϑ is the angle between this direction and the photon direction $\boldsymbol{n}$. Since the radiation field is isotropic, the square of this velocity component averaged over all photon directions is $(h\nu_0/mc)^2/3$. Hence, the velocity fluctuations of a two-level atom are given by

$$\left\langle \frac{\Delta v^2}{\Delta t} \right\rangle = w_1^{\mathrm{B}} \left\langle \frac{\Delta v^2}{\Delta t} \right\rangle_1 + w_2^{\mathrm{B}} \left\langle \frac{\Delta v^2}{\Delta t} \right\rangle_2, \tag{7.2.25}$$

where

$$\left\langle \frac{\Delta v^2}{\Delta t} \right\rangle_1 = B_{12} B_{\nu_0}(T) \frac{1}{3} \left(\frac{h\nu_0}{mc}\right)^2$$

$$= \frac{A_{21}}{3} \left(\frac{h\nu_0}{mc}\right)^2 \frac{\frac{g_2}{g_1}\mathrm{e}^{-h\nu_0/kT}}{1-\mathrm{e}^{-h\nu_0/kT}} \tag{7.2.26}$$

and

$$\left\langle\frac{\Delta v^2}{\Delta t}\right\rangle_2 = [A_{21}+B_{21}B_{\nu_0}(T)]\frac{1}{3}\left(\frac{h\nu_0}{mc}\right)^2$$

$$= \frac{A_{21}}{3}\left(\frac{h\nu_0}{mc}\right)^2 \frac{1}{1-\mathrm{e}^{-h\nu_0/kT}} \ . \tag{7.2.27}$$

Taking (7.2.16) into account, from (7.2.25 – 27)

$$\left\langle\frac{\Delta v^2}{\Delta t}\right\rangle = \frac{2A_{21}}{3}\left(\frac{h\nu_0}{mc}\right)^2 \frac{1}{\left(1+\frac{g_1}{g_2}\mathrm{e}^{h\nu_0/kT}\right)(1-\mathrm{e}^{-h\nu_0/kT})} \ . \tag{7.2.28}$$

Note that $w_2^{\mathrm{B}}\langle\Delta v^2/\Delta t\rangle_2 = w_1^{\mathrm{B}}\langle\Delta v^2/\Delta t\rangle_1$ holds, that is, absorptions and emissions contribute equally to thermal fluctuations. This is a consequence of detailed balance. By contrast, the deceleration of an atom due to absorptions is always larger than the acceleration due to (stimulated) emissions, $w_2^{\mathrm{B}}\langle\Delta v/\Delta t\rangle_2 < |w_1^{\mathrm{B}}\langle\Delta v/\Delta t\rangle_1|$, thus leading to a net deceleration of a moving atom.

Comparison of (7.2.24, 28) now shows that

$$\left\langle\frac{\Delta v^2}{\Delta t}\right\rangle = \frac{2kT}{m}\zeta \tag{7.2.29}$$

is fulfilled. Therefore, on account of (7.2.23, 29), Brownian motion of two-level atoms emitting and absorbing line radiation in a blackbody radiation field obtains. Consequently, the relaxation time of the kinetic degrees of freedom to approach a Maxwell distribution is according to (7.2.5, 24) given by

$$\vartheta_{\mathrm{kin}} = \frac{3}{2A_{21}}\frac{mc^2kT}{(h\nu_0)^2}\left(1+\frac{g_1}{g_2}\mathrm{e}^{h\nu_0/kT}\right)(1-\mathrm{e}^{-h\nu_0/kT}) \ . \tag{7.2.30}$$

As expected, $\vartheta_{\mathrm{exc}}/\vartheta_{\mathrm{kin}} \ll 1$ because of $(h\nu_0)^2/mc^2kT \ll 1$, a fact already used implicitly by assuming a Boltzmann ratio $w_2^{\mathrm{B}}/w_1^{\mathrm{B}}$ during the entire time evolution of the kinetic degrees of freedom.

7.2.2 Thomson Scattering of Electrons

For nonrelativistic free electrons, the radiative process responsible for the approach to thermal equilibrium in a blackbody radiation field is Thomson scattering. In the frame in which the electron is at rest before the photon collision takes place, Thomson scattering is coherent in frequency, that is, the frequency of the outgoing photon is equal to that of the incoming one. The differential cross section for the scattering process $(\nu^{\circ};\boldsymbol{n}^{\circ})\rightarrow(\nu^{\circ};\boldsymbol{n}^{\circ\prime},d\Omega^{\circ\prime})$ is independent of the frequency ν°,

$$\sigma(\psi)\,d\Omega^{\circ\prime}=\frac{1}{2}\left(\frac{e^2}{m_{\mathrm{e}}c^2}\right)^2(1+\cos^2\psi)\,d\Omega^{\circ\prime}\ , \qquad (7.2.31)$$

where ψ is the scattering angle of the photon, i.e., $\cos\psi=\boldsymbol{n}^{\circ}\cdot\boldsymbol{n}^{\circ\prime}$.

We now calculate $\langle\Delta v/\Delta t\rangle$ for an electron of velocity $\boldsymbol{v}$. The scattering process $(\nu^{\circ},\boldsymbol{n}^{\circ})\rightarrow(\nu^{\circ},\boldsymbol{n}^{\circ\prime})$ gives rise to a velocity change of the electron whose component in the direction of $\boldsymbol{v}$ is

$$\frac{h\nu^{\circ}}{m_{\mathrm{e}}c}(\cos\vartheta^{\circ}-\cos\vartheta^{\circ\prime})\ ,$$

where ϑ° and $\vartheta^{\circ\prime}$ denote the angles between the velocity $\boldsymbol{v}$ and the photon directions $\boldsymbol{n}^{\circ}$ and $\boldsymbol{n}^{\circ\prime}$, respectively. Since the flux density of incoming photons $(\nu^{\circ},d\nu^{\circ};\boldsymbol{n}^{\circ},d\Omega^{\circ})$ is $I^{\circ}(\nu^{\circ},\boldsymbol{n}^{\circ})d\nu^{\circ}d\Omega^{\circ}/h\nu^{\circ}$, (1.4.40), where I° is the specific intensity in the rest frame of the electron, the velocity change per unit time due to Thomson scattering is

$$\left\langle\frac{\Delta v}{\Delta t}\right\rangle=\iiint\frac{I^{\circ}(\nu^{\circ},\vartheta^{\circ})}{h\nu^{\circ}}\sigma(\psi)\left[1+\frac{c^2}{2h\nu^{\circ3}}I^{\circ}(\nu^{\circ},\vartheta^{\circ})\right]$$

$$\cdot\frac{h\nu^{\circ}}{m_{\mathrm{e}}c}(\cos\vartheta^{\circ}-\cos\vartheta^{\circ\prime})\,d\nu^{\circ}d\Omega^{\circ}d\Omega^{\circ\prime}\ , \qquad (7.2.32)$$

where induced scattering has been taken into account, and where the intensity I° is given by (7.2.10) with $\boldsymbol{n}^{\circ}\cdot\boldsymbol{v}=v\cos\vartheta^{\circ}$. The angle ψ appearing in the differential cross section (7.2.31) is related to the angles ϑ° and $\vartheta^{\circ\prime}$ through

$$\cos\psi=\cos\vartheta^{\circ}\cos\vartheta^{\circ\prime}+\sin\vartheta^{\circ}\sin\vartheta^{\circ\prime}\cos(\varphi^{\circ}-\varphi^{\circ\prime})\ , \qquad (7.2.33)$$

where φ° and $\varphi^{\circ\prime}$ are the azimuthal angles of $\boldsymbol{n}^{\circ}$ and $\boldsymbol{n}^{\circ\prime}$. This is easily seen by writing the unit vectors $\boldsymbol{n}^{\circ}$ and $\boldsymbol{n}^{\circ\prime}$ in Cartesian coordinates, e.g.,

$$\boldsymbol{n}^{\circ}=(\sin\vartheta^{\circ}\cos\varphi^{\circ},\sin\vartheta^{\circ}\sin\varphi^{\circ},\cos\vartheta^{\circ})\ ,$$

and forming the scalar product $\boldsymbol{n}^{\circ}\cdot\boldsymbol{n}^{\circ\prime}=\cos\psi$. Integrating with respect to $d\Omega^{\circ}$ and $d\Omega^{\circ\prime}$ leads to

$$\left\langle\frac{\Delta v}{\Delta t}\right\rangle=-v\frac{2^5\pi^2}{3^2}\frac{e^4}{m_{\mathrm{e}}^3c^6}\int_0^{\infty}\frac{B_{\nu^{\circ}}(T)}{1-\mathrm{e}^{-h\nu^{\circ}/kT}}\frac{h\nu^{\circ}}{kT}d\nu^{\circ}\ , \qquad (7.2.34)$$

where a term of relativistic order $O(v^2/c^2)$ has been neglected. Inserting here the Planck function according to (7.2.8) and introducing the new variable $\xi = h\nu^{\circ}/kT$, this can be written as

$$\left\langle \frac{\Delta v}{\Delta t} \right\rangle = -v \frac{2^6 \pi^2}{3^2} \frac{e^4 (kT)^4}{m_e^3 h^3 c^8} \int_0^{\infty} \frac{\xi^4 e^{\xi}}{(e^{\xi}-1)^2} d\xi \ . \tag{7.2.35}$$

The value of the integral is $4\pi^4/15$, and one finally obtains

$$\left\langle \frac{\Delta v}{\Delta t} \right\rangle = -\zeta v \tag{7.2.36}$$

with the coefficient of dynamical friction

$$\zeta = \frac{2^8 \pi^6}{3^3 5} \frac{e^4 (kT)^4}{m_e^3 h^3 c^8} \ . \tag{7.2.37}$$

We now proceed to calculate $\langle \Delta v^2/\Delta t \rangle$ for an electron at rest. An electron initially at rest has after the scattering process $(\nu, \boldsymbol{n}) \rightarrow (\nu, \boldsymbol{n}')$ a velocity whose component in an arbitrary but fixed direction, characterized by the unit vector $\boldsymbol{a}$, is

$$\delta v = \frac{h\nu}{m_e c} (\boldsymbol{n} - \boldsymbol{n}') \cdot \boldsymbol{a} \ .$$

Denoting the angle between the vectors $\boldsymbol{a}$ and $\boldsymbol{n} - \boldsymbol{n}'$ by γ, and observing that $|\boldsymbol{n} - \boldsymbol{n}'| = 2\sin(\psi/2)$, where ψ is again the scattering angle from (7.2.31) (see Fig. 3.2, replacing there $\boldsymbol{n} \leftrightarrow \boldsymbol{n}'$ and $\vartheta \rightarrow \psi$), then

$$\delta v(\psi, \gamma) = \frac{2h\nu}{m_e c} \sin\frac{\psi}{2} \cos\gamma \ .$$

For fixed ψ, the average over all directions γ of the square of this expression is

$$\overline{\delta v^2(\psi)} = \frac{1}{3} \left(\frac{2h\nu}{m_e c} \right)^2 \sin^2\frac{\psi}{2} \quad \text{or}$$

$$\overline{\delta v^2(\psi)} = \frac{2}{3} \left(\frac{h\nu}{m_e c} \right)^2 (1 - \cos\psi) \ . \tag{7.2.38}$$

The quantity $\langle \Delta v^2/\Delta t \rangle$ is therefore given by

$$\left\langle \frac{\Delta v^2}{\Delta t} \right\rangle = \iint \frac{4\pi B_\nu(T)}{h\nu} \sigma(\psi) \left[1 + \frac{c^2}{2h\nu^3} B_\nu(T) \right] \overline{\delta v^2(\psi)} \, d\nu \, d\Omega \ , \tag{7.2.39}$$

where $d\Omega = 2\pi \sin\psi \, d\psi$. Inserting here (7.2.8, 31, 38), one can perform the integration $d\Omega$ with the result

$$\left\langle \frac{\Delta v^2}{\Delta t} \right\rangle = \frac{2^7 \pi^2}{3^2} \frac{e^4 (kT)^5}{m_e^4 h^3 c^8} \int_0^\infty \frac{\xi^4 e^\xi}{(e^\xi - 1)^2} d\xi \tag{7.2.40}$$

in terms of the variable $\xi = h\nu/kT$. Again, the value of the integral is $4\pi^4/15$, so that finally

$$\left\langle \frac{\Delta v^2}{\Delta t} \right\rangle = \frac{2^9 \pi^6}{3^3 5} \frac{e^4 (kT)^5}{m_e^4 h^3 c^8} . \tag{7.2.41}$$

Comparing (7.2.37,41) shows that

$$\left\langle \frac{\Delta v^2}{\Delta t} \right\rangle = \frac{2kT}{m_e} \zeta \tag{7.2.42}$$

holds. Thus, (7.2.36,42) show the Brownian motion of free electrons interacting with blackbody radiation via Thomson scattering. According to (7.2.5,37), the corresponding relaxation time for approaching a Maxwell distribution if the electrons are initially at rest is given by

$$\vartheta_{\text{kin}} = \frac{3^3 5}{2^9 \pi^6} \frac{m_e^3 h^3 c^8}{e^4 (kT)^4} , \tag{7.2.43}$$

and is hence proportional to T^{-4}, that is, inversely proportional to the energy density of the blackbody radiation field.

7.3 Radiative Forces

Radiative forces result from momentum exchange between radiation field and gas, owing to emission, absorption, and scattering of photons by particles. In an anisotropic radiation field this momentum transfer leads to a directed radiative force on the material gas. As an important special case, let us consider a gas composed of two different particle types t and f, where t stands for test particle and f for field particle. The t particles interact with the anisotropic radiation field, whereas radiative interactions of the f particles are assumed to be negligible. The stationary gas of f particles is supposed to have a much higher density than the t particles gas ($n_f \gg n_t$) so that collisions among t particles are negligible. If the t particles were free, the momentum exchange with the radiation field would continuously accelerate them. However, owing to collisions with f particles, a constant diffusion velocity of the t particle gas is finally reached such that the momentum gain from the radiation field is exactly counterbalanced by the momentum loss to the surrounding gas through collisions (dynamical friction). The resulting diffusion velocity u_t follows from

$$K_t^{\text{rad}} = (n_f Q_{tf} u_t)(\mu_{tf} kT)^{1/2} . \tag{7.3.1}$$

In words, the mean radiative force K_t^{rad} on a t particle (momentum gain per unit time) equals the momentum loss per unit time, which is given by the product of the "effective collision frequency" $n_f Q_{tf} u_t$, with Q_{tf} being the cross section for elastic tf collisions, and the mean momentum transfer per collision, given by $\mu_{tf}\bar{v}_{\text{rel}}$, where $\mu_{tf}=m_t m_f/(m_t+m_f)$ is the reduced mass and $\bar{v}_{\text{rel}}=(kT/\mu_{tf})^{1/2}$ is the mean relative velocity of two colliding particles. Equation (7.3.1) thus leads to a diffusion velocity of the test particles due to radiative forces

$$u_t=\kappa_{tf}K_t^{\text{rad}}=\frac{D_{tf}}{kT}K_t^{\text{rad}} \tag{7.3.2}$$

since the mobility κ is related to the diffusion constant D through $\kappa=D/kT$; hence from (7.3.1)

$$D_{tf}=\frac{1}{n_f Q_{tf}}\left(\frac{kT}{\mu_{tf}}\right)^{1/2} . \tag{7.3.3}$$

To derive the relation $\kappa=D/kT$, consider thermal equilibrium in a constant gravitational field $-\gamma \boldsymbol{e}_z$. The force per particle $m\gamma$ leads to the particle flow $\Gamma\!\downarrow=\kappa m\gamma n$. On the other hand, the density gradient due to the Boltzmann distribution $n(z)=n_0\exp(-m\gamma z/kT)$ gives rise to the diffusion flow $\Gamma\!\uparrow=-D\,dn/dz=D(m\gamma/kT)\,n$. In equilibrium $\Gamma\!\downarrow=\Gamma\!\uparrow$, and $\kappa=D/kT$ follows at once.

The diffusion constant D_{tf} can also be derived directly by observing that, according to (6.3.37), $D_{tf}\sim\bar{v}_t l$, where $\bar{v}_t$ is the mean thermal speed of a t particle. Writing the mean free path l in terms of an appropriate collision frequency γ, $l\sim\bar{v}_t/\gamma$, gives $D_{tf}\sim\bar{v}_t^2/\gamma\sim kT/m_t\gamma$, see (6.3.38). On the other hand, the collision frequency can be written as $\gamma\sim n_f\bar{v}_{\text{rel}}Q_{\text{eff}}$, where $\bar{v}_{\text{rel}}\sim(kT/\mu_{tf})^{1/2}$ is the mean relative velocity, and $Q_{\text{eff}}\sim(\mu_{tf}/m_t)\,Q_{tf}$ is the effective cross section for momentum transfer, related to the "geometrical" cross section Q_{tf} so that $Q_{\text{eff}}\sim Q_{tf}$ if $m_t\lesssim m_f$, whereas $Q_{\text{eff}}\sim(m_f/m_t)\,Q_{tf}$ if $m_t>m_f$.

Radiative interactions can be divided into two classes. On the one side, there is bremsstrahlung emission and absorption (free-free transitions) and scattering (Rayleigh, Thomson) which leave the particle type unchanged (e.g., $\mathrm{e}+\gamma\rightarrow\mathrm{e}+\gamma$). On the other hand, line absorption and emission (bound-bound transitions), and photoionization and radiative recombination (free-bound transitions) give rise to changes of the particle type [e.g., $\mathrm{H}(1s)+\gamma\rightarrow\mathrm{H}(2p)$, $\mathrm{H}^++\mathrm{e}\rightarrow\mathrm{H}(3d)+\gamma$].

Calculating radiative forces is in principle straightforward for the first class of processes. As an example, we discuss (nonresonant) scattering in Sect. 7.3.1.

However, in cases where the radiative interaction leads to changes of the particle types, one must go back to the general definition of the radiative force as the production rate of particle momentum. Accordingly, the radiative force on the particle type a is given by

$$\boldsymbol{k}_a^{\mathrm{rad}} = \int m_a \boldsymbol{v} \left(\frac{\delta F_a}{\delta t}\right)_{\mathrm{rad}} d^3 v \ , \tag{7.3.4}$$

where m_a is the mass of an a particle, and $(\delta F_a/\delta t)_{\mathrm{rad}}$ is the collision term due to radiative interactions that appears on the right-hand side of the kinetic equation for the distribution function $F_a(\boldsymbol{r}, \boldsymbol{v}, t)$, see (2.1.11).

For example, the total radiative force on a gas of ions in a particular bound level i due to bound-bound and free-bound transitions results from all line absorptions leading to higher levels $k(>i)$ of the same ion, from all line emissions leading to lower levels $j(<i)$, from all photoionizations leading to levels of the ion of the next higher stage of ionization (collectively denoted by $+$), and from all radiative recombinations to levels of the ion of the next lower stage of ionization (collectively denoted by $-$). For instance, if the considered particle type is $O^{5+}(3^2P)$, then examples of levels $k, j, +, -$ are, respectively, $O^{5+}(5^2D)$, $O^{5+}(2^2S)$, $O^{6+}(1^1S)$, $O^{4+}(2^3P)$. Thus, following (7.3.4), the corresponding radiative force density is given by

$$\boldsymbol{k}_i^{\mathrm{rad}} = \int d^3 v\, m \boldsymbol{v} \left[\sum_{k>i} \left(\frac{\delta F_i}{\delta t}\right)_{\mathrm{rad}}^{ik} + \sum_{j<i} \left(\frac{\delta F_i}{\delta t}\right)_{\mathrm{rad}}^{ji} \right.$$

$$\left. + \sum_{+} \left(\frac{\delta F_i}{\delta t}\right)_{\mathrm{rad}}^{i+} + \sum_{-} \left(\frac{\delta F_i}{\delta t}\right)_{\mathrm{rad}}^{-i} \right] , \tag{7.3.5}$$

where m is the mass of the ion considered.

Sections 7.3.2, 3 treat collision terms due to bound-bound and free-bound transitions, taking momentum transfer into account, and discuss the notion of net momentum gain in a given transition, which must be distinguished from the notion of a radiative force.

7.3.1 Scattering

The radiative force density on the gas of a given particle type s due to scattering is given by

$$\boldsymbol{k}_s^{\mathrm{sc}} = \int d^3 v\, F_s(\boldsymbol{v}) \iiint \frac{h\nu}{c} (\boldsymbol{n} - \boldsymbol{n}') \frac{1}{h\nu} I_\nu(\boldsymbol{n})\, \sigma(\nu, \vartheta)$$

$$\cdot \left[1 + \frac{c^2}{2 h \nu^3} I_{\nu'}(\boldsymbol{n}') \right] d\nu\, d\Omega\, d\Omega' \ . \tag{7.3.6}$$

Here $F_s(\boldsymbol{v})$ is the distribution function of the scattering particles, $\sigma(\nu, \vartheta)$ is the differential cross section for the scattering $(\nu, \boldsymbol{n}) \rightarrow (\nu', \boldsymbol{n}')$ where $\cos\vartheta = \boldsymbol{n} \cdot \boldsymbol{n}'$, and the integrations $d\Omega$ and $d\Omega'$ refer to the directions of incoming and outgoing photons, respectively. The frequency ν' denotes the frequency of the outgoing photon scattered by a particle of velocity $\boldsymbol{v}$, (3.3.19),

$$\nu' = \nu + \frac{\nu}{c}(\boldsymbol{n}' - \boldsymbol{n}) \cdot \boldsymbol{v} \ . \tag{7.3.7}$$

According to Sect. 3.3.2, we have set $\nu' = \nu$ everywhere in (7.3.6) except in $I_{\nu'}(\boldsymbol{n}')$. In particular, the momentum gain of a particle of initial velocity $\boldsymbol{v}$ due to the scattering $(\nu, \boldsymbol{n}) \rightarrow (\nu', \boldsymbol{n}')$ has been approximated by

$$\frac{h\nu}{c}\boldsymbol{n} - \frac{h\nu'}{c}\boldsymbol{n}' \simeq \frac{h\nu}{c}(\boldsymbol{n} - \boldsymbol{n}') \ . \tag{7.3.8}$$

To simplify the force density (7.3.6), we introduce the total scattering cross section $\sigma(\nu)$ through, see (3.3.8),

$$\sigma(\nu, \vartheta) = \sigma(\nu)\, g_1(\vartheta) \ , \tag{7.3.9}$$

assuming dipole scattering with the normalized phase function, see (3.3.11a),

$$g_1(\vartheta) = \frac{3}{16\pi}(1 + \cos^2\vartheta) \ . \tag{7.3.10}$$

Since in practically all situations the contribution of stimulated scattering to the radiative force is negligible because of

$$\iint (\boldsymbol{n} - \boldsymbol{n}') I_\nu(\boldsymbol{n}) I_{\nu'}(\boldsymbol{n}')\, g_1(\vartheta)\, d\Omega\, d\Omega' \simeq 0 \ , \tag{7.3.11}$$

the integration in (7.3.6) over the velocities $\boldsymbol{v}$ can be performed yielding the density n_s of the scattering particles. Thus,

$$\begin{aligned} \boldsymbol{k}_s^{\mathrm{sc}} &= \frac{1}{c} n_s \int d\nu\, \sigma(\nu) \iint (\boldsymbol{n} - \boldsymbol{n}') I_\nu(\boldsymbol{n})\, g_1(\vartheta)\, d\Omega\, d\Omega' \\ &= \frac{1}{c} n_s \int d\nu\, \sigma(\nu) \int \boldsymbol{n} I_\nu(\boldsymbol{n})\, d\Omega \end{aligned} \tag{7.3.12}$$

on account of

$$\int g_1(\vartheta)\, d\Omega' = 1 \ , \quad \int \boldsymbol{n}' g_1(\vartheta)\, d\Omega' = 0 \ . \tag{7.3.13}$$

The radiative force density due to nonresonant scattering can finally be written as

$$\boldsymbol{k}_s^{\mathrm{sc}} = \frac{1}{c} n_s \int \sigma(\nu) \boldsymbol{f}_\nu\, d\nu \tag{7.3.14}$$

in terms of the monochromatic flux density of radiant energy, see (4.2.2),

$$f_\nu = \int \boldsymbol{n} I_\nu(\boldsymbol{n})\, d\Omega \ . \tag{7.3.15}$$

7.3.2 Bound-Bound Transitions

As an example of a radiative collision term appearing in the radiative force density (7.3.5) we consider the term corresponding to the bound-bound transition from the lower level 1 to the upper level 2 of an atom. In analogy to (2.4.6)

$$\begin{aligned}\left(\frac{\delta F_1}{\delta t}\right)^{12}_{\text{rad}} &= -F_1(\boldsymbol{v}) B_{12} \iint I_\nu(\boldsymbol{n})\, \alpha_{12}(\xi)\, d\nu \frac{d\Omega}{4\pi} \\ &\quad + A_{21} \iint F_2(\boldsymbol{v}')\, \eta_{21}(\xi,\boldsymbol{n}) \left[1 + \frac{B_{21}}{A_{21}} I_\nu(\boldsymbol{n})\right] d\nu \frac{d\Omega}{4\pi} \ ,\end{aligned} \tag{7.3.16}$$

where the first term describes absorptions and the second, spontaneous and induced emissions. Here and in the following the argument of a collision term is understood to be $\boldsymbol{v}$, $\delta F/\delta t \equiv \delta F(\boldsymbol{v})/\delta t$. Conservation of momentum requires

$$m\boldsymbol{v}' = m\boldsymbol{v} + \frac{h\nu}{c}\boldsymbol{n} \simeq m\boldsymbol{v} + \frac{h\nu_0}{c}\boldsymbol{n} \tag{7.3.17}$$

because $\nu \simeq \nu_0$ for all photons of the spectral line considered, $\nu_0 \equiv \nu_0^{\circ}$ denoting again the frequency of the line center in the atomic rest frame. According to our approximation (2.4.3), in (7.3.16) only the photon frequency has been transformed from the laboratory frame to the atomic rest frame. Thus, in the profile coefficients $\alpha_{12}(\xi)$ and $\eta_{21}(\xi,\boldsymbol{n})$ appears, see (2.4.7),

$$\xi = \nu - \frac{\nu_0}{c}\boldsymbol{n}\cdot\boldsymbol{v} \simeq \nu - \frac{\nu_0}{c}\boldsymbol{n}\cdot\boldsymbol{v}' \tag{7.3.18}$$

which is the rest frame frequency of a photon $(\nu,\boldsymbol{n})$ for an atom of velocity $\boldsymbol{v}$ *and* for an atom of velocity $\boldsymbol{v}'$, up to terms of the order v^2/c^2.

Let us now turn to the *net momentum gain* of the atoms due to radiative transitions $1 \leftrightarrow 2$, which is defined by

$$\hat{\boldsymbol{\pi}}^{12}_{\text{rad}} = \int m\boldsymbol{v} \left[\left(\frac{\delta F_1}{\delta t}\right)^{12}_{\text{rad}} + \left(\frac{\delta F_2}{\delta t}\right)^{12}_{\text{rad}}\right] d^3v \ . \tag{7.3.19}$$

It should be noted that the net momentum gain $\hat{\boldsymbol{\pi}}^{12}_{\text{rad}}$ refers to *two* particle types, namely, atoms A_1 and A_2 in two different excitation states, in contrast to the force densities $\boldsymbol{k}_1^{\text{rad}}$ and $\boldsymbol{k}_2^{\text{rad}}$ which apply to single particle types A_1 and

A_2, respectively. One should further notice that $\hat{\pi}^{12}_{\rm rad} = \boldsymbol{k}^{\rm rad}_1 + \boldsymbol{k}^{\rm rad}_2$ holds only for a simple two-level atom.

To transform (7.3.19) into a more familiar form, we write

$$\hat{\pi}^{12}_{\rm rad} = \hat{\pi}^{12}_{\rm abs} + \hat{\pi}^{12}_{\rm em} \; , \tag{7.3.20}$$

where the terms on the right-hand side refer to absorptions and emissions, respectively. The absorption term is given by

$$\begin{aligned}\hat{\pi}^{12}_{\rm abs} &= \iiint d^3v \, d\nu \frac{d\Omega}{4\pi}(-m\boldsymbol{v}+m\boldsymbol{v}')F_1(\boldsymbol{v})B_{12}I_\nu(\boldsymbol{n})\,\alpha_{12}(\xi)\\ &= \frac{h\nu_0}{c}\frac{B_{12}}{4\pi}\iint d\nu\, d\Omega\, \boldsymbol{n} I_\nu(\boldsymbol{n}) \int d^3v\, F_1(\boldsymbol{v})\,\alpha_{12}(\xi) \; ,\end{aligned} \tag{7.3.21}$$

and the emission term by

$$\begin{aligned}\hat{\pi}^{12}_{\rm em} &= \iiint d^3v' \, d\nu \frac{d\Omega}{4\pi}(m\boldsymbol{v}-m\boldsymbol{v}')F_2(\boldsymbol{v}')A_{21}\eta_{21}(\xi,\boldsymbol{n})\left[1+\frac{B_{21}}{A_{21}}I_\nu(\boldsymbol{n})\right]\\ &= -\frac{h\nu_0}{c}\frac{A_{21}}{4\pi}\iint d\nu\, d\Omega\, \boldsymbol{n}\int d^3v' F_2(\boldsymbol{v}')\,\eta_{21}(\xi,\boldsymbol{n})\\ &\quad -\frac{h\nu_0}{c}\frac{B_{21}}{4\pi}\iint d\nu\, d\Omega\, \boldsymbol{n} I_\nu(\boldsymbol{n}) \int d^3v' F_2(\boldsymbol{v}')\,\eta_{21}(\xi,\boldsymbol{n}) \; ,\end{aligned} \tag{7.3.22}$$

where (7.3.17) has been used. Now, according to (3.2.9, 10), the line emission coefficient is

$$\varepsilon^{21}_\nu(\boldsymbol{n}) = \frac{h\nu_0}{4\pi}A_{21}\int d^3v\, F_2(\boldsymbol{v})\,\eta_{21}(\xi,\boldsymbol{n}) \; , \tag{7.3.23}$$

and according to (3.2.10, 13, 15) the line absorption coefficient is

$$\kappa^{12}_\nu(\boldsymbol{n}) = \frac{h\nu_0}{4\pi}\left[B_{12}\int d^3v\, F_1(\boldsymbol{v})\,\alpha_{12}(\xi) - B_{21}\int d^3v\, F_2(\boldsymbol{v})\,\eta_{21}(\xi,\boldsymbol{n})\right] . \tag{7.3.24}$$

From (7.3.20 – 24) one finally gets

$$\hat{\pi}^{12}_{\rm rad} = \frac{1}{c}\iint \boldsymbol{n}\,[\kappa^{12}_\nu(\boldsymbol{n})I_\nu(\boldsymbol{n}) - \varepsilon^{21}_\nu(\boldsymbol{n})]\, d\nu\, d\Omega \tag{7.3.25}$$

which has the form of a negative production rate of radiant momentum, see (4.2.11), as it should. Note that the meaning of the subscript "rad" in (7.3.25) is opposite to that in (4.2.11). Equation (7.3.25) is the production rate of particle momentum due to radiative interactions, whereas (4.2.11) describes the production of radiant momentum by particles.

7.3.3 Free-Bound Transitions

As a second example of a radiative collision term appearing in the radiative force density (7.3.5), consider the term $(\delta F_0/\delta t)^{0+}_{\mathrm{rad}}$ corresponding to the free-bound transition from the bound level 0 of an ion to the bound level + of the ion of the next higher stage of ionization. Since here, in contrast to bound-bound transitions within the same ion, the cross sections are slowly varying functions of the photon frequency, we shall neglect the Doppler effect. Thus, not only the radiation intensity and the photon direction, but also the photon frequency in the atom's rest frame is set equal to the corresponding quantity in the laboratory frame, see (2.4.3). Then the collision term is given by, see (2.4.21),

$$\left(\frac{\delta F_0}{\delta t}\right)^{0+}_{\mathrm{rad}} = -F_0(\boldsymbol{v}) \iint d\nu\, d\Omega \frac{1}{h\nu} I_\nu(\boldsymbol{n})\, \sigma_{0+}(\nu)$$

$$+ \iint d\nu\, d\Omega \left[1 + \frac{c^2}{2h\nu^3} I_\nu(\boldsymbol{n})\right] \iint{}' d^3v_+ d^3v_e$$

$$\cdot F_+(\boldsymbol{v}_+) F_e(\boldsymbol{v}_e) |\boldsymbol{v}_e - \boldsymbol{v}_+|\, \sigma_{+0}(|\boldsymbol{v}_e - \boldsymbol{v}_+|, \vartheta) \ , \qquad (7.3.26)$$

where the first term describes photoionizations and the second, spontaneous and stimulated radiative recombinations. In the latter, indicated by the dash at the second two integral signs, the integrations must be performed only over those velocities of the recombining ions and electrons that lead to an outgoing photon in the range $(\nu, d\nu; \boldsymbol{n}, d\Omega)$. Conservation of momentum requires

$$m\boldsymbol{v} + \frac{h\nu}{c}\boldsymbol{n} = m_+ \boldsymbol{v}_+ + m_e \boldsymbol{v}_e \ , \qquad (7.3.27)$$

where m_+ and m_e are the ion and electron masses, and

$$m = m_+ + m_e \qquad (7.3.28)$$

is the atom mass. In (7.3.26), $\sigma_{0+}(\nu)$ is the total cross section for photoionization by photons of frequency ν, and $\sigma_{+0}(|\boldsymbol{v}_e - \boldsymbol{v}_+|, \vartheta)$ is the differential cross section for spontaneous recombination, which depends on the modulus of the relative velocity of the electron-ion pair and on the angle between this relative velocity and the direction of the outgoing photon.

We now introduce the center-of-mass velocity $\boldsymbol{V}$ and the relative velocity $\boldsymbol{w}$ of the electron-ion pair,

$$m\boldsymbol{V} = m_+\boldsymbol{v}_+ + m_e\boldsymbol{v}_e \ , \quad \boldsymbol{w} = \boldsymbol{v}_e - \boldsymbol{v}_+ \tag{7.3.29}$$

so that

$$\boldsymbol{v}_+ = \boldsymbol{V} - \frac{m_e}{m}\boldsymbol{w} \ , \quad \boldsymbol{v}_e = \boldsymbol{V} + \frac{m_+}{m}\boldsymbol{w} \ , \tag{7.3.30}$$

m being defined by (7.3.28). Since, in the center-of-mass system, the velocity of the atom is given by $\boldsymbol{v} - \boldsymbol{V} = -(h\nu/mc)\boldsymbol{n}$, energy conservation requires

$$\frac{1}{2}m\left(\frac{h\nu}{mc}\right)^2 + h\nu = \frac{1}{2}\mu_{+e}w^2 + E_{+0} \ ,$$

or, as the first term is negligible,

$$h\nu = \tfrac{1}{2}\mu_{+e}w^2 + E_{+0} \ . \tag{7.3.31}$$

Here $E_{+0} = E_+ - E_0$ is the ionization energy of the considered transition, and

$$\mu_{+e} = \frac{m_+ m_e}{m_+ + m_e} = \frac{m_+ m_e}{m} \tag{7.3.32}$$

is the reduced mass of the electron-ion pair. Note that (7.3.31) is consistent with our neglecting the Doppler effect. On the other hand, (7.3.27) now takes the form

$$\boldsymbol{v} + \frac{h\nu}{mc}\boldsymbol{n} = \boldsymbol{V} \ . \tag{7.3.33}$$

The collision term (7.3.26) can now be written as

$$\begin{aligned}\left(\frac{\delta F_0}{\delta t}\right)^{0+}_{\mathrm{rad}} &= -F_0(\boldsymbol{v}) \iint d\nu\, d\Omega \frac{1}{h\nu} I_\nu(\boldsymbol{n})\, \sigma_{0+}(\nu) \\ &+ \iint d\nu\, d\Omega \left[1 + \frac{c^2}{2h\nu^3} I_\nu(\boldsymbol{n})\right] \iint d^3V\, d^3w \\ &\cdot F_+\left(\boldsymbol{V} - \frac{m_e}{m}\boldsymbol{w}\right) F_e\left(\boldsymbol{V} + \frac{m_+}{m}\boldsymbol{w}\right) w\,\sigma_{+0}(w, \vartheta) \\ &\cdot \delta\left(\boldsymbol{V} - \boldsymbol{v} - \frac{h\nu}{mc}\boldsymbol{n}\right) \frac{1}{w^2}\, \delta\left(w - \left[\frac{2}{\mu_{+e}}(h\nu - E_{+0})\right]^{1/2}\right) \ .\end{aligned} \tag{7.3.34}$$

Here use has been made of (7.3.30) and the relation, see (E.2.10),

$$d^3v_+ d^3v_e = d^3V\, d^3w \ , \tag{7.3.35}$$

and the angle ϑ is defined by

$$\cos\vartheta = \frac{\boldsymbol{n}\cdot\boldsymbol{w}}{w} \ . \tag{7.3.36}$$

In (7.3.34), the delta functions ensure conservation of momentum and energy, (7.3.33, 31), and the second delta function is normalized by the factor $1/w^2$ because $d^3w = w^2 dw\, d\Omega_w = 2\pi w^2 dw \sin\vartheta\, d\vartheta$.

We now consider the *net momentum gain* of the heavy particles (atom and ion) owing to the free-bound transition $0 \leftrightarrow +$, defined by

$$\hat{\boldsymbol{\pi}}^{0+}_{\rm rad} = \int \left[m\boldsymbol{v} \left(\frac{\delta F_0}{\delta t}\right)^{0+}_{\rm rad} + m_+ \boldsymbol{v} \left(\frac{\delta F_+}{\delta t}\right)^{0+}_{\rm rad} \right] d^3v \ . \tag{7.3.37}$$

We write

$$\hat{\boldsymbol{\pi}}^{0+}_{\rm rad} = \hat{\boldsymbol{\pi}}^{0+}_{\rm ph} + \hat{\boldsymbol{\pi}}^{0+}_{\rm rec} \tag{7.3.38}$$

where the terms on the right refer to photoionizations and radiative recombinations, respectively. For simplicity, in the following we neglect stimulated recombinations.

The photoionization term is given by

$$\begin{aligned} \hat{\boldsymbol{\pi}}^{0+}_{\rm ph} &= \iiiint d^3v\, d\nu\, d\Omega\, d\Omega' (-m\boldsymbol{v} + m_+ \boldsymbol{v}_+) F_0(\boldsymbol{v}) \frac{1}{h\nu} I_\nu(\boldsymbol{n})\, \sigma_{0+}(\nu, \vartheta) \\ &= \iiiint d^3v\, d\nu\, d\Omega\, d\Omega' \left(\frac{h\nu}{c}\boldsymbol{n} - m_e \boldsymbol{v}_e\right) F_0(\boldsymbol{v}) \frac{1}{h\nu} I_\nu(\boldsymbol{n})\, \sigma_{0+}(\nu, \vartheta) \ , \end{aligned} \tag{7.3.39}$$

where we have used momentum conservation (7.3.27). Here, $d\Omega$ is the element of solid angle connected with the photon direction $\boldsymbol{n}$, and $\sigma_{0+}(\nu, \vartheta)$ is the differential cross section for photoionization by a photon of frequency ν such that the relative velocity of the electron-ion pair produced makes an angle ϑ with the photon direction. It is related to the total cross section for photoionization $\sigma_{0+}(\nu)$ by

$$\sigma_{0+}(\nu) = \int \sigma_{0+}(\nu, \vartheta)\, d\Omega' \ , \tag{7.3.40}$$

where here and in (7.3.39) $d\Omega' = 2\pi \sin\vartheta\, d\vartheta$. We now make the usual approximation of stationary ions by replacing the relative velocity of the electron-ion pair by the electron velocity, and the reduced mass by the electron mass. Thus, instead of (7.3.31, 36) we now have

$$h\nu = \tfrac{1}{2} m_e v_e^2 + E_{+0} \ , \tag{7.3.41}$$

$$\cos\vartheta = \frac{\boldsymbol{n}\cdot\boldsymbol{v}_e}{v_e} \ . \tag{7.3.42}$$

After integrating over $\boldsymbol{v}$, which yields the atom density n_0, and using (7.3.40), the photoionization term (7.3.39) can be written as

$$\begin{aligned}\hat{\pi}_{\rm ph}^{0+} &= \frac{1}{c} n_0 \iint d\nu\, d\Omega\, \boldsymbol{n} I_\nu(\boldsymbol{n})\, \sigma_{0+}(\nu) \\ &\quad - n_0 \iint d\nu\, d\Omega \frac{1}{h\nu} I_\nu(\boldsymbol{n}) \int d^3 v_e m_e \boldsymbol{v}_e \sigma_{0+}(\nu,\vartheta) \\ &\quad \cdot \frac{1}{v_e^2}\, \delta\left(v_e - \left[\frac{2}{m_e}(h\nu - E_{+0})\right]^{1/2}\right) \ . \end{aligned} \tag{7.3.43}$$

Here the delta function ensures conservation of energy, (7.3.41), and the normalization factor $1/v_e^2$ appears because $d^3 v_e = v_e^2 dv_e d\Omega'$.

On the other hand, neglecting stimulated recombinations, the recombination term of (7.3.38) is given by

$$\begin{aligned}\hat{\pi}_{\rm rec}^{0+} &= \iiint d^3 v_+ d^3 v_e d\Omega (m\boldsymbol{v} - m_+ \boldsymbol{v}_+) F_+(\boldsymbol{v}_+) F_e(\boldsymbol{v}_e) \\ &\quad \cdot |\boldsymbol{v}_e - \boldsymbol{v}_+|\, \sigma_{+0}(|\boldsymbol{v}_e - \boldsymbol{v}_+|, \vartheta) \\ &= \iiint d^3 v_+ d^3 v_e d\Omega \left(-\frac{h\nu}{c}\boldsymbol{n} + m_e \boldsymbol{v}_e\right) F_+(\boldsymbol{v}_+) F_e(\boldsymbol{v}_e) \\ &\quad \cdot |\boldsymbol{v}_e - \boldsymbol{v}_+|\, \sigma_{+0}(|\boldsymbol{v}_e - \boldsymbol{v}_+|, \vartheta) \ , \end{aligned} \tag{7.3.44}$$

where (7.3.27) has been used. Here ν and ϑ are defined by (7.3.31, 36) (where $\boldsymbol{w} \equiv \boldsymbol{v}_e - \boldsymbol{v}_+$), respectively, and $d\Omega = 2\pi \sin\vartheta d\vartheta$ is the element of solid angle connected with the photon direction $\boldsymbol{n}$. Using again the approximation of stationary ions, we integrate over $\boldsymbol{v}_+$, yielding the ion density n_+:

$$\begin{aligned}\hat{\pi}_{\rm rec}^{0+} &= -\frac{1}{c} n_+ \iint d^3 v_e d\Omega\, h\nu \boldsymbol{n} F_e(\boldsymbol{v}_e)\, v_e \sigma_{+0}(v_e, \vartheta) \\ &\quad + n_+ \int d^3 v_e m_e \boldsymbol{v}_e F_e(\boldsymbol{v}_e)\, v_e \sigma_{+0}(v_e) \ , \end{aligned} \tag{7.3.45}$$

where

$$\sigma_{+0}(v) = \int \sigma_{+0}(v, \vartheta)\, d\Omega \tag{7.3.46}$$

is the total cross section for spontaneous radiative recombinations. Now, in view of the energy equation (7.3.41), we may write the first term as an integral over the photon frequency ν and keep the modulus of the electron velocity v_e fixed, rather than as an integral over v_e with fixed ν.

Since

$$d^3v_e = v_e^2 dv_e d\Omega' = v_e^2 \frac{dv_e}{d\nu} d\nu\, d\Omega' = \frac{h v_e}{m_e} d\nu\, d\Omega' \tag{7.3.47}$$

on account of (7.3.41), then

$$\begin{aligned} &\iint d^3v_e d\Omega h\nu \boldsymbol{n} F_e(\boldsymbol{v}_e) v_e \sigma_{+0}(v_e, \vartheta) \\ &= \iint d\nu\, d\Omega\, h\nu \boldsymbol{n} \int d^3v_e \frac{h v_e}{m_e} F_e(\boldsymbol{v}_e) v_e \sigma_{+0}(v_e, \vartheta) \\ &\cdot \frac{1}{v_e^2} \delta\left(v_e - \left[\frac{2}{m_e}(h\nu - E_{+0})\right]^{1/2}\right), \end{aligned} \tag{7.3.48}$$

where the delta function fixes the modulus of the electron velocity in accord with (7.3.41). Thus, finally,

$$\begin{aligned} \hat{\boldsymbol{\pi}}_{\mathrm{rec}}^{0+} = &-\frac{1}{c} n_+ \iint d\nu\, d\Omega \frac{h^2\nu}{m_e} \int d^3v_e F_e(\boldsymbol{v}_e) \sigma_{+0}(v_e, \vartheta) \\ &\cdot \delta\left(v_e - \left[\frac{2}{m_e}(h\nu - E_{+0})\right]^{1/2}\right) \\ &+ n_+ \int d^3v_e m_e \boldsymbol{v}_e F_e(\boldsymbol{v}_e) v_e \sigma_{+0}(v_e) \;. \end{aligned} \tag{7.3.49}$$

Comparing (7.3.43, 49) with (3.2.28, 26) shows that the net momentum gain (7.3.38) can be written in terms of the absorption and emission coefficients of the considered free-bound transition, κ_ν^{0+} and $\varepsilon_\nu^{+0}(\boldsymbol{n})$, as

$$\hat{\boldsymbol{\pi}}_{\mathrm{rad}}^{0+} = \frac{1}{c} \iint \boldsymbol{n} [\kappa_\nu^{0+} I_\nu(\boldsymbol{n}) - \varepsilon_\nu^{+0}(\boldsymbol{n})]\, d\nu\, d\Omega + \text{electron terms} \;, \tag{7.3.50}$$

taking into account that in the absence of stimulated recombinations $\kappa_\nu^{0+} = a_\nu^{0+}$, see (3.2.2). Thus, apart from the terms describing the production of electron momentum [explicitly written in (7.3.43, 49)], the net momentum gain (7.3.50) has the expected form of a negative production rate of radiant momentum, see (4.2.11).

The net momentum gain (7.3.37) refers to the heavy particles only, disregarding the electrons. This definition is appropriate when one is interested, for instance, in the diffusion due to radiative forces of a particular chemical species irrespective of its degree of ionization (say, Be, Be^+,

Be^{2+}, ...). Of course, one can define another net momentum gain which includes the electrons. To the integrand in (7.3.37) must then be added

$$m_e \boldsymbol{v} \left(\frac{\delta F_e}{\delta t} \right)_{\mathrm{rad}}^{0+} .$$

Clearly, this new net momentum gain is given by (7.3.50) *without* the electron terms.

7.4 Compton Scattering

This last section briefly discusses Compton scattering, that is, scattering of photons by relativistic free electrons. However, since the kinetic theory of relativistic gases is no major topic of this book, we limit ourselves to deriving the Compton collision terms for the electron and photon gases, and we refer to the literature for any further aspects of the kinetic theory of Compton scattering [7.8 – 13].

The following discussion considers the distribution functions of electrons and photons as relativistic invariants, see (A.1.23, 26, 2.2),

$$F_e = \mathrm{inv} \ , \quad \phi \propto \frac{I_\nu}{\nu^3} = \mathrm{inv} \ . \tag{7.4.1}$$

As explained in Appendix A, these relations are strictly speaking only approximations, albeit in most cases of practical interest excellent ones.

7.4.1 Kinematics of Compton Scattering

Let us consider Compton scattering of a photon $(\nu, \boldsymbol{n})$ by an electron with momentum $\boldsymbol{p}$,

$$\mathrm{e}(\boldsymbol{p}) + \gamma(\nu, \boldsymbol{n}) \rightarrow \mathrm{e}(\boldsymbol{p}') + \gamma(\nu', \boldsymbol{n}') \ . \tag{7.4.2}$$

Conservation of momentum and energy requires

$$\boldsymbol{p} + \frac{h\nu}{c} \boldsymbol{n} = \boldsymbol{p}' + \frac{h\nu'}{c} \boldsymbol{n}' \ , \tag{7.4.3}$$

$$E + h\nu = E' + h\nu' \ , \tag{7.4.4}$$

where the electron energy is given by, see (A.1.3),

$$E \equiv E(\boldsymbol{p}) = (c^2 p^2 + m_e^2 c^4)^{1/2} \ , \tag{7.4.5}$$

and $E' \equiv E(\boldsymbol{p}')$. Here $\boldsymbol{p}^2 \equiv \boldsymbol{p} \cdot \boldsymbol{p}$ is the squared modulus of the momentum vector $\boldsymbol{p}$, not to be confused with the invariant $\underline{p}^2 \equiv \underline{p} \cdot \underline{p}$ of the energy-momentum four-vector $\underline{p}$ (see below).

For given initial electron and photon momenta $\boldsymbol{p}$ and $\boldsymbol{q} = (h\nu/c)\boldsymbol{n}$, the conservation equations (7.4.3, 4) constitute four constraints for the six components of the final momenta $\boldsymbol{p}'$ and $\boldsymbol{q}' = (h\nu'/c)\boldsymbol{n}'$, which means that the final state has only two degrees of freedom. To specify the final state $(\boldsymbol{p}', \boldsymbol{q}')$ uniquely, one usually specifies the direction of the outgoing photon $\boldsymbol{n}'$ and considers the momentum of the outgoing electron $\boldsymbol{p}'$ and the frequency of the outgoing photon ν' as functions of $\boldsymbol{p}, \nu, \boldsymbol{n}, \boldsymbol{n}'$.

In deriving the functions $\boldsymbol{p}' = \boldsymbol{p}'(\boldsymbol{p}, \nu, \boldsymbol{n}, \boldsymbol{n}')$ and $\nu' = \nu'(\boldsymbol{p}, \nu, \boldsymbol{n}, \boldsymbol{n}')$, explicit handling of Lorentz transformations can be completely avoided by exploiting the calculus of four-vectors [7.14]. In fact, all that is needed here (and in Sect. 7.4.5) is the definition of the scalar product of two four-vectors, $\underline{p} = (\boldsymbol{p}, p^0)$ and $\underline{q} = (\boldsymbol{q}, q^0)$ say, and the fact that it is a relativistic invariant,

$$\underline{p} \cdot \underline{q} \equiv \boldsymbol{p} \cdot \boldsymbol{q} - p^0 q^0 = \text{inv} \ . \tag{7.4.6}$$

Let us first replace the electron momentum $\boldsymbol{p}$ by the electron velocity $\boldsymbol{v}$. From the well-known relations

$$\boldsymbol{p} = \frac{m_e \boldsymbol{v}}{(1 - v^2/c^2)^{1/2}} \ , \tag{7.4.7}$$

$$E = \frac{m_e c^2}{(1 - v^2/c^2)^{1/2}} \ , \tag{7.4.8}$$

one obtains the dimensionless velocity as

$$\boldsymbol{\beta} \equiv \frac{1}{c} \boldsymbol{v} = \frac{c}{E} \boldsymbol{p} \ , \tag{7.4.9}$$

which will be used in the following.

The energy-momentum four-vector of an electron and a photon, respectively, are given by

$$\underline{p} = \left(\boldsymbol{p}, \frac{E}{c} \right) = \frac{E}{c} (\boldsymbol{\beta}, 1) \ , \tag{7.4.10}$$

$$\underline{q} = \left(\frac{h\nu}{c} \boldsymbol{n}, \frac{h\nu}{c} \right) = \frac{h\nu}{c} (\boldsymbol{n}, 1) \ , \tag{7.4.11}$$

leading to the invariants

$$\underline{p}^2 \equiv \underline{p} \cdot \underline{p} = -m_e^2 c^2 \ , \tag{7.4.12}$$

$$\underline{q}^2 \equiv \underline{q} \cdot \underline{q} = 0 \ . \tag{7.4.13}$$

The total energy-momentum four-vector of the electron-photon pair is

$$\underline{P} = \underline{p} + \underline{q} = \left(\boldsymbol{P}, \frac{W}{c}\right) \tag{7.4.14}$$

where

$$\boldsymbol{P} = \frac{1}{c}(E\boldsymbol{\beta} + h\nu\boldsymbol{n}) \ , \quad W = E + h\nu \tag{7.4.15}$$

are the total momentum and the total energy, respectively. Finally, conservation of momentum and energy in a Compton collision (7.4.3, 4) is now expressed by the single equation

$$\underline{P} = \underline{P}' \ . \tag{7.4.16}$$

After these preliminaries we now turn to our task to express the frequency of the outgoing photon ν' in terms of $\boldsymbol{p}, \nu, \boldsymbol{n}, \boldsymbol{n}'$. To this end we form the scalar product $\underline{P} \cdot \underline{q}'$, given by

$$\underline{P} \cdot \underline{q}' = \frac{h\nu'}{c}\left(\boldsymbol{P} \cdot \boldsymbol{n}' - \frac{W}{c}\right) \tag{7.4.17}$$

on account of (7.4.6, 11, 14). On the other hand,

$$\begin{aligned} \underline{P} \cdot \underline{q}' &= \underline{P}' \cdot \underline{q}' = (\underline{p}' + \underline{q}') \cdot \underline{q}' = \underline{p}' \cdot \underline{q}' \\ &= \tfrac{1}{2}[(\underline{p}' + \underline{q}')^2 - \underline{p}'^2 - \underline{q}'^2] \\ &= \tfrac{1}{2}(\underline{P}'^2 - \underline{p}'^2) \\ &= \tfrac{1}{2}(\underline{P}^2 + m_e^2c^2) \ , \end{aligned} \tag{7.4.18}$$

where we have used (7.4.12 – 14, 16). Equating (7.4.17, 18) now leads to

$$\frac{h\nu'}{c} = \frac{1}{2}\,\frac{\underline{P}^2 + m_e^2c^2}{\boldsymbol{P} \cdot \boldsymbol{n}' - \dfrac{W}{c}} \ , \tag{7.4.19}$$

which obviously is the desired relation $\nu' = \nu'(\boldsymbol{p}, \nu, \boldsymbol{n}, \boldsymbol{n}')$. Evaluation of the numerator yields

$$\begin{aligned} \underline{P}^2 + m_e^2c^2 &= \boldsymbol{P}^2 - \frac{W^2}{c^2} + m_e^2c^2 \\ &= \frac{1}{c^2}(E\boldsymbol{\beta} + h\nu\boldsymbol{n})^2 - \frac{1}{c^2}(E + h\nu)^2 + m_e^2c^2 \\ &= -2\frac{Eh\nu}{c^2}(1 - \boldsymbol{\beta} \cdot \boldsymbol{n}) \ , \end{aligned} \tag{7.4.20}$$

where in the last step (7.4.8) has been used. On the other hand, the denominator is given by

$$P\cdot n'-\frac{W}{c}=\frac{1}{c}(E\boldsymbol{\beta}+h\nu n)\cdot n'-\frac{1}{c}(E+h\nu)$$

$$=-\frac{E}{c}(1-\boldsymbol{\beta}\cdot n')-\frac{h\nu}{c}(1-n\cdot n')\ . \tag{7.4.21}$$

Inserting (7.4.20, 21) into (7.4.19) leads finally to

$$\nu'=\nu\frac{1-\boldsymbol{\beta}\cdot n}{1-\boldsymbol{\beta}\cdot n'+\dfrac{h\nu}{E}(1-n\cdot n')}\ . \tag{7.4.22}$$

The explicit form of the momentum of the outgoing electron p' now follows immediately from (7.4.3),

$$p'=p+\frac{h}{c}(\nu n-\nu' n')\ . \tag{7.4.23}$$

Equations (7.4.22, 23) express ν' and p' in terms of p (or $\boldsymbol{\beta}$), ν, n, and n'.

7.4.2 Reciprocity Relation

Consider an electron gas and a photon gas in the laboratory frame interacting via Compton scattering. The number of spontaneous Compton collisions

$$\mathrm{e}(p,d^3p)+\gamma(\nu,d\nu;n,d\Omega)\rightarrow\mathrm{e}(p',d^3p')+\gamma(\nu',d\nu';n',d\Omega') \tag{7.4.24}$$

taking place at the space-time point (r,t) per unit volume and unit time is

$$\frac{1}{h\nu}I_\nu(n)\,d\nu\,d\Omega F_\mathrm{e}(p)\,d^3p\,\sigma(p;\nu,n\,|\,p';\nu',n')\,d\Omega'\ . \tag{7.4.25}$$

Here $I_\nu(n)\equiv I_\nu(n,r,t)$ is the specific intensity of the radiation field, and $F_\mathrm{e}(p)\equiv F_\mathrm{e}(r,p,t)$ the distribution function of the electrons. Recall that in the laboratory frame, $(I_\nu/h\nu)\,d\nu\,d\Omega$ is the flux density of photons with frequencies and directions in the range $(\nu,d\nu;n,d\Omega)$, and $F_\mathrm{e}d^3p$ is the number density of electrons with momenta in the range (p,d^3p). As discussed in the preceding section, one considers the direction of the final photon n' as an independent variable, so that the element of solid angle $d\Omega'$, corresponding to the two degrees of freedom of the final state, is connected with n'. Equation (7.4.25) *defines,* in the laboratory frame, the differential cross section $\sigma(p;\nu,n\,|\,p';\nu',n')$ for the scattering process (7.4.24).

To derive the reciprocity relation that connects the differential cross section $\sigma(p;\nu,n\,|\,p';\nu',n')$ to that of the inverse process $\sigma(p';\nu',n'\,|\,p;\nu,n)$, consider as in Sect. 1.6 detailed balance in thermal equilibrium between the process (7.4.24) and the inverse one:

$$\frac{1}{h\nu} B_\nu(T)\, d\nu\, d\Omega\, F_e^M(p)\, d^3p\, \sigma(p;\nu,\boldsymbol{n}\,|\,p';\nu',\boldsymbol{n}')\, d\Omega' \left[1+\frac{c^2}{2h\nu'^3} B_{\nu'}(T)\right]$$

$$= \frac{1}{h\nu'} B_{\nu'}(T)\, d\nu'\, d\Omega' F_e^M(p')\, d^3p' \sigma(p';\nu',\boldsymbol{n}'|p;\nu,\boldsymbol{n})\, d\Omega \left[1+\frac{c^2}{2h\nu^3} B_\nu(T)\right] . \quad (7.4.26)$$

Here stimulated Compton scattering has been taken into account, whereas the electron gas is considered to be nondegenerate so that there are no fermion factors, (1.3.3b). In (7.4.26), $B_\nu(T)$ denotes the Planck function, and $F_e^M(p)$ is the relativistic Maxwell distribution

$$F_e^M(p) = n_e \xi(T)\, e^{-E(p)/kT} , \quad (7.4.27)$$

where $p \equiv |\boldsymbol{p}|$, $E(p)$ is given by (7.4.5), n_e is the electron density, and the normalization factor $\xi(T)$, whose explicit form will not be needed, is determined through

$$\xi(T) \int_0^\infty e^{-E(p)/kT} 4\pi p^2 dp = 1 . \quad (7.4.28)$$

Inserting into (7.4.26) the Planck function (1.4.45) and the Maxwell distribution (7.4.27), and using conservation of energy (7.4.4), one obtains the reciprocity relation

$$\nu^2 \sigma(p;\nu,\boldsymbol{n}\,|\,p';\nu',\boldsymbol{n}')\, d^3p\, d\nu = \nu'^2 \sigma(p';\nu',\boldsymbol{n}'|p;\nu,\boldsymbol{n})\, d^3p' d\nu' . \quad (7.4.29)$$

Here we do not eliminate the differentials through the appropriate Jacobian [derivable from (7.4.22, 23)], because (7.4.29) is exactly what is needed in the collision terms considered below.

7.4.3 Compton Collision Term (Electrons)

To derive the collision term of the electron distribution function $F_e(\boldsymbol{p}) \equiv F_e(\boldsymbol{r},\boldsymbol{p},t)$, we write the net rate of change per unit volume of electrons $e(\boldsymbol{p}, d^3p)$ as the difference of gain and loss rates due to Compton scattering, respectively,

$$\frac{\delta F_e}{\delta t} d^3p = -F_e(\boldsymbol{p})\, d^3p \iint d\nu\, d\Omega \frac{1}{h\nu} I_\nu(\boldsymbol{n})$$

$$\cdot \int d\Omega' \sigma(p;\nu,\boldsymbol{n}\,|\,p';\nu',\boldsymbol{n}') \left[1+\frac{c^2}{2h\nu'^3} I_{\nu'}(\boldsymbol{n}')\right]$$

$$+\iiiint' d^3p'\,d\nu'\,d\Omega'\,d\Omega\,F_e(p')\frac{1}{h\nu'}I_{\nu'}(n')$$

$$\cdot\,\sigma(p';\nu',n'|p;\nu,n)\left[1+\frac{c^2}{2h\nu^3}I_\nu(n)\right]\,, \tag{7.4.30}$$

where the dash at the second integral signifies that the integrations are to be performed over ranges such that the outgoing electron has a momentum in the range (p, d^3p). Using the reciprocity relation (7.4.29) and canceling the differential d^3p now leads to the collision term

$$\left(\frac{\delta F_e}{\delta t}\right)^{\text{Compt}}_{\text{rad}} = \iiint\left\{\frac{\nu^2}{\nu'^2}\frac{1}{h\nu'}I_{\nu'}(n')\left[1+\frac{c^2}{2h\nu^3}I_\nu(n)\right]F_e(p')\right.$$

$$\left.-\frac{1}{h\nu}I_\nu(n)\left[1+\frac{c^2}{2h\nu'^3}I_{\nu'}(n')\right]F_e(p)\right\}$$

$$\cdot\,\sigma(p;\nu,n|p';\nu',n')\,d\nu\,d\Omega\,d\Omega'\;. \tag{7.4.31}$$

Here ν' and p' are given by (7.4.22, 23), respectively. In terms of the (approximate) relativistic invariants F_e and I_ν/ν^3, see (7.4.1), also, more symmetrically,

$$\left(\frac{\delta F_e}{\delta t}\right)^{\text{Compt}}_{\text{rad}} = \iiint\frac{\nu^2}{h}\left\{\frac{1}{\nu'^3}I_{\nu'}(n')\left[1+\frac{c^2}{2h\nu^3}I_\nu(n)\right]F_e(p')\right.$$

$$\left.-\frac{1}{\nu^3}I_\nu(n)\left[1+\frac{c^2}{2h\nu'^3}I_{\nu'}(n)\right]F_e(p)\right\}$$

$$\cdot\,\sigma(p;\nu,n|p';\nu',n')\,d\nu\,d\Omega\,d\Omega'\;. \tag{7.4.32}$$

Taking energy conservation (7.4.4) into account, one verifies easily that in a blackbody radiation field $I_\nu = B_\nu(T)$, the relativistic Maxwell distribution (7.4.27) is stationary,

$$\left(\frac{\delta F_e^{\mathrm{M}}}{\delta t}\right)^{\text{Compt}}_{\text{rad}} = 0\;. \tag{7.4.33}$$

7.4.4 Compton Collision Term (Photons)

The scattering term appearing on the right-hand side of the radiative transfer equation (3.1.6 or 16) is according to (3.1.7, 8, 10) defined such that

$$\frac{1}{c}\,\frac{\delta I_\nu}{\delta t}\,d\nu\,d\Omega = h\nu \text{ times net creation rate (per unit time and unit volume) of photons } (\nu, d\nu; \boldsymbol{n}, d\Omega) \text{ due to scattering,}$$

where $I_\nu \equiv I_\nu(\boldsymbol{n},\boldsymbol{r},t)$ is the specific intensity. For Compton scattering therefore

$$\begin{aligned}\frac{1}{c}\left(\frac{\delta I_\nu}{\delta t}\right)^{\text{Compt}} d\nu\,d\Omega = &-h\nu\frac{1}{h\nu} I_\nu(\boldsymbol{n})\,d\nu\,d\Omega \int d^3p\,F_e(\boldsymbol{p}) \\ &\cdot \int d\Omega'\sigma(\boldsymbol{p};\nu,\boldsymbol{n}\,|\,\boldsymbol{p}';\nu',\boldsymbol{n}')\left[1+\frac{c^2}{2h\nu'^3}I_{\nu'}(\boldsymbol{n}')\right] \\ &+h\nu\left[1+\frac{c^2}{2h\nu^3}I_\nu(\boldsymbol{n})\right] d\Omega \iiint{}' d^3p'\,d\nu'\,d\Omega' \\ &\cdot \frac{1}{h\nu'} I_{\nu'}(\boldsymbol{n}')F_e(\boldsymbol{p}')\sigma(\boldsymbol{p}';\nu',\boldsymbol{n}'|\boldsymbol{p};\nu,\boldsymbol{n})\ ,\end{aligned} \tag{7.4.34}$$

where the dash at the integral signifies that the integrations are to be performed over regions such that the outgoing photon is in the range $(\nu, d\nu; \boldsymbol{n}, d\Omega)$. Using now the reciprocity relation (7.4.29) again and dropping $d\nu\,d\Omega$ gives the Compton scattering term

$$\begin{aligned}\frac{1}{c}\left(\frac{\delta I_\nu}{\delta t}\right)^{\text{Compt}} = \iint \Bigg\{&\frac{\nu^3}{\nu'^3} I_{\nu'}(\boldsymbol{n}')\left[1+\frac{c^2}{2h\nu^3}I_\nu(\boldsymbol{n})\right]F_e(\boldsymbol{p}') \\ &-I_\nu(\boldsymbol{n})\left[1+\frac{c^2}{2h\nu'^3}I_{\nu'}(\boldsymbol{n}')\right]F_e(\boldsymbol{p})\Bigg\} \\ &\cdot\sigma(\boldsymbol{p};\nu,\boldsymbol{n}\,|\,\boldsymbol{p}';\nu',\boldsymbol{n}')\,d^3p\,d\Omega'\ ,\end{aligned} \tag{7.4.35}$$

where ν' and $\boldsymbol{p}'$ are given by (7.4.22, 23). In terms of the (approximate) relativistic invariants I_ν/ν^3 and F_e, (7.4.1), then, more symmetrically,

$$\begin{aligned}\frac{1}{c}\left(\frac{\delta(I_\nu/\nu^3)}{\delta t}\right)^{\text{Compt}} = \iint \Bigg\{&\frac{1}{\nu'^3} I_{\nu'}(\boldsymbol{n}')\left[1+\frac{c^2}{2h\nu^3}I_\nu(\boldsymbol{n})\right]F_e(\boldsymbol{p}') \\ &-\frac{1}{\nu^3}I_\nu(\boldsymbol{n})\left[1+\frac{c^2}{2h\nu'^3}I_{\nu'}(\boldsymbol{n}')\right]F_e(\boldsymbol{p})\Bigg\} \\ &\cdot\sigma(\boldsymbol{p};\nu,\boldsymbol{n}\,|\,\boldsymbol{p}';\nu',\boldsymbol{n}')\,d^3p\,d\Omega'\ .\end{aligned} \tag{7.4.36}$$

One verifies immediately that blackbody radiation is stationary with respect to Compton scattering by a thermal electron gas of the same temperature, see (7.4.27),

$$\frac{1}{c}\left(\frac{\delta B_\nu(T)}{\delta t}\right)^{\text{Compt}} = 0 \ . \tag{7.4.37}$$

7.4.5 Transformation of Cross Sections

In the previous sections, the Compton collision terms for the electron and photon distribution functions were expressed in terms of the differential cross section

$$\sigma(\boldsymbol{p};\nu,\boldsymbol{n}\,|\,\boldsymbol{p}';\nu',\boldsymbol{n}')\equiv\sigma(\boldsymbol{p};\nu,\boldsymbol{n}\,|\,\boldsymbol{n}') \tag{7.4.38}$$

which applies to electrons that, prior to the Compton collision, have momentum $\boldsymbol{p}$ in the laboratory system. Recall that for given $\boldsymbol{p}, \nu, \boldsymbol{n}$, the quantities $\boldsymbol{p}'$ and ν' are uniquely determined by the direction $\boldsymbol{n}'$ of the outgoing photon, see (7.4.22, 23). The differential cross section (7.4.38), defined through the reaction rate (7.4.25) and which contains only quantities defined in the laboratory system, can be directly calculated in quantum electrodynamics [7.15].

However, it is instructive to derive this cross section from the familiar Klein-Nishina cross section for Compton scattering, which applies to the case where the incoming electron is at rest. Let us denote all quantities in the rest frame of the incoming electron by the superscript $^{\circ}$. Thus, by definition,

$$\beta^{\circ}=0\ ,\quad \boldsymbol{p}^{\circ}=0\ ,\quad E^{\circ}=m_{\mathrm{e}}c^2\ , \tag{7.4.39}$$

while the Klein-Nishina differential cross section is given by [7.15]

$$\sigma^{\circ}(\nu^{\circ},\vartheta^{\circ})=\frac{1}{2}\left(\frac{e^2}{m_{\mathrm{e}}c^2}\right)^2\left(\frac{\nu^{\circ\prime}}{\nu^{\circ}}\right)^2\left\{\frac{\nu^{\circ}}{\nu^{\circ\prime}}+\frac{\nu^{\circ\prime}}{\nu^{\circ}}-[1-(\boldsymbol{n}^{\circ}\cdot\boldsymbol{n}^{\circ\prime})^2]\right\}, \tag{7.4.40}$$

where it is assumed that the incoming particles (electron and photon) are unpolarized, and that the polarization of the outgoing particles is not observed. As in the previous sections, unprimed and primed symbols refer to physical quantities before and after the Compton collision, respectively. In (7.4.40), e is the electron charge in cgs units (and hence $e^2/m_{\mathrm{e}}c^2$ is the classical electron radius), and ϑ° is the scattering angle of the photon, so that $\boldsymbol{n}^{\circ}\cdot\boldsymbol{n}^{\circ\prime} = \cos\vartheta^{\circ}$, $1-(\boldsymbol{n}^{\circ}\cdot\boldsymbol{n}^{\circ\prime})^2=\sin^2\vartheta^{\circ}$. Finally, from (7.4.22, 39) the frequency of the outgoing photon is given by

$$\nu^{\circ\prime}=\frac{\nu^{\circ}}{1+\dfrac{h\nu^{\circ}}{m_{\mathrm{e}}c^2}(1-\boldsymbol{n}^{\circ}\cdot\boldsymbol{n}^{\circ\prime})}\ . \tag{7.4.41}$$

To relate the laboratory-frame cross section (7.4.38) to the Klein-Nishina cross section (7.4.40), one observes that a collision rate per unit time and unit volume is a relativistic invariant,

$$\frac{dN}{d^3r\,dt} = \text{inv} \ . \tag{7.4.42}$$

This is so because dN, being a number of events (namely, of well-defined collisions taking place within a well-defined space-time element), is an invariant, and the space-time element $d^3r\,dt$ is also an invariant,

$$d^3r\,dt = \text{inv} \ . \tag{7.4.43}$$

Indeed, using (A.1.1, 2) it is readily shown that the modulus of the Jacobian

$$|\partial(x', y', z', t')/\partial(x, y, z, t)| = 1 \ ,$$

and hence $dx'dy'dz'dt' = dx\,dy\,dz\,dt$. The physical reason for the invariance of $d^3r\,dt$ is that the Lorentz contraction, proportional to $(1 - v^2/c^2)^{1/2}$, is exactly compensated by the time dilation, proportional to $(1 - v^2/c^2)^{-1/2}$.

Applying the invariance property (7.4.42) to the collision rate per unit time and unit volume (7.4.25), one gets

$$\frac{1}{h\nu} I(\nu,\boldsymbol{n})\,d\nu\,d\Omega\,F_{\rm e}(\boldsymbol{p})\,d^3p\,\sigma\,d\Omega'$$

$$= \frac{1}{h\nu^{\rm o}} I^{\rm o}(\nu^{\rm o},\boldsymbol{n}^{\rm o})\,d\nu^{\rm o}\,d\Omega^{\rm o}\,F^{\rm o}_{\rm e}(\boldsymbol{p}^{\rm o})\,d^3p^{\rm o}\,\sigma^{\rm o}\,d\Omega^{\rm o\prime} \ , \tag{7.4.44}$$

where σ and $\sigma^{\rm o}$ are the differential cross sections (7.4.38, 40), respectively. Now, the following relations hold:

from (A.2.2),

$$F_{\rm e}(\boldsymbol{p}) = F^{\rm o}_{\rm e}(\boldsymbol{p}^{\rm o}) \equiv F^{\rm o}_{\rm e}(0) \ ; \tag{7.4.45}$$

from (A.1.7),

$$\frac{d^3p}{E} = \frac{d^3p^{\rm o}}{E^{\rm o}} \equiv \frac{d^3p^{\rm o}}{m_{\rm e}c^2} \ ; \tag{7.4.46}$$

from (A.1.13),

$$\frac{d\nu}{\nu} = \frac{d\nu^{\rm o}}{\nu^{\rm o}} \ ; \tag{7.4.47}$$

from (A.1.19),

$$\nu^2 d\Omega = \nu^{\rm o2} d\Omega^{\rm o} \ , \quad \nu'^2 d\Omega' = (\nu^{\rm o\prime})^2 d\Omega^{\rm o\prime} \ ; \tag{7.4.48}$$

and from (A.1.26),

$$\frac{I(\nu,\boldsymbol{n})}{\nu^3}=\frac{I^{\circ}(\nu^{\circ},\boldsymbol{n}^{\circ})}{\nu^{\circ 3}} . \tag{7.4.49}$$

Inserting (7.4.45 – 49) into (7.4.44) now yields the desired relation

$$\sigma(\boldsymbol{p};\nu,\boldsymbol{n}\,|\boldsymbol{n}')=\frac{\nu^{\circ}}{\nu}\left(\frac{\nu'}{\nu^{\circ\prime}}\right)^2\frac{m_e c^2}{E}\,\sigma^{\circ}(\nu^{\circ},\vartheta^{\circ}) , \tag{7.4.50}$$

where $E\equiv E(\boldsymbol{p})$ is given by (7.4.5).

It remains to express the quantities

$$\frac{\nu^{\circ}}{\nu} , \quad \frac{\nu^{\circ\prime}}{\nu'} , \quad \frac{\nu^{\circ}}{\nu^{\circ\prime}} , \quad \boldsymbol{n}^{\circ}\cdot\boldsymbol{n}^{\circ\prime}$$

appearing in (7.4.50) and in the Klein-Nishina formula (7.4.40) in terms of the given laboratory-frame quantities $\boldsymbol{\beta}$ [which, in turn, determines $\boldsymbol{p}$ and hence $E=E(\boldsymbol{p})$, (7.4.7 – 9)], $\boldsymbol{n}$, and $\boldsymbol{n}'$. These relations can be derived by the calculus of four-vectors, without having recourse to the explicit formulas of Lorentz transformations [7.14].

Denoting again the energy-momentum four-vectors of the electron and photon by $\underline{\boldsymbol{p}}$ and $\underline{\boldsymbol{q}}$, respectively, then

$$\underline{\boldsymbol{p}}^{\circ}\cdot\underline{\boldsymbol{q}}^{\circ}=\underline{\boldsymbol{p}}\cdot\underline{\boldsymbol{q}} \tag{7.4.51}$$

owing to the invariance of the scalar product of two four-vectors, (7.4.6). Hence, through (7.4.6, 10, 11)

$$-m_e c\frac{h\nu^{\circ}}{c}=\frac{E}{c}\,\frac{h\nu}{c}(\boldsymbol{\beta}\cdot\boldsymbol{n}-1)$$

or

$$\frac{\nu^{\circ}}{\nu}=\frac{E}{m_e c^2}(1-\boldsymbol{\beta}\cdot\boldsymbol{n}) . \tag{7.4.52a}$$

Likewise, from $\underline{\boldsymbol{p}}^{\circ}\cdot\underline{\boldsymbol{q}}^{\circ\prime}=\underline{\boldsymbol{p}}\cdot\underline{\boldsymbol{q}}'$ it follows that

$$\frac{\nu^{\circ\prime}}{\nu'}=\frac{E}{m_e c^2}(1-\boldsymbol{\beta}\cdot\boldsymbol{n}') . \tag{7.4.52b}$$

Equations (7.4.52) are the well-known formulas for the relativistic Doppler effect, see (A.1.12), recalling that

$$\frac{E}{m_e c^2}=(1-\beta^2)^{-1/2} \tag{7.4.53}$$

according to (7.4.8, 9), with $\beta^2\equiv\boldsymbol{\beta}\cdot\boldsymbol{\beta}$.

To calculate the ratio $\nu^{\circ}/\nu^{\circ\prime}$, we write, using (7.4.52),

$$\frac{\nu^{\circ}}{\nu^{\circ\prime}}=\frac{\nu}{\nu'}\,\frac{1-\boldsymbol{\beta}\cdot\boldsymbol{n}}{1-\boldsymbol{\beta}\cdot\boldsymbol{n}'}\ . \tag{7.4.54}$$

But, from (7.4.22),

$$\frac{\nu}{\nu'}=\frac{1-\boldsymbol{\beta}\cdot\boldsymbol{n}'+\dfrac{h\nu}{E}(1-\boldsymbol{n}\cdot\boldsymbol{n}')}{1-\boldsymbol{\beta}\cdot\boldsymbol{n}}\ , \tag{7.4.55}$$

and hence

$$\frac{\nu^{\circ}}{\nu^{\circ\prime}}=1+\frac{h\nu}{E}\,\frac{1-\boldsymbol{n}\cdot\boldsymbol{n}'}{1-\boldsymbol{\beta}\cdot\boldsymbol{n}'}\ . \tag{7.4.56}$$

Finally, the quantity $\boldsymbol{n}^{\circ}\cdot\boldsymbol{n}^{\circ\prime}$ and thus the scattering angle ϑ° follow from considering the invariant scalar product

$$\underline{q}^{\circ}\cdot\underline{q}^{\circ\prime}=\underline{q}\cdot\underline{q}'\ . \tag{7.4.57}$$

According to (7.4.11), this means

$$\frac{h\nu^{\circ}}{c}\,\frac{h\nu^{\circ\prime}}{c}(\boldsymbol{n}^{\circ}\cdot\boldsymbol{n}^{\circ\prime}-1)=\frac{h\nu}{c}\,\frac{h\nu'}{c}(\boldsymbol{n}\cdot\boldsymbol{n}'-1)\ ,$$

so that

$$\boldsymbol{n}^{\circ}\cdot\boldsymbol{n}^{\circ\prime}=1-\frac{\nu}{\nu^{\circ}}\,\frac{\nu'}{\nu^{\circ\prime}}(1-\boldsymbol{n}\cdot\boldsymbol{n}')\ . \tag{7.4.58}$$

Inserting here the Doppler formulas (7.4.52) yields

$$\boldsymbol{n}^{\circ}\cdot\boldsymbol{n}^{\circ\prime}=1-\frac{(m_{\mathrm{e}}c^2/E)^2}{(1-\boldsymbol{\beta}\cdot\boldsymbol{n})(1-\boldsymbol{\beta}\cdot\boldsymbol{n}')}(1-\boldsymbol{n}\cdot\boldsymbol{n}')\ , \tag{7.4.59}$$

which expresses ϑ° in terms of $\boldsymbol{\beta}$, $\boldsymbol{n}$, $\boldsymbol{n}'$.

To sum up: the laboratory-frame cross section (7.4.38) is related to the Klein-Nishina cross section (7.4.40) by (7.4.50), and the various rest-frame quantities are expressed in terms of the laboratory-frame quantities $\boldsymbol{\beta}$, $\boldsymbol{n}$, $\boldsymbol{n}'$ through (7.4.52, 56, 59).

To avoid misunderstandings, let us point out that in the literature still another definition of the differential Compton cross section exists, which, in particular, was adopted by *Pauli* [7.16] in his treatment of Compton scattering by moving electrons. In fact, while a reaction rate per unit volume is a well-defined, relativistically invariant physical quantity, see (7.4.42), it should be realized that the definition of a differential cross section is to a large extent arbitrary. (A clear discussion of relativistic cross sections can be found in [7.14].) As already stressed, the cross section $\sigma(\boldsymbol{p}; \nu, \boldsymbol{n} \,|\, \boldsymbol{n}')$ used earlier in this section is defined through (7.4.25) in which the photon flux density in the laboratory system is used, independently of the considered electron momentum $\boldsymbol{p}$. This definition is very convenient, as the collision terms of the kinetic equations can be written down immediately, see (7.4.30, 34).

Alternatively, one can define another differential cross section for Compton scattering, $\tilde{\sigma}(\boldsymbol{p}; \nu, \boldsymbol{n} \,|\, \boldsymbol{n}')$, such that the cross section differential is a relativistic invariant,

$$d\tilde{\sigma}_{\text{inv}} \equiv \tilde{\sigma}\, d\Omega' = \text{inv} \ , \tag{7.4.60}$$

the solid angle $d\Omega'$ corresponding to the outgoing photon [7.14, 16, 17]. We write $\sigma(\vartheta)$ for a differential cross section (often written $d\sigma/d\Omega$ in the literature), in contradistinction to the cross section differential $d\sigma = \sigma(\vartheta)\, d\Omega = (d\sigma/d\Omega)\, d\Omega$, which includes the solid angle $d\Omega$ of the outgoing particle. The total cross section is $\sigma = \int \sigma(\vartheta)\, d\Omega = \int d\sigma$.

Because of this *postulated* invariance,

$$d\tilde{\sigma}_{\text{inv}} = d\sigma^{\text{o}} \equiv \sigma^{\text{o}}\, d\Omega^{\text{o}\prime} \ , \tag{7.4.61}$$

where σ^{o} is the Klein-Nishina cross section (7.4.40). Using (7.4.48), one obtains from (7.4.60, 61)

$$\tilde{\sigma}(\boldsymbol{p}; \nu, \boldsymbol{n} \,|\, \boldsymbol{n}') = \left(\frac{\nu'}{\nu^{\text{o}\prime}} \right)^2 \sigma^{\text{o}}(\nu^{\text{o}}, \vartheta^{\text{o}}) \ , \tag{7.4.62}$$

which should be contrasted with (7.4.50).

Let us write down the reaction rate per unit time and unit volume for the spontaneous Compton collisions (7.4.24) in terms of this new cross section $\tilde{\sigma}$. We denote this collision rate by dr, and the corresponding number densities of electrons and photons by dn_{e} and dn_{φ}, respectively. Explicitly, see (1.4.39),

$$dn_{\text{e}} = F_{\text{e}}(\boldsymbol{p})\, d^3p \ , \tag{7.4.63}$$

$$dn_{\varphi} = \frac{1}{h\nu c} I_{\nu}(\boldsymbol{n})\, d\nu\, d\Omega \ . \tag{7.4.64}$$

Denoting again quantities in the rest frame of the incoming electron by the superscript ^{o}, one has in view of the invariance property (7.4.42)

$$dr = dr^{\text{o}} \ , \tag{7.4.65}$$

that is, using the definition (7.4.61),

$$dr = c\, d\tilde{\sigma}_{\text{inv}}\, dn_{\text{e}}^{\text{o}}\, dn_{\varphi}^{\text{o}} \ . \tag{7.4.66}$$

On the other hand, on account of (7.4.45 – 49), one shows easily that

$$dn_e^o = \frac{m_e c^2}{E} dn_e \ , \quad dn_\varphi^o = \frac{\nu^o}{\nu} dn_\varphi \ . \tag{7.4.67}$$

Inserting these values into (7.4.66), and taking the Doppler formula (7.4.52a) into account leads finally to

$$dr = c(1 - \boldsymbol{\beta} \cdot \boldsymbol{n}) \, d\tilde{\sigma}_{\text{inv}} dn_e dn_\varphi \ . \tag{7.4.68}$$

This should be compared with collision rate (7.4.25), which in the present notation reads

$$dr = c \, d\sigma \, dn_e \, dn_\varphi \ . \tag{7.4.69}$$

Comparing (7.4.68, 69) and recalling that $d\tilde{\sigma}_{\text{inv}} = \tilde{\sigma} \, d\Omega'$ and $d\sigma = \sigma \, d\Omega'$ yields the following relation between the two differential cross sections:

$$\sigma(\boldsymbol{p}; \nu, \boldsymbol{n} \,|\boldsymbol{n}') = (1 - \boldsymbol{\beta} \cdot \boldsymbol{n}) \tilde{\sigma}(\boldsymbol{p}; \nu, \boldsymbol{n} \,|\boldsymbol{n}') \ , \tag{7.4.70}$$

which can also be obtained directly from (7.4.50, 62) together with the Doppler formula (7.4.52a).

Appendices

A. Transformation Formulas for Radiative Quantities

In this appendix we derive the transformation laws for the frequency of a photon (Doppler effect), for the propagation direction of a photon (aberration), and for the specific intensity of the radiation field when passing from one inertial coordinate system to another. In this book, the main application of these formulas is the transformation of radiative quantities from the laboratory (or observer's) coordinate system to the rest frame of a moving particle and vice versa. These formulas are best derived in the framework of the special theory of relativity [A.1,2], even if one is mainly interested only in their nonrelativistic limit. It is again very convenient to consider the radiation field as a photon gas.

A.1 Transformation Formulas

Consider two inertial frames in relative motion with respect to each other, $\Sigma \equiv [x, y, z, t]$ and $\Sigma' \equiv [x', y', z', t']$, whose axes are parallel and whose origins O and O′ coincide at time $t = t' = 0$. If, when measured in Σ, the point O′ has a constant velocity $\boldsymbol{v}$ parallel to the x axis, the transformation from Σ to Σ' is the familiar Lorentz transformation

$$
\begin{aligned}
x' &= \gamma(x - vt) \ , \\
y' &= y \ , \\
z' &= z \ , \\
t' &= \gamma\left(t - \frac{vx}{c^2}\right) \ ,
\end{aligned}
\tag{A.1.1}
$$

where

$$
\gamma \equiv \gamma(v) = \frac{1}{(1 - v^2/c^2)^{1/2}} \ . \tag{A.1.2}
$$

The inverse transformation is obtained by performing the substitutions $x \leftrightarrow x'$, $t \leftrightarrow t'$, $v \rightarrow -v$.

As always in this book, photons are considered as particles. For a free particle of mass m, the momentum $\boldsymbol{p} = (p_x, p_y, p_z)$ and the energy

$$E \equiv E(\boldsymbol{p}) = c(p_x^2 + p_y^2 + p_z^2 + m^2c^2)^{1/2} \tag{A.1.3}$$

form a four-vector. More precisely, $(p_x, p_y, p_z, E/c)$ transforms like (x, y, z, ct), so that according to (A.1.1)

$$\begin{aligned} p_x' &= \gamma\left(p_x - \frac{vE}{c^2}\right) , \\ p_y' &= p_y , \\ p_z' &= p_z , \\ E' &= \gamma(E - vp_x) , \end{aligned} \tag{A.1.4}$$

where $E' \equiv E(\boldsymbol{p}')$ using (A.1.3).

We now derive the transformation law for the volume element in momentum space $d^3p = dp_x\, dp_y\, dp_z$. Quite generally,

$$d^3p' = |J|\, d^3p \; , \quad \text{where} \tag{A.1.5}$$

$$J = \frac{\partial(p_x', p_y', p_z')}{\partial(p_x, p_y, p_z)}$$

is the Jacobian. Choosing our coordinate system again such that the x axis points in the direction of the velocity $\boldsymbol{v}$, (A.1.4) applies, and the Jacobian reduces to

$$J = \frac{\partial p_x'}{\partial p_x} = \gamma\left(1 - \frac{v}{c^2}\,\frac{\partial E}{\partial p_x}\right) = \gamma\left(1 - \frac{vp_x}{E}\right) = \frac{E'}{E} , \tag{A.1.6}$$

where in the last step the last line of (A.1.4) has been used. It follows from (A.1.5, 6) that

$$\frac{d^3p'}{E'} = \frac{d^3p}{E} = \text{inv} \tag{A.1.7}$$

is a relativistic invariant.

For photons with mass $m = 0$

$$\boldsymbol{p} = \frac{h\nu}{c}\boldsymbol{n} \; , \quad E = h\nu \; , \tag{A.1.8}$$

where ν is the frequency and $\boldsymbol{n}$ the unit vector in the direction of propagation. Substitution of these expressions into (A.1.4) leads to

$$\begin{aligned} \nu' n'_x &= \gamma\nu(n_x - v/c) \ , \\ \nu' n'_y &= \nu n_y \ , \\ \nu' n'_z &= \nu n_z \ , \end{aligned} \tag{A.1.9}$$

and

$$\nu' = \gamma(1 - v n_x/c)\,\nu \ . \tag{A.1.10}$$

Equation (A.1.10) describes the relativistic Doppler effect. On the other hand, inserting (A.1.10) into (A.1.9) gives

$$\begin{aligned} n'_x &= \frac{n_x - v/c}{1 - v n_x/c} \ , \\ n'_y &= \frac{n_y}{\gamma(1 - v n_x/c)} \ , \\ n'_z &= \frac{n_z}{\gamma(1 - v n_x/c)} \ , \end{aligned} \tag{A.1.11}$$

which describe the relativistic aberration of a photon.

Let us rewrite these formulas in a form applicable to an arbitrary velocity $\boldsymbol{v}$ of the coordinate system (Fig. A.1). Since $v n_x = \boldsymbol{n} \cdot \boldsymbol{v}$, (A.1.10) can be written as

$$\nu' = \gamma\left(1 - \frac{\boldsymbol{n} \cdot \boldsymbol{v}}{c}\right)\nu \ . \tag{A.1.12}$$

In particular, it follows that

$$\frac{d\nu'}{\nu'} = \frac{d\nu}{\nu} = \mathrm{inv} \tag{A.1.13}$$

is a relativistic invariant. Furthermore,

$$\boldsymbol{n} = \boldsymbol{n}_{\parallel} + \boldsymbol{n}_{\perp}, \quad \boldsymbol{n}_{\parallel} = \frac{\boldsymbol{n} \cdot \boldsymbol{v}}{v^2}\,\boldsymbol{v} \ , \tag{A.1.14}$$

where $\boldsymbol{n}_{\parallel}$ and $\boldsymbol{n}_{\perp}$ are, respectively, vectors parallel and perpendicular to $\boldsymbol{v}$. Equations (A.1.11) can then be written as

$$\boldsymbol{n}'_{\parallel} = \frac{\boldsymbol{n}_{\parallel} - \boldsymbol{v}/c}{1 - \boldsymbol{n} \cdot \boldsymbol{v}/c} \ , \quad \boldsymbol{n}'_{\perp} = \frac{\boldsymbol{n}_{\perp}}{\gamma(1 - \boldsymbol{n} \cdot \boldsymbol{v}/c)} \ , \tag{A.1.15}$$

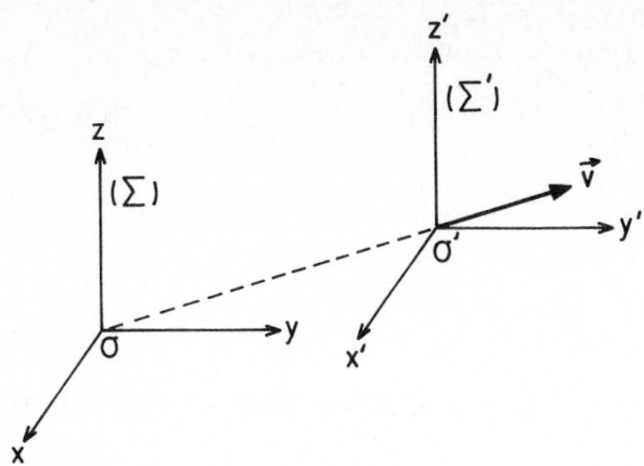

Fig. A.1. Two coordinate frames with parallel axes in relative motion. Frame $\Sigma(x, y, z, t)$ may be the laboratory frame, and frame $\Sigma'(x', y', z', t')$ the rest frame of a particle of velocity $\boldsymbol{v}$, or the local rest frame of a gas moving with macroscopic velocity $\boldsymbol{v}$

so finally, using (A.1.14),

$$\boldsymbol{n}' = \frac{1}{\gamma(1-\boldsymbol{n}\cdot\boldsymbol{v}/c)}\left(\boldsymbol{n}+\frac{\boldsymbol{v}}{c}\left[(\gamma-1)\frac{c}{v^2}\boldsymbol{n}\cdot\boldsymbol{v}-\gamma\right]\right) . \tag{A.1.16}$$

Equations (A.1.12, 16) are the desired Doppler and aberration formulas for an arbitrary velocity $\boldsymbol{v}$. The inverse formulas are obtained by the substitutions $\nu\leftrightarrow\nu'$, $\boldsymbol{n}\leftrightarrow\boldsymbol{n}'$, $\boldsymbol{v}\rightarrow-\boldsymbol{v}$.

To derive the transformation law for the element of solid angle $d\Omega$ connected with the photon direction $\boldsymbol{n}$, observe that

$$d^3p = p^2 dp\, d\Omega , \tag{A.1.17}$$

where $p = |\boldsymbol{p}| = h\nu/c$. We now use (A.1.7) to obtain

$$\frac{d\Omega'}{d\Omega} = \frac{p^2/E}{p'^2/E'}\,\frac{dp}{dp'} = \frac{\nu}{\nu'}\,\frac{d\nu}{d\nu'} = \frac{\nu^2}{\nu'^2} \tag{A.1.18}$$

on account of (A.1.8, 13). Hence

$$\nu'^2 d\Omega' = \nu^2 d\Omega = \mathrm{inv} \tag{A.1.19}$$

is a relativistic invariant.

Let us now turn to the transformation law for the specific intensity $I_\nu(\boldsymbol{n})$. To this end, we first express $I_\nu(\boldsymbol{n})$ in terms of the photon distribution functions $\phi(\boldsymbol{p})$. By definition, $\phi(\boldsymbol{p})\,d^3r\,d^3p$ is the number of photons in the volume element d^3r that have momenta in the range $(\boldsymbol{p}, d^3p)$, or equivalently, $\phi(\boldsymbol{p})\,d^3p$ is the number density of these photons. Therefore, on account of (1.4.39),

$$\phi(\boldsymbol{p})\,d^3p = \frac{1}{h\nu c} I_\nu(\boldsymbol{n})\,d\nu\, d\Omega . \tag{A.1.20}$$

Using now, see (A.1.17),

$$d^3p = p^2 dp\, d\Omega = \frac{h^3\nu^2}{c^3}\, d\nu\, d\Omega \ , \tag{A.1.21}$$

then

$$\phi(\boldsymbol{p}) = \frac{c^2}{h^4}\,\frac{I_\nu(\boldsymbol{n})}{\nu^3} \ , \tag{A.1.22}$$

where $\boldsymbol{p} = (h\nu/c)\boldsymbol{n}$. This is the desired relation between photon distribution function and specific intensity.

The reason for considering the photon distribution function rather than the specific intensity is that the photon distribution function is, to a very good approximation, a relativistic invariant, that is,

$$\phi'(\boldsymbol{p}') \simeq \phi(\boldsymbol{p}) \simeq \text{inv} \ , \tag{A.1.23}$$

where $\boldsymbol{p}'$ and $\boldsymbol{p}$ are connected through a Lorentz transformation, e.g., (A.1.4). The question of the Lorentz invariance of distribution functions is discussed in more detail in Appendix A.2.

For *nonrelativistic* velocities $\boldsymbol{v}$, which are the main concern of this book, (A.1.23) holds rigorously, and can be derived in the following way. First observe that for a Galileo transformation, the time is not transformed so that there is a common time t for two different observers. Now, at a given time t, the number of well-defined photons in a given element of μ space is an invariant because it is a pure number,

$$\phi'(\boldsymbol{r}', \boldsymbol{p}', t)\, d^3r'\, d^3p' = \phi(\boldsymbol{r}, \boldsymbol{p}, t)\, d^3r\, d^3p \ , \tag{A.1.24}$$

where $d^3r' d^3p'$ and $d^3r\, d^3p$ correspond to the same six-dimensional element in μ space. By the same reason, the number of quantum states in this μ space element is an invariant, so that according to (1.4.7, 32)

$$2\frac{d^3r'\, d^3p'}{h^3} = 2\frac{d^3r\, d^3p}{h^3} \ , \tag{A.1.25}$$

where the factor 2 is due to the two polarization states of a photon. Equation (A.1.23) follows immediately from (A.1.24, 25). This proof depends crucially on the use of a common time t in (A.1.24).

We conclude from (A.1.22, 23) that

$$\frac{I'_{\nu'}(\boldsymbol{n}')}{\nu'^3} \simeq \frac{I_\nu(\boldsymbol{n})}{\nu^3} \simeq \text{inv} \tag{A.1.26}$$

is (approximately) a relativistic invariant, where ν', $\boldsymbol{n}'$ are connected to ν, $\boldsymbol{n}$ by a Lorentz transformation, e.g., (A.1.12, 16). For nonrelativistic velocities of the moving coordinate frames, (A.1.26) is an exact relation.

We now rewrite our formulas for the nonrelativistic limit by neglecting terms of the order v^2/c^2, so that, in particular, $\gamma = 1$, see (A.1.2). From (A.1.12) then follows the nonrelativistic Doppler formula

$$\nu' = \left(1 - \frac{1}{c}\boldsymbol{n} \cdot \boldsymbol{v}\right)\nu \ , \tag{A.1.27}$$

and from (A.1.16) the nonrelativistic aberration formula

$$\boldsymbol{n}' = \boldsymbol{n} - \frac{1}{c}[\boldsymbol{v} - (\boldsymbol{n} \cdot \boldsymbol{v})\boldsymbol{n}] \ . \tag{A.1.28}$$

Note that in (A.1.27), $\boldsymbol{n} \cdot \boldsymbol{v} = v_{\|}$ is the velocity component in the direction of photon propagation $\boldsymbol{n}$, and in (A.1.28), $\boldsymbol{v} - (\boldsymbol{n} \cdot \boldsymbol{v})\boldsymbol{n} = \boldsymbol{v}_{\perp}$ is the velocity vector perpendicular to $\boldsymbol{n}$.

To derive the nonrelativistic transformation formula for the specific intensity, (A.1.26) is written in the form

$$I'(\nu', \cos\vartheta', \varphi') = \left(\frac{\nu'}{\nu}\right)^3 I(\nu, \cos\vartheta, \varphi) \ , \tag{A.1.29}$$

where ϑ and φ are the polar and azimuthal angles of the photon direction $\boldsymbol{n}$ with respect to the velocity $\boldsymbol{v}$ as polar axis. Instead of the angle ϑ itself, we may use $\cos\vartheta$ as argument of the intensity because it is a single-valued function in the interval $0 \leqslant \vartheta \leqslant \pi$. The transformation formula for $\cos\vartheta$ follows from (A.1.11) by observing that in the particular coordinate system used in these equations $n_x = \cos\vartheta$, so that

$$\cos\vartheta' = \frac{\cos\vartheta - v/c}{1 - (v/c)\cos\vartheta} \ , \tag{A.1.30}$$

while the transformation formula for $\varphi = \arctan(z/y)$ follows from (A.1.1):

$$\varphi' = \varphi \ . \tag{A.1.31}$$

Thus, according to (A.1.27, 30), the following transformation formulas apply to the nonrelativistic limit,

$$\nu' = \left(1 - \frac{v}{c}\cos\vartheta\right)\nu \ , \tag{A.1.32}$$

$$\cos\vartheta' = \cos\vartheta - \frac{v}{c}(1 - \cos^2\vartheta) \ , \tag{A.1.33}$$

and (A.1.31). The inverse formulas are obtained by the substitutions $\nu' \leftrightarrow \nu$, $\cos\vartheta' \leftrightarrow \cos\vartheta$, $v \rightarrow -v$.

In the nonrelativistic limit, (A.1.29) takes the form

$$I'(\nu', \cos\vartheta', \varphi') = \left(1 - 3\frac{v}{c}\cos\vartheta\right) I(\nu, \cos\vartheta, \varphi)$$

$$= \left(1 - 3\frac{v}{c}\cos\vartheta'\right) I(\nu, \cos\vartheta, \varphi) \qquad \text{(A.1.34)}$$

to first order in v/c, where (A.1.32, 33) have been used. Now, up to first order in v/c,

$$I(\nu, \cos\vartheta, \varphi) = I(0) + (\nu - \nu')\frac{\partial I(0)}{\partial \nu'} + (\cos\vartheta - \cos\vartheta')\frac{\partial I(0)}{\partial \cos\vartheta'}$$

$$= I(0) + \frac{v}{c}\nu'\cos\vartheta'\frac{\partial I(0)}{\partial \nu'} + \frac{v}{c}(1 - \cos^2\vartheta')\frac{\partial I(0)}{\partial \cos\vartheta'}$$

$$= I(0) + \frac{v}{c}\nu'\cos\vartheta'\frac{\partial I(0)}{\partial \nu'} - \frac{v}{c}\sin\vartheta'\frac{\partial I(0)}{\partial \vartheta'}\,, \qquad \text{(A.1.35)}$$

where $I(0) \equiv I(\nu', \cos\vartheta', \varphi')$. In (A.1.35), use has been made of (A.1.31) and of the inverse formulas of (A.1.32, 33). Inserting (A.1.35) into (A.1.34) and neglecting terms of the order v^2/c^2, one obtains the nonrelativistic transformation formula for the specific intensity

$$I' = I + \frac{v}{c}\cos\vartheta\left(\nu\frac{\partial I}{\partial \nu} - 3I\right) - \frac{v}{c}\sin\vartheta\frac{\partial I}{\partial \vartheta} \qquad \text{(A.1.36)}$$

in which I' and I have to be considered as functions of the same variables: $I' \equiv I'(\nu, \vartheta, \varphi)$ and $I \equiv I(\nu, \vartheta, \varphi)$. Recall that ϑ denotes the angle between the velocity $\boldsymbol{v}$ and the photon direction $\boldsymbol{n}$ so that $\boldsymbol{n} \cdot \boldsymbol{v} = v\cos\vartheta$. In the primed coordinate system, to determine the specific intensity that corresponds to the value $I_\nu(\boldsymbol{n})$ in the unprimed frame, one must first determine the function I' from (A.1.36), and then insert as arguments the values ν' and $\boldsymbol{n}'$ obtained from (A.1.32, 33).

A.2 Is the Distribution Function a Relativistic Invariant?

It is often claimed in the literature that the distribution function of a gas of particles or photons is a relativistic invariant. To investigate this point, let us first state explicitly what invariance of the distribution function really means.

The one-particle distribution function F of a gas is defined as follows [A.3]. An observer O finds at time t

$$F(\boldsymbol{r}, \boldsymbol{p}, t)\, \Delta^3 r\, \Delta^3 p \tag{A.2.1a}$$

particles in the small volume $\Delta^3 r$ centered at point $\boldsymbol{r}$ that have momenta in the small range $\Delta^3 p$ around the value $\boldsymbol{p}$. Here we write $\Delta^3 r$, $\Delta^3 p$ rather than $d^3 r$, $d^3 p$ as elsewhere in this book to emphasize particularly that we are dealing with small, finite ranges in ordinary and momentum space. Another observer O′, moving with a relativistic velocity $\boldsymbol{v}$ relative to O, finds at time t'

$$F'(\boldsymbol{r}', \boldsymbol{p}', t')\, \Delta^3 r'\, \Delta^3 p' \tag{A.2.1b}$$

particles in the corresponding ranges $\Delta^3 r'$, $\Delta^3 p'$. More precisely, $(\boldsymbol{r}', t')$ and $(\boldsymbol{p}', E')$ are connected to $(\boldsymbol{r}, t)$ and $(\boldsymbol{p}, E)$, respectively, by the same Lorentz transformation, e.g., (A.1.1,4), $E = E(\boldsymbol{p})$ and $E' = E(\boldsymbol{p}')$ being the energies of the particles, see (A.1.3). The volume $\Delta^3 r'$ to be used by observer O′ at time t' is determined by the intersection of the hyperplane $t' = \text{const}$ with the world lines of the surface points of the volume $\Delta^3 r$, chosen by observer O at time t and supposed to be at rest relative to him (Fig. A.2). Clearly, owing to Lorentz contraction, $\Delta^3 r' = \gamma \Delta^3 r$, where γ is given by (A.1.2), and $\Delta^3 p' = (E'/E)\, \Delta^3 p$ on account of (A.1.7).

Relativistic invariance of the distribution function would mean that

$$F'(\boldsymbol{r}', \boldsymbol{p}', t') = F(\boldsymbol{r}, \boldsymbol{p}, t) \tag{A.2.2}$$

holds quite generally. However, owing to the fact that observers O and O′ use *different events* for defining their respective distribution functions F and F', it is easy to give counterexamples to (A.2.2). For instance, consider Fig. A.2. Observer O observes at his time t in $\Delta^3 r$ three particles with equal momenta $\boldsymbol{p}_1$, say, but observer O′, due to a scattering event at space-time point S and an absorption event at space-time point A, observes at his time t' in $\Delta^3 r'$ two particles with different momenta $\boldsymbol{p}_1'$ and $\boldsymbol{p}_2'$, say, whereas invariance of the distribution function would require the presence of three particles with equal momenta $\boldsymbol{p}_1'$. This simple example shows that the distribution function $F(\boldsymbol{r}, \boldsymbol{p}, t)$ is *not* a relativistic invariant in the usual sense of the word. It follows

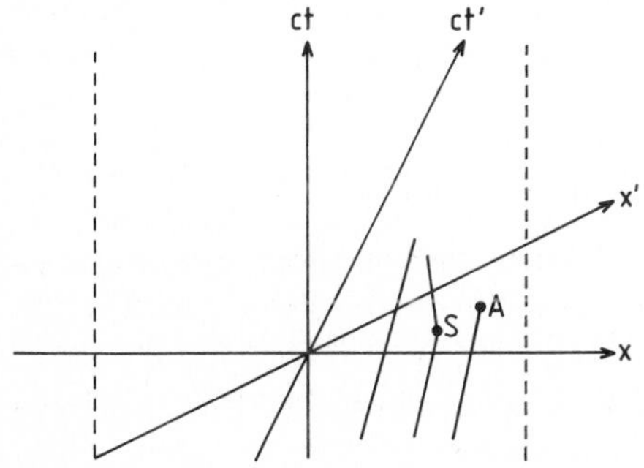

Fig. A.2. Two different observers in relative motion, O and O′, observe different distribution functions owing to the scattering event at S and the absorption event at A. (For electrons, e.g., S may denote collision with another particle, and A radiative recombination with an ion; for photons, S may be Thomson scattering by an electron, and A line absorption by an atom.) Dashed lines indicate the world lines of the surface of volume $\Delta^3 r$ used by observer O

that previously published "proofs" of this invariance must be erroneous. (For a list of such proofs and their refutation, see [A.3].) In particular, demonstrating the invariance of the photon distribution function after (A.1.23) (which, with obvious changes, applies to particles too) is valid only in the non-relativistic case where events that are simultaneous for observer O are likewise simultaneous for observer O′, as can be seen from (A.1.24) where the *same* time t appears in both distribution functions.

Let us now return to Fig. A.2 which illustrated that the distribution function is not a relativistic invariant in an arbitrary situation. However, the distribution function F as defined by (A.2.1) is useful for describing the behavior of a gas only if there is a small, finite linear dimension Δx over which the gas can be considered as homogeneous. If, moreover, temporal changes over time intervals $\Delta t \sim \Delta x/c$ are negligible (which is a very mild restriction), we have the situation depicted in Fig. A.3. For every event that tends to destroy the invariance of F (like those at S and A considered in Fig. A.2), there is another equally probable event which compensates the effect of the first (like those at S' and A' shown in Fig. A.3). This reasoning holds for particles and photons alike, and independently of whether or not the number of particles is conserved. Hence, in such cases, the distribution function is *approximately* a relativistic invariant,

$$F'(\boldsymbol{r}',\boldsymbol{p}',t') \simeq F(\boldsymbol{r},\boldsymbol{p},t) \simeq \text{inv} \ . \tag{A.2.3}$$

It is not possible, in general cases, to go beyond such a statement.

In contrast to the coarse-grained distribution function F considered above, one also defines in statistical physics the so-called fine-grained one-particle distribution function by

$$\varphi(\boldsymbol{r},\boldsymbol{p},t) = \sum_n \delta(\boldsymbol{r}-\boldsymbol{r}_n(t))\,\delta(\boldsymbol{p}-\boldsymbol{p}_n(t)) \, , \tag{A.2.4}$$

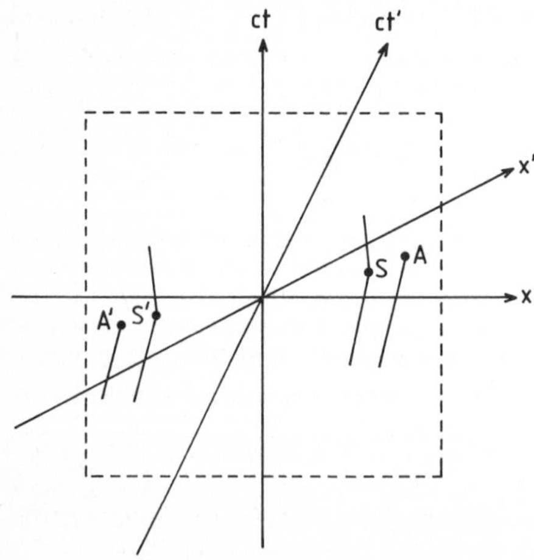

Fig. A.3. If the gas is sufficiently homogeneous within the small space-time region $\Delta^3 r \Delta t$ indicated by dashed lines, the events at S' and A' are on the average as frequent as those at S and A, so that the distribution function behaves as if it were a relativistic invariant

or by an ensemble average of such expressions. Here $\boldsymbol{r}_n(t)$ and $\boldsymbol{p}_n(t)$ are the position and momentum vectors of particle n at time t, and the sum is over all particles of the gas. We shall show presently that $\varphi(\boldsymbol{r}, \boldsymbol{p}, t)$ is a relativistic invariant [A.4].

Obviously, the term "distribution function" for φ is a misnomer because the function φ provides a complete, microscopic description of the gas, or, to put it differently, it describes for an observer O all physical events at all space-time points of the gas. For example, the function φ may describe, among others, the events that at space-time point $(\boldsymbol{r}_a, t_a)$ there is no particle present, at $(\boldsymbol{r}_b, t_b)$ one particle with momentum $\boldsymbol{p}_0$, and at $(\boldsymbol{r}_c, t_c)$ two (just colliding) particles with momenta $\boldsymbol{p}_1$ and $\boldsymbol{p}_2$. Now consider another observer O$'$ whose fine-grained distribution function is defined by

$$\varphi'(\boldsymbol{r}', \boldsymbol{p}', t') = \sum_n \delta(\boldsymbol{r}' - \boldsymbol{r}_n'(t'))\, \delta(\boldsymbol{p}' - \boldsymbol{p}_n'(t')) \ . \tag{A.2.5}$$

For O$'$, the same events are described by saying that at $(\boldsymbol{r}_a', t_a')$ there is no particle present, at $(\boldsymbol{r}_b', t_b')$ one particle with momentum $\boldsymbol{p}_0'$, and at $(\boldsymbol{r}_c', t_c')$ two particles with momenta $\boldsymbol{p}_1'$ and $\boldsymbol{p}_2'$, and these events are contained in φ', as φ and φ' are identical in form. Quite generally, all events contained in φ are likewise contained in φ', merely expressed in another language, and vice versa. It follows that the function φ is invariant,

$$\varphi'(\boldsymbol{r}', \boldsymbol{p}', t') = \varphi(\boldsymbol{r}, \boldsymbol{p}, t) = \mathrm{inv} \ . \tag{A.2.6}$$

To be specific, observer O[O$'$] finds, for example, that for $(\boldsymbol{r}_b, t_b)$ $[(\boldsymbol{r}_b', t_b')]$ the function $\varphi[\varphi']$ vanishes for all $\boldsymbol{p}[\boldsymbol{p}']$ except for $\boldsymbol{p} = \boldsymbol{p}_0 [\boldsymbol{p}' = \boldsymbol{p}_0']$ where precisely one delta function is different from zero. To obtain from the (infinite) delta function the (finite) number of particles, observer O[O$'$] has to integrate $\varphi[\varphi']$ over a small, six-dimensional volume $d^3r\, d^3p[d^3r'\, d^3p']$ with $t = t_b = \mathrm{const}$ $[t' = t_b' = \mathrm{const}]$ and centered at $(\boldsymbol{r}_b, \boldsymbol{p}_0)$ $[(\boldsymbol{r}_b', \boldsymbol{p}_0')]$. In spite of the fact that these integration volumes are different (since they lie in different hyperplanes, $t = \mathrm{const}$ and $t' = \mathrm{const}$, respectively), both observers find the same result (namely unity) because their respective delta functions are different from zero only at the same event point of the seven-dimensional $(\boldsymbol{r}, \boldsymbol{p}, t)$ space which belongs to both $d^3r\, d^3p$ and $d^3r'\, d^3p'$.

Provided that a classical description by means of point-like particles is adequate, the invariance property (A.2.6) holds for particles and photons alike. Clearly, the validity of (A.2.6) is not affected by averaging over a statistical ensemble of independent systems. If the particles have spin (electric moment, magnetic moment, etc.) $\boldsymbol{s}$, the μ space must be enlarged from $(\boldsymbol{r}, \boldsymbol{p})$ to $(\boldsymbol{r}, \boldsymbol{p}, \boldsymbol{s})$, and in the fine-grained distribution functions, the factors $\delta(\boldsymbol{s} - \boldsymbol{s}_n(t))$ must be included in order to make them invariant.

As shown, the validity of (A.2.6) is independent of any special property of Lorentz transformations, which are needed only for calculating explicitly the primed quantities from the unprimed ones, and vice versa. It follows, then, that the invariance property (A.2.6) holds quite generally for arbitrary transformations of space-time coordinates.

Let us summarize this discussion. The fine-grained distribution function φ deals with physical events at strictly localized space-time points, and is therefore a relativistic invariant. By contrast, the coarse-grained distribution function F is merely a numerical label attached to a space-time point, whose evaluation involves counting events over extended regions of μ space. While the particles involved in a given event are the same for observers O and O$'$, the particles observed by O at his time t in the finite region $\Delta^3 r \Delta^3 p$ are not necessarily the same as those observed by O$'$ at his time t' in the finite region $\Delta^3 r' \Delta^3 p'$ because the particles move and possibly interact with other particles or fields during the time between both observations. Therefore, the distribution function F is, in general, not a relativistic invariant.

B. Atomic Absorption and Emission Profiles

Appendix B treats some aspects of atomic absorption and emission profiles in view of the importance of these quantities for the theory of radiation transport in spectral lines. It must be emphasized from the outset that this topic is far from being settled, and that many questions remain to be answered. We do not intend to review the present state of the art in this field, rather, we discuss in some detail the semi-classical Weisskopf-Woolley model which from an intuitive point of view is very appealing. As shown, this model completely agrees in some simple situations with the exact quantum mechanical calculations. Whatever its limitations in more complex situations may be, the Weisskopf-Woolley model, like the Bohr model of the atom, will always serve as a reference point for the physical interpretation of more realistic quantum mechanical results. Appendix B.6 discusses the more general approach to atomic absorption and emission profiles via generalized redistribution functions.

B.1 Definitions and General Remarks

The atomic absorption and emission profiles for a particular spectral line are defined in the rest frame of the atom. In order not to burden our notation, in this appendix the superscript ^{o} is dropped which refers to the atomic rest frame elsewhere in this book, and the rest frame values of photon frequency, photon direction, and specific radiation intensity are written simply

$$\nu \equiv \nu^{\mathrm{o}} \;, \quad \boldsymbol{n} \equiv \boldsymbol{n}^{\mathrm{o}} \;, \quad I \equiv I^{\mathrm{o}} \;.$$

Consider a bound-bound transition with a lower atomic level 1 and an upper level 2. The *atomic absorption profile* $\alpha_{12}(\nu)$ for this transition is defined as follows: for an atom in level 1, $\alpha_{12}(\nu)d\nu$ is the probability of absorbing a photon whose frequency is in the range $(\nu, d\nu)$ if the atom is irradiated with white light, that is, light having constant intensity over the frequency range of the considered spectral line. Since this book considers only unpolarized atoms, we always assume the atomic absorption profile to be isotropic. On the other hand, the *atomic emission profile* $\eta_{21}(\nu,\boldsymbol{n})$ of an atom in level 2 is defined such that $\eta_{21}(\nu,\boldsymbol{n})\,d\nu\,d\Omega/4\pi$ is the probability of emitting a photon $(\nu, d\nu; \boldsymbol{n}, d\Omega)$, provided that a *spontaneous* radiative transition $2\rightarrow 1$ takes place. Note that the atomic emission profile can be anisotropic even if the atomic absorption profile is isotropic, since for atoms excited by absorbing a photon $(\nu', \boldsymbol{n}')$, the probability of spontaneously emitting a photon $(\nu, \boldsymbol{n})$ in general depends on the angle between the directions $\boldsymbol{n}$ and $\boldsymbol{n}'$, thus leading in an anisotropic radiation field to anisotropic spontaneous reemission. By definition, both profiles are normalized,

$$\int \alpha_{12}(\nu)\,d\nu = 1 \ , \tag{B.1.1a}$$

$$\iint \eta_{21}(\nu,\boldsymbol{n})\,d\nu \frac{d\Omega}{4\pi} = 1 \ . \tag{B.1.1b}$$

Recall that in this book, the radiation field is always supposed to be unpolarized. The profiles $\alpha_{12}(\nu)$ and $\eta_{21}(\nu,\boldsymbol{n})$ as defined above are therefore appropriate mean values of the corresponding more general quantities whose definitions include photon polarization.

The above definitions of atomic absorption and emission profiles suppose the atom to be in either the lower or the upper level of the considered transition. Therefore, in the general case where the quantum mechanical state of the atom is a superposition of the lower and upper quantum states, these definitions break down. However, one assumes that for the purposes of plasma spectroscopy, such quantum states can be disregarded, and that atomic absorption and emission profiles are always well-defined physical quantities.

As stated above, the atomic emission profile is defined for spontaneous emissions. To take stimulated emissions into account, one proceeds in the usual way by first calculating the spontaneous emission of photons $(\nu, \boldsymbol{n})$ alone, and then multiplying the result by the factor

$$1 + \frac{c^2}{2h\nu^3} I(\nu,\boldsymbol{n}) \simeq 1 + \frac{B_{21}}{A_{21}} I(\nu,\boldsymbol{n}) \tag{B.1.2}$$

on account of (1.3.3a, 4.43, 6.31), where A_{21} and B_{21} are the Einstein coefficients for spontaneous and stimulated emission, respectively. The emission profile η_{21} thus applies to spontaneous and stimulated emissions alike. It should be noted, however, that stimulated emissions may alter the atomic emission profile, for example through the additional broadening of the upper atomic level owing to the shortening of its lifetime, as compared to a situation without stimulated emissions. In other words, the atomic emission profile, while identical for spontaneous and stimulated emissions, depends in general on the actual rate of stimulated emissions (see Appendices B.3, 6).

The cross section $\sigma_{12}(\nu)$ of an atom in level 1 for absorbing a photon with frequency ν in a transition $1 \rightarrow 2$ can be readily expressed in terms of the Einstein coefficient for absorption B_{12} and the atomic absorption profile $\alpha_{12}(\nu)$. To this end, consider a photon beam $I(\nu,\boldsymbol{n})\,d\nu\,d\Omega$. Recalling that the photon flux density of this beam is $I(\nu,\boldsymbol{n})\,d\nu\,d\Omega/h\nu$, see (1.4.40), one can write the absorption rate per atom in the following two equivalent forms,

$$\frac{1}{h\nu}I(\nu,\boldsymbol{n})\,d\nu\,d\Omega\,\sigma_{12}(\nu) = B_{12}I(\nu,\boldsymbol{n})\,\alpha_{12}(\nu)\,d\nu\frac{d\Omega}{4\pi} \ ,$$

where the right-hand side follows from (1.6.29) and the definition of $\alpha_{12}(\nu)$. Hence,

$$\sigma_{12}(\nu) = \frac{h\nu}{4\pi}B_{12}\alpha_{12}(\nu) \simeq \frac{h\nu_0}{4\pi}B_{12}\alpha_{12}(\nu) \ , \quad \text{and} \tag{B.1.3}$$

$$\int \sigma_{12}(\nu)\,d\nu = \frac{h\nu_0}{4\pi}B_{12} \ , \tag{B.1.4}$$

where ν_0 denotes the frequency of the line center as measured in the atomic rest frame. Thus, up to a numerical factor, the Einstein coefficient for absorption is the absorption cross section integrated over the whole spectral line.

Let us now consider detailed balance in thermal equilibrium between absorptions and emissions of photons $(\nu, d\nu; \boldsymbol{n}, d\Omega)$. Taking stimulated emissions into account according to (B.1.2), detailed balance requires that

$$\begin{aligned} & n_1 B_{12} B_{\nu_0}(T)\,\alpha_{12}(\nu)\,d\nu\frac{d\Omega}{4\pi} \\ & \quad = n_2 A_{21}\,\eta_{21}(\nu,\boldsymbol{n})\,d\nu\frac{d\Omega}{4\pi}\left[1 + \frac{B_{21}}{A_{21}}B_{\nu_0}(T)\right] \ , \end{aligned} \tag{B.1.5}$$

where $B_{\nu_0}(T)$ is the Planck function. Using the Einstein relations (1.6.31, 32) and inserting into (B.1.5) the Boltzmann distribution (1.4.19) for the ratio n_2/n_1 of the occupation numbers [or, more simply, using (1.6.29) directly, which is the integrated form of (B.1.5)] leads to

$$\eta_{21}(\nu,\boldsymbol{n}) = \alpha_{12}(\nu) \ . \tag{B.1.6}$$

Thus, in thermal equilibrium, the atomic emission profile is equal to the atomic absorption profile[1].

The above derivation shows that the validity of (B.1.6) is independent of the particular value of the quantity $h\nu_0/kT$ appearing in the Planck function. It follows, then, that (B.1.6) holds quite generally in any isotropic radiation field whose intensity is constant within the frequency range of the spectral line, because, for the spectral line considered, such a radiation field is equivalent to blackbody radiation of a temperature T^* defined through $B_{\nu_0}(T^*) = I(\nu) = I(\nu_0)$.

If the atoms are excited by inelastic collisions, the spectrum of spontaneously emitted photons is likewise governed by (B.1.6). Excitation by isotropic white light or by inelastic collisions is usually referred to as *natural excitation*. Hence, for natural excitation of the upper level 2, the atomic emission profile is equal to the atomic absorption profile. If broadening of the lower level 1 cannot be neglected, the absorption profile to be used in (B.1.6) must likewise correspond to a naturally populated level 1 (Appendices B.3, 6).

In the literature on plasma spectroscopy, the atomic absorption profile is usually referred to as the atomic line profile (apart from the additional Doppler broadening of the line in the laboratory frame), and it is calculated on the basis of (B.1.6). Supposing the upper atomic level to be naturally populated, one determines the frequency spectrum of spontaneously emitted photons which, according to (B.1.6), coincides with the atomic absorption profile [B.1 – 5].

In the framework of the Weisskopf-Woolley model, described in greater detail below, two distinct mechanisms giving rise to the broadening of an atomic level are to be distinguished. First, there is level broadening caused via the time-energy uncertainty relation by the finite lifetime of the level due to emissions, absorptions, and inelastic collisions. And second, there is so-called

---

[1] Detailed balance in thermal equilibrium applies to the laboratory frame. Equation (B.1.5) tacitly assumes the atom mass to be infinite so that all atoms are at rest, and the atomic rest frame coincides with the laboratory frame. Adopting for simplicity approximation (2.4.3) (i.e., transforming only the photon frequency, but neither the photon direction nor the radiation intensity), for atoms of finite mass detailed balance in the laboratory frame is described by (B.1.5) if one substitutes there for the number densities

$$n_i \to n_i f^{\mathrm{M}}(\boldsymbol{v})\, d^3v \quad (i = 1, 2) \ ,$$

where $f^{\mathrm{M}}(\boldsymbol{v})$ is a normalized Maxwell distribution, and for the frequency

$$\nu \to \xi = \nu - \frac{\nu_0}{c} \boldsymbol{n} \cdot \boldsymbol{v} \ ,$$

where ξ is the frequency of a photon $(\nu, \boldsymbol{n})$ as measured in the rest frame of an atom of velocity $\boldsymbol{v}$, (2.4.7). One now obtains from (B.1.5)

$$\eta_{21}(\xi, \boldsymbol{n}) = \alpha_{12}(\xi) \ ,$$

which is again (B.1.6), recalling that there $\nu \equiv \nu^{\mathrm{o}}$

collision broadening. These two broadening mechanisms differ with respect to the frequency distribution of spontaneously re-emitted photons following a previous absorption of a photon in the same spectral line. An atom in a lifetime-broadened upper level tends to re-emit coherently, that is, the frequencies of the absorbed and re-emitted photons are approximately equal, up to an uncertainty determined solely by the width of the lower level of the transition. On the other hand, an atom in a collision-broadened upper level re-emits according to complete redistribution, that is, it behaves like a naturally excited atom.

It should be noted that the above picture ascribes a well-defined width to every atomic *level.* However, while it is possible in the framework of quantum mechanics to speak of the width of a lifetime-broadened level, this notion becomes meaningless for collision broadening. This can be easily understood. In a classical picture, collision broadening of an emitted spectral line is due to perturbations of the phase of the oscillating electric dipole as a result of collisions with neighboring particles. Now, in quantum mechanics, this oscillating atomic dipole is formed by the wave functions of the upper and lower levels, so that a broadening can be attributed only to a specific *transition* involving *two* atomic levels. Thus, in the Weisskopf-Woolley model, one must suppose that the quantum mechanical collision widths of the various atomic transitions can be reproduced by a suitable choice of collision widths of the individual atomic levels. It should be noted further that a clear-cut distinction between inelastic and elastic (broadening) collisions is an idealization insofar as a nonvanishing inelastic cross section necessarily entails a nonvanishing elastic cross section; in the important special case of close strong collisions, these two cross sections are in fact of the same order of magnitude. Finally, the approximation of infinitely short duration of inelastic collisions, implicitly contained in the picture given above, may break down for transitions between levels of nearly the same energy, and in the far wings of a spectral line.

Section 6.3.4 introduced the distinction between the two limiting cases of correlated and uncorrelated re-emission of radiation-excited atoms. By definition, uncorrelated re-emission takes place if, during the lifetime of the excited level, elastic collisions occur that give rise to velocity changes of the excited atom. Owing to this strong collisional interaction, the excited atom will behave like a naturally excited one, so that the atomic emission profile is equal to the atomic absorption profile. Therefore, in this appendix, we need to consider only the opposite limiting case of correlated re-emission where the velocity of the atom does not change during the lifetime of the excited level. Recalling that we neglect velocity changes of an atom caused by elastic broadening collisions, inelastic collisions, and photon emission and absorption, we see that atoms, which in the laboratory frame have velocities in the range $(\boldsymbol{v}, d^3v)$, can be considered independently of all the other atoms as far as their atomic absorption and emission profiles are concerned.

The following three sections discuss atomic absorption and emission profiles for two-level and three-level atoms from the point of view of a semiclassical model which describes broadened atomic levels in terms of a continuous distribution of "sublevels", so that the actual state of the atom is characterized by the occupation of a particular sublevel. For radiation-broadened levels, this model emerged quite naturally from *Weisskopf's* quantum mechanical calculations [B.6–8] which followed *Weisskopf* and *Wigner's* fundamental investigation of the natural linewidth [B.9]. Essentially the same results were obtained by *Woolley* [B.10, 11] using purely semiclassical reasoning based on the principle of detailed balance (see also [B.12, 13]). On the other hand, it follows from the important work by *Omont* et al. [B.14] that in many situations, the Weisskopf-Woolley model provides a reasonable picture even in the presence of collisions (see also [B.15]).

B.2 Two-Level Atom (I): One Level Broadened

We first consider the simplest case, namely, a two-level atom with an unbroadened lower level 1 and a broadened upper level 2. Here, the atomic absorption profile exhibits the broadening of the upper atomic level directly. Use of the terminology of the Weisskopf-Woolley model here does not mean that this model is already applied. Rather, the above statement simply means that the lifetime broadening of level 1 can be neglected, whereas the collision broadening pertaining to the transition $1 \leftrightarrow 2$ may be pictorially attributed to the upper level 2 without loss of generality and without any further physical consequences.

Since the atomic emission profile of atoms excited by inelastic collisions is equal to the atomic absorption profile (Appendix B.1), the general atomic emission profile is the sum of two terms corresponding to atoms excited through absorptions and excitation collisions, respectively,

$$\eta_{21}(\nu, \boldsymbol{n}) = \lambda \rho_{21}(\nu, \boldsymbol{n}) + (1 - \lambda)\, \alpha_{12}(\nu) \ . \tag{B.2.1}$$

Here λ and $1-\lambda$ denote, respectively, the fractions of radiation- and collision-excited atoms. [In the laboratory frame, $\lambda = \lambda(\boldsymbol{v})$ depends in general on the velocity $\boldsymbol{v}$ of the atom.] The quantity $\rho_{21}(\nu, \boldsymbol{n})$ is hence the normalized emission profile of radiation-excited atoms,

$$\iint \rho_{21}(\nu, \boldsymbol{n})\, d\nu \frac{d\Omega}{4\pi} = 1 \ . \tag{B.2.2}$$

To obtain an explicit expression for the emission profile $\rho_{21}(\nu, \boldsymbol{n})$, the *atomic redistribution function* $r(\nu', \boldsymbol{n}'; \nu, \boldsymbol{n})$ is defined such that

$$r(\nu', \boldsymbol{n}'; \nu, \boldsymbol{n})\, d\nu' \frac{d\Omega'}{4\pi}\, d\nu \frac{d\Omega}{4\pi}$$

is the probability that an atom in an isotropic radiation field with frequency-independent intensity absorbs a photon $(\nu', d\nu'; \boldsymbol{n}', d\Omega')$ and re-emits a photon

$(\nu, d\nu; \boldsymbol{n}, d\Omega)$ in the same spectral line, provided that the atom absorbs from a *naturally populated* lower level, and that spontaneous re-emission takes place. (Strictly speaking, in isotropic white radiation, the restriction to spontaneous emissions is unnecessary.) By definition, the atomic redistribution function is normalized,

$$\iiiint r(\nu', \boldsymbol{n}'; \nu, \boldsymbol{n})\, d\nu' \frac{d\Omega'}{4\pi} d\nu \frac{d\Omega}{4\pi} = 1 \ , \tag{B.2.3}$$

and it is related to the atomic absorption profile through

$$\iint r(\nu', \boldsymbol{n}'; \nu, \boldsymbol{n})\, d\nu \frac{d\Omega}{4\pi} = \alpha_{12}(\nu') \ . \tag{B.2.4}$$

The angle-averaged atomic redistribution function is defined by

$$\begin{aligned} r(\nu', \nu) &= \iint r(\nu', \boldsymbol{n}'; \nu, \boldsymbol{n}) \frac{d\Omega'}{4\pi} \frac{d\Omega}{4\pi} \\ &= \int r(\nu', \boldsymbol{n}'; \nu, \boldsymbol{n}) \frac{d\Omega}{4\pi} = \int r(\nu', \boldsymbol{n}'; \nu, \boldsymbol{n}) \frac{d\Omega'}{4\pi} \ , \end{aligned} \tag{B.2.5}$$

where in the second line we have used the fact that for unpolarized atoms in an unpolarized radiation field, the redistribution function depends only on the angle between the photon directions $\boldsymbol{n}'$ and $\boldsymbol{n}$. The normalization is clearly

$$\iint r(\nu', \nu)\, d\nu' d\nu = 1 \ , \tag{B.2.6}$$

and from (B.2.4, 5) it follows that

$$\int r(\nu', \nu)\, d\nu = \alpha_{12}(\nu') \ . \tag{B.2.7}$$

In the simplest case,

$$r(\nu', \boldsymbol{n}'; \nu, \boldsymbol{n}) = \gamma(\boldsymbol{n}', \boldsymbol{n})\, r(\nu', \nu) \tag{B.2.8}$$

with a phase function $\gamma(\boldsymbol{n}', \boldsymbol{n})$ which is independent of the photon frequencies ν' and ν. The phase function $\gamma(\boldsymbol{n}', \boldsymbol{n})$ is related to the phase function $g(\boldsymbol{n}', \boldsymbol{n})$ sometimes used in this book, e.g., (3.3.8 – 11), through $\gamma(\boldsymbol{n}', \boldsymbol{n}) = 4\pi g(\boldsymbol{n}', \boldsymbol{n})$. In the present context, the function γ is more convenient. On the other hand, the function $g(\vartheta)$ directly relates the differential scattering cross section $\sigma(\nu, \vartheta)$ to the total cross section $\sigma(\nu)$ through $\sigma(\nu, \vartheta) = g(\vartheta)\, \sigma(\nu)$.

As the phase function is only a function of the angle ϑ between the two photon directions ($\cos\vartheta = \boldsymbol{n}' \cdot \boldsymbol{n}$), it is symmetric in its arguments,

$$\gamma(\boldsymbol{n}', \boldsymbol{n}) = \gamma(\boldsymbol{n}, \boldsymbol{n}') \ , \tag{B.2.9}$$

and is normalized

$$\int \gamma(\boldsymbol{n}',\boldsymbol{n})\frac{d\Omega}{4\pi} = \int \gamma(\boldsymbol{n}',\boldsymbol{n})\frac{d\Omega'}{4\pi} = 1 \ . \tag{B.2.10}$$

The special cases of isotropic and dipole scattering correspond to

$$\gamma_0(\boldsymbol{n}',\boldsymbol{n}) = 1 \quad \text{and} \tag{B.2.11a}$$

$$\gamma_1(\boldsymbol{n}',\boldsymbol{n}) = \tfrac{3}{4}(1+\cos^2\vartheta) \ , \tag{B.2.11b}$$

respectively.

One often writes the angle-averaged redistribution function in a form that displays explicitly the atomic absorption profile $\alpha_{12}(\nu')$,

$$r(\nu',\nu) = \alpha_{12}(\nu')\, p_{21}(\nu'\to\nu) \ . \tag{B.2.12}$$

Because of the normalization

$$\int p_{21}(\nu'\to\nu)\,d\nu = 1 \tag{B.2.13}$$

which follows on account of (B.1.1a, 2.6), $p_{21}(\nu'\to\nu)\,d\nu$ is to be interpreted as the (conditional) probability that the atom, after absorbing a photon of frequency ν', re-emits a photon with a frequency in the range $(\nu, d\nu)$, provided that spontaneous emission takes place.

Let us now consider detailed balance in a thermal radiation field $B_\nu(T)$ between the process that consists of photon absorption $(\nu', d\nu'; \boldsymbol{n}', d\Omega')$ followed by photon emission $(\nu, d\nu; \boldsymbol{n}, d\Omega)$, and the inverse process that consists of photon absorption $(\nu, d\nu; \boldsymbol{n}, d\Omega)$ followed by photon emission $(\nu', d\nu'; \boldsymbol{n}', d\Omega')$. Assuming for simplicity stationary atoms of infinite mass, [the final result (B.2.15) is independent of this assumption; see the footnote after (B.1.6)], and taking stimulated emissions into account according to (B.1.2), detailed balance is expressed by

$$n_1 B_{12} B_{\nu_0}(T)\, r(\nu',\boldsymbol{n}';\nu,\boldsymbol{n})\,d\nu'\frac{d\Omega'}{4\pi}\,d\nu\frac{d\Omega}{4\pi}\left[1+\frac{B_{21}}{A_{21}}B_{\nu_0}(T)\right]$$

$$= n_1 B_{12} B_{\nu_0}(T)\, r(\nu,\boldsymbol{n};\nu',\boldsymbol{n}')\,d\nu\frac{d\Omega}{4\pi}\,d\nu'\frac{d\Omega'}{4\pi}\left[1+\frac{B_{21}}{A_{21}}B_{\nu_0}(T)\right] \ , \tag{B.2.14}$$

where B_{12}, B_{21}, A_{21} are the usual Einstein coefficients, n_1 is the number density of atoms in the lower level 1, and ν_0 is the frequency of the line center. From this equation one obtains at once the following reciprocity relation:

$$r(\nu',\boldsymbol{n}';\nu,\boldsymbol{n}) = r(\nu,\boldsymbol{n};\nu',\boldsymbol{n}') \ . \tag{B.2.15}$$

It follows further from (B.2.8, 9) that

$$r(\nu', \nu) = r(\nu, \nu') \;, \tag{B.2.16a}$$

$$\alpha_{12}(\nu')\, p_{21}(\nu' \to \nu) = \alpha_{12}(\nu)\, p_{21}(\nu \to \nu') \;, \tag{B.2.16b}$$

and from (B.2.7, 16a)

$$\int r(\nu', \nu)\, d\nu' = \alpha_{12}(\nu) \;, \tag{B.2.17a}$$

$$\int \alpha_{12}(\nu')\, p_{21}(\nu' \to \nu)\, d\nu' = \alpha_{12}(\nu) \;. \tag{B.2.17b}$$

Equation (B.2.14) calls for a comment, for resonance scattering is treated there like ordinary nonresonant scattering. Indeed, excited atoms do not appear explicitly in (B.2.14), and neither the Einstein relations (1.6.31, 32), relating the Einstein coefficients for absorption and emission, nor the thermal Boltzmann distribution of the occupation numbers of the atomic levels were required to arrive at the reciprocity relation (B.2.15). To justify this procedure it is necessary to show that a quantum mechanical treatment of resonance scattering leads indeed to the same reciprocity relation (B.2.15). This was shown to be the case by *Weisskopf* [B.6, 7] for natural broadening (i.e., radiative lifetime broadening) of the atomic levels, and by *Omont* et al. [B.14] for general lifetime broadening (in the absence of stimulated emissions) as well as for collision broadening in the so-called impact approximation. In this book, we assume the reciprocity relation (B.2.15) to be generally valid.

From the point of view of quantum mechanics, the detailed balance equation (B.2.14) may be justified by observing that resonance scattering is a single quantum process during which the atom cannot be said to be in the upper or lower level of the considered transition [B.16]. If, however, an ensemble of scattering atoms in a blackbody radiation field is probed by a beam of electrons, say, the correct elastic and inelastic collision rates are obtained by using the Boltzmann distribution for the occupation numbers of the atomic levels. Therefore, a semiclassical picture which assumes that the atom is always in one of its energy levels suffices for all purposes of plasma spectroscopy. It also underlies Einstein's treatment of the approach to thermal equilibrium in a blackbody radiation field discussed in Sect. 7.2. Adopting this picture, one may say that the symbol $r(\nu', \boldsymbol{n}'; \nu, \boldsymbol{n})$ used in (B.2.14) is too compact insofar as it hides the fact that absorption and re-emission events are separated by a finite time interval during which the atom is in the excited level.

After this long digression, let us return to the emission profile $\rho_{21}(\nu, \boldsymbol{n})$ of radiation-excited atoms. In terms of the atomic redistribution function, it can be written as

$$N \rho_{21}(\nu, \boldsymbol{n}) = \iint I(\nu', \boldsymbol{n}')\, r(\nu', \boldsymbol{n}'; \nu, \boldsymbol{n})\, d\nu' \frac{d\Omega'}{4\pi} \;. \tag{B.2.18}$$

Here N is a normalization constant obtained by integrating (B.2.18) over all ν and $\boldsymbol{n}$,

$$N = \iiiint I(\nu',\boldsymbol{n}')\,r(\nu',\boldsymbol{n}';\nu,\boldsymbol{n})\,d\nu'\frac{d\Omega'}{4\pi}\,d\nu\frac{d\Omega}{4\pi}$$

$$= \iint I(\nu',\boldsymbol{n}')\,\alpha_{12}(\nu')\,d\nu'\frac{d\Omega'}{4\pi}\,, \tag{B.2.19}$$

where (B.2.2,4) have been used. Defining

$$I_{12} = \iint I(\nu,\boldsymbol{n})\,\alpha_{12}(\nu)\,d\nu\frac{d\Omega}{4\pi} = \int \bar{I}(\nu)\,\alpha_{12}(\nu)\,d\nu\;, \tag{B.2.20}$$

with

$$\bar{I}(\nu) = \int I(\nu,\boldsymbol{n})\frac{d\Omega}{4\pi} \tag{B.2.21}$$

being the mean radiation intensity in the atomic rest frame, one obtains from (B.2.18 – 20) the emission profile of radiation-excited atoms:

$$\rho_{21}(\nu,\boldsymbol{n}) = \frac{1}{I_{12}}\iint I(\nu',\boldsymbol{n}')\,r(\nu',\boldsymbol{n}';\nu,\boldsymbol{n})\,d\nu'\frac{d\Omega'}{4\pi}\;. \tag{B.2.22}$$

According to (B.2.1), we must now determine the fractions λ and $1-\lambda$ of radiation- and collision-excited atoms, respectively. Up to a normalization factor N, clearly

$$\begin{aligned} N\lambda &= B_{12}\iiiint I(\nu',\boldsymbol{n}')\,r(\nu',\boldsymbol{n}';\nu,\boldsymbol{n})\,d\nu'\frac{d\Omega'}{4\pi}\,d\nu\frac{d\Omega}{4\pi} = B_{12}I_{12}\;,\\ N(1-\lambda) &= S_{12}\;, \end{aligned} \tag{B.2.23}$$

where S_{12} denotes the collisional excitation rate per atom (which, elsewhere in this book, is written as $S_{12} = n_e C_{12}$, n_e being the electron density). Hence,

$$\lambda = \frac{B_{12}I_{12}}{B_{12}I_{12}+S_{12}}\;,\quad 1-\lambda = \frac{S_{12}}{B_{12}I_{12}+S_{12}}\;. \tag{B.2.24}$$

Using these values, (B.2.1) takes the form

$$\eta_{21}(\nu,\boldsymbol{n}) = \frac{B_{12}I_{12}\rho_{21}(\nu,\boldsymbol{n})+S_{12}\alpha_{12}(\nu)}{B_{12}I_{12}+S_{12}}\;, \tag{B.2.25}$$

or, on account of (B.2.22),

$$\eta_{21}(\nu,\boldsymbol{n}) = \frac{1}{B_{12}I_{12}+S_{12}}\left[B_{12}\iint I(\nu',\boldsymbol{n}')\,r(\nu',\boldsymbol{n}';\nu,\boldsymbol{n})\,d\nu'\frac{d\Omega'}{4\pi}+S_{12}\alpha_{12}(\nu)\right]. \tag{B.2.26}$$

This is the atomic emission profile of a two-level atom in terms of the atomic redistribution function.

The actual form of the atomic redistribution function must be obtained from quantum mechanics. For our present case of a two-level atom with a sharp lower level, *Omont* et al. [B.14] obtained the following results. The atomic absorption profile is a Lorentz profile,

$$\alpha_{12}(\nu) = \frac{\delta_2/\pi}{(\nu-\nu_0)^2+\delta_2^2}\,, \tag{B.2.27}$$

where ν_0 is again the frequency of the line center (including the collisional shift of the line). The half halfwidth δ_2 of this profile is given by

$$4\pi\delta_2 = A_{21}+S_{21}+S_2\,, \tag{B.2.28}$$

where A_{21} is the Einstein coefficient for spontaneous emission, S_{21} is the rate of de-excitation collisions per atom (which, elsewhere in this book, is written as $S_{21}=n_e C_{21}$, n_e being the electron density), and S_2 is the effective frequency of elastic broadening collisions. The width of the atomic absorption profile is thus a superposition of radiative and collisional lifetime broadening ($A_{21}+S_{21}$) and of collision broadening (S_2). On the other hand, the conditional redistribution function, defined through (B.2.12), is given by

$$p_{21}(\nu'\to\nu) = \beta\delta(\nu-\nu')+(1-\beta)\,\alpha_{12}(\nu)\,. \tag{B.2.29}$$

Thus, spontaneous re-emission following the absorption of a photon is a superposition of coherent re-emission (the emitted photon having the same frequency as the absorbed photon), and of completely redistributed re-emission (the frequency distribution of emitted photons being governed by the atomic absorption profile). The branching ratio between coherent and completely redistributed re-emission is determined by the ratio of lifetime broadening to collision broadening. Explicitly,

$$\beta = \frac{A_{21}+S_{21}}{A_{21}+S_{21}+S_2}\,,\quad 1-\beta = \frac{S_2}{A_{21}+S_{21}+S_2}\,. \tag{B.2.30}$$

Apart from some other simplifying assumptions (e.g., assuming isolated lines, or neglecting spontaneous emission during collisions), the main approximation made by *Omont* et al. is the so-called impact approximation, which restricts the validity of (B.2.27 – 30) to the core of the spectral line, defined by $|\nu-\nu_0|\lesssim 1/\tau_{coll}$, where τ_{coll} is the duration of a typical collision. However, for simplicity, we assume that (B.2.27 – 30) apply quite generally.

It should be noted that in (B.2.28), the contribution of stimulated emissions to the lifetime broadening of the excited level 2 has been neglected. This is consistent with the assumption of an unbroadened lower level 1 of the atom. Indeed, owing to the equality of the Einstein coefficients B_{21} and B_{12}, see (1.6.32), the rate of stimulated emissions per atom is of the same order of magnitude as the rate of absorptions. Hence, if the lifetime broadening of level 1 due to absorptions can be neglected, that of level 2 due to stimulated emissions can be neglected, too.

The conditional redistribution function (B.2.29) now leads via (B.2.8, 12) to the following form of the atomic emission profile (B.2.26):

$$\eta_{21}(\nu,\boldsymbol{n}) = \frac{\alpha_{12}(\nu)}{B_{12}I_{12}+S_{12}}\left[\beta B_{12}\int I(\nu,\boldsymbol{n}')\,\gamma(\boldsymbol{n}',\boldsymbol{n})\frac{d\Omega'}{4\pi}\right.$$
$$\left.+(1-\beta)B_{12}\iint I(\nu',\boldsymbol{n}')\,\gamma(\boldsymbol{n}',\boldsymbol{n})\,\alpha_{12}(\nu')\,d\nu'\frac{d\Omega'}{4\pi}+S_{12}\right]. \quad \text{(B.2.31)}$$

In the special case of isotropic re-emission, $\gamma=1$, see (B.2.11a), the emission profile is isotropic and takes the simple form

$$\eta_{21}(\nu) = \frac{\alpha_{12}(\nu)}{B_{12}I_{12}+S_{12}}\left[\beta B_{12}\bar{I}(\nu)+(1-\beta)B_{12}I_{12}+S_{12}\right] . \quad \text{(B.2.32)}$$

Here $\bar{I}(\nu)$ is the mean intensity in the atomic rest frame (B.2.21), and the quantity I_{12} is given by (B.2.20). Recall that in (B.2.31, 32), the first term on the right-hand side is due to coherent re-emission of radiation-excited atoms, the second term due to completely redistributed re-emission of radiation-excited atoms, and the third term due to emission of collision-excited atoms.

It is readily seen that the emission profiles (B.2.31) and (B.2.32) reduce to the absorption profile, $\eta_{21}(\nu)=\alpha_{12}(\nu)$, if the radiation field corresponds to isotropic white light, i.e., $I(\nu,\boldsymbol{n})=I$ in the spectral line, or if radiative excitation is negligible compared to collisional one, i.e., $B_{12}\bar{I}(\nu)$, $B_{12}I_{12} \ll S_{12}$, in agreement with the discussion in Appendix B.1.

In general, however, the atomic emission profile is not equal to the atomic absorption profile. In fact, owing to the correlation between absorbed and emitted frequencies provided by the coherent part of the re-emission, the emission profile depends on the spectral distribution of the radiation intensity locally present.

We now turn definitively to the isotropic emission profile (B.2.32), and show that it allows a very suggestive interpretation in terms of the *Weisskopf-Woolley model.*

First of all, let us describe the broadened excited level 2 of the atom (Fig. B.1) by a normalized weight function $w_2(\mu)$,

$$\int w_2(\mu)\,d\mu = 1 , \quad \text{(B.2.33)}$$

such that the atomic absorption profile can be written as

$$\alpha_{12}(\nu) = w_2(\nu-\nu_0) . \quad \text{(B.2.34)}$$

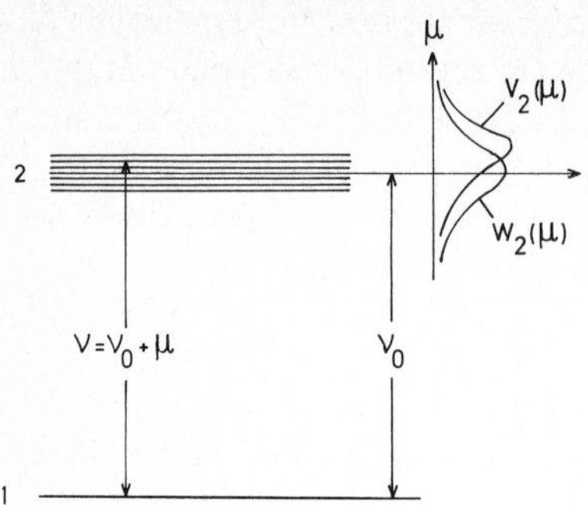

Fig. B.1. Two-level atom with sharp lower and broadened upper level. The broadening of the upper level is described by a continuous distribution of "sublevels" with weight function $w_2(\mu)$. If the population density of atoms in level 2, $v_2(\mu)$, differs from $w_2(\mu)$ (as shown in the figure), the atomic emission profile differs from the atomic absorption profile

Following *Weisskopf* and *Woolley*, one may interpret the function $w_2(\mu)$ as the distribution function of "sublevels" within level 2, i.e., $w_2(\mu)\,d\mu$ is proportional to the "number of sublevels" in the frequency range $(\mu, d\mu)$. According to (B.2.34), the center of level 2 corresponds to $\mu = 0$. On the other hand, since the ground level 1 is supposed to be infinitely sharp, its weight function is a delta function.

Let us now define another normalized function $v_2(\mu)$,

$$\int v_2(\mu)\,d\mu = 1 \;, \qquad \text{(B.2.35)}$$

which for an ensemble of excited atoms describes the population of the sublevels of the atomic level 2, and thus determines the spontaneous emission spectrum. More precisely, $v_2(\mu)\,d\mu$ is proportional to the number of excited atoms which occupy sublevels in the frequency range $(\mu, d\mu)$ of the atomic level 2, so that the atomic emission profile can be written as

$$\eta_{21}(\nu) = v_2(\nu - \nu_0) \;, \qquad \text{(B.2.36)}$$

in analogy to (B.2.34) for the atomic absorption profile. If the sublevels of atomic level 2 are uniformly occupied, $v_2(\mu) = w_2(\mu)$.

Comparing (B.2.32) with (B.2.36) shows that the distribution function $v_2(\mu)$ of excited atoms over the sublevels of level 2 can be written as

$$N v_2(\mu) = \beta B_{12}\bar{I}(\nu_0 + \mu)\,w_2(\mu) + (1-\beta)\,B_{12}I_{12}\,w_2(\mu) + S_{12}\,w_2(\mu) \;, \qquad \text{(B.2.37)}$$

N being a normalization factor. The three terms on the right-hand side allow the following interpretation. The first term describes atoms excited into a particular sublevel μ by the absorption of a photon, and which remain in this sublevel until the instant of re-emission. As the probability of reaching a sublevel in the range $(\mu, d\mu)$ is proportional to the number of photons having the appropriate frequency $\nu = \nu_0 + \mu$, and to the number of sublevels $w_2(\mu)\,d\mu$, this term is proportional to the product $\bar{I}(\nu_0 + \mu)\,w_2(\mu)$. The second term describes atoms that are also excited through photon absorption but that, at the instant of re-emission, occupy the sublevels uniformly due to the "shuffling effect" of elastic broadening collisions. Finally, the last term describes atoms excited by inelastic collisions, thus occupying the sublevels uniformly.

The normalization factor N in (B.2.37) is easily determined by integrating this equation over all μ. Taking (B.2.33, 35) into account, and observing that

$$I_{12} = \int \bar{I}(\nu_0 + \mu)\, w_2(\mu)\, d\mu \tag{B.2.38}$$

on account of (B.2.20, 34), the distribution function of excited atoms over the sublevels of level 2 corresponding to the atomic emission profile (B.2.32) is given by

$$v_2(\mu) = \frac{w_2(\mu)}{B_{12}I_{12} + S_{12}} \left[\beta B_{12}\bar{I}(\nu_0 + \mu) + (1-\beta) B_{12} I_{12} + S_{12}\right] . \tag{B.2.39}$$

In the terminology of the Weisskopf-Woolley model, complete redistribution in the atomic rest frame corresponds to uniform occupation of the sublevels of the excited level, so that the distribution of excited atoms over the sublevels is equal to the distribution of the sublevels, $v_2(\mu) = w_2(\mu)$. If, however, $v_2(\mu) \neq w_2(\mu)$ owing to the first term on the right of (B.2.39), the atomic emission profile differs from the atomic absorption profile.

In the next two sections we apply the Weisskopf-Woolley model to more complex situations in order to get some feeling for the physics involved in atomic line profiles.

B.3 Two-Level Atom (II): Two Levels Broadened

We now turn to the case where broadening of the ground state of the two-level atom is not neglected. The redistribution function for a two-level atom with two broadened levels, derived by *Omont* et al. [B.14], looks at first sight very complicated. However, using a clever trick, *Heinzel* and *Hubený* [B.17] have shown that in fact it corresponds again to a simple superposition of quasi-coherent re-emission and completely redistributed re-emission, where "quasi-coherent re-emission" means that the frequencies of the absorbed and re-emitted photons are approximately equal, up to an uncertainty determined solely by the width of the lower atomic level. This is precisely the situation expected on the basis of the Weisskopf-Woolley model. In other words, the Weisskopf-Woolley model agrees also in this more general case with the quantum mechanical calculations, at least in the framework of the impact approximation.

Let us therefore use the Weisskopf-Woolley model to derive the atomic absorption and emission profiles of a two-level atom with two broadened levels. However, in contrast to the quantum mechanical calculations of redistribution functions mentioned above, we take stimulated emissions into account because, according to the Weisskopf-Woolley model, they should give rise to interesting new features that have not yet been investigated in the framework of quantum mechanics.

Let $w_1(\mu_1)$ and $w_2(\mu_2)$ be the normalized weight functions describing the distribution of sublevels within the atomic levels 1 and 2, respectively, with $\mu_1 = 0$ and $\mu_2 = 0$ corresponding to the level centers (Fig. B.2). Apart from collision broadening, the upper atomic level has a lifetime broadening due to spontaneous emissions, stimulated emissions, and de-excitation collisions, while the lifetime broadening of the lower level is due to absorptions and excitation collisions. Explicitly, for Lorentzian weight functions

$$w_i(\mu_i) = \frac{\delta_i/\pi}{\mu_i^2 + \delta_i^2} \tag{B.3.1}$$

the half halfwidths δ_i are respectively given by

$$4\pi\delta_1 = B_{12}I_{12} + S_{12} + S_1 \ , \tag{B.3.2a}$$

$$4\pi\delta_2 = A_{21} + B_{21}I_{21} + S_{21} + S_2 \ . \tag{B.3.2b}$$

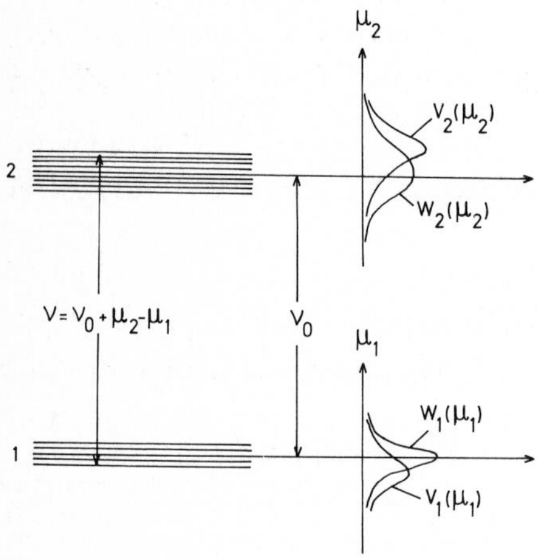

Fig. B.2. Two-level atom with broadened upper and lower levels

Here S_1 and S_2 are the effective frequencies of elastic broadening collisions corresponding to the atomic levels 1 and 2. [To keep the formulas symmetric, we always ascribe collision broadening to each atomic level. It is assumed that a set of values S_i $(i = 1, 2, \ldots)$ can be found to reproduce the actual line broadenings.] On the other hand, S_{12} and S_{21} are the rates per atom of excitation and de-excitation collisions, B_{12}, A_{21}, B_{21} are the usual Einstein coefficients, and the quantities I_{12} and I_{21} are given in terms of the atomic absorption and emission profiles by

$$I_{12} = \int \bar{I}(\nu)\,\alpha_{12}(\nu)\,d\nu \ , \tag{B.3.3a}$$

$$I_{21} = \int \bar{I}(\nu)\,\eta_{21}(\nu)\,d\nu \ , \tag{B.3.3b}$$

with $\bar{I}(\nu)$ being the mean radiation intensity in the atomic rest frame, (B.2.21), see also (B.2.20).

In the Weisskopf-Woolley model, the actual distribution of an ensemble of atoms over the atomic sublevels is described by normalized functions $v_1(\mu_1)$ and $v_2(\mu_2)$. Since the frequency ν of a transition connecting the sublevels μ_1 (in level 1) and μ_2 (in level 2) is given by

$$\nu = \nu_0 + \mu_2 - \mu_1 \ , \tag{B.3.4}$$

where ν_0 is the frequency of the line center including collisional level shifts (Fig. B.2), the atomic absorption and emission profiles can be written in terms of the functions $w(\mu)$ and $v(\mu)$ as

$$\alpha_{12}(\nu) = \int v_1(\mu_1)\, w_2(\nu - \nu_0 + \mu_1)\, d\mu_1 = \int v_1(\nu_0 - \nu + \mu_2)\, w_2(\mu_2)\, d\mu_2 \ , \tag{B.3.5}$$

$$\eta_{21}(\nu) = \int v_2(\mu_2)\, w_1(\nu_0 - \nu + \mu_2)\, d\mu_2 = \int v_2(\nu - \nu_0 + \nu_1)\, w_1(\mu_1)\, d\mu_1 \ . \tag{B.3.6}$$

Notice, in particular, that the absorption profile is now no longer a quantity whose shape can be specified a priori as Lorentzian, because it depends on the as yet unknown function $v_1(\mu_1)$. For a sharp lower level, $w_1(\mu_1) = v_1(\mu_1) = \delta(\mu_1)$, (B.3.5, 6) reduce to (B.2.34, 36), respectively, as they should. The quantities I_{12} and I_{21}, (B.3.3), now take the form

$$I_{12} = \iint \bar{I}(\nu_0 + \mu_2 - \mu_1)\, v_1(\mu_1)\, w_2(\mu_2)\, d\mu_1\, d\mu_2 \ , \tag{B.3.7a}$$

$$I_{21} = \iint \bar{I}(\nu_0 + \mu_2 - \mu_1)\, v_2(\mu_2)\, w_1(\mu_1)\, d\mu_1\, d\mu_2 \ . \tag{B.3.7b}$$

Following the procedure of the preceding section, (B.2.37), the distribution function of excited atoms over the sublevels of level 2 is determined by

$$\begin{aligned} N v_2(\mu_2) = {} & \beta_2 B_{12} \int \bar{I}(\nu_0 + \mu_2 - \mu_1)\, v_1(\mu_1)\, d\mu_1\, w_2(\mu_2) \\ & + (1 - \beta_2) B_{12} I_{12} w_2(\mu_2) + S_{12} w_2(\mu_2) \end{aligned} \tag{B.3.8}$$

with the branching ratio between lifetime and collision broadening of level 2, see (B.2.30),

$$\beta_2 = \frac{A_{21} + B_{21} I_{21} + S_{21}}{A_{21} + B_{21} I_{21} + S_{21} + S_2} \ . \tag{B.3.9}$$

The normalization factor N is determined by integrating (B.3.8) over μ_2, so finally

$$\begin{aligned} v_2(\mu_2) = {} & \frac{w_2(\mu_2)}{B_{12} I_{12} + S_{12}} \left[\beta_2 B_{12} \int \bar{I}(\nu_0 + \mu_2 - \mu_1)\, v_1(\mu_1)\, d\mu_1 \right. \\ & \left. + (1 - \beta_2) B_{12} I_{12} + S_{12}\right] \ . \end{aligned} \tag{B.3.10}$$

On the other hand, the distribution function of atoms over the sublevels of the ground state 1 follows from

$$N v_1(\mu_1) = \beta_1 \left[A_{21} w_1(\mu_1) + B_{21} \int \bar{I}(\nu_0 + \mu_2 - \mu_1) v_2(\mu_2) d\mu_2 w_1(\mu_1)\right] + (1-\beta_1)(A_{21} + B_{21} I_{21}) w_1(\mu_1) + S_{21} w_1(\mu_1) \ , \qquad \text{(B.3.11)}$$

where

$$\beta_1 = \frac{B_{12} I_{12} + S_{12}}{B_{12} I_{12} + S_{12} + S_1} \qquad \text{(B.3.12)}$$

is the branching ratio between lifetime and collision broadening of level 1. It is perhaps useful to indicate how the first two terms on the right-hand side of (B.3.11) are obtained:

$$\iint v_2(\mu_2) A_{21} \left[1 + \frac{c^2}{2h\nu^3} I(\nu, \boldsymbol{n})\right] w_1(\mu_1) d\nu \frac{d\Omega}{4\pi}$$

$$= w_1(\mu_1) \int v_2(\mu_2) A_{21} \left[1 + \frac{B_{21}}{A_{21}} \bar{I}(\nu)\right] d\nu$$

$$= w_1(\mu_1) \int v_2(\mu_2) [A_{21} + B_{21} \bar{I}(\nu_0 + \mu_2 - \mu_1)] d\mu_2$$

$$= A_{21} w_1(\mu_1) + B_{21} \int \bar{I}(\nu_0 + \mu_2 - \mu_1) v_2(\mu_2) d\mu_2 w_1(\mu_1) \ .$$

Notice, in particular, that the first term on the right-hand side of (B.3.11) means that spontaneous emissions lead to uniform occupation of the sublevels of the lower level of the transition, which is very plausible. Determining the normalization factor N by integrating (B.3.11) over all μ_1, then finally

$$v_1(\mu_1) = \frac{w_1(\mu_1)}{A_{21} + B_{21} I_{21} + S_{21}} \{\beta_1 [A_{21} + B_{21} \int \bar{I}(\nu_0 + \mu_2 - \mu_1) v_2(\mu_2) d\mu_2] + (1-\beta_1)(A_{21} + B_{21} I_{21}) + S_{21}\} \ . \qquad \text{(B.3.13)}$$

The functions $v_2(\mu_2)$ and $v_1(\mu_1)$, (B.3.10,13), together with the level profiles $w_2(\mu_2)$ and $w_1(\mu_1)$, (B.3.1), determine in the framework of the Weisskopf-Woolley model the atomic absorption and emission profiles through (B.3.5,6). Recall that in our discussion, velocity-changing elastic collisions with the atoms have been ignored. In many cases, the atomic ground state can be considered as naturally populated even in the presence of stimulated emissions, due to the occurrence of elastic collisions during its lifetime.

The atomic absorption and emission profiles are functions of the radiation field locally present, because all four quantities v_1, v_2, w_1, w_2 depend on the radiation field. On the one hand, the distribution functions v_1 and v_2 depend explicitly on the mean radiation intensity through the terms describing induced emissions and absorptions. On the other hand, the level profiles w_1 and w_2 are indirectly affected by the radiation field via the contributions of absorptions and induced emissions to the lifetime broadening of the atomic levels.

It should be realized that apart from their dependence on the radiation intensity, the four functions w_1, w_2, v_1, v_2 have been determined above only in an implicit form. Indeed, they are related to each other in a rather complicated way. From (B.3.1, 2a, 7a)

$$w_1 = w_1\,[I_{12}(v_1, w_2)]\ ,$$

from (B.3.1, 2b, 7b)

$$w_2 = w_2\,[I_{21}(w_1, v_2)]\ ,$$

from (B.3.13, 7b) [and (B.3.12, 7a) if $S_1 \neq 0$, i.e., $\beta_1 \neq 1$],

$$v_1 = v_1\,\{w_1, v_2, I_{21}(w_1, v_2), \beta_1\,[I_{12}(v_1, w_2)]\}\ ,$$

and from (B.3.10, 7a, 9, 7b),

$$v_2 = v_2\,\{w_2, v_1, I_{12}(v_1, w_2), \beta_2\,[I_{21}(w_1, v_2)]\}\ .$$

For example, v_2 as given by (B.3.10) depends explicitly on w_2 and v_1, and implicitly via I_{12} on v_1 and w_2, and via β_2, which depends on I_{21}, on w_1 and on v_2 itself.

Therefore, w_1, w_2, v_1, v_2 and the mean intensity $\bar{I}(\nu)$ must be determined self-consistently. This self-consistency problem is, however, still more complicated in the laboratory frame because it includes in addition the determination of the atomic distribution functions $F_i(\boldsymbol{v})$.

Let us briefly consider the special case when stimulated emissions are negligible. Then, according to (B.3.13), the ground state is naturally populated,

$$v_1(\mu_1) = w_1(\mu_1)\ , \tag{B.3.14}$$

and the atomic absorption profile (B.3.5) becomes the folding of two Lorentzians,

$$\alpha_{12}(\nu) = \int w_1(\mu_1)\, w_2(\nu - \nu_0 + \mu_1)\, d\mu_1\ . \tag{B.3.15}$$

On the other hand, using (B.3.4, 10, 14, 15) the atomic emission profile (B.3.6) now takes the form

$$\eta_{21}(\nu) = \frac{\beta_2 B_{12}}{B_{12} I_{12} + S_{12}} \iint \bar{I}(\nu')\, w_1(\mu_1)\, w_2(\nu' - \nu_0 + \mu_1)\, w_1(\nu' - \nu + \mu_1)\, d\mu_1\, d\nu'$$
$$+ \frac{(1-\beta_2) B_{12} I_{12} + S_{12}}{B_{12} I_{12} + S_{12}}\, \alpha_{12}(\nu)\ . \tag{B.3.16}$$

Here the second term on the right-hand side corresponds to complete redistribution in the atomic rest frame, while the first term describes quasi-coherent re-emission, showing the characteristic folding of three Lorentzians (Appendix B.5).

B.4 Three-Level Atom

Our last example applies the Weisskopf-Woolley model to an atom with three bound levels. The new feature here, in comparison with the two-level atom, is the (easily understandable) fact that the absorption and emission profiles of an atom of velocity $\boldsymbol{v}$ depend explicitly on the distribution functions $F_i(\boldsymbol{v})$ of atoms in the various bound levels i. Recall that in our approximation, the velocity of an atom is supposed to be unaltered by radiative processes, inelastic collisions, and all interactions leading to collision broadening of the atomic levels, and that no velocity-changing elastic collisions occur during time intervals of the order of the lifetimes of the excited levels (Appendix B.1).

For simplicity, we make the following assumptions:

(1) stimulated emissions are neglected;
(2) broadening of the atomic ground state is neglected;
(3) contributions to the lifetime broadening of the excited levels due to absorptions and inelastic collisions are neglected.

In the Weisskopf-Woolley model, the two excited levels 2 and 3 of the three-level atom are described by normalized functions $w_2(\mu_2)$, $v_2(\mu_2)$ and $w_3(\mu_3)$, $v_3(\mu_3)$, with the level centers corresponding to $\mu_2=0$ and $\mu_3=0$, respectively. The level profiles $w_2(\mu_2)$ and $w_3(\mu_3)$ are again Lorentzians (B.3.1), with half halfwidths δ_2 and δ_3 given by

$$4\pi\delta_2 = A_{21}+S_2 \ , \tag{B.4.1a}$$

$$4\pi\delta_3 = A_{31}+A_{32}+S_3 \ , \tag{B.4.1b}$$

where A_{ji} is the Einstein A coefficient for the transition $j \to i$, and S_i is the effective frequency of broadening collisions of level i. For ground state 1, $w_1(\mu_1) = v_1(\mu_1) = \delta(\mu_1)$, by assumption.

Let ν_{21}^0, ν_{31}^0, ν_{32}^0 be the frequencies of the centers of the three spectral lines corresponding to the transitions $2 \leftrightarrow 1$, $3 \leftrightarrow 1$, $3 \leftrightarrow 2$, respectively. In terms of the functions w_i and v_i, the atomic absorption and emission profiles are given by, see (B.3.5,6),

$$\alpha_{ij}(\nu) = \int v_i(\mu_i)\, w_j(\nu-\nu_{ji}^0+\mu_i)\, d\mu_i \ , \tag{B.4.2}$$

$$\eta_{ji}(\nu) = \int v_j(\mu_j)\, w_i(\nu_{ji}^0-\nu+\mu_j)\, d\mu_j \ , \tag{B.4.3}$$

where $j>i$ is always understood here and in the following. As the ground state is sharp, then simply, see (B.2.34,36),

$$\alpha_{1j}(\nu) = w_j(\nu-\nu_{j1}^0) \ , \quad \eta_{j1}(\nu) = v_j(\nu-\nu_{j1}^0) \quad (j=2,3) \ .$$

Furthermore, we define the quantity, see (B.3.3a, 7a),

$$I_{ij} = \int \bar{I}(\nu)\,\alpha_{ij}(\nu)\,d\nu = \iint \bar{I}(\nu_{ji}^{0} + \mu_j - \mu_i)\,v_i(\mu_i)\,w_j(\mu_j)\,d\mu_i\,d\mu_j \ . \tag{B.4.4}$$

Since induced emissions are neglected, the analogous quantity I_{ji}, see (B.3.3b, 7b), is not needed in the following.

Turning now to the determination of the occupation density $v_2(\mu_2)$ of sublevels within atomic level 2, we must realize that we are considering atoms whose velocities, measured in the laboratory frame, are in the range $(\boldsymbol{v}, d^3v)$. Since, by assumption, radiative and inelastic collision processes do not alter the velocity of an atom, the creation of atoms in level 2 in the velocity range $(\boldsymbol{v}, d^3v)$ through excitation from ground state 1 is proportional to the number of atoms in level 1 with velocities in the same range, that is, it is proportional to $F_1(\boldsymbol{v})\,d^3v$ with $F_1(\boldsymbol{v})$ being the distribution function of atoms in level 1. Likewise, the corresponding creation rate of atoms in level 2 owing to de-excitation of the higher level 3 is proportional to $F_3(\boldsymbol{v})\,d^3v$. In the following, the velocity $\boldsymbol{v}$, which is defined in the laboratory frame, should be simply considered as a label of a particular group of atoms. All other quantities refer as before to the atomic rest frame. Note also that $v_2(\mu_2) \equiv v_2(\mu_2; \boldsymbol{v})$, etc.

Following our by now familiar procedure, the function $v_2(\mu_2)$ is hence determined by

$$\begin{aligned} N v_2(\mu_2) = {} & F_1(\boldsymbol{v})\,d^3v\,[\beta_2 B_{12}\bar{I}(\nu_{21}^{0} + \mu_2)\,w_2(\mu_2) \\ & + (1-\beta_2)\,B_{12} I_{12}\,w_2(\mu_2) + S_{12}\,w_2(\mu_2)] \\ & + F_3(\boldsymbol{v})\,d^3v\,[A_{32}\,w_2(\mu_2) + S_{32}\,w_2(\mu_2)] \end{aligned} \tag{B.4.5}$$

[see (B.2.37, 3.11) with $B_{21} = 0$ there], where N is a normalization factor, and

$$\beta_2 = \frac{A_{21}}{A_{21} + S_2} \tag{B.4.6}$$

is the branching ratio of lifetime to collision broadening of level 2, see (B.4.1a). In (B.4.5) we have used the fact that spontaneous emissions and inelastic collisions populate an atomic level naturally. Integration of (B.4.5) over μ_2 yields the normalization factor

$$N = F_1(\boldsymbol{v})\,d^3v\,(B_{12} I_{12} + S_{12}) + F_3(\boldsymbol{v})\,d^3v\,(A_{32} + S_{32}) \tag{B.4.7}$$

so that finally

$$\begin{aligned} v_2(\mu_2) = {} & \frac{w_2(\mu_2)}{(B_{12} I_{12} + S_{12}) + [F_3(\boldsymbol{v})/F_1(\boldsymbol{v})]\,(A_{32} + S_{32})} \\ & \cdot \{[\beta_2 B_{12}\bar{I}(\nu_{21}^{0} + \mu_2) + (1-\beta_2)\,B_{12} I_{12} + S_{12}] \\ & + [F_3(\boldsymbol{v})/F_1(\boldsymbol{v})]\,(A_{32} + S_{32})\} \ . \end{aligned} \tag{B.4.8}$$

The occupation density $v_2(\mu_2)$ of sublevels of level 2 is hence a function of the ratio $F_3(\boldsymbol{v})/F_1(\boldsymbol{v})$ of the distribution functions of atoms in the other two levels 1 and 3. The arbitrary differential d^3v has dropped out, as it should.

Similarly, for the occupation density $v_3(\mu_3)$ of sublevels of the highest atomic level 3

$$\begin{aligned} N v_3(\mu_3) = {} & F_1(\boldsymbol{v})\, d^3v\, [\beta_3 B_{13} \bar{I}(v_{31}^0 + \mu_3)\, w_3(\mu_3) \\ & + (1-\beta_3) B_{13} I_{13}\, w_3(\mu_3) + S_{13}\, w_3(\mu_3)] \\ & + F_2(\boldsymbol{v})\, d^3v\, [\beta_3 B_{23} \int \bar{I}(v_{32}^0 + \mu_3 - \mu_2)\, v_2(\mu_2)\, d\mu_2\, w_3(\mu_3) \\ & + (1-\beta_3) B_{23} I_{23}\, w_3(\mu_3) + S_{23}\, w_3(\mu_3)] \ , \end{aligned} \tag{B.4.9}$$

where

$$\beta_3 = \frac{A_{31} + A_{32}}{A_{31} + A_{32} + S_3} \tag{B.4.10}$$

is the branching ratio of level 3, see (B.4.1b). The normalization factor N is given by

$$N = F_1(\boldsymbol{v})\, d^3v\, (B_{13} I_{13} + S_{13}) + F_2(\boldsymbol{v})\, d^3v\, (B_{23} I_{23} + S_{23}) \ , \tag{B.4.11}$$

so that

$$\begin{aligned} v_3(\mu_3) = {} & \frac{w_3(\mu_3)}{(B_{13} I_{13} + S_{13}) + [F_2(\boldsymbol{v})/F_1(\boldsymbol{v})](B_{23} I_{23} + S_{23})} \\ & \cdot \{[\beta_3 B_{13} \bar{I}(v_{31}^0 + \mu_3) + (1-\beta_3) B_{13} I_{13} + S_{13}] \\ & + [F_2(\boldsymbol{v})/F_1(\boldsymbol{v})]\, [\beta_3 B_{23} \int \bar{I}(v_{32}^0 + \mu_3 - \mu_2)\, v_2(\mu_2)\, d\mu_2 \\ & + (1-\beta_3) B_{23} I_{23} + S_{23}]\} \ . \end{aligned} \tag{B.4.12}$$

Again, the occupation density $v_3(\mu_3)$ depends explicitly on the ratio $F_2(\boldsymbol{v})/F_1(\boldsymbol{v})$ of the distribution functions of atoms in the two other levels. However, it also depends on the ratio $F_3(\boldsymbol{v})/F_1(\boldsymbol{v})$, namely implicitly via the function $v_2(\mu_2)$ which occurs in (B.4.12) because absorptions $2 \rightarrow 3$ proceed from level 2 whose sublevels are populated according to $v_2(\mu_2)$. Note that the function $v_3(\mu_3)$ likewise occurs in the explicit expression of the function $v_2(\mu_2)$ if induced emissions are taken into account. The occupation densities v_2 and v_3 (B.4.8, 12) together with the level profiles w_2 and w_3, (B.3.1), determine via (B.4.2, 3) the various absorption and emission profiles of the three-level atom.

We remark that temporal variations of the particle and photon distribution functions $F_i(\boldsymbol{v})$ and $\bar{I}(v)$ have been assumed to be negligible over time intervals of the order of the lifetimes of the excited atomic levels, otherwise distribution functions referring to the same time t could not be used.

As shown, the explicit dependence of the atomic absorption and emission profiles on the distribution functions $F_i(\boldsymbol{v})$ (and hence, a fortiori, on the number densities n_i) has a straightforward origin which is not linked to the semiclassical model used, and thus will survive in more refined descriptions. On the other hand, the Weisskopf-Woolley model predicts that the atomic absorption and emission profiles corresponding to a particular spectral line should be influenced by radiative transitions occurring in other spectral lines. For example, the absorption profile $\alpha_{23}(\nu)$ for the spectral line at ν_{32}^0 depends on $v_2(\mu_2)$, which in turn is determined in part by absorptions $1\to 2$ due to the spectral line at ν_{21}^0.

We do not intend to discuss in detail the absorption and emission profiles obtained in this section, as they are based on a semiclassical and hence hypothetical model. In any case, profiles of such complexity must always be simplified before they can be used in practical line transfer calculations.

B.5 Atomic Redistribution Functions

This section discusses briefly some atomic redistribution functions for two-level atoms as defined in Appendix B.2, which in many cases also describe the scattering in a subordinate line of a multilevel atom. For simplicity, we consider only angle-averaged redistribution functions $r(\nu', \nu)$.

Let us first recall some properties of Lorentz distributions needed in the following. We denote by

$$L_i(t) = \frac{\delta_i/\pi}{t^2+\delta_i^2}, \quad L_{ij}(t) = \frac{\delta_{ij}/\pi}{t^2+\delta_{ij}^2} \tag{B.5.1}$$

normalized Lorentz distributions with half halfwidths δ_i and δ_{ij}, respectively, and the sums of two and three half halfwidths by

$$\delta_{ij} = \delta_i + \delta_j, \quad \delta_{ijk} = \delta_i + \delta_j + \delta_k . \tag{B.5.2}$$

The folding of two Lorentzians with half halfwidths δ_1 and δ_2 is a Lorentzian with the half halfwidth $\delta_{12} = \delta_1 + \delta_2$,

$$\int_{-\infty}^{\infty} L_1(t-x_1) L_2(t-x_2)\, dt = L_{12}(x_1-x_2) . \tag{B.5.3}$$

Furthermore, the folding of three Lorentzians is given by Henyey's formula [B.13, 18],

$$\int_{-\infty}^{\infty} L_1(t-x_1) L_2(t-x_2) L_3(t-x_3)\, dt$$

$$= 4\pi \frac{\delta_1\delta_2\delta_3\delta_{123}}{\delta_{12}\delta_{23}\delta_{31}} L_{12}(x_1-x_2) L_{23}(x_2-x_3) L_{31}(x_3-x_1)$$

$$+\frac{\delta_1\delta_2}{\delta_{23}\delta_{31}}L_{23}(x_2-x_3)L_{31}(x_3-x_1)$$

$$+\frac{\delta_2\delta_3}{\delta_{31}\delta_{12}}L_{31}(x_3-x_1)L_{12}(x_1-x_2)$$

$$+\frac{\delta_3\delta_1}{\delta_{12}\delta_{23}}L_{12}(x_1-x_2)L_{23}(x_2-x_3)\ . \qquad \text{(B.5.4)}$$

Let us now consider a two-level atom with a lower level 1 and an upper level 2, neglecting stimulated emissions. The following atomic redistribution functions, corresponding to various special cases, are found in the literature:

(1) Both atomic levels 1 and 2 are sharp, so that the re-emission is necessarily coherent,

$$r_{\mathrm{I}}(\nu',\nu)=\delta(\nu'-\nu_0)\,\delta(\nu-\nu_0)\ . \qquad \text{(B.5.5)}$$

(2) Level 1 is sharp, level 2 is broadened, and the re-emission is coherent,

$$r_{\mathrm{II}}(\nu',\nu)=L_2(\nu'-\nu_0)\,\delta(\nu-\nu')\ . \qquad \text{(B.5.6)}$$

(3) Level 1 is sharp, level 2 is broadened, and the re-emission is completely redistributed,

$$r_{\mathrm{III}}(\nu',\nu)=L_2(\nu'-\nu_0)\,L_2(\nu-\nu_0)\ . \qquad \text{(B.5.7)}$$

(4) Heitler's redistribution function [B.16]

$$r_{\mathrm{IV}}(\nu',\nu)=L_1(\nu-\nu')\,L_2(\nu-\nu_0) \qquad \text{(B.5.8)}$$

does *not* correctly describe resonance fluorescence by a two-level atom with two lifetime-broadened levels (in contrast to the Weisskopf-Woolley redistribution function r_{V} to be discussed presently). Indeed, writing (B.5.8) in the form, see (B.5.11) below,

$$r_{\mathrm{IV}}(\nu',\nu)=\int_{-\infty}^{\infty}L_1(t-\nu')\,L_2(t-\nu_0)\,\delta(t-\nu)\,dt\ ,$$

one notices that the level 1 of the absorbing atom is considered to be broadened, whereas it is considered sharp for the re-emitting atom, which are two contradictory assumptions. However, Heitler's function r_{IV}, mutatis mutandis, describes correctly the radiative cascade $3\rightarrow2\rightarrow1$ in a multilevel atom, leading from a naturally populated level 3 via a lifetime-broadened level 2 into the sharp ground level 1 (Appendix B.6).

(5) Both levels are lifetime broadened (in the absence of inelastic collisions, the finite lifetime of the lower level is due to absorptions, and that of the upper level is due to spontaneous emissions). According to the Weisskopf-

Woolley model (which in this case is exact, Appendix B.3), the atomic redistribution function is given by

$$r_{\mathrm{V}}(\nu', \nu) = \int w_1(\mu_1)\, w_2(\mu_2)\, w_1(\mu_1')\, d\mu_1 \tag{B.5.9}$$

with frequencies μ_2 and μ_1' defined through, see (B.3.4) and Fig. B.2,

$$\nu' = \nu_0 + \mu_2 - \mu_1 \ , \quad \nu = \nu_0 + \mu_2 - \mu_1' \ . \tag{B.5.10}$$

In (B.5.9) we used the fact that in the absence of stimulated emissions, the ground state of the absorbing atom is naturally populated, see (B.3.14). Recalling that the profiles of lifetime-broadened levels are Lorentzians, $w_i(\mu_i) = L_i(\mu_i)$, and introducing the new variable $t = \nu' + \mu_1$, (B.5.9) takes the symmetrical form

$$r_{\mathrm{V}}(\nu', \nu) = \int_{-\infty}^{\infty} L_1(t-\nu') L_2(t-\nu_0) L_1(t-\nu)\, dt \ , \tag{B.5.11}$$

which is the redistribution function describing quasi-coherent re-emission.

(6) Both atomic levels 1 and 2 are broadened and naturally populated, so that the re-emission is completely redistributed,

$$r_{\mathrm{VI}}(\nu', \nu) = L_{12}(\nu' - \nu_0) L_{12}(\nu - \nu_0) \ . \tag{B.5.12}$$

Notice that in contrast to the other redistribution functions, Heitler's function r_{IV} does not satisfy the reciprocity relation $r(\nu', \nu) = r(\nu, \nu')$, (B.2.16a), which also shows that it is not a true redistribution function.

The notation $r_{\mathrm{I}} \ldots r_{\mathrm{IV}}$ for the redistribution functions corresponding to cases $1-4$ is from *Hummer* [B.19]. The notation r_{V} for the Weisskopf-Woolley redistribution function (B.5.11) is from *Heinzel* [B.20], who showed that also the functions $r_{\mathrm{I}} \ldots r_{\mathrm{IV}}$ can be written formally as convolution integrals of three Lorentz distributions, thus providing a unified and powerful starting point for carrying out the averages over Maxwellian velocity distributions. On the other hand, the redistribution function r_{VI} cannot be written as a single convolution integral involving L_1 and L_2, but it is formally identical with r_{III}, connected to it by the simple substitution $\delta_2 \rightarrow \delta_{12}$.

As pointed out in Appendix B.3, the Weisskopf-Woolley model is exact for a two-level atom in the framework of the usual approximations (impact approximation, no stimulated emissions). Hence, the most general atomic redistribution function of a two-level atom is of the form

$$r(\nu', \nu) = \beta r_{\mathrm{V}}(\nu', \nu) + (1-\beta) r_{\mathrm{VI}}(\nu', \nu) \tag{B.5.13}$$

with $0 \leqslant \beta \leqslant 1$. This function also applies to resonance scattering in a subordinate line of a multilevel atom, provided the lower level of the transition is naturally populated and the approximation of isolated lines is valid.

In (B.5.13), the r_{VI} term is trivial. It gives rise to the emission of a Lorentzian spectral line with half halfwidth $\delta_{12} = \delta_1 + \delta_2$ independently of the frequency ν' of the previously absorbed photon.

On the other hand, using (B.5.1,4) with $\delta_3 = \delta_1$ and dropping for simplicity the index V, the redistribution function r_V, (B.5.11), can be written as

$$r(\nu',\nu) = \frac{4\delta_1^2\delta_2(2\delta_1+\delta_2)}{\pi^2} \frac{1}{(\nu'-\nu_0)^2+\delta_{12}^2} \frac{1}{(\nu-\nu_0)^2+\delta_{12}^2} \frac{1}{(\nu-\nu')^2+4\delta_1^2}$$
$$+\frac{\delta_1\delta_2}{\pi^2} \frac{1}{(\nu-\nu_0)^2+\delta_{12}^2} \frac{1}{(\nu-\nu')^2+4\delta_1^2}$$
$$+\frac{\delta_1\delta_2}{\pi^2} \frac{1}{(\nu-\nu')^2+4\delta_1^2} \frac{1}{(\nu'-\nu_0)^2+\delta_{12}^2}$$
$$+\frac{\delta_1^2}{\pi^2} \frac{1}{(\nu'-\nu_0)^2+\delta_{12}^2} \frac{1}{(\nu-\nu_0)^2+\delta_{12}^2} \,. \qquad \text{(B.5.14)}$$

According to (B.2.7), the corresponding atomic absorption profile is then given by

$$\alpha(\nu') = \int r(\nu',\nu)\,d\nu = \int_{-\infty}^{\infty} L_1(t-\nu')L_2(t-\nu_0)\,dt = L_{12}(\nu'-\nu_0) \qquad \text{(B.5.15)}$$

on account of (B.5.11,3), that is,

$$\alpha(\nu') = \frac{\delta_{12}/\pi}{(\nu'-\nu_0)^2+\delta_{12}^2} \,, \qquad \text{(B.5.16)}$$

as expected.

To understand the physics contained in (B.5.16) better, let us consider the special case of irradiation with monochromatic light in the far line wing where $|\nu'-\nu_0| \gg \delta_1, \delta_2$. Equations (B.5.16,14) then reduce to [B.13]

$$\alpha(\nu') \simeq \frac{\delta_{12}/\pi}{(\nu'-\nu_0)^2} \quad \text{and} \qquad \text{(B.5.17)}$$

$$r(\nu',\nu) \simeq \frac{4\delta_1^2\delta_2(2\delta_1+\delta_2)}{\pi^2} \frac{1}{(\nu'-\nu_0)^2} \frac{1}{[(\nu-\nu_0)^2+\delta_{12}^2][(\nu-\nu')^2+4\delta_1^2]}$$
$$+\frac{\delta_1\delta_2}{\pi^2} \frac{1}{[(\nu-\nu_0)^2+\delta_{12}^2][(\nu-\nu')^2+4\delta_1^2]}$$
$$+\frac{\delta_1\delta_2}{\pi^2} \frac{1}{(\nu'-\nu_0)^2} \frac{1}{(\nu-\nu')^2+4\delta_1^2}$$
$$+\frac{\delta_1^2}{\pi^2} \frac{1}{(\nu'-\nu_0)^2} \frac{1}{(\nu-\nu_0)^2+\delta_{12}^2} \,. \qquad \text{(B.5.18)}$$

The function $r(\nu',\nu)$, (B.5.18), has two strong maxima at $\nu=\nu'$ and $\nu=\nu_0$, respectively. If $\nu\simeq\nu'$, only the second and third term contribute, so that, setting $\nu=\nu'+\varepsilon$ with a small ε,

$$r(\nu',\nu'+\varepsilon)\simeq\frac{\delta_2/\pi}{(\nu'-\nu_0)^2}\,\frac{2\delta_1/\pi}{\varepsilon^2+4\delta_1^2}=\frac{\delta_2}{\delta_{12}}\,\alpha(\nu')\,\frac{2\delta_1/\pi}{\varepsilon^2+4\delta_1^2}\;. \tag{B.5.19a}$$

On the other hand, if $\nu\simeq\nu_0$, only the second and fourth term contribute, so that, setting $\nu=\nu_0+\varepsilon$ and using the fact that $(\nu-\nu')^2\simeq(\nu'-\nu_0)^2$,

$$r(\nu',\nu_0+\varepsilon)\simeq\frac{\delta_1/\pi}{(\nu'-\nu_0)^2}\,\frac{\delta_{12}/\pi}{\varepsilon^2+\delta_{12}^2}=\frac{\delta_1}{\delta_{12}}\,\alpha(\nu')\,\frac{\delta_{12}/\pi}{\varepsilon^2+\delta_{12}^2}\;. \tag{B.5.19b}$$

The total redistribution function is approximately equal to the sum of (B.5.19a, b). Hence, the conditional redistribution function, defined through (B.2.12), is in the limit $|\nu'-\nu_0|\gg\delta_1,\delta_2$ given by

$$p(\nu'\to\nu)=\frac{\delta_2}{\delta_{12}}\,\frac{2\delta_1/\pi}{(\nu-\nu')^2+4\delta_1^2}+\frac{\delta_1}{\delta_{12}}\,\frac{\delta_{12}/\pi}{(\nu-\nu_0)^2+\delta_{12}^2}\;. \tag{B.5.20}$$

We note that it is normalized, see (B.2.13), and thus provides a consistent approximation.

The redistribution function (B.5.20) shows at first sight unexpected behavior: the atoms re-emit not only around the incident frequency ν' (as expected for "quasi-coherent re-emission"), but also around the line center ν_0. To explain this behavior, one observes that the former spectral line is due to absorptions *starting* near the center of the lower level followed by quasi-

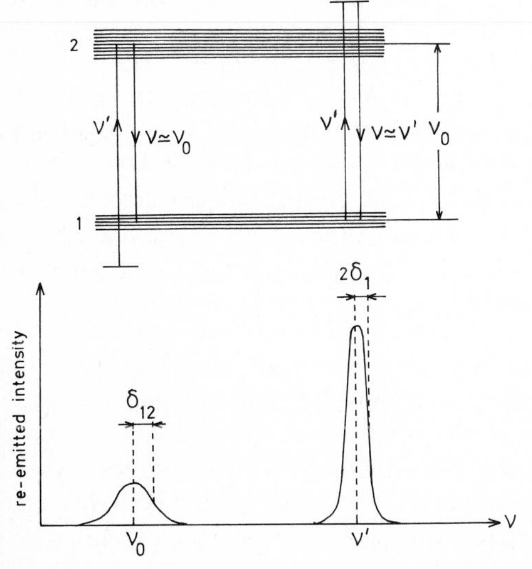

Fig. B.3. Re-emission of two spectral lines by atoms with two lifetime-broadened levels, after monochromatic excitation by photons of frequency ν' in the far line wing coherent re-emission, while the latter is due to absorptions *ending* near the center of the upper level followed by re-emission near the line center, Fig. B.3. For a sharp lower level ($\delta_1 = 0$, $\delta_{12} = \delta_2$), there is no re-emission at $\nu = \nu_0$, and one has true coherent re-emission $p(\nu' \to \nu) = \delta(\nu - \nu')$, in agreement with (B.2.29) with $\beta = 1$. This effect should be kept in mind when using the notion "quasi-coherent re-emission".

B.6 Absorption and Emission Profiles in Terms of Generalized Redistribution Functions

In Appendices B.2 – 4, determining atomic absorption and emission profiles in the framework of the Weisskopf-Woolley model provided some feeling for the physics involved in such profiles. However, while essentially correct for two-level atoms, the semiclassical Weisskopf-Woolley model cannot be expected to describe correctly all quantum mechanical phenomena of these processes, in particular in multilevel atoms. We therefore discuss now an alternative approach to atomic absorption and emission profiles through generalized atomic redistribution functions which, formally, is quite general and independent of any particular model [B.21]. It leads to closed (albeit in general implicit) expressions for the atomic absorption and emission profiles if stimulated emissions are neglected, and to an iterative approximation scheme if stimulated emissions are taken into account.

In order to have something specific at hand, we consider a three-level atom, Fig. B.4. Any frequency of a photon involved in the radiative transition $i \to j$ is denoted by ν_{ij}. Thus, if $i < j$, ν_{ij} is the frequency of an absorbed photon, whereas if $i > j$, it is the frequency of an emitted photon, Fig. B.4. The frequencies of the line centers are denoted by $\nu_{ij}^0 \equiv \nu_{ji}^0$. Again, as in the previous sections, we consider only the limiting case where the velocity of an atom is supposed to be constant while the processes under investigation, e.g., absorption followed by re-emission, take place (Appendices B.1, 4).

Let us now define *generalized atomic redistribution functions* r_{ij}, r_{ijk}, r_{ijkl}, ... of an atom, thought to be immersed in an isotropic white radiation field, for consecutive radiative transitions $i \to j$, $i \to j \to k$, $i \to j \to k \to l$, ... *that start from a naturally populated initial level i,* restricting ourselves for simplicity to the isotropic (angle-averaged) functions:

$r_{ij}(\nu_{ij})\, d\nu_{ij}$ is the probability that a photon with frequency in the range $(\nu_{ij}, d\nu_{ij})$ is involved (that is, absorbed if $i < j$, or emitted if $i > j$) in the radiative transition $i \to j$;

$r_{ijk}(\nu_{ij}, \nu_{jk})\, d\nu_{ij}\, d\nu_{jk}$ is the probability that two photons with frequencies in the ranges $(\nu_{ij}, d\nu_{ij})$ and $(\nu_{jk}, d\nu_{jk})$, respectively, are involved in the consecutive radiative transitions $i \to j \to k$;

$r_{ijkl}(\nu_{ij}, \nu_{jk}, \nu_{kl})\, d\nu_{ij}\, d\nu_{jk}\, d\nu_{kl}$ is the probability that three photons with frequencies in the ranges $(\nu_{ij}, d\nu_{ij})$, $(\nu_{jk}, d\nu_{jk})$, $(\nu_{kl}, d\nu_{kl})$, respectively, are involved in the consecutive radiative transitions $i \to j \to k \to l$, and so on.

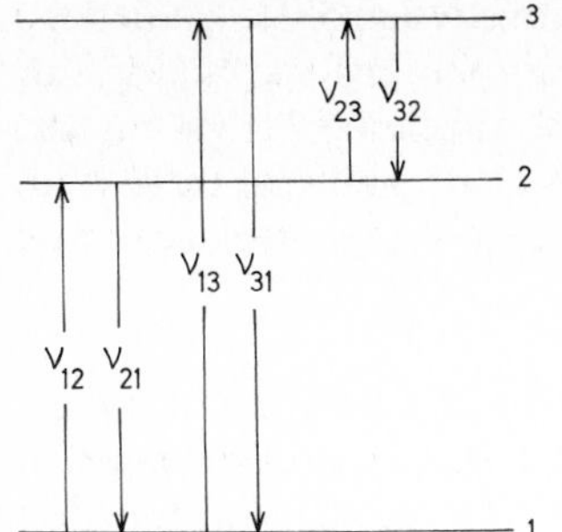

Fig. B.4. Energy level diagram of a three-level atom with the notation of absorbed and emitted frequencies used in the text

In these definitions it does not matter whether or not stimulated emissions are taken into account, apart from the minor effect of additional level broadening due to stimulated emissions.

By definition, the generalized redistribution functions are normalized,

$$\int r_{ij}(\nu_{ij})\,d\nu_{ij} = \iint r_{ijk}(\nu_{ij},\nu_{jk})\,d\nu_{ij}\,d\nu_{jk}$$

$$= \iiint r_{ijkl}(\nu_{ij},\nu_{jk},\nu_{kl})\,d\nu_{ij}\,d\nu_{jk}\,d\nu_{kl} = \cdots = 1 \ . \tag{B.6.1}$$

Moreover, among others, the following relations hold:

$$\int r_{ijk}(\nu_{ij},\nu_{jk})\,d\nu_{jk} = r_{ij}(\nu_{ij}) \ , \tag{B.6.2a}$$

$$\int r_{ijk}(\nu_{ij},\nu_{jk})\,d\nu_{ij} = r_{jk}(\nu_{jk}) \ , \tag{B.6.2b}$$

$$\int r_{ijkl}(\nu_{ij},\nu_{jk},\nu_{kl})\,d\nu_{kl} = r_{ijk}(\nu_{ij},\nu_{jk}) \ , \tag{B.6.2c}$$

$$\int r_{ijkl}(\nu_{ij},\nu_{jk},\nu_{kl})\,d\nu_{ij} = r_{jkl}(\nu_{jk},\nu_{kl}) \ , \tag{B.6.2d}$$

$$\int r_{ijkl}(\nu_{ij},\nu_{jk},\nu_{kl})\,d\nu_{jk} = r_{ij}(\nu_{ij})\,r_{kl}(\nu_{kl}) \ . \tag{B.6.2e}$$

Let us consider some examples. The function $r_{23}(\nu_{23})$ describes the absorption probability in the transition $2\to3$ from a naturally populated level 2, that is, it is the absorption profile for the transition $2\to3$ provided level 2 is *naturally populated.* Likewise, the function $r_{32}(\nu_{32})$ is the emission profile for the transition $3\to2$ provided level 3 is *naturally populated.* The function $r_{121}(\nu_{12},\nu_{21})$ is the ordinary redistribution function $r(\nu',\nu)$ for the resonance transition $1\leftrightarrow2$ as defined in Appendix B.2, with $\nu_{12}=\nu'$ and $\nu_{21}=\nu$. The function $r_{1232}(\nu_{12},\nu_{23},\nu_{32})$ describes two successive absorptions, $1\to2\to3$, followed by emission $3\to2$, etc.

Hubený [B.22] has shown explicitly that all generalized two-photon redistribution functions $r_{ijk}(\nu_{ij},\nu_{jk})$ are correctly described by the Weisskopf-Woolley model if the assumptions made by *Omont* et al. [B.14] are met (impact approximation, isolated lines). For generalized redistribution functions involving three photons, see [B.23, 24].

As a simple application, let us determine the emission profiles η_{32} and η_{21} of a three-level atom which is optically pumped by photons ν_{13}, that is, for which the only excitation mechanism is the absorption of light in the transition $1 \to 3$. Assuming that the ground state is sharp and that there is no collision broadening of the other levels, we consider the two special cases of excitation by white and monochromatic light, respectively.

Case I: Excitation by White Light. This leads to natural population of level 3, so that the radiative cascade $3 \to 2 \to 1$ is described by the generalized redistribution function r_{321}. According to Hubený's result just quoted, this function can be determined using the Weisskopf-Woolley model. Denoting normalized Lorentz distributions by $L_i(\mu_i)$, (B.5.1), therefore

$$r_{321}(\nu_{32}, \nu_{21}) = \int L_3(\mu_3) L_2(\mu_2) \delta(\mu_1) d\mu_1 \qquad \text{(B.6.3a)}$$

with, see (B.3.4),

$$\nu_{32} = \nu_{32}^0 + \mu_3 - \mu_2 \ , \quad \nu_{21} = \nu_{21}^0 + \mu_2 - \mu_1 \ , \qquad \text{(B.6.3b)}$$

so that

$$r_{321}(\nu_{32}, \nu_{21}) = L_3(\nu_{32} + \nu_{21} - \nu_{31}^0) L_2(\nu_{21} - \nu_{21}^0) \ . \qquad \text{(B.6.4)}$$

The function r_{321} corresponds to Heitler's function r_{IV}, (B.5.8). One obtains (B.6.4) from (B.5.8) by setting $\nu_0 = \nu_{21}^0$ and by substituting absorbing level 1 $\to$level 3, absorbed frequency $\nu' \to \nu_{31}^0 - \nu_{31}$, emitted frequency $\nu \to \nu_{21}$. Equation (B.6.4) shows that the two successively emitted photons are correlated: owing to energy conservation, $\nu_{32} + \nu_{21}$ must be equal to the frequency of the absorbed photon, so that the probability of emitting ν_{21} depends on the previously emitted frequency ν_{32}. On the other hand, the two emission profiles are given by

$$\eta_{32}^{\text{W}}(\nu_{32}) = \int r_{321}(\nu_{32}, \nu_{21}) d\nu_{21} = L_{32}(\nu_{32} - \nu_{32}^0) \ , \qquad \text{(B.6.5a)}$$

$$\eta_{21}^{\text{W}}(\nu_{21}) = \int r_{321}(\nu_{32}, \nu_{21}) d\nu_{32} = L_2(\nu_{21} - \nu_{21}^0) \ , \qquad \text{(B.6.5b)}$$

where (B.5.3) has been used in (B.6.5a).

Case II: Excitation by Monochromatic Light of Frequency ν_{31}^0. Here only the "sublevel $\mu_3 = 0$" of level 3 is populated, so that this case is formally described by setting $L_3(\mu_3) = \delta(\mu_3)$ in the preceding formulas. Hence, from (B.6.4)

$$r_{321}^{\delta}(\nu_{32}, \nu_{21}) = \delta(\nu_{32} + \nu_{21} - \nu_{31}^0) L_2(\nu_{21} - \nu_{21}^0) \ , \qquad \text{(B.6.6)}$$

which is not a generalized redistribution function of the three-level atom considered, but which corresponds to a generalized redistribution function of a fictitious three-level atom with a sharp level 3. Here the correlation between two successively emitted photons simply means $\nu_{32}+\nu_{21}=\nu_{31}^0$ (conservation of energy), and the emission profiles are now given by

$$\eta_{32}^{\delta}(\nu_{32})=\int r_{321}^{\delta}(\nu_{32},\nu_{21})\,d\nu_{21}=L_2(\nu_{32}-\nu_{32}^0)\ , \tag{B.6.7a}$$

$$\eta_{21}^{\delta}(\nu_{21})=\int r_{321}^{\delta}(\nu_{32},\nu_{21})\,d\nu_{32}=L_2(\nu_{21}-\nu_{21}^0)\ . \tag{B.6.7b}$$

Thus, in spite of the existence of correlations between two successively emitted photons, the emission profiles (B.6.5,7), being statistical averages, show the simple forms expected from a naive point of view. In particular,

$$\eta_{21}^{W}(\nu_{21})=\eta_{21}^{\delta}(\nu_{21})=L_2(\nu_{21}-\nu_{21}^0)$$

shows very clearly that spontaneous emissions 3→2 lead to natural population of level 2, quite independently of the particular "excitation state" of level 3 [which is given by $v_3(\mu_3)=w_3(\mu_3)=L_3(\mu_3)$ in Case I, and by $v_3(\mu_3)=\delta(\mu_3)$ in Case II, using the terminology of the Weisskopf-Woolley model]. This fact is, of course, already contained in (B.6.3a).

We now turn to our main topic, the determination of the atomic absorption and emission profiles of a three-level atom in terms of generalized redistribution functions. We first consider a three-level atom *in the absence of stimulated emissions.*

Recalling that by definition the initial level of a generalized redistribution function is naturally populated, to determine a particular profile one must first look for all independent sequences of consecutive radiative transitions that start from a naturally populated level and terminate with the particular radiative transition under investigation. Then one must add the contributions of all of these sequences, weighted with their respective probabilities of occurrence.

Consider a particular atomic level i. We denote by

$$\rightarrow i^*$$

the ensemble of all processes that create atoms with a naturally populated level i^* (the asterisk standing for natural population). Explicitly, these processes are excitation collisions, de-excitation collisions, and spontaneous emissions creating atoms in level i. Note that therefore, in the absence of stimulated emissions, the ground state of an atom is always naturally populated, (B.3.14). By contrast, as they may give rise to deviations from natural population of level i, absorptions from lower levels j are denoted

$$j\Rightarrow i\quad (j<i)\ .$$

The probabilities of such processes leading to level i are

$$\text{prob}\{\rightarrow i^*\}\ ,\quad \text{prob}\{j\Rightarrow i\}\ ,\ \ldots\ .$$

Using these definitions, the various absorption and emission profiles $\alpha_{ij}(\nu_{ij})$ and $\eta_{ij}(\nu_{ij})$ of a three-level atom can at once be written down in terms of generalized redistribution functions. Below, 1* reminds the reader that the ground state of the atom is naturally populated. Then the following relations, interpreted presently, arise:

$$\alpha_{12}(\nu_{12}) = r_{12}(\nu_{12}) \ , \tag{B.6.8}$$

$$\alpha_{13}(\nu_{13}) = r_{13}(\nu_{13}) \ , \tag{B.6.9}$$

$$\alpha_{23}(\nu_{23}) = \text{prob}\{\rightarrow 2^*\} r_{23}(\nu_{23}) + \text{prob}\{1^* \Rightarrow 2\} j_{123}(\nu_{23}) \ , \tag{B.6.10}$$

$$\eta_{21}(\nu_{21}) = \text{prob}\{\rightarrow 2^*\} r_{21}(\nu_{21}) + \text{prob}\{1^* \Rightarrow 2\} j_{121}(\nu_{21}) \ , \tag{B.6.11}$$

$$\begin{aligned}\eta_{31}(\nu_{31}) = {} & \text{prob}\{\rightarrow 3^*\} r_{31}(\nu_{31}) + \text{prob}\{1^* \Rightarrow 3\} j_{131}(\nu_{31}) \\ & + \text{prob}\{\rightarrow 2^* \Rightarrow 3\} j_{231}(\nu_{31}) + \text{prob}\{1^* \Rightarrow 2 \Rightarrow 3\} j_{1231}(\nu_{31}) \ ,\end{aligned} \tag{B.6.12}$$

$$\begin{aligned}\eta_{32}(\nu_{32}) = {} & \text{prob}\{\rightarrow 3^*\} r_{32}(\nu_{32}) + \text{prob}\{1^* \Rightarrow 3\} j_{132}(\nu_{32}) \\ & + \text{prob}\{\rightarrow 2^* \Rightarrow 3\} j_{232}(\nu_{32}) + \text{prob}\{1^* \Rightarrow 2 \Rightarrow 3\} j_{1232}(\nu_{32}) \ .\end{aligned} \tag{B.6.13}$$

Here we have introduced the quantities

$$j_{ijk}(\nu_{jk}) = \frac{\int \bar{I}(\nu_{ij}) r_{ijk}(\nu_{ij}, \nu_{jk}) d\nu_{ij}}{\int \bar{I}(\nu_{ij}) r_{ij}(\nu_{ij}) d\nu_{ij}} \ , \tag{B.6.14a}$$

$$j_{ijkl}(\nu_{kl}) = \frac{\iint \bar{I}(\nu_{ij}) \bar{I}(\nu_{jk}) r_{ijkl}(\nu_{ij}, \nu_{jk}, \nu_{kl}) d\nu_{ij} d\nu_{jk}}{\iint \bar{I}(\nu_{ij}) \bar{I}(\nu_{jk}) r_{ijk}(\nu_{ij}, \nu_{jk}) d\nu_{ij} d\nu_{jk}} \ , \tag{B.6.14b}$$

with $\bar{I}(\nu)$ being the mean radiation intensity in the atomic rest frame, and where the denominators have been added to obtain normalized quantities, see (B.6.2a, c),

$$\int j_{ijk}(\nu_{jk}) d\nu_{jk} = \int j_{ijkl}(\nu_{kl}) d\nu_{kl} = 1 \ . \tag{B.6.15}$$

Let us now discuss the absorption and emission profiles (B.6.8 – 13). Equations (B.6.8, 9) follow immediately from the definition of the function r_{ij} together with the fact that level 1 is naturally populated (formally: $\text{prob}\{\rightarrow 1^*\} = 1$). Next, the absorption profile α_{23}, (B.6.10), is composed of two parts, the first due to atoms created with a naturally populated level 2*, and the second due to atoms in level 2 created by a previous absorption from the ground state. It is these latter atoms which, when absorbing a photon ν_{23}, perform two consecutive absorptions $1^* \Rightarrow 2 \Rightarrow 3$, and which are thus responsible for the fact that the absorption profile α_{23} in general differs from the function r_{23}. The emission profile η_{21}, (B.6.11), is interpreted similarly. Note that the j_{121} term describes the usual redistribution of photons in the resonance transition $1 \leftrightarrow 2$ in a radiation field $\bar{I}(\nu)$, where the (ordinary) redistribution function r_{121} occurring in j_{121} is generally composed of two parts corresponding to coherent (or quasi-coherent) and completely redistributed re-emission, respectively. The probability that an atom in level 2 *emits* a photon ν_{21} according to a naturally populated level 2* is greater than the probability prob$\{\rightarrow 2^*\}$ to *create* an atom in a naturally populated level 2*, owing to the corresponding contribution of completely redistributed re-emission of the j_{121} term due to elastic broadening collisions. This remark holds, mutatis mutandis, for the other profiles too, and explains the arrow in the notation prob$\{\rightarrow i^*\}$. Finally, the emission profiles η_{31} and η_{32}, (B.6.12, 13), are readily interpreted along the same lines, taking into account all independent sequences of radiative transitions that start from a naturally populated initial level.

It remains to calculate the various probabilities occurring in Eqs. (B.6.8 – 13). We use our standard notation with S_{ji} being the rate per atom of inelastic collisions $j \rightarrow i$, A_{ji} the Einstein coefficient for spontaneous emissions $j \rightarrow i$, and $B_{ji} I_{ji}$ the rate per atom of absorptions $j \rightarrow i$, where B_{ji} is the Einstein coefficient for absorption, and, see (B.2.20),

$$I_{ji} = \int \bar{I}(\nu)\, \alpha_{ji}(\nu)\, d\nu \quad (j<i) \ , \tag{B.6.16a}$$

formed with the *true* absorption profile $\alpha_{ji}(\nu)$. The corresponding quantity formed with the purely atomic quantity $r_{ji}(\nu)$ is

$$I_{ji}^* = \int \bar{I}(\nu)\, r_{ji}(\nu)\, d\nu \quad (j<i) \ . \tag{B.6.16b}$$

Further let χ_i be the number density of atoms in level i under consideration. As mentioned above, the velocity $\boldsymbol{v}$ of an excited atom in the laboratory frame is supposed to be constant. Hence, if we consider a group of atoms with velocities in the range $(\boldsymbol{v}, d^3 v)$ in the laboratory frame, then

$$\chi_i = F_i(\boldsymbol{v})\, d^3 v \ , \tag{B.6.17}$$

where $F_i(\boldsymbol{v})$ is the distribution function of atoms in level i.

The probabilities occurring in (B.6.8 – 13) can now be written as

$$\text{prob}\{\rightarrow 2^*\} = \frac{\chi_1 S_{12} + \chi_3 (A_{32} + S_{32})}{\chi_1 (B_{12} I_{12}^* + S_{12}) + \chi_3 (A_{32} + S_{32})} \ , \tag{B.6.18}$$

$$\text{prob}\{1^* \Rightarrow 2\} = \frac{\chi_1 B_{12} I_{12}^*}{\chi_1 (B_{12} I_{12}^* + S_{12}) + \chi_3 (A_{32} + S_{32})} \ , \tag{B.6.19}$$

$$\text{prob}\{\rightarrow 3^*\} = \frac{\chi_1 S_{13} + \chi_2 S_{23}}{\chi_1 (B_{13} I_{13}^* + S_{13}) + \chi_2 (B_{23} I_{23} + S_{23})} \ , \tag{B.6.20}$$

$$\text{prob}\{1^* \Rightarrow 3\} = \frac{\chi_1 B_{13} I_{13}^*}{\chi_1 (B_{13} I_{13}^* + S_{13}) + \chi_2 (B_{23} I_{23} + S_{23})} \ , \tag{B.6.21}$$

$$\mathrm{prob}\{\rightarrow 2^* \Rightarrow 3\} = \frac{\chi_2 B_{23} I_{23}}{\chi_1 (B_{13} I_{13}^* + S_{13}) + \chi_2 (B_{23} I_{23} + S_{23})} \cdot \frac{\chi_1 S_{12} + \chi_3 (A_{32} + S_{32})}{\chi_1 (B_{12} I_{12}^* + S_{12}) + \chi_3 (A_{32} + S_{32})} , \tag{B.6.22}$$

$$\mathrm{prob}\{1^* \Rightarrow 2 \Rightarrow 3\} = \frac{\chi_2 B_{23} I_{23}}{\chi_1 (B_{13} I_{13}^* + S_{13}) + \chi_2 (B_{23} I_{23} + S_{23})} \cdot \frac{\chi_1 B_{12} I_{12}^*}{\chi_1 (B_{12} I_{12}^* + S_{12}) + \chi_3 (A_{32} + S_{32})} . \tag{B.6.23}$$

Here (B.6.22, 23) have been obtained by observing that

$$\mathrm{prob}\{\rightarrow 2^* \Rightarrow 3\} + \mathrm{prob}\{1^* \Rightarrow 2 \Rightarrow 3\} = \frac{\chi_2 B_{23} I_{23}}{\chi_1 (B_{13} I_{13}^* + S_{13}) + \chi_2 (B_{23} I_{23} + S_{23})} \tag{B.6.24}$$

and taking (B.6.18, 19) into account. Furthermore, we have used the fact that

$$I_{12} = I_{12}^* , \quad I_{13} = I_{13}^* \tag{B.6.25}$$

according to (B.6.8, 9, 16).

One readily verifies that

$$\mathrm{prob}\{\rightarrow 2^*\} + \mathrm{prob}\{1^* \Rightarrow 2\} = 1 , \tag{B.6.26a}$$

$$\mathrm{prob}\{\rightarrow 3^*\} + \mathrm{prob}\{1^* \Rightarrow 3\} + \mathrm{prob}\{\rightarrow 2^* \Rightarrow 3\} + \mathrm{prob}\{1^* \Rightarrow 2 \Rightarrow 3\} = 1 . \tag{B.6.26b}$$

On account of (B.6.1, 15, 26), all absorption and emission profiles (B.6.8 – 13) are normalized,

$$\int \alpha_{ij}(\nu_{ij})\, d\nu_{ij} = \int \eta_{ij}(\nu_{ij})\, d\nu_{ij} = 1 . \tag{B.6.27}$$

The profiles α_{23}, η_{21}, η_{31}, and η_{32} depend on the radiation intensity $\bar{I}(\nu)$ through the $j_{ijk\ldots}$ terms, and through the probabilities prob$\{\ldots\}$ which contain $B_{ij} I_{ij}$ terms. Furthermore, they also depend on the ratios $\chi_3/\chi_1 = F_3(\boldsymbol{v})/F_1(\boldsymbol{v})$ and $\chi_2/\chi_1 = F_2(\boldsymbol{v})/F_1(\boldsymbol{v})$ of the atomic distribution functions which enter the probabilities (B.6.18 – 23). Moreover, (B.6.10 – 13) are only implicit equations for determining these profiles, because the absorption profile α_{23} appears on their right-hand sides via the $B_{23} I_{23}$ terms in the various probabilities. However, this implicit dependence is very weak, and in most cases $I_{23} \simeq I_{23}^*$, thus making (B.6.10 – 13) explicit expressions in terms of the distribution functions $\bar{I}(\nu)$ and $F_i(\boldsymbol{v})$. (Compare the analogous discussions in Appendices B.2 – 4.)

By suppressing the atomic level 3, one obtains the relations for a *two-level atom.* Equations (B.6.8, 11) remain unchanged,

$$\alpha_{12}(\nu_{12}) = r_{12}(\nu_{12}) \ , \tag{B.6.28}$$

$$\eta_{21}(\nu_{21}) = \text{prob}\{\to 2^*\} r_{21}(\nu_{21}) + \text{prob}\{1^* \Rightarrow 2\} j_{121}(\nu_{21}) \ , \tag{B.6.29}$$

where, see (B.6.14a, 16b),

$$j_{121}(\nu_{21}) = \frac{1}{I^*_{12}} \int \bar{I}(\nu_{12}) r_{121}(\nu_{12}, \nu_{21}) d\nu_{12} \ . \tag{B.6.30}$$

Equations (B.6.18, 19) now reduce to

$$\text{prob}\{\to 2^*\} = \frac{S_{12}}{B_{12}I^*_{12} + S_{12}} \ , \quad \text{prob}\{1^* \Rightarrow 2\} = \frac{B_{12}I^*_{12}}{B_{12}I^*_{12} + S_{12}} \ , \tag{B.6.31}$$

so that

$$\eta_{21}(\nu_{21}) = \frac{1}{B_{12}I^*_{12} + S_{12}} [S_{12} r_{21}(\nu_{21}) + B_{12} \int \bar{I}(\nu_{12}) r_{121}(\nu_{12}, \nu_{21}) d\nu_{12}] \ , \tag{B.6.32}$$

which is identical with (B.2.26) on account of (B.6.28), using the evident relation

$$r_{21}(\nu_{21}) = r_{12}(\nu_{21}) \ . \tag{B.6.33}$$

Equation (B.6.33) simply expresses the fact that for natural excitation, the atomic emission profile is equal to the atomic absorption profile (Appendix B.1).

Let us now consider a three-level atom *taking stimulated emissions into account.* As the discussion of the two-level atom in the framework of the Weisskopf-Woolley model in Appendix B.3 has already shown, the problem becomes much more complicated by the presence of stimulated emissions. Indeed, it turns out that in this case one must be content with a cumbersome iteration procedure to obtain approximations for the atomic absorption and emission profiles in terms of generalized redistribution functions.

A first, rough approximation consists in simply keeping the absorption and emission profiles as given by (B.6.8 – 13), merely changing the probabilities (B.6.18 – 23) by substituting everywhere

$$A_{ji} \to A_{ji} + B_{ji} I_{ji} \quad (j > i) \ . \tag{B.6.34}$$

Here B_{ji} is the Einstein coefficient for stimulated emission, and, see (B.3.3b),

$$I_{ji} = \int \bar{I}(\nu) \eta_{ji}(\nu) d\nu \quad (j > i) \tag{B.6.35a}$$

formed with the *true* emission profile $\eta_{ji}(\nu)$, while, in analogy to (B.6.16b),

$$I_{ji}^* = \int \bar{I}(\nu)\, r_{ji}(\nu)\, d\nu \quad (j>i) \tag{B.6.35b}$$

denotes the corresponding quantity formed with the generalized redistribution function $r_{ji}(\nu)$. This approximation accounts for the alterations of the emission rates owing to the presence of stimulated emissions, but disregarding possible additional deviations from natural population of the atomic levels due to stimulated emissions.

To see how a better approximation can be obtained, let us consider as an example the emission profile η_{21} of a three-level atom in greater detail. For simplicity, suppose the ground state of the atom to be naturally populated. (In particular, this assumption is valid when ground-state broadening can be neglected altogether.)

Since now absorptions *and* emissions can give rise to deviations from natural population, it is useful to change the notation. We now denote radiative transitions (through absorptions or emissions) from level j to level i by

$$j \Rightarrow i \quad (j \lessgtr i) \ ,$$

and collisional transitions (through excitation or de-excitation collisions) from level j to level i by

$$j \rightarrow i \quad (j \lessgtr i) \ .$$

The corresponding probabilities are denoted by

$$\mathrm{p}\{j \Rightarrow i\} \ , \quad \mathrm{p}\{j \rightarrow i\} \ ,$$

normalized such that

$$\sum_j (\mathrm{p}\{j \Rightarrow i\} + \mathrm{p}\{j \rightarrow i\}) = 1 \ .$$

Explicitly, for levels 2 and 3 of a three-level atom

$$\mathrm{p}\{1 \Rightarrow 2\} = \frac{\chi_1 B_{12} I_{12}^*}{Z_2} \ , \quad \mathrm{p}\{1 \rightarrow 2\} = \frac{\chi_1 S_{12}}{Z_2} \ ,$$

$$\mathrm{p}\{3 \Rightarrow 2\} = \frac{\chi_3 (A_{32} + B_{32} I_{32})}{Z_2} \ , \quad \mathrm{p}\{3 \rightarrow 2\} = \frac{\chi_3 S_{32}}{Z_2} \ , \tag{B.6.36a}$$

$$Z_2 \equiv \chi_1 (B_{12} I_{12}^* + S_{12}) + \chi_3 (A_{32} + B_{32} I_{32} + S_{32}) \ ;$$

$$\mathrm{p}\{1 \Rightarrow 3\} = \frac{\chi_1 B_{13} I_{13}^*}{Z_3} \ , \quad \mathrm{p}\{1 \rightarrow 3\} = \frac{\chi_1 S_{13}}{Z_3} \ ,$$

$$\mathrm{p}\{2 \Rightarrow 3\} = \frac{\chi_2 B_{23} I_{23}}{Z_3} \ , \quad \mathrm{p}\{2 \rightarrow 3\} = \frac{\chi_2 S_{23}}{Z_3} \ , \tag{B.6.36b}$$

$$Z_3 \equiv \chi_1 (B_{13} I_{13}^* + S_{13}) + \chi_2 (B_{23} I_{23} + S_{23}) \ ,$$

where $\chi_i = F_i(\boldsymbol{v})d^3v$, see (B.6.17). Since the ground state 1 is naturally populated, (B.6.8,9) are still valid, so that in (B.6.36) $I_{12} = I_{12}^*$, $I_{13} = I_{13}^*$.

To obtain the emission profile η_{21}, one must consider all sequences of atomic transitions ending with

$$\ldots 2 \Rightarrow 1 \ .$$

There are four possibilities:

$$\ldots 1 \rightarrow 2^* \Rightarrow 1 \ , \tag{a}$$

$$\ldots 3 \rightarrow 2^* \Rightarrow 1 \ , \tag{b}$$

$$\ldots 1^* \Rightarrow 2 \Rightarrow 1 \ , \tag{c}$$

$$\ldots 3 \Rightarrow 2 \Rightarrow 1 \ , \tag{d}$$

where at relevant places in the reaction chains we have indicated natural population of a level by an asterisk. Hence,

$$\eta_{21}(\nu_{21}) = \mathrm{p}\{1 \rightarrow 2\} r_{21}(\nu_{21}) + \mathrm{p}\{3 \rightarrow 2\} r_{21}(\nu_{21}) + \mathrm{p}\{1 \Rightarrow 2\} j_{121}(\nu_{21}) + \mathrm{p}\{3 \Rightarrow 2\} X_{321}(\nu_{21}) \ . \tag{B.6.37}$$

Here the first three terms on the right correspond to the sequences (a–c) above, and are readily interpreted, see (B.6.11). On the other hand, the normalized quantity X_{321},

$$\int X_{321}(\nu_{21})\, d\nu_{21} = 1 \ , \tag{B.6.38}$$

describes the emission $2 \Rightarrow 1$ of atoms whose level 2 is populated by radiative transitions $3 \Rightarrow 2$. To determine X_{321}, one must consider the various possibilities of creating atoms in level 3. The reaction chain (d) above thus splits up into four sequences:

$$\ldots 1 \rightarrow 3^* \Rightarrow 2 \Rightarrow 1 \ , \tag{d1}$$

$$\ldots 2 \rightarrow 3^* \Rightarrow 2 \Rightarrow 1 \ , \tag{d2}$$

$$\ldots 1^* \Rightarrow 3 \Rightarrow 2 \Rightarrow 1 \ , \tag{d3}$$

$$\ldots 2 \Rightarrow 3 \Rightarrow 2 \Rightarrow 1 \ . \tag{d4}$$

Accordingly,

$$X_{321}(\nu_{21}) = (\mathrm{p}\{1 \rightarrow 3\} + \mathrm{p}\{2 \rightarrow 3\}) \frac{\int r_{321}(\nu_{32}, \nu_{21})[1 + \gamma_{32}\bar{I}(\nu_{32})]\, d\nu_{32}}{\iint r_{321}(\nu_{32}, \nu_{21})[1 + \gamma_{32}\bar{I}(\nu_{32})]\, d\nu_{32} d\nu_{21}} + \mathrm{p}\{1 \Rightarrow 3\} \frac{\iint \bar{I}(\nu_{13}) r_{1321}(\nu_{13}, \nu_{32}, \nu_{21})[1 + \gamma_{32}\bar{I}(\nu_{32})]\, d\nu_{13} d\nu_{32}}{\iiint \bar{I}(\nu_{13}) r_{1321}(\nu_{13}, \nu_{32}, \nu_{21})[1 + \gamma_{32}\bar{I}(\nu_{32})]\, d\nu_{13} d\nu_{32} d\nu_{21}} + \mathrm{p}\{2 \Rightarrow 3\} Y_{2321}(\nu_{21}) \ . \tag{B.6.39}$$

Here stimulated emissions in the transition $3 \Rightarrow 2$ have been taken into account by the terms containing

$$\gamma_{32} \equiv \frac{B_{32}}{A_{32}} ,$$

i.e., the ratio of the two Einstein coefficients, (B.1.2). The denominators in (B.6.39) ensure the correct normalizations, and the still unknown quantity $Y_{2321}(\nu_{21})$ must be normalized too, (B.6.38).

The procedure is now clear. To determine Y_{2321}, one encounters in the next round another unknown quantity $Z_{32321}(\nu_{21})$, and so on, owing to the infinite sequence of transitions

$$\ldots \Rightarrow 2 \Rightarrow 3 \Rightarrow 2 \Rightarrow 3 \Rightarrow \ldots .$$

The corresponding infinite chain of equations must therefore be cut somewhere. For example, let us make the (very mild) approximation that for determining η_{21}, the contribution of the sequence

$$\ldots 2 \Rightarrow 3 \Rightarrow 2 \Rightarrow 1$$

may be replaced by that of the sequence

$$\ldots 2 \Rightarrow 3^* \Rightarrow 2 \Rightarrow 1$$

with a naturally populated level 3*. Then the Y_{2321} term behaves like the first term on the right-hand side of (B.6.39). Grouping these terms together, integrating with the aid of (B.6.2), and taking definitions (B.6.16b, 35b) into account leads to

$$\begin{aligned} X_{321}(\nu_{21}) = {} & (\mathrm{p}\{1 \to 3\} + \mathrm{p}\{2 \to 3\} + \mathrm{p}\{2 \Rightarrow 3\}) \\ & \cdot \frac{r_{21}(\nu_{21}) + \gamma_{32} \int \bar{I}(\nu_{32}) r_{321}(\nu_{32}, \nu_{21}) d\nu_{32}}{1 + \gamma_{32} I_{32}^*} \\ & + \mathrm{p}\{1 \Rightarrow 3\} \frac{I_{13}^* r_{21}(\nu_{21}) + \gamma_{32} \iint \bar{I}(\nu_{13}) \bar{I}(\nu_{32}) r_{1321}(\nu_{13}, \nu_{32}, \nu_{21}) d\nu_{13} d\nu_{32}}{I_{13}^* + \gamma_{32} \iint \bar{I}(\nu_{13}) \bar{I}(\nu_{32}) r_{132}(\nu_{13}, \nu_{32}) d\nu_{13} d\nu_{32}} . \end{aligned} \tag{B.6.40}$$

In the considered approximation, (B.6.37, 40) together determine the atomic emission profile η_{21} in the presence of stimulated emissions. If stimulated emissions are neglected, then $\gamma_{32} = 0$, and (B.6.40) reduces to

$$X_{321}(\nu_{21}) = r_{21}(\nu_{21}) ,$$

so that (B.6.37) reduces to (B.6.11), as it should.

Note that determining the emission profile η_{21} would have been even more tedious if we had allowed for possible deviations from natural population of the atomic ground level. Let us further recall that the broadening of all atomic levels depends in principle on the radiation field locally present through the contributions of absorptions and stimulated emissions to the lifetime broadening of the levels.

Before closing this section, let us indicate the following symmetry relations for generalized redistribution functions:

$$r_{ij}(\nu) = r_{ji}(\nu) \ , \tag{B.6.41a}$$

$$r_{ijk}(\nu', \nu) = r_{kji}(\nu, \nu') \ , \tag{B.6.41b}$$

$$r_{ijkl}(\nu'', \nu', \nu) = r_{lkji}(\nu, \nu', \nu'') \ , \tag{B.6.41c}$$

and so on. They follow at once from the definition of these functions, together with the symmetry of radiative processes with respect to time reversal. Equation (B.6.41a) has already been encountered in (B.6.33). On the other hand, (B.6.41b) with $i = k = 1$, $j = 2$ is the familiar reciprocity relation (B.2.16a) for the ordinary redistribution function $r(\nu', \nu)$. Furthermore, it follows from (B.6.41a) that

$$I_{ij}^* = I_{ji}^* \tag{B.6.42}$$

on account of the definitions (B.6.16b, 35b).

Up to now it has been tacitly assumed that different atomic transitions correspond to different spectral lines. In the case of substantial overlap of spectral lines, the definition of generalized atomic redistribution functions has to be changed. As a specific example, consider a three-level atom with $E_2 - E_1 = E_3 - E_2$, so that the spectral lines at ν_{21}^0 and ν_{32}^0 coincide. Then the function r_{1231}, say, must be defined such that

$$r_{1231}([\nu', \nu''], \nu_{31})$$

is the probability density for absorbing photons ν' and ν'' in $1^* \Rightarrow 2 \Rightarrow 3$ *irrespective of their relative time ordering* (that is, irrespective of whether ν' is absorbed in $1 \rightarrow 2$ and ν'' in $2 \rightarrow 3$, or vice versa), and that a photon ν_{31} is emitted in $3 \rightarrow 1$.

The case just considered provides a lucid example of the fact that quantum mechanical interference is not treated correctly by the Weisskopf-Woolley model (even in the collisionless case where otherwise this model is exact). Indeed, even assuming purely radiation-broadened levels 2 and 3, quantum mechanics yields

$$r_{1231}([\nu', \nu''], \nu_{31}) \propto |A_1 + A_2|^2$$

where $A_1(A_2)$ denotes the corresponding probability amplitude when ν' is absorbed first and ν'' second (ν'' first and ν' second), whereas in the Weisskopf-Woolley model, the probabilities of these processes are simply added, so that

$$r_{1231}^{\mathrm{WW}}([\nu', \nu''], \nu_{31}) \propto |A_1|^2 + |A_2|^2 \, .$$

C. The Boltzmann Equation

This appendix discusses some general features of the Boltzmann equation [C.1 – 4]. In particular, it will be shown that the Boltzmann collision term has the three properties of elastic collision terms stated in Sect. 2.2.3.

The Boltzmann equation is a kinetic equation which applies to dilute gases of neutral particles. For simplicity, we consider here only a one-component gas. The Boltzmann collision term describes the effect of elastic collisions among the particles, which are assumed to be spherically symmetric. The internal structure of the particles, and hence the possibility of inelastic collisions, are ignored in the following.

According to (2.2.1), the Boltzmann collision term is given by

$$\left(\frac{\delta F}{\delta t}\right)_{\mathrm{B}} = \iint (F_1 F_1' - FF')\, w q(w, \vartheta)\, d\Omega\, d^3v' \ . \tag{C.1}$$

Here all distribution functions refer to the same space-time point $(\boldsymbol{r}, t)$, and $F = F(\boldsymbol{v}) \equiv F(\boldsymbol{r}, \boldsymbol{v}, t)$, $F' = F(\boldsymbol{v}')$, $F_1 = F(\boldsymbol{v}_1)$, and $F_1' = F(\boldsymbol{v}_1')$. Furthermore, $q(w, \vartheta)$ is the differential cross section for the elastic collision $(\boldsymbol{v}, \boldsymbol{v}' \rightarrow \boldsymbol{v}_1, \boldsymbol{v}_1')$, $w = |\boldsymbol{w}|$ is the modulus of the relative velocity of the collision partners before the collision $\boldsymbol{w} = \boldsymbol{v} - \boldsymbol{v}'$, ϑ is the angle between $\boldsymbol{w}$ and the relative velocity after the collision $\boldsymbol{w}_1 = \boldsymbol{v}_1 - \boldsymbol{v}_1'$, and finally $d\Omega = \sin\vartheta\, d\vartheta\, d\varphi$ is the element of solid angle.

To interpret the collision term (C.1), let us first look at the elastic collision $(\boldsymbol{v}, \boldsymbol{v}' \rightarrow \boldsymbol{v}_1, \boldsymbol{v}_1')$ where $\boldsymbol{v}, \boldsymbol{v}'$ are the particle velocities before, and $\boldsymbol{v}_1, \boldsymbol{v}_1'$ those after the collision. Conservation of momentum and energy requires

$$\boldsymbol{v} + \boldsymbol{v}' = \boldsymbol{v}_1 + \boldsymbol{v}_1' \ , \tag{C.2}$$

$$v^2 + v'^2 = v_1^2 + v_1'^2 \ . \tag{C.3}$$

These are four equations for the six components of $\boldsymbol{v}_1$ and $\boldsymbol{v}_1'$, thus leaving two components still unspecified. Introducing the center-of-mass velocities

$$\boldsymbol{V} = \tfrac{1}{2}(\boldsymbol{v} + \boldsymbol{v}') \ , \quad \boldsymbol{V}_1 = \tfrac{1}{2}(\boldsymbol{v}_1 + \boldsymbol{v}_1') \ , \tag{C.4}$$

and the relative velocities

$$\boldsymbol{w} = \boldsymbol{v} - \boldsymbol{v}' \ , \quad \boldsymbol{w}_1 = \boldsymbol{v}_1 - \boldsymbol{v}_1' \ , \tag{C.5}$$

it follows from (C.2, 3) that

$$\boldsymbol{V} = \boldsymbol{V}_1 \ , \tag{C.6}$$

$$|\boldsymbol{w}| = |\boldsymbol{w}_1| \ , \tag{C.7}$$

i.e., the center-of-mass velocity and the modulus of the relative velocity do not change. Hence the only effect of an elastic collision consists in turning the vector of the relative velocity in a new direction, leaving its length unaffected,

$$\boldsymbol{w}_1 = w\boldsymbol{s} \ , \tag{C.8}$$

where $w = |\boldsymbol{w}|$, $\boldsymbol{s}$ being a unit vector ($|\boldsymbol{s}| = 1$). An elastic collision is thus uniquely specified by the two independent coordinates of this unit vector $\boldsymbol{s}$, for instance by its polar and azimuthal angles ϑ and φ relative to the vector $\boldsymbol{w}$ as polar axis.

For a given collision (that is, for given values of ϑ and φ),

$$d^3v\, d^3v' = d^3v_1\, d^3v_1' \ . \tag{C.9}$$

This can be shown in the following way. Computing the Jacobians from (C.4, 5), one finds

$$\left|\frac{\partial(\boldsymbol{V},\boldsymbol{w})}{\partial(\boldsymbol{v},\boldsymbol{v}')}\right| = \left|\frac{\partial(\boldsymbol{V}_1,\boldsymbol{w}_1)}{\partial(\boldsymbol{v}_1,\boldsymbol{v}_1')}\right| = 1$$

so that

$$d^3v\, d^3v' = d^3V\, d^3w \ , \quad d^3v_1\, d^3v_1' = d^3V_1\, d^3w_1 \ . \tag{C.10}$$

Now, from (C.6),

$$d^3V = d^3V_1 \ . \tag{C.11}$$

On the other hand, also

$$d^3w = d^3w_1 \ , \tag{C.12}$$

which can be seen most easily by introducing spherical coordinates with respect to an arbitrary, fixed polar axis, writing $d^3w = w^2 dw\, d\omega$ and $d^3w_1 = w_1^2 dw_1\, d\omega_1$. [The elements of solid angle $d\omega$ and $d\omega_1$ should not be confused with $d\Omega = \sin\vartheta\, d\vartheta\, d\varphi$ used in (C.1).] Now, $w = w_1$ and $dw = dw_1$ according to (C.7), and $d\omega = d\omega_1$ because the relative orientation of the vectors $\boldsymbol{w}$ and $\boldsymbol{w}_1$ is fixed as ϑ and φ are fixed. This proves (C.9).

Finally, the cross sections for the collision ($\boldsymbol{v}, \boldsymbol{v}' \to \boldsymbol{v}_1, \boldsymbol{v}_1'$) and the inverse collision ($\boldsymbol{v}_1, \boldsymbol{v}_1' \to \boldsymbol{v}, \boldsymbol{v}'$) are equal,

$$q(w,\vartheta)\, d\Omega = q_1(w_1,\vartheta_1)\, d\Omega_1 \ , \tag{C.13}$$

or, recalling that $w_1 = w$ and $\vartheta_1 = \vartheta$, and hence $d\Omega_1 = d\Omega$,

$$q(w,\vartheta) = q_1(w_1,\vartheta_1) = q_1(w,\vartheta) \ , \tag{C.14}$$

which can be written more explicitly as

$$q(\boldsymbol{v},\boldsymbol{v}';\boldsymbol{v}_1,\boldsymbol{v}_1') = q(\boldsymbol{v}_1,\boldsymbol{v}_1';\boldsymbol{v},\boldsymbol{v}') \ . \tag{C.15}$$

Indeed, this reciprocity relation follows from the invariance of the particle interaction with respect to space reflection P and time reversal T. For spherically symmetric particles, invariance under P yields, see (1.5.9),

$$q(\boldsymbol{v},\boldsymbol{v}';\boldsymbol{v}_1,\boldsymbol{v}_1') = q(-\boldsymbol{v},-\boldsymbol{v}';-\boldsymbol{v}_1,-\boldsymbol{v}_1') \ , \tag{C.16}$$

and invariance under T, see (1.5.10),

$$q(\boldsymbol{v},\boldsymbol{v}';\boldsymbol{v}_1,\boldsymbol{v}_1') = q(-\boldsymbol{v}_1,-\boldsymbol{v}_1';-\boldsymbol{v},-\boldsymbol{v}') \ . \tag{C.17}$$

Hence, the combined symmetry operation TP yields (C.15).

We are now able to interpret the collision term (C.1) in the following way. The Boltzmann collision term is the difference between a creation and a destruction term. Each collision $(\boldsymbol{v},\boldsymbol{v}'\to\boldsymbol{v}_1,\boldsymbol{v}_1')$ removes a particle from the velocity range $(\boldsymbol{v},d^3v)$ considered, and the corresponding destruction rate per unit volume is

$$d^3v \iint F(\boldsymbol{v})F(\boldsymbol{v}')\,w\,q(w,\vartheta)\,d\Omega\,d^3v' \ ,$$

where the integrations must be performed over all unit vectors $\boldsymbol{s}$, i.e., over $d\Omega$, for fixed $\boldsymbol{v}'$, and then over all velocities $\boldsymbol{v}'$. On the other hand, each collision $(\boldsymbol{v}_1,\boldsymbol{v}_1'\to\boldsymbol{v},\boldsymbol{v}')$ creates a particle in the velocity range $(\boldsymbol{v},d^3v)$ considered, and the corresponding creation rate per unit volume is

$$\begin{aligned}&\iiint F(\boldsymbol{v}_1)F(\boldsymbol{v}_1')\,w_1q_1(w_1,\vartheta_1)\,d\Omega_1d^3v_1d^3v_1'\\&\quad = d^3v\iint F(\boldsymbol{v}_1)F(\boldsymbol{v}_1')\,w\,q(w,\vartheta)\,d\Omega\,d^3v' \ .\end{aligned}$$

Here (C.7,9,13) have been used, taking into account that the integrations $d^3v_1d^3v_1'$ in the first line have to be carried out over such a region that one of the outgoing particles is in the velocity range $(\boldsymbol{v},d^3v)$. Taking the difference of these two rates leads to the net production of particles $(\boldsymbol{v},d^3v)$ due to elastic collisions, that is, to the Boltzmann collision term (C.1).

The decisive statistical assumption underlying the Boltzmann collision term (C.1) is the so-called assumption of *molecular chaos* according to which any two collision partners are considered uncorrelated, which means that the probability of a collision $(\boldsymbol{v}_1,d^3v_1;\boldsymbol{v}_2,d^3v_2)\to(\boldsymbol{v}_3,d^3v_3;\boldsymbol{v}_4,d^3v_4)$ is proportional to the product of the particle numbers $F(\boldsymbol{v}_1)\,d^3v_1F(\boldsymbol{v}_2)\,d^3v_2$. This assumption introduces irreversibility into the Boltzmann equation and, therefore, is responsible for the validity of the H theorem (see below).

Let us now show that the Boltzmann collision term has the three properties of elastic collision terms stated in Sect. 2.2.3.

(1) The distribution function always remains positive, that is, if $F(\boldsymbol{r},\boldsymbol{v},0)\geqslant 0$ at time $t=0$, then $F(\boldsymbol{r},\boldsymbol{v},t)\geqslant 0$ for all later times $t>0$.

To prove this, suppose that contrary to the statement F becomes negative, for the first time at time t_0 at the point $(\boldsymbol{r}_0,\boldsymbol{v}_0)$ of μ space. This point must correspond to a minimum of F, hence

$$F = 0 \ , \quad \frac{\partial F}{\partial \boldsymbol{r}} = \frac{\partial F}{\partial \boldsymbol{v}} = 0 \ , \quad \frac{\partial F}{\partial t} < 0 \ , \tag{C.18}$$

where $F \equiv F(\boldsymbol{r}_0, \boldsymbol{v}_0, t_0)$. At this time and at this point of μ space, the Boltzmann equation [see (2.1.7)]

$$\frac{\partial F}{\partial t} + \boldsymbol{v} \cdot \frac{\partial F}{\partial \boldsymbol{r}} + \frac{\boldsymbol{K}}{m} \cdot \frac{\partial F}{\partial \boldsymbol{v}} = \left(\frac{\delta F}{\delta t}\right)_{\mathrm{B}} \tag{C.19}$$

reduces to

$$\frac{\partial F}{\partial t} = \iint F_1 F_1' w q(w, \vartheta) \, d\Omega \, d^3 v' \tag{C.20}$$

using (C.1, 18). But (C.20) contains a contradiction: according to (C.18) its left-hand side is negative, whereas its right-hand side is nonnegative since $F_1 \geqslant 0$ and $F_1' \geqslant 0$ by assumption. It follows that $F(\boldsymbol{r}, \boldsymbol{v}, t) \geqslant 0$ for all $t > 0$.

(2) The Boltzmann collision term conserves mass, momentum, and kinetic energy at every space-time point $(\boldsymbol{r}, t)$,

$$\int m \left(\frac{\delta F}{\delta t}\right)_{\mathrm{B}} d^3 v = 0 \ , \tag{C.21a}$$

$$\int m \boldsymbol{v} \left(\frac{\delta F}{\delta t}\right)_{\mathrm{B}} d^3 v = 0 \ , \tag{C.21b}$$

$$\int \frac{1}{2} m v^2 \left(\frac{\delta F}{\delta t}\right)_{\mathrm{B}} d^3 v = 0 \ . \tag{C.21c}$$

We first show that the following general relation holds:

$$\int \psi \left(\frac{\delta F}{\delta t}\right)_{\mathrm{B}} d^3 v = \frac{1}{4} \int (\psi + \psi' - \psi_1 - \psi_1') \left(\frac{\delta F}{\delta t}\right)_{\mathrm{B}} d^3 v \ . \tag{C.22}$$

Here $\psi = \psi(\boldsymbol{v})$ is an arbitrary function of velocity $\boldsymbol{v}$, and $\psi' = \psi(\boldsymbol{v}')$, $\psi_1 = \psi(\boldsymbol{v}_1)$, $\psi_1' = \psi(\boldsymbol{v}_1')$ where the four velocities $\boldsymbol{v}, \boldsymbol{v}', \boldsymbol{v}_1, \boldsymbol{v}_1'$ correspond to the elastic collision $(\boldsymbol{v}, \boldsymbol{v}' \to \boldsymbol{v}_1, \boldsymbol{v}_1')$. According to (C.1),

$$\begin{aligned}
\int \psi \left(\frac{\delta F}{\delta t}\right)_{\mathrm{B}} d^3 v &= \iiint \psi (F_1 F_1' - F F') w q(w, \vartheta) \, d\Omega \, d^3 v \, d^3 v' \\
&= \iiint \psi' (F_1 F_1' - F F') w q(w, \vartheta) \, d\Omega \, d^3 v \, d^3 v' \\
&= -\iiint \psi_1 (F_1 F_1' - F F') w q(w, \vartheta) \, d\Omega \, d^3 v \, d^3 v' \\
&= -\iiint \psi_1' (F_1 F_1' - F F') w q(w, \vartheta) \, d\Omega \, d^3 v \, d^3 v' \ .
\end{aligned} \tag{C.23}$$

Here the second line is obtained from the first by the replacements $\boldsymbol{v}\leftrightarrow\boldsymbol{v}'$, $\boldsymbol{v}_1\leftrightarrow\boldsymbol{v}_1'$, the third line by $\boldsymbol{v}\leftrightarrow\boldsymbol{v}_1$, $\boldsymbol{v}'\leftrightarrow\boldsymbol{v}_1'$, and the fourth line by $\boldsymbol{v}\leftrightarrow\boldsymbol{v}_1'$, $\boldsymbol{v}'\leftrightarrow\boldsymbol{v}_1$. In writing the last two lines, use has been made of (C.7,9,13). Equation (C.23) now leads immediately to (C.22).

In particular, if one takes for ψ an additive collision invariant χ, for which by definition

$$\chi+\chi'=\chi_1+\chi_1' \tag{C.24}$$

holds, then from (C.22)

$$\int\chi\left(\frac{\delta F}{\delta t}\right)_{\mathrm{B}} d^3v=0\ . \tag{C.25}$$

There are five additive collision invariants for elastic collisions: the mass, the three components of the momentum, and the kinetic energy,

$$\chi=m\ ,\quad mv_x\ ,\quad mv_y\ ,\quad mv_z\ ,\quad \tfrac{1}{2}mv^2\ , \tag{C.26}$$

so that (C.21) is proved.

(3) The Boltzmann collision term obeys the H theorem according to which

$$\frac{\delta H}{\delta t}\leqslant 0 \tag{C.27}$$

generally and $\delta H/\delta t=0$ only in thermal equilibrium, where the H function is defined by

$$H=\int F\ln F\,d^3v\ , \tag{C.28}$$

and $\delta/\delta t$ denotes the time variation due to collisions. The H theorem (C.27) expresses the fact that the local entropy production due to elastic collisions is positive definite, Sect. 5.2.1.

From (C.28),

$$\frac{\delta H}{\delta t}=\int(1+\ln F)\left(\frac{\delta F}{\delta t}\right)_{\mathrm{B}} d^3v=\int\ln F\left(\frac{\delta F}{\delta t}\right)_{\mathrm{B}} d^3v\ , \tag{C.29}$$

where in the last step we have applied (C.25) with the collision invariant $\chi=1$. Using now (C.22) with $\psi=\ln F$, then

$$\begin{aligned}\frac{\delta H}{\delta t}&=\frac{1}{4}\int\ln\left(\frac{FF'}{F_1F_1'}\right)\left(\frac{\delta F}{\delta t}\right)_{\mathrm{B}} d^3v\\ &=\frac{1}{4}\iiint\ln\left(\frac{FF'}{F_1F_1'}\right)(F_1F_1'-FF')\,wq(w,\vartheta)\,d\Omega\,d^3v\,d^3v'\ ,\end{aligned} \tag{C.30}$$

taking (C.1) into account. Here the integrand is, apart from positive factors, of the form $(x-y)\ln(y/x)$ with $x, y \geqslant 0$ (since $F \geqslant 0$), and is hence $\leqslant 0$. This proves (C.27).

Equation (C.30) shows that $\delta H/\delta t = 0$ if and only if

$$FF' = F_1 F_1' \quad \text{or} \tag{C.31}$$

$$\ln F + \ln F' = \ln F_1 + \ln F_1' \tag{C.32}$$

for all $\boldsymbol{v}, \boldsymbol{v}', \boldsymbol{v}_1, \boldsymbol{v}_1'$, that is, for all collisions. This means that here $\ln F$ must be a linear combination of the five additive collision invariants (C.26), thus

$$\ln F = A + \boldsymbol{B} \cdot \boldsymbol{v} + Cv^2 \quad \text{or} \tag{C.33}$$

$$F = \alpha \exp[-\beta(\boldsymbol{v}-\boldsymbol{u})^2] \ . \tag{C.34}$$

As (C.34) is a Maxwell distribution around the local mean velocity $\boldsymbol{u}$, this completes the proof of the H theorem.

Equation (C.31) may be written as

$$FF'wq(w,\vartheta)\,d\Omega\,d^3v\,d^3v' = F_1F_1'wq(w,\vartheta)\,d\Omega\,d^3v\,d^3v'$$

or, using (C.7,9,13), as

$$FF'wq(w,\vartheta)\,d\Omega\,d^3v\,d^3v' = F_1F_1'w_1q_1(w_1,\vartheta_1)\,d\Omega_1 d^3v_1 d^3v_1' \ , \tag{C.35}$$

which expresses detailed balance of elastic collisions $(\boldsymbol{v}, \boldsymbol{v}' \rightleftarrows \boldsymbol{v}_1, \boldsymbol{v}_1')$. Hence we have the result that vanishing entropy production (C.31), thermal equilibrium (C.34), and detailed balance (C.35) are equivalent statements, see Sect. 5.2.1.

D. Brownian Motion and Fokker-Planck Equation

D.1 Brownian Motion

This section considers Brownian motion [D.1 – 5] of a free particle in a homogeneous and isothermal medium; for instance, of a colloidal particle immersed in a fluid, or of an atom emitting and absorbing photons in a black-body radiation field. More precisely, of the true three-dimensional motion of the particle we shall consider only its projection on an arbitrary, fixed direction (one-dimensional Brownian motion).

Our starting point is the *Langevin equation*

$$m\dot{v} = -\frac{1}{\mu}v + \tilde{K}(t) \ , \tag{D.1.1}$$

where m is the mass and v the velocity of the particle. By Langevin's equation, the force on the particle by the surrounding thermal medium is decomposed into a systematic friction force $-v/\mu$, μ being the mobility, and a fluctuating force $\tilde{K}(t)$ which accounts for the thermal fluctuations of the particle. This fluctuating force is specified only by its statistical properties, so that (D.1.1) is a stochastic equation rather than a differential equation in the usual sense. We write (D.1.1) in the form

$$\dot{v}+\zeta v=A(t) \quad \text{with} \tag{D.1.2}$$

$$\zeta=\frac{1}{m\mu}\ , \quad A(t)=\frac{1}{m}\tilde{K}(t)\ . \tag{D.1.3}$$

The quantities ζ and $A(t)$ are supposed to be independent of the particle velocity v. Since the systematic deceleration due to friction is accounted for by the term ζv, one assumes quite naturally that the mean value of the fluctuating force vanishes,

$$\langle A(t)\rangle=0\ , \tag{D.1.4}$$

independently of the instantaneous particle velocity v. Here and in the following, $\langle\dots\rangle$ denote averages over an ensemble of equivalent, independent Brownian particles that do not interact with each other (test particles).

The basic assumption of the theory of Brownian motion is the assumption of *weak collisions*. This means that there is a time interval τ which is small on the macroscopic time scale (determined by the slowing-down time ζ^{-1}), but large on the microscopic time scale (determined by the reciprocal collision frequency γ^{-1}), such that during the time τ many collisions of particles of the surrounding medium with the Brownian particle take place, but the velocity of the latter remaining nearly constant. The collisions are assumed to occur statistically independently one after the other, which implies that the duration of a collision t_c must be much smaller than the mean time between successive collisions γ^{-1}. Hence

$$t_c\ll\gamma^{-1}\ll\tau\ll\zeta^{-1}\ . \tag{D.1.5}$$

The time scale of Brownian motion is fixed by τ, the individual collisions forming an unresolved background.

Following these hypotheses, let us assume that the random acceleration $A(t)$ is approximately described by a Gaussian random process with infinitely short correlation time,

$$\langle A(t)A(t')\rangle=\alpha\delta(t-t')\ . \tag{D.1.6}$$

In words, the duration of a collision is neglected ($t_c\to 0$), and successive collisions are statistically independent. The proportionality constant α in (D.1.6) can be determined by the equipartition theorem, according to which a free particle in thermal equilibrium undergoes velocity fluctuations such that $\langle mv^2/2\rangle=kT/2$. Now, Langevin's equation has for $t>0$ the solution

$$v(t) = v(0)\mathrm{e}^{-\zeta t} + \mathrm{e}^{\zeta t}\int_0^t A(u)\mathrm{e}^{\zeta u}du \; . \tag{D.1.7}$$

Squaring and taking the ensemble average leads to

$$\begin{aligned}\langle v^2(t)\rangle &= \langle v^2(0)\rangle \mathrm{e}^{-2\zeta t} + \mathrm{e}^{-2\zeta t}\int_0^t du \int_0^t du' \langle A(u)A(u')\rangle \mathrm{e}^{\zeta(u+u')} \\ &= \langle v^2(0)\rangle \mathrm{e}^{-2\zeta t} + \frac{\alpha}{2\zeta}(1-\mathrm{e}^{-2\zeta t})\end{aligned} \tag{D.1.8}$$

using (D.1.6). Moreover, $\langle v(0)A(t)\rangle = \langle v(0)\rangle\langle A(t)\rangle = 0$, see (D.1.4), as $A(t)$ is independent of v by assumption. The equipartition theorem now requires

$$\langle v^2(t)\rangle = \langle v^2(0)\rangle = \frac{kT}{m} \; , \tag{D.1.9}$$

and hence

$$\alpha = \frac{2kT}{m}\zeta \; . \tag{D.1.10}$$

Thus, finally, from (D.1.6, 10)

$$\langle A(t)A(t')\rangle = \frac{2kT}{m}\zeta\delta(t-t') \; , \tag{D.1.11}$$

which characterizes the fluctuating force on the Brownian particle.

Let us now turn to the approach to thermal equilibrium. More precisely, we wish to determine the probability $w(v, t; v_0)dv$ that, at time $t>0$, a Brownian particle has a velocity in the range (v, dv) if its velocity was v_0 at time $t=0$. One may view $w(v, t; v_0)$ as the velocity distribution function at time $t>0$ corresponding to the initial distribution $w(v, 0; v_0) = \delta(v-v_0)$. Since the Brownian particles are in contact with a thermal medium (a heat bath), $w(v, t; v_0)$ must approach a Maxwell distribution for $t\to\infty$. The connection between $w(v, t; v_0)$ and the true velocity distribution of Brownian particles $f(v, t)$ is obviously

$$f(v, t) = \int w(v, t; v_0) f(v_0, 0)\, dv_0 \; . \tag{D.1.12}$$

Thus, f's approach to equilibrium is known if w's is known.

To study the approach to equilibrium of the function $w(v, t; v_0)$, we first remark that from (D.1.7, 4)

$$\langle v(t)\rangle = v_0\mathrm{e}^{-\zeta t} \tag{D.1.13}$$

since $v(0) = v_0$, showing the expected exponential decrease of the mean velocity. Now making use of the existence of the time interval τ as defined by (D.1.5), (D.1.7) can be written in the form

$$v(t) - v_0 e^{-\zeta t} = e^{-\zeta t} \int_0^t A(u) e^{\zeta u} du$$

$$\simeq e^{-\zeta t} \sum_{j=0}^{(t/\tau)-1} e^{j\zeta\tau} G_j \,, \tag{D.1.14}$$

approximating the integral by a sum of $t/\tau\,(\gg 1)$ terms, where

$$G_j = \int_{j\tau}^{(j+1)\tau} A(u)\, du \tag{D.1.15}$$

denotes the net change of the velocity of the Brownian particle during the time interval $j\tau < t < (j+1)\tau$. Clearly, according to (D.1.4),

$$\langle G_i \rangle = 0 \ . \tag{D.1.16}$$

Furthermore,

$$\langle G_i G_j \rangle = \langle G_\tau^2 \rangle\, \delta_{ij} \tag{D.1.17}$$

because the G_i corresponding to different time intervals are statistically independent, and where

$$\langle G_\tau^2 \rangle = \int_0^\tau du \int_0^\tau du' \langle A(u) A(u') \rangle = \frac{2kT}{m} \zeta\tau \tag{D.1.18}$$

on account of (D.1.11). Using (D.1.16, 17), from (D.1.14)

$$\langle [v(t) - v_0 e^{-\zeta t}]^2 \rangle = e^{-2\zeta t} \langle G_\tau^2 \rangle \sum_{j=0}^{(t/\tau)-1} e^{2j\zeta\tau}$$

$$= e^{-2\zeta t} \langle G_\tau^2 \rangle \frac{e^{2\zeta t} - 1}{e^{2\zeta\tau} - 1} \ . \tag{D.1.19}$$

Inserting here (D.1.13, 18), and using the fact that $\zeta\tau \ll 1$, see (D.1.5), so that

$$e^{2\zeta\tau} - 1 \simeq 2\zeta\tau \,, \tag{D.1.20}$$

one finally obtains

$$\langle [v(t) - \langle v(t) \rangle]^2 \rangle = \frac{kT}{m} (1 - e^{-2\zeta t}) \ . \tag{D.1.21}$$

Equations (D.1.13, 21) are the mean value and the mean square deviation, respectively, of the velocity $v(t)$ of the Brownian particle. Now, it is plausible (and can indeed be proved) that the random variable $v(t)$ is distributed according to a Gaussian distribution. Hence, in view of (D.1.13,21), the distribution function for $v(t)$, which is nothing but the probability density $w(v, t; v_0)$, is given by

$$w(v, t; v_0) = \left(\frac{m}{2\pi kT(1-e^{-2\zeta t})}\right)^{1/2} \exp\left(-\frac{m(v-v_0e^{-\zeta t})^2}{2kT(1-e^{-2\zeta t})}\right) . \quad \text{(D.1.22)}$$

Equation (D.1.22) shows that for $t = 0$, $w(v, t; v_0)$ reduces to $\delta(v - v_0)$ as required, and that for $t \to \infty$ it approaches a Maxwell distribution. The mean velocity slows down from v_0 to zero with a time constant $1/\zeta$, while at the same time the velocity fluctuations increase from zero to their thermal value with a time constant $1/2\zeta$.

The following remark concludes this section. By integrating the Langevin equation (D.1.2) from $t = 0$ to $t = \tau$, one obtains to lowest order in τ

$$v(\tau) - v(0) = -\zeta\tau v(0) + G_\tau \quad \text{(D.1.23)}$$

or

$$\Delta v = -\zeta\tau v + G_\tau , \quad \text{(D.1.24)}$$

writing $\Delta v = v(\tau) - v(0)$. Neglecting again terms of the order τ^2 and using (D.1.16, 18), one obtains from (D.1.24) for the average changes of Δv and $(\Delta v)^2$ per unit time, see (D.1.16, 18),

$$\frac{\langle \Delta v \rangle}{\tau} \equiv \left\langle \frac{\Delta v}{\Delta t} \right\rangle = -\zeta v , \quad \text{(D.1.25)}$$

$$\frac{\langle (\Delta v)^2 \rangle}{\tau} \equiv \left\langle \frac{\Delta v^2}{\Delta t} \right\rangle = \frac{2kT}{m}\zeta . \quad \text{(D.1.26)}$$

These two relations are characteristic of Brownian motion in a thermal medium of temperature T.

D.2 Fokker-Planck Equation

Another approach to Brownian motion is provided by the Fokker-Planck equation [D.6]. Let $W(v, \Delta v)\, d\Delta v$ denote the probability that a Brownian particle of initial velocity v has, after time τ, undergone a velocity change in the range $(\Delta v, d\Delta v)$, that is, the particle velocity at time $t + \tau$ is in the range $(v + \Delta v, d\Delta v)$ if it was v at time t. The quantity $W(v, \Delta v)$ is normalized,

$$\int W(v, \Delta v)\, d\Delta v = 1 . \quad \text{(D.2.1)}$$

As in the preceding section, we assume again that a time interval τ can be chosen such that there are many collisions during time τ while the velocity of the Brownian particle remains approximately constant, see (D.1.5). This means that $W(v, \Delta v)$ differs appreciably from zero only for very small values of Δv.

For the velocity distribution $f(v, t)$ of Brownian particles up to terms of the order of $(\Delta v)^2 \equiv \Delta v^2$ therefore

$$f(v, t+\tau) = \int f(v-\Delta v, t)\, W(v-\Delta v, \Delta v)\, d\Delta v$$

$$\simeq \int \left(f - \Delta v \frac{\partial f}{\partial v} + \frac{1}{2} \Delta v^2 \frac{\partial^2 f}{\partial v^2} \right)$$

$$\cdot \left(W - \Delta v \frac{\partial W}{\partial v} + \frac{1}{2} \Delta v^2 \frac{\partial^2 W}{\partial v^2} \right) d\Delta v$$

$$\simeq f(v, t) - \frac{\partial}{\partial v} (\langle \Delta v \rangle f) + \frac{1}{2} \frac{\partial^2}{\partial v^2} (\langle \Delta v^2 \rangle f) \; , \tag{D.2.2}$$

where $f = f(v, t)$ and $W = W(v, \Delta v)$. Here $\langle \Delta v \rangle$ and $\langle \Delta v^2 \rangle$ denote the ensemble averages for Brownian particles of velocity v,

$$\langle \Delta v \rangle_v \equiv \langle \Delta v \rangle = \int \Delta v \, W(v, \Delta v)\, d\Delta v = \tau \left\langle \frac{\Delta v}{\Delta t} \right\rangle \; , \tag{D.2.3}$$

$$\langle \Delta v^2 \rangle_v \equiv \langle \Delta v^2 \rangle = \int \Delta v^2 \, W(v, \Delta v)\, d\Delta v = \tau \left\langle \frac{\Delta v^2}{\Delta t} \right\rangle \; , \tag{D.2.4}$$

using in the last step the notation introduced in (D.1.25, 26). Setting now

$$f(v, t+\tau) - f(v, t) \simeq \tau \frac{\partial f}{\partial t} \; , \tag{D.2.5}$$

we finally obtain from (D.2.2) the one-dimensional *Fokker-Planck equation* for a homogeneous system without external forces,

$$\frac{\partial f}{\partial t} = -\frac{\partial}{\partial v} \left(\left\langle \frac{\Delta v}{\Delta t} \right\rangle f \right) + \frac{1}{2} \frac{\partial^2}{\partial v^2} \left(\left\langle \frac{\Delta v^2}{\Delta t} \right\rangle f \right) . \tag{D.2.6}$$

The right-hand side is the one-dimensional form of the general Fokker-Planck collision term (2.2.4). Notice that the approximation (D.2.5) shows clearly that the time scale of Brownian motion is determined by the time interval τ, as already pointed out in the preceding section.

The specific form of the Fokker-Planck collision term depends on the particular interaction that determines $\langle \Delta v / \Delta t \rangle$ and $\langle \Delta v^2 / \Delta t \rangle$. For Brownian particles in a thermal medium of temperature T, these quantities are given by (D.1.25, 26), so that in this case, the Fokker-Planck equation takes the form

$$\frac{\partial f}{\partial t} = \zeta \frac{\partial}{\partial v} \left(v f + \frac{kT}{m} \frac{\partial f}{\partial v} \right) . \tag{D.2.7}$$

The solution of this equation, corresponding to the initial condition $f(v,0)=\delta(v-v_0)$, is given by

$$f(v,t)=\left(\frac{m}{2\pi kT(1-\mathrm{e}^{-2\zeta t})}\right)^{1/2}\exp\left(-\frac{m(v-v_0\mathrm{e}^{-\zeta t})^2}{2kT(1-\mathrm{e}^{-2\zeta t})}\right), \qquad \text{(D.2.8)}$$

as may be verified by direct substitution into (D.2.7). This particular distribution function was called $w(v,t;v_0)$ in Appendix D.1, and comparison with (D.1.22) shows that the Fokker-Planck equation correctly describes the approach to thermal equilibrium of Brownian particles.

E. Reciprocity Relations for Inelastic Collisions with Heavy Particles

In Chap. 1, only inelastic collisions with free electrons have been considered. However, inelastic collisions with heavy particles, such as protons or neutral atoms, are sometimes important in plasma spectroscopy. In this appendix, we derive the reciprocity relations for inelastic collisions with heavy particles [E.1].

E.1 Excitation and De-Excitation

Consider detailed balance between excitation and de-excitation collisions of an atom A with a heavy particle S,

$$A_1(\boldsymbol{v}_1,d^3v_1)+S(\boldsymbol{v}_S,d^3v_S)\rightleftarrows A_2(\boldsymbol{v}_2,d^3v_2)+S(\boldsymbol{v}_S',d^3v_S') \ . \qquad \text{(E.1.1)}$$

The energies of the lower (1) and upper (2) bound level of atom A are E_1 and E_2, respectively. Conservation of momentum and energy then requires

$$m_0\boldsymbol{v}_1+m_S\boldsymbol{v}_S=m_0\boldsymbol{v}_2+m_S\boldsymbol{v}_S' \ , \qquad \text{(E.1.2)}$$

$$\tfrac{1}{2}m_0v_1^2+\tfrac{1}{2}m_Sv_S^2=\tfrac{1}{2}m_0v_2^2+\tfrac{1}{2}m_Sv_S'^2+E_{21} \ , \qquad \text{(E.1.3)}$$

where m_0 and m_S are the masses of particles A and S, respectively, and $E_{21}=E_2-E_1$ is the excitation energy in question. Let us introduce center-of-mass velocities $\boldsymbol{V}_1$, $\boldsymbol{V}_2$ and relative velocities $\boldsymbol{w}_1$, $\boldsymbol{w}_2$ through

$$(m_0+m_S)\boldsymbol{V}_1=m_0\boldsymbol{v}_1+m_S\boldsymbol{v}_S \ ; \quad (m_0+m_S)\boldsymbol{V}_2=m_0\boldsymbol{v}_2+m_S\boldsymbol{v}_S' \ , \qquad \text{(E.1.4)}$$

$$\boldsymbol{w}_1=\boldsymbol{v}_S-\boldsymbol{v}_1 \ ; \quad \boldsymbol{w}_2=\boldsymbol{v}_S'-\boldsymbol{v}_2 \ . \qquad \text{(E.1.5)}$$

Momentum conservation (E.1.2) now takes the form

$$V_1 = V_2 \ , \tag{E.1.6}$$

and energy conservation (E.1.3)

$$\tfrac{1}{2}\mu_{0S} w_1^2 = \tfrac{1}{2}\mu_{0S} w_2^2 + E_{21} \ , \quad \text{where} \tag{E.1.7}$$

$$\mu_{0S} = \frac{m_0 m_S}{m_0 + m_S} \tag{E.1.8}$$

is the reduced mass. (For the following, compare the analogous considerations for elastic collisions in Appendix C.) The effect of an excitation collision then consists simply in turning the vector of the relative velocity in a new direction, specified by a unit vector $\boldsymbol{s}$, combined with a shortening of its length,

$$\boldsymbol{w}_2 = \left(w_1^2 - \frac{2E_{21}}{\mu_{0S}} \right)^{1/2} \boldsymbol{s} \ . \tag{E.1.9}$$

An excitation collision is hence uniquely specified by the two independent components of $\boldsymbol{s}$, for which one may choose its polar and azimuthal angles ϑ and φ relative to $\boldsymbol{w}_1$ as polar axis, i.e., $\boldsymbol{w}_1 \cdot \boldsymbol{w}_2 = w_1 w_2 \cos\vartheta$.

Calculating the Jacobians from (E.1.4, 5) gives

$$\left| \frac{\partial(\boldsymbol{V}_1, \boldsymbol{w}_1)}{\partial(\boldsymbol{v}_1, \boldsymbol{v}_S)} \right| = \left| \frac{\partial(\boldsymbol{V}_2, \boldsymbol{w}_2)}{\partial(\boldsymbol{v}_2, \boldsymbol{v}_S')} \right| = 1 \ ,$$

so that

$$d^3v_1\, d^3v_S = d^3V_1\, d^3w_1 \ ; \quad d^3v_2\, d^3v_S' = d^3V_2\, d^3w_2 \ . \tag{E.1.10}$$

On the other hand, from (E.1.6, 7) follows

$$d^3V_1 = d^3V_2 \ , \tag{E.1.11}$$

$$w_1\, dw_1 = w_2\, dw_2 \ . \tag{E.1.12}$$

Using spherical coordinates with respect to an arbitrary polar axis, then $d^3w_1 = w_1^2\, dw_1\, d\omega_1$, $d^3w_2 = w_2^2\, dw_2\, d\omega_2$, $d\omega_1$ and $d\omega_2$ being elements of solid angle. For a well-defined excitation collision (that is, for a given unit vector $\boldsymbol{s}$), $d\omega_1 = d\omega_2$ since the relative orientation of $\boldsymbol{w}_1$ and $\boldsymbol{w}_2$ is held fixed. Hence, using (E.1.12),

$$\frac{d^3w_1}{d^3w_2} = \frac{w_1^2 dw_1 d\omega_1}{w_2^2 dw_2 d\omega_2} = \frac{w_1}{w_2} \ . \tag{E.1.13}$$

The rate equation for the reactions (E.1.1) in detailed balance is

$$n_1 f_1(\boldsymbol{v}_1)\, d^3 v_1 n_S f_S(\boldsymbol{v}_S)\, d^3 v_S w_1 q_{12}(w_1, \vartheta)\, d\Omega$$

$$= n_2 f_2(\boldsymbol{v}_2)\, d^3 v_2 n_S f_S(\boldsymbol{v}_S')\, d^3 v_S' w_2 q_{21}(w_2, \vartheta)\, d\Omega \ . \qquad \text{(E.1.14)}$$

Here q_{12} and q_{21} are the differential cross sections for excitation and de-excitation collisions, which depend on the modulus of the relative velocity and the scattering angle ϑ ($\boldsymbol{w}_1 \cdot \boldsymbol{w}_2 = w_1 w_2 \cos\vartheta$), but which are, of course, independent of the center-of-mass velocity. Furthermore, $d\Omega = \sin\vartheta\, d\vartheta\, d\varphi$ is the element of solid angle corresponding to the unit vector $\boldsymbol{s}$, see (E.1.9). Inserting into (E.1.14) Maxwell distributions for f_1, f_2, and f_S, and a Boltzmann distribution for n_2/n_1, and using the various relations indicated above, one obtains the reciprocity relation

$$g_1 w_1^2 q_{12}(w_1, \vartheta) = g_2 w_2^2 q_{21}(w_2, \vartheta) \ , \qquad \text{(E.1.15)}$$

g_1, g_2 being again the statistical weights of the atomic levels, where w_1 and w_2 are related to each other by (E.1.7). Equation (E.1.15) is the analogue to (1.6.7). It contains as a special case the reciprocity relation for elastic collisions $q_{12}(w, \vartheta) = q_{21}(w, \vartheta)$, see (C.14), which is obtained from (E.1.15) by setting there $w_1 = w_2$ and $g_1 = g_2$.

E.2 Ionization and Three-Body Recombination. Dissociation and Recombination

We now turn to detailed balance between collisional ionization of an atom A by a heavy particle S, and the corresponding three-body recombination of the ion A^+ with an electron e,

$$A(\boldsymbol{v}_0, d^3 v_0) + S(\boldsymbol{v}_S, d^3 v_S) \rightleftarrows A^+(\boldsymbol{v}_+, d^3 v_+) + \mathrm{e}(\boldsymbol{v}_\mathrm{e}, d^3 v_\mathrm{e}) + S(\boldsymbol{v}_S', d^3 v_S') . \qquad \text{(E.2.1)}$$

The energies of the atomic and ionic bound levels are E_0 and E_+, respectively. The conservation equations for momentum and energy are then

$$m_0 \boldsymbol{v}_0 + m_S \boldsymbol{v}_S = m_+ \boldsymbol{v}_+ + m_\mathrm{e} \boldsymbol{v}_\mathrm{e} + m_S \boldsymbol{v}_S' \ , \qquad \text{(E.2.2)}$$

$$\tfrac{1}{2} m_0 v_0^2 + \tfrac{1}{2} m_S v_S^2 = \tfrac{1}{2} m_+ v_+^2 + \tfrac{1}{2} m_\mathrm{e} v_\mathrm{e}^2 + \tfrac{1}{2} m_S v_S'^2 + E_{+0} \ , \qquad \text{(E.2.3)}$$

where $E_{+0} = E_+ - E_0$ is the ionization energy in question. Introducing again center-of-mass velocities and relative velocities by

$$M\boldsymbol{V} = m_0 \boldsymbol{v}_0 + m_S \boldsymbol{v}_S \ ; \quad M\boldsymbol{V}' = m_+ \boldsymbol{v}_+ + m_\mathrm{e} \boldsymbol{v}_\mathrm{e} + m_S \boldsymbol{v}_S' \ , \qquad \text{(E.2.4)}$$

$$\boldsymbol{w} = \boldsymbol{v}_S - \boldsymbol{v}_0 \ ; \quad \boldsymbol{w}_+ = \boldsymbol{v}_\mathrm{e} - \boldsymbol{v}_+ \ ; \quad \boldsymbol{w}_S = \boldsymbol{v}_\mathrm{e} - \boldsymbol{v}_S' \ , \qquad \text{(E.2.5)}$$

where

$$M = m_0 + m_S = m_+ + m_e + m_S \tag{E.2.6}$$

is the total mass, the conservation equations (E.2.2, 3) become

$$V = V', \tag{E.2.7}$$

$$\frac{1}{2}\mu_{0S} w^2 = \frac{1}{2M}[m_+(m_S + m_e) w_+^2 + m_S(m_+ + m_e) w_S^2$$

$$- 2 m_+ m_S \boldsymbol{w}_+ \cdot \boldsymbol{w}_S] + E_{+0} , \tag{E.2.8}$$

with the reduced mass

$$\mu_{0S} = \frac{m_0 m_S}{m_0 + m_S} . \tag{E.2.9}$$

Equation (E.2.8) takes a simple form if the electron mass is neglected ($m_e = 0$). Then $m_+ = m_0$, and hence

$$\tfrac{1}{2}\mu_{0S} w^2 = \tfrac{1}{2}\mu_{0S} |\boldsymbol{w}_+ - \boldsymbol{w}_S|^2 + E_{+0} .$$

However, we shall not use this approximation in the following.

Again, one verifies readily from (E.2.4, 5) that the Jacobians have modulus unity,

$$\left|\frac{\partial(\boldsymbol{V}, \boldsymbol{w})}{\partial(\boldsymbol{v}_0, \boldsymbol{v}_S)}\right| = \left|\frac{\partial(\boldsymbol{V}', \boldsymbol{w}_+, \boldsymbol{w}_S)}{\partial(\boldsymbol{v}_+, \boldsymbol{v}_e, \boldsymbol{v}_S')}\right| = 1$$

so that

$$d^3v_0 d^3v_S = d^3V d^3w \; ; \quad d^3v_+ d^3v_e d^3v_S' = d^3V' d^3w_+ d^3w_S . \tag{E.2.10}$$

Furthermore, according to (E.2.7)

$$d^3V = d^3V' . \tag{E.2.11}$$

Let us now write down the detailed balance equation for the reactions (E.2.1) in analogy to (1.6.13),

$$n_0 f_0(\boldsymbol{v}_0) d^3v_0 n_S f_S(\boldsymbol{v}_S) d^3v_S w \omega_{0+}(\boldsymbol{w}; \boldsymbol{w}_+, \boldsymbol{w}_S) d^3w_+ d^3w_S$$

$$= n_+ f_+(\boldsymbol{v}_+) d^3v_+ n_e f_e(\boldsymbol{v}_e) d^3v_e n_S f_S(\boldsymbol{v}_S') d^3v_S'$$

$$\cdot w_+ w_S \omega_{+0}(\boldsymbol{w}_+, \boldsymbol{w}_S; \boldsymbol{w}) d^3w , \tag{E.2.12}$$

where we have indicated the obvious fact that ω_{0+} and ω_{+0} are independent of the center-of-mass motion. We now insert in the usual manner thermal distribution functions into (E.2.12). However, since we have kept a finite electron mass, we cannot use the approximate Saha distribution as given by (1.4.25), rather, we must use its exact form

$$\left(\frac{n_e n_+}{n_0}\right)^{\mathrm{S}} = \frac{2 g_+}{g_0} \frac{(2\pi\mu_{+e} kT)^{3/2}}{h^3} \mathrm{e}^{-E_{+0}/kT} , \tag{E.2.13}$$

where

$$\mu_{+e} = \frac{m_+ m_e}{m_+ + m_e} = \frac{m_+ m_e}{m_0} \tag{E.2.14}$$

is the reduced mass of the electron-ion pair. Thus from (E.2.13) the reciprocity relation

$$g_0 w\, \omega_{0+}(w; w_+, w_S) = 2 g_+ \frac{\mu_{+e}^3}{h^3} w_+ w_S\, \omega_{+0}(w_+, w_S; w) \tag{E.2.15}$$

is derived in complete analogy to (1.6.14).

As its derivation shows, (E.2.15) applies likewise to dissociation and its inverse recombination, for example,

$$A + S \rightleftarrows B + C + S , \tag{E.2.16}$$

where S need not be a heavy particle, but may also be an electron. Here, it is perhaps more natural to introduce instead of (E.2.5) the relative velocities

$$\boldsymbol{w}_A = \boldsymbol{v}_S - \boldsymbol{v}_A , \quad \boldsymbol{w}_B = \boldsymbol{v}'_S - \boldsymbol{v}_B , \quad \boldsymbol{w}_C = \boldsymbol{v}'_S - \boldsymbol{v}_C , \tag{E.2.17}$$

and, accordingly, to define quantities $\omega_{A,BC}$ and $\omega_{BC,A}$ such that the reaction rates for $A + S \rightarrow B + C + S$ and $B + C + S \rightarrow A + S$ are proportional to $w_A \omega_{A,BC}$ and $w_B w_C \omega_{BC,A}$, respectively. Then, substituting in (E.2.15)

$$g_0 \rightarrow g_A , \quad 2 g_+ \rightarrow g_B g_C , \quad \mu_{+e} \rightarrow \mu_{BC}$$

where

$$\mu_{BC} = \frac{m_B m_C}{m_B + m_C} = \frac{m_B m_C}{m_A} , \tag{E.2.18}$$

one obtains the reciprocity relation

$$g_A w_A\, \omega_{A,BC}(w_A; w_B, w_C) = g_B g_C \frac{\mu_{BC}^3}{h^3} w_B w_C\, \omega_{BC,A}(w_B, w_C; w_A) . \tag{E.2.19}$$

F. Elastic Collision Term of the Standard Problem

Appendix F derives the elastic collision term for the standard problem of non-LTE line transfer as given by (6.4.4). Recall that this is the elastic collision term of excited two-level atoms immersed in a thermal gas of nonexited two-level atoms, in the approximation where excited and nonexcited atoms are considered as hard spheres. Only collisions of excited with nonexcited atoms are taken into account, while collisons between excited atoms or with other particle species, e.g., with free electrons, are ignored.

We denote excited and nonexcited atoms by A_2 and A_1, respectively, and the corresponding distribution functions by $F_2(\boldsymbol{v})$ and $F_1(\boldsymbol{v})$. The elastic collisions to be considered are of the type

$$A_2(\boldsymbol{v})+A_1(\boldsymbol{v}_1)\rightarrow A_2(\boldsymbol{v}')+A_1(\boldsymbol{v}_1') \ . \tag{F.1}$$

By a straightforward generalization of (C.1), the Boltzmann collision term for the distribution function $F_2(\boldsymbol{v})$ of the excited atoms is given by

$$\left(\frac{\delta F_2}{\delta t}\right)_{\mathrm{el}} = \int d^3v_1 \int d\Omega\, Wq(W,\vartheta)\left[F_2(\boldsymbol{v}')F_1(\boldsymbol{v}_1')-F_2(\boldsymbol{v})F_1(\boldsymbol{v}_1)\right] \tag{F.2}$$

where the second term in the brackets describes the collisions of type (F.1), while the first term describes the corresponding inverse collisions. We denote by $\boldsymbol{W}$ the relative velocity of a nonexcited atom A_1 with respect to a colliding excited atom A_2, thus

$$\boldsymbol{W}=\boldsymbol{v}_1-\boldsymbol{v} \ , \quad \boldsymbol{W}'=\boldsymbol{v}_1'-\boldsymbol{v}' \ . \tag{F.3}$$

Here $\boldsymbol{W}$ rather than $\boldsymbol{w}$ as in Appendix C denotes the relative velocity in order to avoid confusion with the thermal velocity w, (F.18). In (F.2), $W=|\boldsymbol{W}|$ is the modulus of the relative velocity, ϑ is the deflection angle of the relative velocity (i.e., $\boldsymbol{W}\cdot\boldsymbol{W}'=W\,W'\cos\vartheta$), $d\Omega=\sin\vartheta\, d\vartheta\, d\varphi$ is the element of solid angle, and $q(W,\vartheta)$ is the differential cross section for the collision (F.1).

In the Boltzmann collision term (F.2), use has been made of the following relations for elastically colliding particles of equal mass:

$$W'=W \tag{F.4}$$

according to (C.7);

$$d^3v'd^3v_1'=d^3v\, d^3v_1 \tag{F.5}$$

according to (C.9); and, for spherically symmetric atoms A_1 and A_2,

$$q(\boldsymbol{v}',\boldsymbol{v}_1';\boldsymbol{v},\boldsymbol{v}_1)=q(\boldsymbol{v},\boldsymbol{v}_1;\boldsymbol{v}',\boldsymbol{v}_1')=q(W,\vartheta) \tag{F.6}$$

according to Eqs. (C.13,15).

We now proceed to express the velocities $\boldsymbol{v}_1$, $\boldsymbol{v}'$, $\boldsymbol{v}_1'$ in terms of the velocities $\boldsymbol{v}$ and $\boldsymbol{W}$. To this end we use the unit vector $\boldsymbol{a}$ ($|\boldsymbol{a}| = 1$) as defined in Fig. F.1. Taking (F.4) into account, one readily derives from Fig. F.1

$$\boldsymbol{W}' = \boldsymbol{W} - 2(\boldsymbol{W} \cdot \boldsymbol{a})\boldsymbol{a} \ , \tag{F.7}$$

and hence, from (F.3),

$$\begin{aligned} \boldsymbol{v}_1 &= \boldsymbol{v} + \boldsymbol{W} \ , \\ \boldsymbol{v}' &= \boldsymbol{v} + (\boldsymbol{W} \cdot \boldsymbol{a})\boldsymbol{a} \ , \\ \boldsymbol{v}_1' &= \boldsymbol{v} + \boldsymbol{W} - (\boldsymbol{W} \cdot \boldsymbol{a})\boldsymbol{a} \ . \end{aligned} \tag{F.8}$$

To express the solid angle $d\Omega = \sin\vartheta\, d\vartheta\, d\varphi$ referring to $\boldsymbol{W}'$ by the solid angle $d\Omega_a = \sin\Theta\, d\Theta\, d\varphi$ referring to $\boldsymbol{a}$, Fig. F.1 shows that

$$\vartheta = \pi - 2\Theta \ , \tag{F.9}$$

and hence

$$\sin\vartheta = \sin(\pi - 2\Theta) = 2\sin\Theta\cos\Theta \ ,$$

$$d\vartheta\, d\varphi = \left|\frac{\partial(\vartheta,\varphi)}{\partial(\Theta,\varphi)}\right| d\Theta\, d\varphi = 2\, d\Theta\, d\varphi \ ,$$

$$\sin\vartheta\, d\vartheta\, d\varphi = 4\sin\Theta\cos\Theta\, d\Theta\, d\varphi \ ,$$

that is,

$$d\Omega = 4\cos\Theta\, d\Omega_a \ . \tag{F.10}$$

Observing that $0 \leqslant \Theta \leqslant \pi/2$ (in contrast to $0 \leqslant \vartheta \leqslant \pi$), then

$$\begin{aligned} \int d\Omega &= \int_0^{2\pi} d\varphi \int_0^{\pi/2} 4\cos\Theta\sin\Theta\, d\Theta \\ &= \frac{1}{2}\int_0^{2\pi} d\varphi \int_0^{\pi} 4|\cos\Theta|\sin\Theta\, d\Theta \\ &= 2\int |\cos\Theta|\, d\Omega_a \ , \end{aligned} \tag{F.11}$$

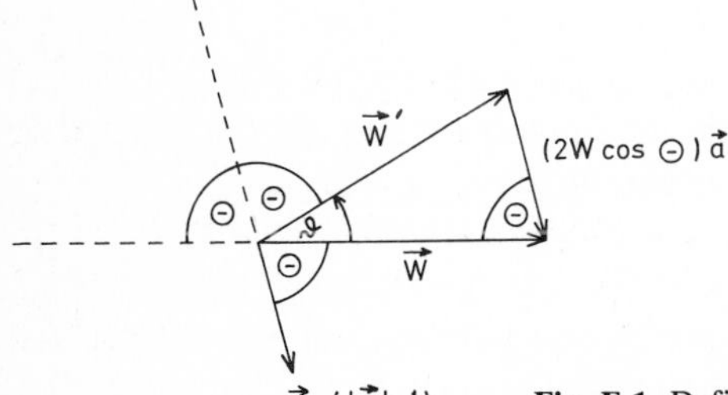

Fig. F.1. Definitions of the unit vector $\boldsymbol{a}$ and the angle Θ

where in the last line the integration is now again over the full sphere $0 \leqslant \Theta \leqslant \pi$, $0 \leqslant \varphi \leqslant 2\pi$.

Finally, from $\boldsymbol{v}_1 = \boldsymbol{W} + \boldsymbol{v}$, (F.8),

$$\int d^3 v_1 = \int d^3 W \tag{F.12}$$

as $\boldsymbol{v}$ is constant for the integration considered.

Using (F.8, 9, 11, 12), the Boltzmann collision term (F.2) can be written as

$$\begin{aligned}\left(\frac{\delta F_2}{\delta t}\right)_{\mathrm{el}} &= 2 \int d^3 W \int d\Omega_a |\cos\Theta| \, W q(W, \pi - 2\Theta) \\ &\quad \cdot [F_2(\boldsymbol{v} + (\boldsymbol{W}\cdot\boldsymbol{a})\boldsymbol{a}) F_1(\boldsymbol{v} + \boldsymbol{W} - (\boldsymbol{W}\cdot\boldsymbol{a})\boldsymbol{a}) \\ &\quad - F_2(\boldsymbol{v}) F_1(\boldsymbol{v} + \boldsymbol{W})] \ .\end{aligned} \tag{F.13}$$

We now introduce our two main assumptions. First, we consider the atoms A_1 and A_2 as hard spheres of radii r_1 and r_2, respectively, so that the differential cross section becomes the constant

$$q = \tfrac{1}{4} r^2 \ , \tag{F.14}$$

with $r = r_1 + r_2$, and the total cross section becomes the constant

$$Q = \int q \, d\Omega = 4\pi q = \pi r^2 \ . \tag{F.15}$$

Second, we assume that the distribution function of the nonexcited atoms is Maxwellian of temperature T,

$$F_1(\boldsymbol{v}) = n_1 f^{\mathrm{M}}(\boldsymbol{v}) \ , \tag{F.16}$$

where n_1 is the density of nonexcited atoms, and

$$f^{\mathrm{M}}(\boldsymbol{v}) = \frac{1}{\pi^{3/2} w^3} \exp\left(-\frac{\boldsymbol{v}^2}{w^2}\right) \tag{F.17}$$

is a normalized Maxwell distribution. Here

$$w = \left(\frac{2kT}{m}\right)^{1/2} \tag{F.18}$$

is the thermal velocity of the atoms, and m is the mass of an atom.

We use the thermal velocity w to define a characteristic collision frequency of excited atoms A_2 with nonexcited atoms A_1 by

$$\gamma_2 = n_1 Q w = n_1 \pi r^2 w \ . \tag{F.19}$$

On account of (F.14, 16), and writing the collision term (F.13) as the difference of creation and destruction terms, one gets

$$\left(\frac{\delta F_2}{\delta t}\right)_{\mathrm{el}} = \Gamma_2^+ - \Gamma_2^- \ , \qquad \text{where} \tag{F.20}$$

$$\Gamma_2^+(\boldsymbol{v}) = \tfrac{1}{2} n_1 r^2 \int d^3W \int d\Omega_a W|\cos\Theta| \cdot f^{\mathrm{M}}(\boldsymbol{v}+\boldsymbol{W}-(\boldsymbol{W}\cdot\boldsymbol{a})\boldsymbol{a}) F_2(\boldsymbol{v}+(\boldsymbol{W}\cdot\boldsymbol{a})\boldsymbol{a}) \tag{F.21}$$

and

$$\Gamma_2^-(\boldsymbol{v}) = \tfrac{1}{2} n_1 r^2 \int d^3W \int d\Omega_a W|\cos\Theta| f^{\mathrm{M}}(\boldsymbol{v}+\boldsymbol{W}) F_2(\boldsymbol{v}) \ , \tag{F.22}$$

where $f^{\mathrm{M}}(\boldsymbol{v})$ is the Maxwell distribution (F.17).

Evaluation of the destruction term (F.22) is straightforward, yielding

$$\Gamma_2^-(\boldsymbol{v}) = \gamma_2 \left[\left(\frac{v}{w}+\frac{w}{2v}\right)\operatorname{erf}\left(\frac{v}{w}\right)+\frac{1}{\pi^{1/2}}\exp\left(-\frac{v^2}{w^2}\right)\right] F_2(\boldsymbol{v}) \ , \tag{F.23}$$

where $v = |\boldsymbol{v}|$, γ_2 is the collision frequency (F.19), and $\operatorname{erf} x$ denotes the error function

$$\operatorname{erf} x = \frac{2}{\pi^{1/2}} \int_0^x e^{-t^2} dt \ . \tag{F.24}$$

To evaluate the creation term (F.21), we decompose the vector $\boldsymbol{W}$ into its components parallel and perpendicular to the unit vector $\boldsymbol{a}$ (Fig. F.2),

$$\begin{aligned} \boldsymbol{W} &= \boldsymbol{W}_{\|} + \boldsymbol{W}_{\perp} \ , \\ \boldsymbol{W}_{\|} &= (\boldsymbol{W}\cdot\boldsymbol{a})\boldsymbol{a} = (W\cos\Theta)\boldsymbol{a} \ , \\ \boldsymbol{W}_{\perp} &= \boldsymbol{W} - (\boldsymbol{W}\cdot\boldsymbol{a})\boldsymbol{a} \ . \end{aligned} \tag{F.25}$$

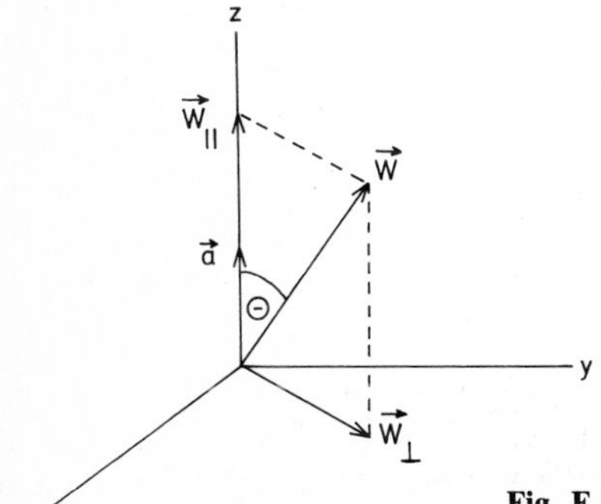

Fig. F.2. Decompositon of the relative velocity $\boldsymbol{W}$ into $\boldsymbol{W}_{\|}$ and $\boldsymbol{W}_{\perp}$

Since $W\,|\cos\Theta| = |W_{\|}| \geqslant 0$, and writing $\int d^3W = \int dW_{\|} \int d^2W_{\perp}$, (F.21) can now be written as

$$\Gamma_2^+ = \frac{1}{2} n_1 r^2 \int d\Omega_a \int\limits_{-\infty}^{\infty} dW_{\|}\,|W_{\|}| F_2(\boldsymbol{v}+\boldsymbol{W}_{\|}) \iint\limits_{-\infty}^{\infty} d^2W_{\perp} f^{\mathrm{M}}(\boldsymbol{v}+\boldsymbol{W}_{\perp}) \ . \tag{F.26}$$

To combine here the integrations $d\Omega_a$ and $dW_{\|}$, we define the vector [F.1]

$$\boldsymbol{u} = |W_{\|}|\,\boldsymbol{a} \tag{F.27}$$

which is always parallel (and never antiparallel) to the vector $\boldsymbol{a}$ as $|W_{\|}| \geqslant 0$. With

$$d^3u = u^2 du\, d\Omega_a \quad (u = |W_{\|}| \geqslant 0) \tag{F.28}$$

then

$$\int d\Omega_a \int\limits_{-\infty}^{\infty} dW_{\|} = 2 \int d\Omega_a \int\limits_{0}^{\infty} dW_{\|} = 2 \int \frac{d^3u}{u^2} \tag{F.29}$$

because the first integrations run over the $W_{\|}$ space twice. Equation (F.26) now takes the form

$$\Gamma_2^+ = n_1 r^2 \int d^3u \frac{1}{u} F_2(\boldsymbol{v}+\boldsymbol{u})\, g(\boldsymbol{v},\boldsymbol{u}) \ , \tag{F.30}$$

where

$$g(\boldsymbol{v},\boldsymbol{u}) = \iint\limits_{-\infty}^{\infty} d^2W_{\perp} f^{\mathrm{M}}(\boldsymbol{v}+\boldsymbol{W}_{\perp}) \ . \tag{F.31}$$

The function g is easily calculated in the coordinate frame shown in Fig. F.2. Indeed, through (F.17, 31)

$$\begin{aligned} g(\boldsymbol{v},\boldsymbol{u}) &= \frac{1}{\pi^{3/2} w^3} \exp\left(-\frac{v_z^2}{w^2}\right) \int\limits_{-\infty}^{\infty} dW_x \exp\left(-\frac{(v_x+W_x)^2}{w^2}\right) \\ &\quad \cdot \int\limits_{-\infty}^{\infty} dW_y \exp\left(-\frac{(v_y+W_y)^2}{w^2}\right) \\ &= \frac{1}{\pi^{1/2} w} \exp\left(-\frac{v_z^2}{w^2}\right) \end{aligned} \tag{F.32}$$

since the value of each of the integrals on the right-hand side is $\pi^{1/2} w$. Recalling that

$$v_z = \boldsymbol{v} \cdot \boldsymbol{a} = \frac{\boldsymbol{v} \cdot \boldsymbol{u}}{u} , \tag{F.33}$$

one obtains

$$g(\boldsymbol{v},\boldsymbol{u}) = \frac{1}{\pi^{1/2} w} \exp\left(-\left[\frac{\boldsymbol{v} \cdot \boldsymbol{u}}{wu}\right]^2\right) . \tag{F.34}$$

Inserting (F.34) into (F.30) finally yields the creation term

$$\Gamma_2^+(\boldsymbol{v}) = \gamma_2 \frac{1}{\pi^{3/2} w^2} \int d^3u \frac{1}{u} \exp\left(-\left[\frac{\boldsymbol{v} \cdot \boldsymbol{u}}{wu}\right]^2\right) F_2(\boldsymbol{v}+\boldsymbol{u}) \tag{F.35}$$

with γ_2 being the collision frequency as defined by (F.19). Equations (F.20, 23, 35) are our final result, see also (6.4.4).

It can be checked that for a Maxwellian distribution $F_2(\boldsymbol{v}) = n_2 f^{\mathrm{M}}(\boldsymbol{v})$ of the same temperature T as the distribution function $F_1(\boldsymbol{v})$, the relation

$$\Gamma_2^+(\boldsymbol{v}) = \Gamma_2^-(\boldsymbol{v})$$

holds for all velocities $\boldsymbol{v}$ as required by the principle of detailed balance, and hence

$$\left(\frac{\delta F_2}{\delta t}\right)_{\mathrm{el}} = n_2 \left(\frac{\delta f^{\mathrm{M}}}{\delta t}\right)_{\mathrm{el}} = 0$$

as it should.

General Bibliography

Griem, H. R.: *Plasma Spectroscopy* (McGraw-Hill, New York 1964)

Mihalas, D.: *Stellar Atmospheres,* 2nd ed. (Freeman, San Francisco 1978)

Mihalas, D., Weibel Mihalas, B.: *Foundations of Radiation Hydrodynamics* (Oxford U. Press, Oxford 1985)

Mitchner, M., Kruger C. H.: *Partially Ionized Gases* (Wiley, New York 1974)

Pomraning, G. C.: *The Equations of Radiation Hydrodynamics* (Pergamon, Oxford 1973)

Sampson, D. H.: *Radiative Contribution to Energy and Momentum Transport in a Gas* (Wiley-Interscience, New York 1965)

Sobelman, I. I.: *Atomic Spectra and Radiative Transitions,* Springer Ser. Chem. Phys., Vol. 1 (Springer, Berlin, Heidelberg 1979)

Sobelman, I. I., Vainshtein, L. A., Yukov, E. A.: *Excitation of Atoms and Broadening of Spectral Lines,* Springer Ser. Chem. Phys., Vol. 7 (Springer, Berlin, Heidelberg 1981)

References

Chapter 1

1.1 Fowler, R. H.: *Statistical Mechanics,* 2nd ed. (Cambridge U. Press, Cambridge 1936)
1.2 Feynman, R. P., Leighton, R. B., Sands, M.: *The Feynman Lectures on Physics,* Vol. 3 (Addison-Wesley, Reading, MA 1965)
1.3 Lipkin, H. J.: *Quantum Mechanics* (North Holland, Amsterdam 1973)
1.4 Oster, L.: Am. J. Phys. **30**, 754 (1970)
1.5 Frauenfelder, H., Henley, E. M.: *Nuclear and Particle Physics,* Part A (Benjamin, Reading, MA 1975)
1.6 Williams, W. S. C.: *An Introduction to Elementary Particles,* 2nd ed. (Academic, New York 1971)
1.7 Pert, G. J.: J. Phys. B**4**, L 72 (1971)

Chapter 2

2.1 Balescu, R.: *Equilibrium and Nonequilibrium Statistical Mechanics* (Wiley, New York 1975)
2.2 Chapman, S., Cowling, T. G.: *The Mathematical Theory of Non-Uniform Gases,* 3rd ed. (Cambridge U. Press, Cambridge 1970)
2.3 Ecker, G.: *Theory of Fully Ionized Plasmas* (Academic, New York 1972)
2.4 Montgomery, D. C., Tidman, T. A.: *Plasma Kinetic Theory* (McGraw-Hill, New York 1964)
2.5 Résibois, P., De Leener, M.: *Classical Kinetic Theory of Fluids* (Wiley, New York 1977)
2.6 Moreau, E., Salmon, J.: J. Phys. Radium **21**, 217 (1960)
2.7 Bernstein, I. B.: "Electron Distribution Functions in Weakly Ionized Plasmas", in *Advances in Plasma Physics,* Vol. 3, ed. by Simon, A., Thompson, W. B. (Interscience, New York 1969)
2.8 Risken, H.: *The Fokker-Planck Equation,* Springer Ser. Syn., Vol. 18 (Springer, Berlin, Heidelberg 1984)
2.9 Bates, D. R., Dalgarno, A.: "Electronic Recombination", in *Atomic and Molecular Processes,* ed. by Bates, D. R. (Academic, New York 1962)
2.10 Seaton, M. J., Story, P. J.: "Di-electronic Recombination", in *Atomic Processes and Applications,* ed. by Burke, P. G., Moiseiwitsch, B. L. (North-Holland, Amsterdam 1976)
2.11 Bekefi, G.: *Radiation Processes in Plasmas* (Wiley, New York 1966)
2.12 Jackson, J. D.: *Classical Electrodynamics,* 2nd ed. (Wiley, New York 1975)

Chapter 3

3.1 Acquista, C., Anderson, J. L.: Ann. Phys. (N.Y.) **106**, 435 (1977)
3.2 Cooper, J., Zoller, P.: Astrophys. J. **277**, 813 (1984)
3.3 Landi Degl'innocenti, E., Landi Degl'innocenti, M.: Nuovo Cimento **27B**, 134 (1975)
3.4 Gelinas, R. J., Ott, R. L.: Ann. Phys. (N. Y.) **59**, 323 (1970)

3.5 Wolf, E.: Phys. Rev. D**13**, 869 (1976)
3.6 Bekefi, G.: *Radiation Processes in Plasmas* (Wiley, New York 1966)
3.7 Enomé, S.: Publ. Astron. Soc. Jpn. **21**, 367 (1969)
3.8 Harris, E. G.: Phys. Rev. **138**, B 479 (1965)
3.9 Pomraning, G. C.: *The Equations of Radiation Hydrodynamics* (Pergamon, Oxford 1973)
3.10 Zheleznyakov, V. V.: Astrophys. J. **148**, 849 (1967)
3.11 Acquista, C., Anderson, J. L.: Astrophys. J. **191**, 567 (1974)
3.12 Chandrasekhar, S.: *Radiative Transfer* (Dover, New York 1960)
3.13 Weber, A. (ed): *Raman Spectroscopy of Gases and Liquids,* Topics Current Phys., Vol. 11 (Springer, Berlin, Heidelberg 1979)
3.14 Jackson, J. D.: *Classical Electrodynamics,* 2nd ed. (Wiley, New York 1975)

Chapter 4

4.1 Balescu, R.: *Equilibrium and Nonequilibrium Statistical Mechanics* (Wiley, New York 1975)
4.2 Chapman, S., Cowling, T. G.: *The Mathematical Theory of Non-Uniform Gases,* 3rd ed. (Cambridge U. Press, Cambridge 1970)
4.3 Ecker, G.: *Theory of Fully Ionized Plasmas* (Academic, New York 1972)
4.4 Ferziger, J. H., Kaper, H. G.: *Mathematical Theory of Transport Processes in Gases* (North-Holland, Amsterdam 1972)
4.5 Hirschfelder, J. O., Curtis, C. F., Bird, R. B.: *The Molecular Theory of Gases and Liquids* (Wiley, New York 1954)
4.6 Résibois, P., De Leener, M.: *Classical Kinetic Theory of Fluids* (Wiley, New York 1977)
4.7 Waldmann, L.: "Transporterscheinungen in Gasen von mittlerem Druck", in *Encyclopedia of Physics,* Vol. XII, ed. by Flügge, S. (Springer, Berlin, Göttingen, Heidelberg 1958)
4.8 Van Bladel, J.: *Relativity and Engineering,* Springer Ser. Electrophys., Vol. 15 (Springer, Berlin, Heidelberg 1984)
4.9 Alfvén, H., Fälthammar, C. G.: *Cosmical Electrodynamics,* 2nd ed. (Clarendon, Oxford 1963)
4.10 Thomas, L. H.: Q. J. Math. **1**, 239 (1930)
4.11 Simon, R.: J. Quant. Spectrosc. Radiat. Transfer **3**, 1 (1963)
4.12 Anderson, J. L., Spiegel, E. A.: Astrophys. J. **171**, 127 (1972)
4.13 Hazlehurst, J., Sargent, W. L. W.: Astrophys. J. **130**, 276 (1959)
4.14 Hsieh, S. H., Spiegel, E. A.: Astrophys. J. **207**, 244 (1976)
4.15 Pomraning, G. C.: *The Equations of Radiation Hydrodynamics* (Pergamon, Oxford 1973)

Chapter 5

5.1 Ter Haar, D.: *Elements of Statistical Mechanics* (Constable, London 1954)
5.2 Landau, L. D., Lifshitz, E. M.: *Statistical Physics Part 1,* 3rd ed. (Pergamon, Oxford 1980)
5.3 Toda, M., Kubo, R., Saitô, N.: *Statistical Physics I,* Springer Ser. Solid-State Sci., Vol. 30 (Springer, Berlin, Heidelberg 1983)
5.4 Planck, M.: *The Theory of Heat Radiation* (Dover, New York 1959)
5.5 Ore, A.: Phys. Rev. **98**, 887 (1955)
5.6 Rosen, R.: Phys. Rev. **96**, 555 (1954)
5.7 Wildt, R.: Astrophys. J. **123**, 107 (1956)
5.8 Pauli, W.: "Über das H-Theorem vom Anwachsen der Entropie vom Standpunkt der neuen Quantenmechanik", in *Collected Scientific Papers,* Vol. 1 (Interscience, New York 1964)
5.9 Wu, T. Y.: *Kinetic Equations of Gases and Plasmas* (Addison-Wesley, Reading, MA 1966)
5.10 Oxenius, J.: J. Quant. Spectrosc. Radiat. Transfer **6**, 65 (1966)
5.11 Stückelberg, E. C. G.: Helv. Phys. Acta **25**, 577 (1952)
5.12 Morimoto, T.: J. Phys. Soc. Jpn. **18**, 328 (1963)

5.13 Kröll, W.: J. Quant. Spectrosc. Radiat. Transfer **7**, 715 (1967)
5.14 Glansdorff, P., Prigogine, I.: Physica **20**, 773 (1954); **30**, 351 (1964)
5.15 Glansdorff, P., Prigogine, I.: *Thermodynamic Theory of Structure, Stability and Fluctuations* (Wiley-Interscience, London 1971)

Chapter 6

6.1 Sobelman, I. I., Vainshtein, L. A., Yukov, E. A.: *Excitation of Atoms and Broadening of Spectral Lines,* Springer Ser. Chem. Phys., Vol. 7 (Springer, Berlin, Heidelberg 1981)
6.2 Düchs, D. F., Post, D. E., Rutherford, P. H.: Nucl. Fusion **17**, 565 (1977)
6.3 Drawin, H. W.: Ann. Physik (Leipzig) **14**, 262 (1964)
6.4 Drawin, H. W., Emard, F.: Physica **85C**, 333 (1977); **94C**, 134 (1978)
6.5 Drawin, H. W., Emard, F.: Report EUR-CEA-FC 697 (Fontenay-aux-Roses, France 1973)
6.6 Burgess, A., Summers, H. P.: Astrophys. J. **157**, 1007 (1969)
6.7 Burgess, A., Summers, H. P.: Mon. Not. R. Astron. Soc. **174**, 345 (1976)
6.8 Bates, D. R., Kingston, A. E., McWhirter, R. W. P.: Proc. R. Soc. A**267**, 297 (1962)
6.9 Athay, R. G.: *Radiation Transport in Spectral Lines* (Reidel, Dordrecht 1972)
6.10 Hummer, D. G., Rybicki, G.: Ann. Rev. Astron. Astrophys. **9**, 237 (1971)
6.11 Ivanov, V. V.: *Transfer of Radiation in Spectral Lines,* NBS Special Publication 385 (Government Printing Office, Washington, DC 1973)
6.12 Jefferies, J. T.: *Spectral Line Formation* (Blaisdell, Waltham, MA 1968)
6.13 Mihalas, D.: *Stellar Atmospheres,* 2nd ed. (Freeman, San Francisco 1978)
6.14 Thomas, R. N.: *Some Aspects of Non-Equilibrium Thermodynamics in the Presence of a Radiation Field* (U. Colorado Press, Boulder 1965)
6.15 Oxenius, J.: J. Quant. Spectrosc. Radiat. Transfer **5**, 771 (1965)
6.16 Hubený, I., Oxenius, J., Simonneau, E.: J. Quant. Spectrosc. Radiat. Transfer **29**, 477, 495 (1983)
6.17 Hummer, D. G.: Mon. Not. R. Astron. Soc. **125**, 21 (1962)
6.18 Heinzel, P.: J. Quant. Spectrosc. Radiat. Transfer **25**, 483 (1981)
6.19 Baschek, B., Mihalas, D., Oxenius, J.: Astron. Astrophys. **97**, 43 (1981)
6.20 Hubený, I.: Bull. Astron. Inst. Czech. **32**, 271 (1981)
6.21 Cooper, J., Ballagh, R. J., Burnett, K., Hummer, D. G.: Astrophys. J. **260**, 299 (1982)
6.22 Cooper, J., Hubený, I., Oxenius, J.: Astron. Astrophys. **127**, 224 (1983)
6.23 Feautrier, P.: Ann. Astrophys. **30**, 347 (1967)
6.24 Avrett, E. H., Hummer, D. G.: Mon. Not. R. Astron. Soc. **130**, 295 (1965)
6.25 Düchs, D. F., Oxenius, J.: Z. Naturforsch. **32a**, 124 (1978)
6.26 Düchs, D. F., Rehker, S., Oxenius, J.: Z. Naturforsch. **33a**, 124 (1978)
6.27 Cipolla, J. W., Morse, T. F.: J. Quant. Spectrosc. Radiat. Transfer **22**, 365 (1979)
6.28 Borsenberger, J., Oxenius, J., Simonneau, E.: J. Quant. Spectrosc. Radiat. Transfer (to be published)
6.29 Hummer, D. G.: Mon. Not. R. Astron. Soc. **145**, 95 (1969)
6.30 Biberman, L. M., Vorobev, V. S., Yakubov, I. T.: High Temp. (USSR) **6**, 359 (1968)
6.31 Peyraud, N.: J. Physique **29**, 201, 747, 997 (1968)
6.32 Shaw, J. F., Mitchner, M., Kruger, C. H.: Phys. Fluids **13**, 325, 339 (1970)
6.33 Oxenius, J.: Z. Naturforsch. **25a**, 1302 (1970)
6.34 Shoub, E. C.: Astrophys. J. Suppl. **34**, 259, 277 (1977)
6.35 Biberman, L. M., Vorobev, V. S., Yakubov, I. T.: Sov. Phys. Usp. **22**, 411 (1979)
6.36 Claaßen, H. A.: Z. Naturforsch. **28a**, 1875 (1973); **30a**, 451 (1975)
6.37 Oxenius, J.: J. Quant. Spectrosc. Radiat. Transfer **14**, 731 (1974)

Chapter 7

7.1 Field, G. B.: Astrophys. J. **129**, 551 (1959)
7.2 Lam, J. F., Berman, P. R.: Phys. Rev. A**14**, 1683 (1976)

7.3 Einstein, A.: Physik. Z. **18**, 121 (1917)
7.4 Milne, E. A.: Proc. Cambridge Philos. Soc. **23**, 465 (1926)
7.5 Fowler, R. H.: *Statistical Mechanics,* 2nd ed. (Cambridge U. Press, Cambridge 1936)
7.6 Oxenius, J.: J. Quant. Spectrosc. Radiat. Transfer **7**, 837 (1967)
7.7 Pauli, W.: Z. Physik **18**, 272 (1923)
7.8 Dreicer, H.: Phys. Fluids **7**, 735 (1964)
7.9 Kompaneets, A. S.: Sov. Phys. JETP **4**, 730 (1957)
7.10 Peyraud, J.: J. Physique **29**, 88, 306 (1968)
7.11 Pomraning, G. C.: *The Equations of Radiation Hydrodynamics* (Pergamon, Oxford 1973)
7.12 Weymann, R.: Phys. Fluids **8**, 2112 (1965)
7.13 Wienke, B. R.: J. Quant. Spectrosc. Radiat. Transfer **15**, 151 (1975)
7.14 Terrall, J. R.: Am. J. Phys. **38**, 1460 (1970)
7.15 Jauch, J. M., Rohrlich, F.: *The Theory of Photons and Electrons,* Texts and Monogr. in Phys., 2nd ed. (Springer, New York 1976)
7.16 Pauli, W.: Helv. Phys. Acta **6**, 279 (1933)
7.17 Gould, R. J.: Am. J. Phys. **38**, 1460 (1970)

Appendices

A.1 Pauli, W.: *Theory of Relativity* (Pergamon, London 1958)
A.2 Van Bladel, J.: *Relativity and Engineering,* Springer Ser. Electrophys., Vol. 15 (Springer, Berlin, Heidelberg 1984)
A.3 Van Kampen, N. G.: Physica **43**, 244 (1969)
A.4 Oxenius, J.: Z. Naturforsch. **34a**, 895 (1979)
B.1 Baranger, M.: "Spectral Line Broadening in Plasmas", in *Atomic and Molecular Processes,* ed. by Bates, D. R. (Academic, New York 1962)
B.2 Griem, H. R.: *Plasma Spectroscopy* (McGraw-Hill, New York 1964)
B.3 Griem, H. R.: *Spectral Line Broadening by Plasmas* (Academic, New York 1974)
B.4 Van Regemorter, H.: "Spectral Line Broadening", in *Atoms and Molecules in Astrophysics,* ed. by Carson, T. R., Roberts, M. J. (Academic, London 1972)
B.5 Sobelman, I. I., Vainshtein, L. A., Yukov, E. A.: *Excitation of Atoms and Broadening of Spectral Lines,* Springer Ser. Chem. Phys., Vol. 7 (Springer, Berlin, Heidelberg 1981)
B.6 Weisskopf, V.: Ann. Physik (Leipzig) **9**, 23 (1931)
B.7 Weisskopf, V.: Z. Physik **85**, 451 (1933)
B.8 Weisskopf, V.: Observatory **56**, 291 (1933)
B.9 Weisskopf, V., Wigner, E.: Z. Physik **63**, 54 (1930); **65**, 18 (1930)
B.10 Woolley, R. v. d. R.: Mon. Not. R. Astron. Soc. **91**, 977 (1931)
B.11 Woolley, R. v. d. R.: Mon. Not. R. Astron. Soc. **98**, 624 (1938)
B.12 Spitzer, L.: Mon. Not. R. Astron. Soc. **96**, 794 (1936)
B.13 Woolley, R. v. d. R., Stibbs, D. W. N.: *The Outer Layers of a Star* (Clarendon, Oxford 1953)
B.14 Omont, A., Smith, E. W., Cooper, J.: Astrophys. J. **175**, 185 (1972)
B.15 Zanstra, H.: Mon. Not. R. Astron. Soc. **101**, 273 (1941)
B.16 Heitler, W.: *The Quantum Theory of Radiation,* 3rd ed. (Clarendon, Oxford 1954)
B.17 Heinzel, P., Hubený, I.: J. Quant. Spectrosc. Radiat. Transfer **27**, 1 (1982)
B.18 Henyey, L. G.: Astrophys. J. **103**, 332 (1946)
B.19 Hummer, D. G.: Mon. Not. R. Astron. Soc. **125**, 21 (1962)
B.20 Heinzel, P.: J. Quant. Spectrosc. Radiat. Transfer **25**, 483 (1981)
B.21 Hubený, I., Oxenius, J., Simonneau, E.: J. Quant. Spectrosc. Radiat. Transfer **29**, 477 (1983)
B.22 Hubený, I.: J. Quant. Spectrosc. Radiat. Transfer **27**, 593 (1982)
B.23 Magnan, C.: Astron. Astrophys. **142**, 117 (1985)
B.24 Hubený, I., Oxenius, J.: J. Quant. Spectrosc. Radiat. Transfer (to be published)

C.1 Balescu, R.: *Equilibrium and Nonequilibrium Statistical Mechanics* (Wiley, New York 1975)
C.2 Chapman, S., Cowling, T. G.: *The Mathematical Theory of Non-Uniform Gases,* 3rd ed. (Cambridge U. Press, Cambridge 1970)
C.3 Huang, K.: *Statistical Mechanics* (Wiley, New York 1963)
C.4 Résibois, P., De Leener, M.: *Classical Kinetic Theory of Fluids* (Wiley, New York 1977)
D.1 Balescu, R.: *Equilibrium and Nonequilibrium Statistical Mechanics* (Wiley, New York 1975)
D.2 Becker, R.: *Theory of Heat,* 2nd ed., revised by G. Leibfried (Springer, Berlin, Heidelberg 1967)
D.3 Chandrasekhar, S.: Rev. Mod. Phys. **15**, 1 (1943)
D.4 Résibois, P., De Leener, M.: *Classical Kinetic Theory of Fluids* (Wiley, New York 1977)
D.5 Toda, M., Kubo, R., Saitô, N.: *Statistical Physics I,* Springer Ser. Solid-State Sci., Vol. 30 (Springer, Berlin, Heidelberg 1983)
D.6 Risken, M.: *The Fokker-Planck Equation,* Springer Ser. Syn., Vol. 18 (Springer, Berlin, Heidelberg 1984)
E.1 Fowler, R. H.: *Statistical Mechanics,* 2nd ed. (Cambridge U. Press, Cambridge 1936)
F.1 Cercignani, C.: *Mathematical Methods in Kinetic Theory* (Plenum, New York 1969)

Subject Index

Springer Series in Chemical Physics

Editors: **V. J. Goldanskii, R. Gomer, F. P. Schäfer, J. P. Toennies**

A selection:

Volume 1
I. I. Sobelman

Atomic Spectra and Radiative Transitions

1979. 21 figures, 46 tables. XII, 306 pages. ISBN 3-540-09082-7

Contents: Elementary Information on Atomic Spectra: The Hydrogen Spectrum. Systematics of the Spectra of Multielectron Atoms. Spectra of Multielectron Atoms. – Theory of Atomic Spectra: Angular Momenta. Systematics of the Levels of Multielectron Atoms. Hyperfine Structure of Spectral Lines. The Atom in an External Electric Field. The Atom in an External Magnetic Field. Radiative Transitions. References. – List of Symbols. – Subject Index.

Volume 7
I. I. Sobelman, L. A. Vainshtein, E. A. Yukov

Excitation of Atoms and Broadening of Spectral Lines

1981. 34 figures, 40 tables. X, 315 pages. ISBN 3-540-09890-9

Contents: Elementary Processes Giving Rise to Spectra. – Theory of Atomic Collisions. – Approximate Methods for Calculating Cross Sections. – Collisions Between Heavy Particles. – Some Problems of Excitation Kinetics. – Tables and Formulas for the Estimation of Effective Cross Sections. – Broadening of Spectral Lines. – References. – List of Symbols. – Subject Index. – Erratra for volume 1 of this series.

Volume 13
I. Lindgren, J. Morrison

Atomic Many-Body Theory

1982. 96 figures. XIII, 469 pages. ISBN 3-540-10504-2

Contents: Angular-Momentum Theory and the Independent-Particle Model: Introduction. Angular-Momentum and Spherical Tensor Operators. Angular-Momentum Graphs. Further Developments of Angular-Momentum Graphs. Applications to Physical Problems. The Independent-Particle Model. The Central-Field Model. The Hartree-Fock Model. Many-Electron Wave Functions. – Perturbation Theory and the Treatment of Atomic Many-Body Effects: Perturbation Theory. First-Order Perturbation for Closed-Shell Atoms. Second Quantization and the Particle-Hole Formalism. Application of Perturbation Theory to Closed-Shell Systems. Application of Perturbation Theory to Open-Shell Systems. The Hyperfine Interaction. The Pair-Correlation Problem and the Coupled-Cluster Approach. – Appendices A–D. – References. – Author Index. – Subject Index.

Springer-Verlag
Berlin Heidelberg
New York Tokyo